interactions between
ring theory and
representations of algebras

PURE AND APPLIED MATHEMATICS

A Program of Monographs, Textbooks, and Lecture Notes

LECTURE NOTES IN PURE AND APPLIED MATHEMATICS

Additional Volumes in Preparation

interactions between ring theory and representations of algebras

proceedings of the conference held in Murcia, Spain

edited by

Freddy Van Oystaeyen
University of Antwerp/UIA
Wilrijk, Belgium

Manuel Saorin
University of Murcia
Murcia, Spain

MARCEL DEKKER, INC.   NEW YORK · BASEL

ISBN: 0-8247-0367-7

This book is printed on acid-free paper

Headquarters
Marcel Dekker, Inc
270 Madison Avenue. New York, NY 10016
tel 212-696-9000, fax 212-685-4540

Eastern Hemisphere Distribution
Marcel Dekker AG
Hutgasse 4, Postfach 812, CH-4001 Basel, Switzerland
tel 41-61-261-8482, fax 41-61-261-8896

World Wide Web
http //www dekker com

The publisher offers discounts on this book when ordered in bulk quantities For more information,
write to Special Sales/Professional Marketing at the headquarters address above

Current printing (last digit)
10 9 8 7 6 5 4 3 2 1

PRINTED IN THE UNITED STATES OF AMERICA

Preface

The meeting at Murcia, Spain, on which this book is based, was part of a series of four Eurocongresses organized within the framework of a European Commission Training and Mobility of Researchers (TMR) Programme. These proceedings present the set of lectures and also contain some invited papers. We would like to thank everybody involved in making the meeting a success, in particular, all participants, the lecturers, and most of all, the local committee.

We also acknowledge the support of the following organizations:
DGES of the Spanish Ministry of Education
Fundacion Seneca of the Autonomous Government of Murcia
The European Commission TMR Programme

Freddy Van Oystaeyen
Manuel Saorin

Contents

Contributors

Toma Albu   University of Bucharest, Bucharest, Romania

L'Moufadal Ben Yakoub   Faculty of Sciences, Tetouan, Morocco

Lu Chen   Shaanxi Normal University, Xian, P. R. China

Claude Cibils   Université de Montpellier 2, Montpellier, France

Flávio Ulhoa Coelho   IME, University of Sao Paolo, Sao Paolo, Brazil

Miriam Cohen   Ben Gurion University of the Negev, Beer-Sheva, Israel

Riccardo Colpi   Università degli Studi di Padova, Padova, Italy

Willem A. de Graaf   School of Computational Sciences, St-Andrews, Scotland

Alberto Del Valle   University of Murcia, Murcia, Spain

Yuri A. Drozd   Kiev Taras Shevchenko University, Kiev, Ukraine

Moha Helmy Ibrahim El Baroudy   Ain Shams University, Cairo, Egypt

Edgar E. Enochs   University of Kentucky, Lexington, Kentucky

Miguel Ferrero   Universidade Federal do Rio Grande do Sul, Porto Allegre, Brazil

Odile Garotta   Institut Fourier, St-Martin-d'Héres, France

Yuval Ginosar   Brandeis University, Waltham, Massachusetts

José L. Gómez Pardo   University of Santiago, Santiago de Compostela, Spain

K. R. Goodearl   University of California, Santa Barbara, California

Pedro A. Guil Asensio   University of Murcia, Murcia, Spain

Gábor Ivanyos   Computer and Automation Institute, Hungarian Academy of Sciences, Budapest, Hungary

Overtoun M. G. Jenda   Auburn University, Auburn, Alabama

Eric Jespers   Vrije Universiteit Brussel, Brussels, Belgium

T. H. Lenagan   J.C.M.B., Edinburgh, Scotland

Huishi Li   Shaanxi Normal University, Xian, P. R. China

Shao Xue Liu   Beijing Normal University, Beijing, P. R. China

Leandro Marin   University of Murcia, Murcia, Spain

Jerzy Matczuk   Warsaw University, Warsaw, Poland

Jan Okniński   Warsaw University, Warsaw, Poland

Uri Onn   Technion, Haifa, Mt. Carmel, Israel

Barbara L. Osofsky   Hill Center for Mathematics, Rutgers University, Piscataway, New Jersey

Richard Rossmanith   Friedrich Schiller University, Jena, Germany

Manuel Saorin   University of Murcia, Murcia, Spain

Peter Schauenburg   Mathematics Institute of the University of Munich, Munich, Germany

Daniel Simson   Universytet Mikotaya Korpernika u Torciniu, Toruń, Poland

Boris Širola   University of Zagreb, Zagreb, Yugoslavia

Patrick F. Smith   University of Glasgow, Glasgow, Scotland

Yorck Sommerhäuser   Mathematics Institute of the University of Munich, Munich, Germany

Alfons Van Daele   University of Leuven, Heverlee, Belgium

Freddy Van Oystaeyen   University of Antwerp, Antwerp, Belgium

Robert Wisbauer   Mathematical Institute of the University of Düsseldorf, Düsseldorf, Germany

Yinhuo Zhang   University of Antwerp, Antwerp, Belgium

Global Krull Dimension

TOMA ALBU

Facultatea de Matematică
Universitatea Bucuresti
Str. Academiei 14
Ro-70109 Bucharest 1, Romania

PATRICK F. SMITH

Department of Mathematics
University of Glasgow
Glasgow, G12 8QW, Scotland, UK

Associated with any ring R are two ordinals α and β. The ordinal α (respectively, β) is the supremium of the (dual) Krull dimensions of the right R-modules which have Krull dimension. In this paper we investigate these ordinals and relationships between them. We denote α by $r.gl.k(R)$, β by $r.gl.k^0(R)$ and refer to them as the right global Krull dimension and right global dual Krull dimension of R, respectively.

If R is a right Artinian, or more generally a semiprimary ring, then $r.gl.k(R) = r.gl.k^0(R) = 0$. On the other hand if R is a commutative Noetherian ring then $r.gl.k^0(R)$ is the supremum of the heights of the prime ideals of R and $r.gl.k(R)$ is the Krull dimension of R. If R is a right Loewy ring then $r.gl.k(R) = 0$. For a right V-ring $R, r.gl.k^0(R) = 0$. Also if R is a right perfect ring then $r.gl.k^0(R) = 0$ and $l.gl.k(R) = 0$.

Given an ordinal $\alpha \geq 0$, a ring R satisfies $r.gl.k(R) \leq \alpha$ if and only if every right R-module M which does not have Krull dimension at most α has a proper homomorphic image with the same property. There is a dual result for $r.gl.k^0(R) \leq \alpha$. A commutative ring R satisfies $gl.k(R) = 0$ if and only if every Noetherian R-module is Artinian. Dually, a commutative ring satisfies $gl.k^0(R) = 0$ if and only if every Artinian R-module is Noetherian.

1

In consequence, if R is commutative and $gl.k^0(R) = 0$ then $gl.k(R) = 0$. This last fact is not true in general.

1   Definitions and examples

All rings are associative with identity and all modules are unital. Let R be a ring. It is well known that a right R-module M has Krull dimension if and only if M has dual Krull dimension [14, Corollaire 6], and in this case $k(M)$ and $k^0(M)$ will be used to denote the Krull dimension and dual Krull dimension of M, respectively. For the definition and basic facts about Krull dimension, see [11] or [17]. The ring R has right (respectively, left) Krull dimension if the right (left) R-module R has Krull dimension. In case R has right Krull dimension, $k(R_R)$ and $k^0(R_R)$ will denote the Krull and dual Krull dimensions of the right R-module R, respectively.

For any ring R, we define the *right global Krull dimension*, $r.gl.k(R)$, and *right global dual Krull dimension*, $r.gl.k^0(R)$, of R by

$r.gl.k(R) = \sup\{k(M) : M$ is a right R-module with Krull dimension$\}$, and
$r.gl.k^0(R) = \sup\{k^0(M) : M$ is a right R-module with Krull dimension$\}$.

The left global Krull dimension and left global dual Krull dimensions of R are defined analogously and are denoted by $l.gl.k(R)$ and $l.gl.k^0(R)$, respectively. In [1] it is proved that, unlike Krull dimension, $r.gl.k(R)$ and $r.gl.k^0(R)$ exist for *any* ring R. On the other hand we know of no general relation between $r.gl.k(R)$ and $l.gl.k(R)$.

Given a ring R and an ordinal $\alpha \geq 0$, a right R-module M is called α-*critical* if M has Krull dimension α and $k(M/N) < \alpha$ for every non-zero submodule N of M. Dually, M is called α-*cocritical* if M has dual Krull dimension α and $k^0(N) < \alpha$ for every proper submodule N of M. The module M is called *critical* (respectively, *cocritical*) if M is α-critical (α-cocritical) for some ordinal $\alpha \geq 0$. The module M is called *cocyclic* if M is non-zero and M contains an essential simple submodule. In the literature cocyclic modules are given different names; for example, some authors call cocyclic modules *monolithic* or *subdirectly irreducible*.

Proposition 1.1 *For any non-zero ring R, $r.gl.k(R) = \sup\{k(M) : M$ is a cyclic critical right R-module$\}$, and $r.gl.k^0(R) = \sup\{k^0(M) : M$ is a cocyclic cocritical right R-module$\}$.*

Proof. By [18, Corollary 3.2.12] and [5, Proposition 10],

$$r.gl.k(R) = \sup\{k(M) : M \text{ is a critical right } R - \text{ module }\}, \text{ and}$$
$$r.gl.k^0(R) = \sup\{k^0(M) : M \text{ is a cocritical right } R - \text{ module }\}.$$

But if M is an α-critical right R-module, for some ordinal $\alpha \geq 0$, then so too is mR for any $0 \neq m \in M$ by [11, Proposition 2.3]. On the other hand, if M is α-cocritical then there exists a proper submodule N of M such that M/N is cocyclic and, by [5, Proposition 3], M/N is α-cocritical. The result follows easily.

Next we consider some special cases.

Proposition 1.2 *Let R be a ring with right Krull dimension. Then $r.gl.k(R) = k(R_R)$.*

Proof. By [11, Corollary 4.4].

Corollary 1.3 *Let R be a right Artinian ring. Then $r.gl.k(R) = r.gl.k^0(R) = 0$.*

Proof By Proposition 1.2 and [2, Corollary 15.21].

We shall extend Corollary 1.3 to semiprimary rings. To do so we first prove the following result.

Proposition 1.4 *Let I be an ideal of a ring R. Then $r.gl.k(R/I) \leq r.gl.k(R)$ and $r.gl.k^0(R/I) \leq r.gl.k^0(R)$, with equality in each case if I is nilpotent.*

Proof. The first part is obvious. Now suppose that $I^t = 0$ for some positive integer t. Let M be any right R-module with Krull dimension. Consider the chain

$$M = MI^0 \geq MI \geq MI^2 \geq \ldots \geq MI^t = 0.$$

For each $1 \leq j \leq t$, $k(MI^{j-1}/MI^j) \leq r.gl.k(R/I)$ and hence $k(M) \leq r.gl.k(R/I)$ by [11, Lemma 1.1]. It follows that $r.gl.k(R) \leq r.gl.k(R/I)$ and hence, by the first part, $r.gl.k(R/I) = r.gl.k(R)$. Similarly $r.gl.k^0(R/I) = r.gl.k^0(R)$.

Corollary 1.5 *Let R be a semiprimary ring. Then* $r.gl.k(R) = r.gl.k^0(R) = 0$.

Proof. By Corollary 1.3 and Proposition 1.4.

Corollary 1.6 *Let R be a ring which satisfies acc on right annihilatiors and on left annihilators and let N be the prime radical of R. Then* $r.gl.k(R) = r.gl.k(R/N)$ *and* $r.gl.k^0(R) = r.gl.k^0(R/N)$.

Proof. By Proposition 1.4 and [6, Theorem 1.34].

Corollary 1.5 can be extended to right perfect rings, as we shall show in Section 2. For *PI* rings (i.e. rings which satisfy a polynomial identity) we have the following result.

Proposition 1.7 *Let R be a PI ring with prime radical N. Then* $r.gl.k(R) = r.gl.k(R/N)$ *and* $r.gl.k^0(R) = r.gl.k^0(R/N)$.

Proof. Let M be any right R-module with Krull dimension. If $A = \{r \in R : Mr = 0\}$ then A is an ideal of R and M is a faithful right (R/A)-module with Krull dimension. By [16], the ring R/A has nilpotent prime radical and hence $N^t \subseteq A$ for some positive integer t. By the proof of Proposition 1.4, $k(M) \leq r.gl.k(R/N)$. It follows that $r.gl.k(R) \leq r.gl.k(R/N)$. By Proposition 1.4, $r.gl.k(R) = r.gl.k(R/N)$. Similarly $r.gl.k^0(R) = r.gl.k^0(R/N)$.

Next we consider commutative Noetherian rings. If P is a prime ideal of a commutative Noetherian ring R then the height of P will be denoted by $ht(P)$; i.e. $ht(P)$ is the greatest positive integer n such that there exists a chain $P = P_0 \supset P_1 \supset \ldots \supset P_n$ of distinct prime ideals $P_i(0 \leq i \leq n)$ of R. The *dimension*, $\dim(R)$, of R is defined by

$$\dim(R) = \sup\{ht(P) : P \text{ is a prime ideal of } R\}.$$

Because $ht(P)$ is a non-negative integer for every prime ideal P of R, $\dim(R)$ is a non-negative integer or $\dim R = \omega$, the first infinite ordinal. For the commutative ring R, $r.gl.k(R) = l.gl.k(R)$, of course, and we denote their common value by $gl.k(R)$. Similarly, $gl.k^0(R)$ will denote $r.gl.k^0(R)$ for the commutative ring R.

Theorem 1.8 *Let R be a commutative Noetherian ring. Then $gl.k^0(R) = \dim R \leq gl.k(R) = k(R)$. Moreover, if R has finite Krull dimension then $gl.k^0(R) = k(R)$.*

Proof. By [15, Corollaire 4.5] or [5, Proposition 12].

For direct products of rings we have the following result.

Proposition 1.9 *Let R_i $(1 \leq i \leq n)$ be a finite collection of rings and let R be the ring $R_1 \times \ldots \times R_n$. Then*

$$\begin{aligned} r.gl.k(R) &= \max\{r.gl.k(R_i) : 1 \leq i \leq n\}, \quad and \\ r.gl.k^0(R) &= \max\{r.gl.k^0(R_i) : 1 \leq i \leq n\}. \end{aligned}$$

Proof. Let M be a right R-module. For each $1 \leq i \leq n$, MR_i is an R-submodule of M and a subset N of MR_i is an R-submodule of MR_i if and only if N is an R_i-submodule of MR_i. Thus MR_i has the same Krull dimension when considered as an R-module or as an R_i-module. Now $M = (MR_1) \oplus \ldots \oplus (MR_n)$. It follows that $k(M) \leq \max\{r.gl.k(R_i) : 1 \leq i \leq n\}$ by [11, Lemma 1.1]. Hence $r.gl.k(R) \leq \max\{r.gl.k(R_i) : 1 \leq i \leq n\}$. By Proposition 1.4, $r.gl.k(R) = \max\{r.gl.k(R_i) : 1 \leq i \leq n\}$. Similarly, $r.gl.k^0(R) = \max\{r.gl.k^0(R_i) : 1 \leq i \leq n\}$.

Corollary 1.10 *Let R be a ring and let $T_n(R)$ be the ring of all $n \times n$ upper triangular matrices with entries in R, for some positive integer n. Then $r.gl.k(T_n(R)) = r.gl.k(R)$ and $r.gl.k^0(T_n(R)) = r.gl.k^0(R)$.*

Proof. Let $T = T_n(R)$ and let I denote the ideal of T consisting of all matrices with diagonal entries 0. Then I is a nilpotent ideal of T and $T/I \cong R^{(n)}$. Apply Propositions 1.4 and 1.9.

Now let S and T be rings and let M be a left S-, right T-bimodule. We form the ring $[S, M; 0, T]$ consisting of all "matrices"

$$\begin{bmatrix} s & m \\ 0 & t \end{bmatrix}$$

with $s \in S$, $m \in M$, $t \in T$, where addition and multiplication are the usual matrix operations. For such a ring $R = [S, M; 0, T]$ the ideal $I = [0, M; 0, 0]$ satisfies $I^2 = 0$ and $R/I \cong S \oplus T$. Propositions 1.4 and 1.9 can be combined to give the next result without further proof.

Proposition 1.11 *With the above notation, let $R = [S, M; 0, T]$. Then $r.gl.k(R) = \max\{r.gl.k(S), r.gl.k(T)\}$ and $r.gl.k^0(R) = \max\{r.gl.k^0(S), r.gl.k^0(T)\}$.*

Corollary 1.12 *Let T be any ring with right Krull dimension and let M be a right T-module. Let $R = [\mathbb{Z}, M; 0, T]$. Then the ring R has right Krull dimension if and only if the right T-module M has Krull dimension. In general, $r.gl.k(R) = \max\{1, k(T_T)\}$ and $r.gl.k^0(R) = \max\{1, r.gl.k^0(T)\}$.*

Proof. The first part follows by [11, Lemma 1.1]. Moreover, by Propositions 1.2 and 1.11, $r.gl.k(R) = \max\{1, k(T_T)\}$ On the other hand, by Theorem 1.8 and Proposition 1.11, $r.gl.k^0(R) = \max\{1, r.gl.k^0(T)\}$.

It is natural to question how global (dual) Krull dimension behaves under Morita equivalence. This is dealt with next.

Proposition 1.13 *Let R and R' be rings which are Morita equivalent. Then $r.gl.k(R) = r.gl.k(R')$ and $r.gl.k^0(R) = r.gl.k^0(R')$.*

Proof. By [2, Proposition 21.7].

In particular, Proposition 1.13 shows that if R is a ring and $M_n(R)$ the ring of all $n \times n$ matrices with entries in R, for any positive integer n, then $r.gl.k(M_n(R)) = r.gl.k(R)$ and $r.gl.k^0(M_n(R)) = r.gl.k^0(R)$.

2 Generalized Loewy modules

Let R be a ring and let M be a right R-module. Given an ordinal $\alpha \geq 0$, it will be convenient to write "$k(M) \leq \alpha$" to mean that the module M has Krull dimension β and $\beta \leq \alpha$. On the other hand, "$k(M) \not\leq \alpha$" will mean that either M does not have Krull dimension or M does have Krull dimension β and $\alpha < \beta$. For a general module M and ordinal $\alpha \geq 0$, we define the α-*Loewy series* of M to be the series of submodules

$$0 = S_0^\alpha(M) \subseteq S_1^\alpha(M) \subseteq \ldots \subseteq S_\tau^\alpha(M) \subseteq S_{\tau+1}^\alpha(M) \subseteq \ldots$$

where $S_{\tau+1}^\alpha(M)/S_\tau^\alpha(M)$ is the sum of all submodules X of the factor module $M/S_\tau^\alpha(M)$ such that $k(X) \leq \alpha$, and

$$S_\tau^\alpha(M) = \bigcup_{0 \leq \sigma < \tau} S_\sigma^\alpha(M)$$

if τ is a limit ordinal. The module M is called an α-*Loewy module* if $S_\rho^\alpha(M) = M$ for some ordinal $\rho \geq 0$, and, in this case, the least ordinal ρ such that $S_\rho^\alpha(M) = M$ is called the α-*Loewy length* of M. The ring R is called a *right α-Loewy ring* if the right R-module R is an α-Loewy module and, in this case, the Loewy length of the right R-module R is called the *right α-Loewy length* of R. In case $\alpha = 0$, the 0-Loewy series of M is just the usual Loewy series of M, 0-Loewy modules are called *Loewy modules* or *semi-artinian modules*, right 0-Loewy rings are called *right Loewy rings* or *right semi-artinian rings*, and in this case the right 0-Loewy length of R is called simply the *right Loewy length*.

Lemma 2.1 *The following statements are equivalent for an R-module M and ordinal $\alpha \geq 0$.*

(i) M *is an α-Loewy module.*

(ii) *There exists a chain* $0 = N_0 \subseteq N_1 \subseteq \ldots \subseteq N_\lambda \subseteq N_{\lambda+1} \subseteq \ldots \subseteq N_\mu = M$ *of submodules of M such that $k(N_{\lambda+1}/N_\lambda) \leq \alpha$ for all $0 \leq \lambda < \mu$ and $N_\lambda = \bigcup_{0 \leq \sigma < \lambda} N_\sigma$ if λ is a limit ordinal.*

(iii) *For every proper submodule L of M there exists a submodule N properly containing L such that $k(N/L) \leq \alpha$.*

Proof. (i) $\Rightarrow$ (iii) There exists an ordinal ρ such that $S_\rho^\alpha(M) = M$. Let L be a proper submodule of M. Then $S_0^\alpha(M) \subseteq L$ but $S_\rho^\alpha(M) \not\subseteq L$. Let τ be the least ordinal such that $S_\tau^\alpha(M) \not\subseteq L$. Clearly τ is not a limit ordinal. Let $S = S_{\tau-1}^\alpha(M)$. Note that $S \subseteq L$. There exists a submodule N containing S such that $k(N/S) \leq \alpha$ and $N \not\subseteq L$. Then L is properly contained in the submodule $N + L$ and $k((N + L)/L) = k(N/(N \cap L)) \leq k(N/S) \leq \alpha$.

(iii) $\Rightarrow$ (ii) Suppose that $M \neq 0$. There exists a non-zero submodule N_1 such that $k(N_1) \leq \alpha$. Suppose that $N_1 \neq M$. There exists a submodule N_2 properly containing N_1 such that $k(N_2/N_1) \leq \alpha$. Repeat this process to obtain a chain of submodules $0 = N_0 \subset N_1 \subset N_2 \subset \ldots \subset N_\lambda \subset N_{\lambda+1} \subset \ldots$ such that $k(N_{\lambda+1}/N_\lambda) \leq \alpha$ for each ordinal $\lambda \geq 0$ and $N_\lambda = \cup_{0 \leq \sigma < \lambda} N_\sigma$ if λ is a limit ordinal. Clearly there exists an ordinal $\mu \geq 0$ such that $N_\mu = M$.

(ii) $\Rightarrow$ (i) Suppose that $N_\lambda \subseteq S_\lambda^\alpha(M) = T$ for some ordinal $\lambda \geq 0$. Then

$$k((N_{\lambda+1} + T)/T) = k(N_{\lambda+1}/(N_{\lambda+1} \cap T)) \leq k(N_{\lambda+1}/N_\lambda) \leq \alpha,$$

so that $N_{\lambda+1} \subseteq S_{\lambda+1}^\alpha(M)$. It is now clear that $N_\lambda \subseteq S_\lambda^\alpha(M)$ for all $\lambda \geq 0$ and hence $S_\mu^\alpha(M) = M$.

Corollary 2.2 *For any ring R and ordinal $\alpha \geq 0$, the class of α-Loewy right R-modules is closed under taking submodules, homomorphic images, extensions and direct sums.*

Proof. By Lemma 2.1.

Note that Corollary 2.2 shows that for any ordinal $\alpha \geq 0$ and ring R, the collection of α-Loewy right R-modules forms the torsion class of an hereditary torsion theory in mod $-R$, the category of right R-modules (see, for example, [23]). Another consequence of Lemma 2.1 is the following result.

Corollary 2.3 *Given an ordinal $\alpha \geq 0$, an R-module M satisfies $k(M) \leq \alpha$ if and only if M is an α-Loewy module and every homomorphic image of M has finite uniform dimension.*

Proof. The necessity follows by [11, Proposition 1.4]. Conversely, suppose that M is an α-Loewy module such that M/N has finite uniform dimension for every submodule N. Suppose that $M \neq 0$. There exist a positive integer n and independent uniform submodules $U_i (1 \leq i \leq n)$ such that $U_1 \oplus \ldots \oplus U_n$ is an essential submodule of M. For each $1 \leq i \leq n$, Corollary 2.2 gives that U_i is an α-Loewy module and hence U_i contains a non-zero submodule V_i with $k(V_i) \leq \alpha$ by Lemma 2.1. Then $V = V_1 \oplus \ldots \oplus V_n$ is an essential submodule of M and $k(V) \leq \alpha$. In this way, every non-zero homomorphic image of M contains an essential submodule with Krull dimension at most α. By [7, 6.3], $k(M) \leq \alpha$.

A third consequence of Lemma 2.1 is the next result.

Proposition 2.4 *The following statements are equivalent for a ring R and ordinal $\alpha \geq 0$.*
(i) R is a right α-Loewy ring.
(ii) Every non-zero right R-module contains a non-zero submodule having Krull dimension at most α.
(iii Every right R-module is an α-Loewy module.
Moreover, in this case, the α-Loewy length of any right R-module is not greater than the right α-Loewy length of R.

Proof. (i) $\Rightarrow$ (iii) By Corollary 2.2.
(ii) $\Rightarrow$ (iii) By Lemma 2.1.

(iii) $\Rightarrow$ (i) Clear.

For the last part, suppose that R is a right α-Loewy ring and M is a right R-module. Let Ω denote the (non-empty) collection of right ideals C of R such that $k(C) \leq \alpha$. For each $m \in M$ and $C \in \Omega$, the module mC is a homomorphic image of C and hence $k(mC) \leq k(C) \leq \alpha$, i.e. $mC \leq S_1^\alpha(M)$. Now

$$MS_1^\alpha(R) = \sum_{m \in M} \sum_{C \in \Omega} mC \leq S_1^\alpha(M).$$

By a similar argument it can be shown that if $MS_\tau^\alpha(R) \leq S_\tau^\alpha(M)$ for some ordinal $\tau \geq 0$ then $MS_{\tau+1}^\alpha(R) \subseteq S_{\tau+1}^\alpha(M)$. Moreover, if τ is a limit ordinal such that $MS_\sigma^\alpha(R) \subseteq S_\sigma^\alpha(M)$ for all ordinals $0 \leq \sigma < \tau$ then

$$MS_\tau^\alpha(R) = \bigcup_{0 \leq \sigma < \tau} MS_\sigma^\alpha(R) \subseteq \bigcup_{0 \leq \sigma < \tau} S_\sigma^\alpha(M) = S_\tau^\alpha(M).$$

It follows that $MS_\tau^\alpha(R) \subseteq S_\tau^\alpha(M)$ for all ordinals $\tau \geq 0$. If R has right α-Loewy length ρ then $M = MR = MS_\rho^\alpha(R) \subseteq S_\rho^\alpha(M)$, so that $S_\rho^\alpha(M) = M$ and the α-Loewy length of M is at most ρ.

The next result shows why right α-Loewy rings are of interest in the present study.

Theorem 2.5 *For any ordinal $\alpha \geq 0$, $r.gl.k(R) \leq \alpha$ for any right α-Loewy ring R.*

Proof. Let R be a right α-Loewy ring. Let M be any right R-module with Krull dimension. By [11, Proposition 1.4] every homomorphic image of M has finite uniform dimension. Hence $k(M) \leq \alpha$ by Corollary 2.3 and Proposition 2.4. It follows that $r.gl.k(R) \leq \alpha$.

It is easy to produce examples of α-Loewy rings and modules, for any ordinal $\alpha \geq 0$. Note first that if R is a ring with right Krull dimension α then R is a right α-Loewy ring and hence every right R-module is an α-Loewy module. Let T be an algebra over a field K such that the ring T has right Krull dimension α. Let M be any right T-module and let R denote the ring $[K, M; 0, T]$. Then R is a right α-Loewy ring but R does not have right Krull dimension unless the T-module M has Krull dimenson. With the same ring T, consider the direct product $B = \prod_{n \in \mathbb{N}} T_n$, where $T_n = T$ for every $n \in \mathbb{N}$. Let A be the subring of B consisting of all sequences $\{t_n\}$ such that $t_n \in T$ $(n \in \mathbb{N})$ and there exists $k \in \mathbb{N}$ such that $t_k = t_{k+1} = t_{k+2} = \ldots$ and $t_k \in K$. Then $I = \oplus_{n \in \mathbb{N}} T_n$ is an ideal of A and $A/I \cong K$. If E is a right ideal of A and E has Krull dimension then $E \subseteq I$. It follows that $I = Soc_1^\alpha(A)$ and $A = Soc_2^\alpha(A)$. Thus A is a right α-Loewy ring. Clearly A does not have right Krull dimension (see, for example, [11, Proposition 1.4]).

The most widely studied class of right α-Loewy rings is the case $\alpha = 0$, in which case R is called a right Loewy ring. Any left perfect ring is right Loewy by [2, Theorem 28.4]. Osofsky [19] proved that if λ and ρ are infinite ordinals then there exists a right and left perfect ring with right Loewy length $\rho + 1$ and left Loewy length $\lambda + 1$. (Clearly the right Loewy length of a right Loewy ring cannot be a limit ordinal.) Moreover, Osofsky also showed that for any infinite ordinal ρ there exists a left perfect ring R of right Loewy length $\rho + 1$ but having zero left socle. This is in contrast to the following result of Camillo and Fuller [4, Theorem 1.3].

Proposition 2.6 *Any right Loewy ring of finite right Loewy length is a left Loewy ring of finite left Loewy length.*

Loewy modules can be dualised. For a module M and ordinal $\alpha \geq 0$, "$k^0(M) \leq \alpha$" and "$k^0(M) \not\leq \alpha$" will have the analogous meanings to "$k(M) \leq \alpha$" and "$k(M) \not\leq \alpha$", respectively. By the α-*Hamsher series* of a module M we mean the series

$$M = S_\alpha^0(M) \supseteq S_\alpha^1(M) \supseteq \ldots \supseteq S_\alpha^\tau(M) \supseteq S_\alpha^{\tau+1}(M) \supseteq \ldots$$

where $S_\alpha^{\tau+1}(M)$ is the intersection of all submodules N of $S_\alpha^\tau(M)$ such that $k^0(S_\alpha^\tau(M)/N) \leq \alpha$, for each ordinal $\tau \geq 0$, and

$$S_\alpha^\tau(M) = \bigcap_{0 \leq \sigma < \tau} S_\alpha^\sigma(M)$$

if τ is a limit ordinal. The module M will be called an α-*Hamsher module* if $S_\alpha^\rho(M) = 0$ for some ordinal $\rho \geq 0$. The ring R is called a *right α-Hamsher ring* if the right R-module R is an α-Hamsher module. In case $\alpha = 0$ we call 0-Hamsher modules and right 0-Hamsher rings simply *Hamsher modules* and *right Hamsher rings*, respectively.

The dual of Lemma 2.1 is the following result. We omit the proof.

Lemma 2.7 *The following statements are equivalent for an R-module M and ordinal $\alpha \geq 0$.*

    (i) M is an α-Hamsher module.

    (ii) There exists a chain $M = N_0 \supseteq N_1 \supseteq \cdots \supseteq N_\lambda \supseteq N_{\lambda+1} \supseteq \cdots \supseteq N_\mu = 0$ of submodules of M such that $k^0(N_\lambda/N_{\lambda+1}) \leq \alpha$ for all $0 \leq \lambda < \mu$ and $N_\lambda = \bigcap_{0 \leq \sigma < \lambda} N_\sigma$ if λ is a limit ordinal.

    (iii) For every non-zero submodule N of M there exists a proper submodule L of N such that $k^0(N/L) \leq \alpha$.

Corollary 2.8 *A right R-module M is a Hamsher module if and only if every non-zero submodule of M contains a maximal submodule.*

    Proof. Let M be a Hamsher module and let N be a non-zero submodule of M. By Lemma 2.7 there exists a proper submodule L of N such that $k^0(N/L) \leq 0$, i.e. N/L is a non-zero Noetherian module. Thus N/L, and hence also N, contains a maximal submodule. Conversely, if every non-zero submodule contains a maximal submodule then M is a Hamsher module by Lemma 2.7.

    Corollary 2.8 is the motivation for defining Hamsher, and more generally α-Hamsher, modules as we do (see [10]). Another consequence of Lemma 2.7 is the following result.

Corollary 2.9 *For any ring R and ordinal $\alpha \geq 0$, the class of α-Hamsher right R-modules is closed under taking submodules, extensions and direct products.*

    Proof. By Lemma 2.7.

    It is not the case that any homomorphic image of a Hamsher module is also a Hamsher module. For example, any free $\mathbb{Z}$-module F of infinite rank is a Hamsher module by Corollary 2.8 because every non-zero submodule of F is free. However there exists a submodule G of F such that $F/G \cong \mathbb{Q}$ and hence F/G is not a Hamsher module.

    For any R-module M, the injective hull of M will be denoted by $E(M)$. Corollary 2.2 shows that if $\mathrm{mod} - R$ contains an α-Loewy generator G then every right R-module is α-Loewy (cf. Proposition 2.4). The analogue of this fact is given in the next result.

Theorem 2.10 *The following statements are equivalent for a ring R and ordinal $\alpha \geq 0$.*

(i)   There exists an α-Hamsher cogenerator for $\mathrm{mod} - R$.

(ii)   The injective hull of every simple right R-module is an α-Hamsher module.

(iii)   Every non-zero right R-module M contains a proper submodule N such that $k^0(M/N) \leq \alpha$.

(iv)   Every right R-module is α-Hamsher.

Proof. (i) $\Rightarrow$ (iv) By Corollary 2.9.

(iv) $\Rightarrow$ (ii) Clear.

(ii) $\Rightarrow$ (i) Let S_i $(i \in I)$ be representatives of the isomorphism classes of simple right R-modules. For each $i \in I$ let E_i denote the injective hull of S_i. Let $E = \prod_{i \in I} E_i$. It is well known that E is a cogenerator for $\mathrm{mod} - R$. By (ii) and Corollary 2.9, E is an α-Hamsher module.

(iii) $\Leftrightarrow$ (iv) By Lemma 2.7.

Given an ordinal $\alpha \geq 0$ and a ring R, we shall say that $\mathrm{mod} - R$ is α-Hamsher (respectively, Hamsher) if every right R-module is α-Hamsher (Hamsher). Theorem 2.10 generalises [10, Theorem 1]. It has the following consequence.

Corollary 2.11 *Let R be a ring such that $\mathrm{mod} - R$ is α-Hamsher for some ordinal $\alpha \geq 0$. Then $r.gl.k^0(R) \leq \alpha$.*

Proof. Let M be a non-zero cocritical right R-module. There exists a proper submodule N of M such that $k^0(M/N) \leq \alpha$ (Theorem 2.10). By [5, Proposition 3] $k^0(M) = k^0(M/N) \leq \alpha$. Therefore $r.gl.k^0(R) \leq \alpha$ by Proposition 1.1.

An ideal I of a ring R is *right T-nilpotent* if for any sequence $a_1, a_2, a_3, \ldots$ of elements of I there exists a positive integer n such that $a_n a_{n-1} \ldots a_2 a_1 = 0$. It is well known that an ideal I of R is right T-nilpotent if and only if $M \neq MI$ for any non-zero right R-module M (see, for example, [2, Lemma 28.3]).

Proposition 2.12 *Let I be a right T-nilpotent ideal of a ring R. Then*

(i) $r.gl.k^0(R) = r.gl.k^0(R/I)$.

(ii)   Given an ordinal $\alpha \geq 0$, $\mathrm{mod} - R$ is α-Hamsher if and only if $\mathrm{mod} - (R/I)$ is α-Hamsher.

Proof. (i) By Proposition 1.4, $r.gl.k^0(R/I) \leq r.gl.k^0(R)$. Conversely, let M be a non-zero cocritical right R-module. Then $M \neq MI$. By [5, Proposition 3], $k^0(M) = k^0(M/MI) \leq r.gl.k^0(R/I)$. Now $r.gl.k^0(R) \leq r.gl.k^0(R/I)$ by Proposition 1.1, and hence $r.gl.k^0(R) = r.gl.k^0(R/I)$.

(ii) The necessity is obvious. Conversely, suppose that $\mathrm{mod}-(R/I)$ is α-Hamsher. Let N be any non-zero right R-module. Then $N \neq NI$. By Theorem 2.10, the right (R/I)-module N/NI contains a proper submodule L/NI, for some proper submodule L of N containing NI, such that $k^0((N/NI)/(L/NI)) \leq \alpha$. Note that $k^0(N/L) \leq \alpha$. By Theorem 2.10, $\mathrm{mod}-R$ is α-Hamsher.

For the ring $\mathbb{Z}$ of integers, $\mathrm{mod}-\mathbb{Z}$ is 1-Hamsher because $k^0(E(S)) = 1$ for every simple $\mathbb{Z}$-module S. However $\mathrm{mod}-\mathbb{Z}$ is not 0-Hamsher because the $\mathbb{Z}$-module $\mathbb{Q}$ does not contain a maximal submodule (Corollary 2.8). More generally we have the following result.

Theorem 2.13 *Let R be a commutative ring which contains a T-nilpotent ideal I such that the ring R/I is Noetherian. Then $\mathrm{mod}-R$ is α-Hamsher, where α is the supremum of the heights of the prime ideals of R.*

Proof. By Proposition 2.12, Theorem 2.10 and [5, Proposition 12].

In particular, Theorem 2.13 shows that $\mathrm{mod}-R$ is ω-Hamsher for any commutative Noetherian ring R. But what happens for other commutative rings? In particular, given any ordinal $\alpha \geq 0$ does there exist a (commutative) ring R with $k^0(R_R) = \alpha$ or one with $r.gl.k^0(R) = \alpha$? For a commutative ring R, $\mathrm{mod}-R$ is Hamsher if and only if the Jacobson radical J of R is T-nilpotent and the ring R/J is von Neumann regular (see [12], [13]). Note also that Nastasescu and Popescu [19, Theorem 3.1] proved that $\mathrm{mod}-R$ is Hamsher for any commutative Loewy ring R. On the other hand, $\mathrm{mod}-R$ is Hamsher for any right Loewy ring R with finite right Loewy length. For, let M be a non-zero right R-module. By Proposition 2.4, M has finite Loewy length and hence there exists a proper submodule N of M such that M/N is a semisimple module and hence has a maximal submodule. By Corollary 2.8, $\mathrm{mod}-R$ is Hamsher.

Proposition 2.14 *Let R be a right Loewy ring with finite right Loewy length. Then $r.gl.k(R) = l.gl.k(R) = r.gl.k^0(R) = l.gl.k^0(R) = 0$.*

Proof. By the above remarks, mod $-R$ is Hamsher and by Proposition 2.6 R is a left Loewy ring with finite left Loewy length. Now apply Theorem 2.5 and Corollary 2.11.

Another case of interest is that of right perfect rings.

Proposition 2.15 *Let R be a right perfect ring. Then* mod $-R$ *is Hamsher and $r.gl.k^0(R) = l.gl.k(R) = 0$.*

Proof. By [2, Theorem 28.4] mod $-R$ is Hamsher and R is left Loewy. The result follows by Theorem 2.5 and Corollary 2.11.

A ring R is called a *right V-ring* if every simple right R-module is injective. A classical result of Kaplansky states that a commutative ring is a V-ring if and only if R is von Neumann regular, and this has been generalized to PI-rings by Armendariz and Fisher [3]. More generally, a ring R is called a *right GV-ring* if every singular simple right R-module is injective.

Proposition 2.16 *Let R be a right GV-ring. Then* mod $-R$ *is Hamsher and $r.gl.k^0(R) = 0$. Moreover, if R is commutative then $gl.k(R) = 0$.*

Proof. The first part follows by Corollary 2.8 and [7, Proposition 16.3]. Now suppose that R is commutative. Because R is a GV-ring it is not hard to prove that R is von Neumann regular. Let M be a critical R-module. Then M is uniform. Let $m \in M$, $r \in R$ such that $mr \neq 0$. Then $rR = eR$ for some idempotent e in R and $meR \cap m(1 - e)R = 0$. Thus $m(1 - e) = 0$ and $m = me \in mrR$. It follows that mR is a simple R-module and hence is injective by Kaplansky's Theorem. Thus $M = mR$. It follows that M is simple. By Proposition 1.1, $gl.k(R) = 0$.

Let F be a field and let R denote the ring $\prod_{n \in \mathbb{N}} F_n$, where $F_n = F$ ($n \in \mathbb{N}$). Let I be the ideal $\oplus_{n \in \mathbb{N}} F_n$ and let S denote the factor ring R/I. Then S is a commutative von Neumann regular ring with zero socle and $gl.k(R) = gl.k^0(R) = 0$. Moreover, if C is a Cozzens domain then C is a right and left Noetherian right and left V-domain and $k(R_R) = 1 = k(_RR)$, $k^0(R_R) = k^0(_RR) = r.gl.k^0(R) = l.gl.k^0(R) = 0$ (see [9, Theorems 7.42 and 7.45]).

3 Rings with $r.gl.k(R) \leq \alpha$ or $r.gl.k^0(R) \leq \alpha$.

Let R be a ring and let M be a right R-module. For any ordinal $\alpha \geq 0$, we let $\mathbf{L}_\alpha(M)$ denote the collection of submodules N of M such that $k(M/N) \not\leq \alpha$, i.e. M/N does not have Krull dimension or M/N has Krull dimension β where $\beta \not\leq \alpha$. Of course, it may happen that $\mathbf{L}_\alpha(M)$ is the empty set for some ordinal α, for example if $k(M) \leq \alpha$.

Theorem 3.1 *Given a ring R and ordinal $\alpha \geq 0$, the following statements are equivalent.*

(i) $r.gl.k(R) \leq \alpha$.

(ii) $\mathbf{L}_\alpha(R_R)$ does not contain a maximal member.

(iii) $\mathbf{L}_\alpha(G)$ does not contain a maximal number for every generator G of $\mathrm{mod} - R$.

(iv) For every right R-module M with $k(M) \not\leq \alpha$ there exists a non-zero submodule N such that $k(M/N) \not\leq \alpha$.

Proof. (i) $\Rightarrow$ (iii) Let G be a generator for $\mathrm{mod} - R$ such that $\mathbf{L}_\alpha(G)$ contains a maximal member F. Then $k(G/F) \not\leq \alpha$. Let $G = H_0 \geq H_1 \geq H_2 \geq \ldots$ be any descending chain of submodules of G containing F. If $H_n = F$ for some $n \geq 1$ then $H_n = H_{n+1} = H_{n+2} = \ldots$ and hence $k(H_i/H_{i+1}) = -1 \leq \alpha$ for all $i \geq n$. Otherwise, for all $n \geq 1$, $H_n \neq F$ and hence $k(H_{n-1}/H_n) \leq k(G/H_n) \leq \alpha$, by the choice of F. In any case, we see that $k(G/F) \leq \alpha + 1$ and hence $k(G/F) = \alpha + 1$. Thus $r.gl.k(R) \not\leq \alpha$.

(iii) $\Rightarrow$ (ii) Clear.

(ii) $\Rightarrow$ (iv) Let M be a right R-module with $k(M) \not\leq \alpha$. Suppose that $k(M/N) \leq \alpha$ for every non-zero submodule N of M. Then $k(M) \leq \alpha + 1$, by the proof of (i) $\Rightarrow$ (iii). Thus $k(M) = \alpha + 1$ and M is $(\alpha + 1)$-critical. Let $0 \neq m \in M$ and let $A = \{r \in R : mr = 0\}$. Then $R/A \cong mR$ and the right R-module R/A is $(\alpha + 1)$-critical. Clearly A is a maximal member of $\mathbf{L}_\alpha(R_R)$.

(iv) $\Rightarrow$ (i) Suppose that $r.gl.k(R) \not\leq \alpha$. Then there exists a right R-module L such that L has Krull dimension β for some $\beta \geq \alpha + 1$. By induction on β it is easy to prove that there exist submodules $V \subseteq U$ of L such that $k(U/V) = \alpha + 1$. Another easy induction argument shows that there exist submodules $Y \subseteq X$ of L such that $V \subseteq Y \subseteq X \subseteq U$ and X/Y is $(\alpha + 1)$-critical. Then the module $Z = X/Y$ satisfies $k(Z) \not\leq \alpha$ and $k(Z/T) \leq \alpha$ for every non-zero submodule T of Z.

Corollary 3.2 *The following statements are equivalent for a ring R.*
(i) $r.gl.k(R) = 0$.
(ii) *For every non-Artinian right R-module M there exists a non-zero submodule N of M such that the module M/N is not Artinian.*
(iii) *Every critical right R-module is simple.*

Proof. (i) $\Leftrightarrow$ (ii) By the theorem with $\alpha = 0$.
(i) $\Rightarrow$ (iii) Clear.
(iii) $\Rightarrow$ (i) By Proposition 1.1.

Given a ring R, a right R-module M and an ordinal $\alpha \geq 0$, we let $\mathbf{L}^0_\alpha(M)$ denote the collection of submodules N of M such that $k^0(N) \not\leq \alpha$. The analogue for dual Krull dimension of Theorem 3.1 is the following result. It can be proved by dualizing the proof of Theorem 3.1 and so the proof is omitted.

Theorem 3.3 *Given a ring R and ordinal $\alpha \geq 0$, the following statements are equivalent.*
(i) $r.gl.k^0(R) \leq \alpha$.
(ii) *There exists a cogenerator E for $\mathrm{mod} - R$ such that $\mathbf{L}^0_\alpha(E)$ does not contain a minimal member.*
(iii) $\mathbf{L}_\alpha(H)$ *does not contain a minimal member for every cogenerator H of $\mathrm{mod} - R$.*
(iv) *For every right R-module M with $k^0(M) \not\leq \alpha$ there exists a proper submodule N such that $k^0(N) \not\leq \alpha$.*

Corollary 3.4 *The following statements are equivalent for a ring R.*
(i) $r.gl.k^0(R) = 0$.
(ii) *Every non-Noetherian right R-module contains a non-Noetherian proper submodule.*
(iii) *Every cocritical right R-module is simple.*

Proof. See the proof of Corollary 3.2.

In Corollary 3.4, (i) $\Leftrightarrow$ (ii) is due to Sarath [21, Theorem 2.7]. Now we turn to the case of commutative rings. The first result is well known but is included for completeness. Given a ring R, an R-module M and a non-empty subset X of M we set $\mathbf{r}(X) = \{r \in R : xr = 0 \text{ for all } x \in X\}$.

Lemma 3.5 *Let R be a commutative ring.*

(i) An R-module M is a cyclic critical R-module if and only if $M \cong R/P$ for some prime ideal P of R such that R/P has Krull dimension.

(ii) For any cocritical R-module N there exists a prime ideal Q of R such that $Q = \mathbf{r}(N)$.

Proof. (i) It is clear that the R-module R/P is cyclic critical for every prime ideal P of R such that R/P has Krull dimension. Conversely, suppose that $M = mR$ is a cyclic critical R-module. Let $P = \mathbf{r}(m)$. Then P is a proper ideal of R and $M \cong R/P$. Let $a, b \in R$ such that $ab \in P$. Suppose that $a \notin P$. Then $ma \neq 0$. Define a mapping $\phi : M \to M$ by $\phi(mr) = mar$ for all $r \in R$. Then ϕ is a homomorphism and $M/\ker \phi \cong maR$. Thus $k(M/\ker \phi) = k(maR) = k(M)$ and hence $\ker \phi = 0$. But $mb \in \ker \phi$, so that $mb = 0$ and $b \in P$. It follows that P is a prime ideal of R.

(ii) Let N be a cocritical R-module and let $Q = \mathbf{r}(N)$. Then Q is a proper ideal of R. Let $a, b \in R$ such that $ab \in Q$. Then $Nab = 0$. Define a mapping $\theta : N \to N$ by $\theta(x) = xa$ for all $x \in N$. Then θ is a homomorphism with image Na. Suppose that $Nb \neq 0$. Then $Na \neq N$ and hence $k^0(N/\ker \theta) = k^0(Na) < k^0(N)$, so that $\ker \theta = N$, i.e. $Na = 0$ and $a \in Q$. It follows that Q is a prime ideal of R.

Corollary 3.6 *Let R be a commutative ring. Then*

(i) $gl.k(R) = \sup\{k(R/P) : P$ is a prime ideal of R such that the ring R/P has Krull dimension$\}$,

(ii) $gl.k^0(R) = \sup\{gl.k^0(R/P) : P$ is a prime ideal of $R\}$.

Proof. By Proposition 1.1 and Lemma 3.5.

Corollary 3.7 *Let R be a commutative ring such that every prime ideal is maximal. Then $gl.k(R) = gl.k^0(R) = 0$.*

Proof. By Corollary 3.6.

Proposition 3.8 *Let R be a commutative ring such that $\dim R = 1$. Then $gl.k(R) \leq 1$. Moreover, if R is a domain then $gl.k(R) = 0$ or R is a Noetherian ring and $gl.k(R) = 1$.*

Proof. Let P be a prime ideal of R such that the ring R/P has Krull dimension. Suppose that P is not a maximal ideal of R. For every ideal I properly containing P the ring R/I has Krull dimension and every prime ideal is maximal, so that R/I is an Artinian ring by [11, Corollary 7.5]. It follows that $k(R/P) = 1$. Moreover, R/I is a Noetherian ring for every ideal I properly containing P, so that the ring R/P is Noetherian. The result follows.

Proposition 3.9 *Let R be a commutative ring and let $\alpha \geq 0$ be an ordinal. Then*

(i) $gl.k(R) \leq \alpha$ if and only if there does not exist a prime ideal P of R such that the domain R/P has Krull dimension $\alpha + 1$.

(ii) $gl.k^0(R) \leq \alpha$ if and only if there does not exist a prime ideal Q of R such that $gl.k^0(R/Q) = \alpha + 1$.

Proof. (i) The necessity is clear by Corollary 3.6. Conversely, suppose tht $gl.k(R) \not\leq \alpha$. By the proof of Theorem 3.1 (iv) $\Rightarrow$ (i), there exists an $(\alpha + 1)$-critical R-module M. Let $0 \neq m \in M$ and let $P = \mathbf{r}(mR)$. By Lemma 3.5 (i), P is a prime ideal of R and $k(R/P) = \alpha + 1$.

(ii) Similar to (i).

Theorem 3.10 *Let R be a commutative ring. Then*

(i) $gl.k(R) = 0$ if and only if every Noetherian R-module is Artinian.

(ii) $gl.k^0(R) = 0$ if and only if every Artinian R-module is Noetherian. Moreover, if $gl.k^0(R) = 0$ then $gl.k(R) = 0$.

Proof. (i) The necessity is clear by [11, Proposition 1.3]. Conversely, suppose that every Noetherian R-module is Artinian. Let P be a prime ideal of R such that $k(R/P) = 1$. For every ideal I properly containing P, $k(R/I) = 0$, i.e. R/I is an Artinian ring and hence a Noetherian ring. It follows that R/P is a Noetherian ring. Thus R/P is a Noetherian R-module. By hypothesis, R/P is an Artinian R-module and hence $k(R/P) = 0$, a contradiction. By Proposition 3.9, $gl.k(R) = 0$.

(ii) The necessity is clear. Conversely, suppose that every Artinian R-module is Noetherian. Let M be a cyclic critical R-module. We shall show that M is simple. Suppose that M is not simple. There exists $m \in M$ such that $M = mR$. If $P = \mathbf{r}(m) = \mathbf{r}(M)$ then $R/P \cong mR$ and hence R/P is a domain with Krull dimension, which is not a field. By [11, Theorem

7.1] there exists a prime ideal Q of R with $P \subseteq Q$ and Q is maximal with respect to not being maximal. For any ideal J properly containing Q, the prime ideals of R/J are all maximal and hence the ring R/J is Artinian by [11, Corollary 7.5]. Thus R/J is a Noetherian ring for every ideal J properly containing Q and it follows that R/Q is a Noetherian ring.

There exists a maximal ideal B of R such that $Q \subseteq B$. By [22, Theorem 4.30] the injective hull E of the simple (R/Q)-module R/B is Artinian and hence Noetherian by hypothesis. Let A denote the maximal ideal B/Q of the ring R/Q. Then $EA^n = 0$ for some positive integer n by [22, Proposition 4.23] and $A = 0$ by [22, Proposition 2.6], i.e. $B = Q$, a contradiction.

Thus M is simple. Hence every cyclic critical R-module is simple and $gl.k(R) = 0$ by Proposition 1.1. Moreover, if V is an R-module with Krull dimension then V is Artinian and hence V is Noetherian. Thus $gl.k^0(R) = 0$. This completes the proof of the theorem.

Corollary 3.11 *A commutative ring R satisfies $gl.k^0(R) = 0$ if and only if no homomorphic image of R is isomorphic to a dense subring of a complete local Noetherian domain of dimension 1.*

Proof. By Theorem 3.10 and [8, Proposition 4.4].

From Theorem 3.10, we see that for a commutative ring R, $gl.k^0(R) = 0$ implies that $gl.k(R) = 0$. This is not true for non-commutative rings. If R is a Cozzens domain then $r.gl.k^0(R) = l.gl.k^0(R) = 0$ but $r.gl.k(R) = l.gl.k(R) = 1$. It is not the case that if R is a commutative ring with $gl.k(R) = 0$ then $gl.k^0(R) = 0$ (although it *is* true if R has Krull dimension by Proposition 1.2 and Corollary 1.3), as the following example shows.

Example 3.12 *There exists a commutative domain R with $\dim R = 1$, $gl.k(R) = 0$ and $gl.k^0(R) = 1$.*

Proof. In [8, Example 4.3] an example is given of a commutative domain R such that $\dim R = 1$ and R has only two maximal ideals P and Q with P idempotent and Q a principal ideal. Since $P \neq 0$ it follows that P is not finitely generated and hence R is not a Noetherian ring. By Proposition 3.8, $gl.k(R) = 0$.

Let M be a cocyclic R-module with Krull dimension. Because $gl.k(R) = 0$ it follows that M is Artinian. Let S be the unique simple submodule of M.

Let $0 \neq m \in M$. Then mR is a cyclic Artinian module over the commutative ring R, so that mR has finite length. It follows that $mP^sQ^t = 0$ for some non-negative integers s, t. Suppose that $P = \mathbf{r}(S)$. Then $mP^s = 0$ and hence $mP = 0$, i.e. $m \in S$. In this case $M = S$ and $k^0(M) = 0$. Otherwise $Q = \mathbf{r}(S)$. Hence $mQ^t = 0$. Thus M is Q-torsion. In the usual way it follows that M is a module over the local ring R_Q. Because Q is a principal ideal, the ring R_Q is a DVR and, by Theorem 1.8, $gl.k^0(R_Q) = 1$. But the submodules of the R-module M coincide with the submodules of the R_Q-module M. Therefore $k^0(M) \leq gl.k^0(R_Q) = 1$. In any case, $k^0(M) \leq 1$. By Proposition 1.1, $gl.k^0(R) \leq 1$. But $E(R/Q)$ is an Artinian R-module which is not Noetherian, so that $gl.k^0(R) = 1$.

References

[1]   T. Albu and P.F. Smith, Dual Krull dimension and duality, Rocky Mtn. J., to appear.

[2]   F.W. Anderson and K.R. Fuller, Rings and categories of modules (Springer-Verlag, New York 1974).

[3]   E.P. Armendariz and J.W. Fisher, Regular PI-rings, Proc. Amer. Math. Soc. **39** (1973), 247-251.

[4]   V.P. Camillo and K.R. Fuller, On Loewy length of rings, Pacific J. Math. **53** (1974), 347-354.

[5]   L. Chambless, N-dimension and N-critical modules, Comm. Algebra **8** (1980), 1561-1592.

[6]   A.W. Chatters and C.R. Hajarnavis, Rings with chain conditions (Pitman, London 1980).

[7]   Nguyen Viet Dung, Dinh van Huynh, P.F. Smith and R. Wisbauer, Extending modules (Longman, Harlow 1994).

[8]   A. Facchini, Loewy and Artinian modules over commutative rings, Ann. Mat. Pura Appl. (4) **128** (1981), 359-374.

[9]   C. Faith, Algebra: Rings, modules and categories I (Springer-Verlag, Berlin 1973).

[10]   C. Faith, Rings whose modules have maximal submodules, Publ. Mat. **39** (1995), 201-214.

[11]   R. Gordon and J.C. Robson, Krull dimension, Mem. Amer. Math. Soc. **133** (1973).

[12]   R. Hamsher, Commutative rings over which every module has a maximal submodule, Proc. Amer. Math. Soc. **18** (1967), 1133-1137.

[13]   L.A. Koifman, Rings over which every module has a maximal submodule, Math. Notes **7** (1970), 215-219.

[14]   B. Lemonnier, Déviation des ensembles et groupes, abéliens totalement ordonnés, Bull. Sc. Math. **96** (1972), 289-303.

[15]   B. Lemonnier, Dimension de Krull et codeviation. Application au theoreme d'Eakin, Comm. Algebra **6** (1978), 1647-1665.

[16]   V.T. Markov, On *PI*-rings having a faithful module with Krull dimension, Fundam. i Prikl. Math. 1 (1995), 557-560 (in Russian).

[17]   J.C. McConnell and J.C. Robson, Noncommutative Noetherian rings (Wiley-Interscience, Chichester 1987).

[18]   C. Năstăsescu and F. van Oystaeyen, Dimensions of ring theory (Reidel, Dordrecht 1987).

[19]   C. Năstăsescu and N. Popescu, Anneaux semi-artiniens, Bull. Soc. Math. France 96 (1968), 357-368.

[20]   B.L. Osofsky, Loewy length of perfect rings, Proc. Amer. Math. Soc. **28** (1971), 352-354.

[21]   B. Sarath, Krull dimension and Noetherianness, Illinois J. Math. **20** (1976), 329-353.

[22]   D.W. Sharpe and P. Vamos, Injective modules (Cambridge Univ. Press, Cambridge 1972).

[23]   B. Stenström, Rings of quotients (Springer-Verlag, Berlin 1975).

Dérivations de la *n*-ième algèbre de Weyl quantique multiparametrée

L'Moufadal Ben Yakoub
Faculté des Sciences
Département de mathématiques
B. P. 2121 TETOUAN · MAROC.

Abstract

In [4], we show that each derivations of the quantized Weyl algebra $A_1(K,q)$ is a sum of an inner derivation and a well specified derivation. In this paper, we generalize this results to the n-th quantized Weyl algebra $A_n^{\bar{q},\Lambda}$ introduced by Maltsiniotis in [6].

1 Introduction.

Soient K un corps commutatif algébriquement clos de caractéristique nulle. Pour tout entier strictement positif n. on note $I\!N_n = \{1, 2, ..., n\}$; on prendra $I\!N_0 = \emptyset$. D'autre part. $M_n'(K^*)$ désigne l'ensemble des matrices $(\lambda_{i,j}) \in M_n(K)$ telles que: pour tout $(i,j) \in I\!N_n^2, \lambda_{j,i} = \lambda_{i,j}^{-1}$ et $\lambda_{i,i} = 1$.

1-1 Définition

Soit $n \in I\!N^*$, on note $q(n) = (q_1, q_2, ..., q_n)$ un élément de $(K^*)^n$ et $\Lambda(n) = (\lambda_{i,j}) \in M_n'(K^*)$. La **n-ième algèbre de Weyl quantique multiparametrée** associée à $\bar{q}(n)$ et $\Lambda(n)$. notée $A_n^{\bar{q},\Lambda}$, est l'algèbre à $2n$ générateurs sur K:

$$y_1, y_2, ..., y_n, x_1, x_2, ..., x_n$$

soumis aux relations:

1) Pour tout $(i,j) \in I\!N_n^2$ tel que $i < j$:

$$x_i x_j = \lambda_{i,j} q_i x_j x_i \qquad y_i x_j = \lambda_{i,j}^{-1} q_i^{-1} x_j y_i$$
$$y_i y_j = \lambda_{i,j} y_j y_i \qquad x_i y_j = \lambda_{i,j}^{-1} y_j x_i.$$

2) Pour tout $i \in I\!N_n$:

$$x_i y_i - q_i y_i x_i = 1 + \sum_{j=1}^{i-1} (q_j - 1) y_j x_j.$$

Si $n = 1$, $A_n^{q,\Lambda}$ notée $A_1(K,q)$ est l'algèbre de Weyl quantique [5], si de plus $q = 1$ on retrouve la définition de l'algèbre de Weyl classique notée $A_1(K)$.

Concernant la n-ième algèbre de Weyl quantiques on trouve dans [7] une description complète du spectre premier de $A_n^{q,\Lambda}$, et comme application, il détermine, sous certaines conditions sur q et Λ, les automorphismes de $A_n^{q,\Lambda}$.

Si A est une K-algèbre, on appelle **dérivation** de A, toute application K-linéaire de A dans A telle que:

$$D(ab) = D(a)b + aD(b), \quad \text{pour tout } a, b \in A$$

Par exemple si $x \in A$, l'application ad_x de A dans A telle que $ad_x(a) = xa - ax$ pour tout $a \in A$, est une dérivation de A, appelée la **dérivation intérieure** associée à x. Etant donné $a_1, a_2, ..., a_n \in K$, il est facile de vérifier que l'application K-linéaire δ de $A_n^{q,\Lambda}$ telle que $\delta(x_i) = a_i x_i$ et $\delta(y_i) = -a_i y_i$, est une dérivation de $A_n^{q,\Lambda}$. Le but de cet article est d'étudier les dérivations de $A_n^{\bar{q},\Lambda}$. Si Γ_n le sous-groupe de K^* engendré par les λ_{ij} et les q_i pour $(i,j) \in \mathbb{N}^2$ et $i < j$, ($\Gamma_0 = \{1\}$), est libre de rang $\frac{n(n+1)}{2}$, on montre que toute dérivation de $A_n^{\bar{q},\Lambda}$ est somme d'une dérivation intérieure de $A_n^{\bar{q},\Lambda}$ et d'une dérivation bien définie (Théorème 2.3).

2 Dérivations de $A_n^{\bar{q},\Lambda}$.

2-1 Proposition [7].

1) Si pour $i \in \mathbb{N}_n$, on note $z_i = x_i y_i - y_i x_i$ et $z_0 = 1$, on obtient:

$$z_i = 1 + \sum_{j=1}^{i}(q_j - 1)y_j x_j, \qquad z_i = z_{i-1} + (q_i - 1)y_i x_i, \qquad z_{i-1} = x_i y_i - q_i y_i x_i.$$

2) Pour tout $i \in \mathbb{N}_n$, z_i est normal dans $A_n^{q,\Lambda}$. Plus précisément, pour tout $j \in \mathbb{N}_n$ on a:

$$z_i x_j = x_j z_i \text{ et } z_i y_j = y_j z_i \text{ si } i < j,$$

$$z_i x_j = q_j^{-1} x_j z_i \text{ et } z_i y_j = q_j y_j z_i \text{ si } i \geq j.$$

3) Pour tout $j \in \mathbb{N}^*$, on a:

$$x_n y_n^j = q_n^j y_n^j x_n + (j)_{q_n} z_{n-1} y_n^{j-1} \text{ et } x_n^j y_n = q_n^j y_n x_n^j + (j)_{q_n} z_{n-1} x_n^{j-1}.$$

où $(j)_{q_n} = q_n^{j-1} + q_n^{j-2} + ... + q_n + 1$.

2-2 Proposition [7].

L'algèbre $A_n^{\bar{q},\Lambda}$ est une extension de Ore itérée, intègre et noethérienne, la famille:

$$\{y_1^{i_1} x_1^{j_1} y_2^{i_2} x_2^{j_2} ... y_n^{i_n} x_n^{j_n}; \quad (i_1, j_1, i_2, j_2, ..., i_n, j_n) \in \mathbb{N}^{2n}\}$$

est une base de Poincaré - Birkhof - Witt de $A_n^{\bar{q},\Lambda}$.

2-3 Théorème.

Supposons que Γ_n est libre de rang $\frac{n(n+1)}{2}$. Si D une dérivation de $A_n^{\bar{q},\Lambda}$. Alors, il existe une dérivation δ avec $\delta(x_i) = a_i x_i$ et $\delta(y_i) = -a_i y_i$, où $a_i \in K$ pour tout $i \in \mathbb{N}_n$ telle que:

$$D = ad_P + \delta \quad avec \quad P \in A_n^{\bar{q},\Lambda}.$$

Nous procédons par récurrence sur n. Pour $n = 0$, $A_0^{\bar{q},\Lambda} = K$, dans ([4] Théorème 2.1) on a à établir le cas $n = 1$. Pour le reste de la démonstration, on a besoin de toute la série de lemmes suivante.

Puisque x_n (resp: y_n) normalise x_i et y_i pour tout $i < n$, alors x_n (resp: y_n) normalise tout élément de $A_{n-1}^{\bar{q},\Lambda}$. Si $P \in A_{n-1}^{\bar{q},\Lambda}$, notons $P^{(n)}$, $P^{(1)}$ et $P^{(2)}$ les éléments de $A_{n-1}^{\bar{q},\Lambda}$ tels que $z_n P = P^{(n)} z_n$, $y_n P = P^{(1)} y_n$ et $x_n P = P^{(2)} x_n$.

2-4 Lemme.

Soit $P \in A_{n-1}^{\eta,\Lambda}$.
1) Pour tout $i \in \mathbb{N}$. si $P^{(n)} - q_n^i P$ (resp: $P^{(1)} - q_n^i P$, $P^{(2)} - q_n^i P$) centralise z_n ou bien y_n ou bien x_n. Alors, P centralise z_n, y_n et x_n.
2) Pour tout $i \in \mathbb{N}^*$. si $P^{(n)} = q_n^i P$ (resp: $P^{(1)} = q_n^i P$, $P^{(2)} = q_n^i P$). Alors, $P = 0$.

Preuve. Soit $P_\nu = \alpha_\nu y_1^{i_1} x_1^{j_1} ... y_{n-1}^{i_{n-1}} x_{n-1}^{j_{n-1}}$; avec $\nu = (i_1, j_1, ..., i_{n-1}, j_{n-1}) \in \mathbb{N}^{2(n-1)}$ et $\alpha_\nu \in K$. un monôme dans l'expression de P on a:

$$z_n P_\nu = \alpha_\nu q_1^{i_1 - j_1} ... q_{n-1}^{i_{n-1} - j_{n-1}} y_1^{i_1} x_1^{j_1} ... y_{n-1}^{i_{n-1}} x_{n-1}^{j_{n-1}} z_n,$$

ce qui implique que:

$$P_\nu^{(n)} = q_1^{i_1 - j_1} ... q_{n-1}^{i_{n-1} - j_{n-1}} P_\nu \quad \text{et} \quad P_\nu^{(n)} - q_n^i P_\nu = (q_1^{i_1 - j_1} ... q_{n-1}^{i_{n-1} - j_{n-1}} - q_n^i) P_\nu.$$

i) Si par exemple $P_\nu^{(n)} - q_n^i P_\nu$ centralise z_n on a:

$$(q_1^{i_1 - j_1} ... q_{n-1}^{i_{n-1} - j_{n-1}} - q_n^i) P_\nu^{(n)} = (q_1^{i_1 - j_1} ... q_{n-1}^{i_{n-1} - j_{n-1}} - q_n^i) P_\nu$$

ce qui implique que:

$$(q_1^{i_1 - j_1} ... q_{n-1}^{i_{n-1} - j_{n-1}} - q_n^i)(q_1^{i_1 - j_1} ... q_{n-1}^{i_{n-1} - j_{n-1}}) P_\nu = (q_1^{i_1 - j_1} ... q_{n-1}^{i_{n-1} - j_{n-1}} - q_n^i) P_\nu$$

c'est à dire: $P_\nu = 0$ ou $i_1 - j_1 = i_2 - j_2 = ... = i_{n-1} - j_{n-1} = 0$ ceci entraine que P_ν centralise x_n, y_n et z_n. celà prouve que P l'est aussi.

ii) Si $P_\nu^{(n)} - q_n^i P_\nu$ centralise x_n on a:

$$(q_1^{i_1 - j_1} ... q_{n-1}^{i_{n-1} - j_{n-1}} - q_n^i) P_\nu^{(2)} = (q_1^{i_1 - j_1} ... q_{n-1}^{i_{n-1} - j_{n-1}} - q_n^i) P_\nu$$

ce qui implique que:

$$(q_1^{i_1 - j_1} ... q_{n-1}^{i_{n-1} - j_{n-1}} - q_n^i)((\lambda_{1,n} q_1)^{i_1 - j_1} ... (\lambda_{n-1,n} q_{n-1})^{i_{n-1} - j_{n-1}} - 1) P_\nu = 0$$

celà entraine que P centralise x_n, y_n et z_n.

iii) De même si $P_\nu^{(n)} - q_n^\iota P_\nu$ centralise y_n on a:

$$(q_1^{\iota_1 - \jmath_1}...q_{n-1}^{\iota_{n-1} - \jmath_{n-1}} - q_n^\iota)P_\nu^{(1)} = (q_1^{\iota_1 - \jmath_1}...q_{n-1}^{\iota_{n-1} - \jmath_{n-1}} - q_n^\iota)P_\nu$$

ce qui implique que:

$$(q_1^{\iota_1 - \jmath_1}...q_{n-1}^{\iota_{n-1} - \jmath_{n-1}} - q_n^\iota)((\lambda_{1,n})^{\jmath_1 - \iota_1}...(\lambda_{n-1,n})^{\jmath_{n-1} - \iota_{n-1}} - 1)P_\nu = 0.$$

On en déduit que P centralise x_n, y_n et z_n. On fait la même chose pour $P_\nu^{(2)} - q_n^\iota P_\nu$ et $P_\nu^{(1)} - q_n^\iota P_\nu$.

2) Pour tout $\iota \in I\!N^*$, on a: $P_\nu^{(n)} = q_n^\iota P_\nu$ signifie que $(q_1^{\iota_1 - \jmath_1}...q_{n-1}^{\iota_{n-1} - \jmath_{n-1}} - q_n^\iota)P_\nu = 0$, ce qui entraine que $P_\nu = 0$.

2-5 Lemme

Modulo une dérivation intérieure on peut supposer que $D(x_n)$ est de la forme:

$$D(x_n) = a + \sum_{\iota \geq 1} b_\iota x_n^\iota + \sum_{\iota \geq 1} c_\iota y_n^\iota,$$

avec $a, b_\iota, c_\iota \in \Lambda_{n-1}^{\tilde{q},\Lambda}$, tel que b_ι centralise z_n, y_n et x_n, pour tout $\iota \geq 1$.

Preuve. Si on prend $D(x_n) = a + \sum_{\iota \geq 1} b_\iota x_n^\iota + \sum_{\iota \geq 1} c_\iota y_n^\iota + \sum_{\iota \geq 1} d_{\iota,\jmath} y_n^\iota x_n^\jmath$, avec $a, b_\iota, c_\iota, d_{\iota,\jmath} \in \Lambda_{n-1}^{\tilde{q},\Lambda}$. Notons $d_{\iota,\jmath} = \sum_\nu \alpha_\nu y_1^{\iota_1} x_1^{\jmath_1}...y_{n-1}^{\iota_{n-1}} x_{n-1}^{\jmath_{n-1}}$ et soit $P_\nu = \alpha_\nu y_1^{\iota_1} x_1^{\jmath_1}...y_{n-1}^{\iota_{n-1}} x_{n-1}^{\jmath_{n-1}}$, on a:

$$ad_{P_\nu y_n^\iota x_n^{\jmath-1}}(x_n) = P_\nu y_n^\iota x_n^\jmath - P_\nu^{(2)}(x_n y_n^\iota)x_n^{\jmath-1} = P_\nu y_n^\iota x_n^\jmath - P_\nu^{(2)}(q_n^\iota y_n^\iota x_n + (\iota)_{q_n} z_{n-1} y_n^{\iota-1})x_n^{\jmath-1} =$$

$$(P_\nu - q_n^\iota P_\nu^{(2)})y_n^\iota x_n^\jmath - (\iota)_{q_n} P_\nu^{(2)} z_{n-1} y_n^{\iota-1} x_n^{\jmath-1} =$$

$$(1 - q_n^\iota(\lambda_{1,n}q_1)^{\iota_1 - \jmath_1}...(\lambda_{n-1,n}q_{n-1})^{\iota_{n-1} - \jmath_{n-1}})P_\nu y_n^\iota x_n^\jmath - (\iota)_{q_n} P_\nu^{(2)} z_{n-1} y_n^{\iota-1} x_n^{\jmath-1}.$$

D'où:

$$P_\nu y_n^\iota x_n^\jmath = \frac{ad_{P_\nu y_n^\iota x_n^{\jmath-1}}(x_n) + (\iota)_{q_n} P_\nu^{(2)} z_{n-1} y_n^{\iota-1} x_n^{\jmath-1}}{1 - [q_n^\iota(\lambda_{1,n}q_1)^{\iota_1 - \jmath_1}...(\lambda_{n-1,n}q_{n-1})^{\iota_{n-1} - \jmath_{n-1}}]}$$

Donc, modulo des dérivations intérieures de la forme:

$$\delta_\nu = \frac{ad_{P_\nu y_n^\iota x_n^{\jmath-1}}}{1 - [q_n^\iota(\lambda_{1,n}q_1)^{\iota_1 - \jmath_1}...(\lambda_{n-1,n}q_{n-1})^{\iota_{n-1} - \jmath_{n-1}}]}$$

et si on répète le processus un nombre suffisant de fois on aura la première partie du lemme.

D'autre part, pour tout $\iota \geq 1$, soit $P_\nu = \alpha_\nu y_1^{\iota_1} x_1^{\jmath_1}...y_{n-1}^{\iota_{n-1}} x_{n-1}^{\jmath_{n-1}}$ un monôme dans l'expression de b_ι. On a:

$$ad_{P_\nu x_n^{\iota-1}}(x_n) = P_\nu x_n^\iota - x_n P_\nu x_n^{\iota-1} = (P_\nu - P_\nu^{(2)})x_n^\iota.$$

On:

$$P_\nu - P_\nu^{(2)} = \alpha_\nu(1 - (\lambda_{1,n}q_1)^{\imath_1-\jmath_1}...(\lambda_{n-1,n}q_{n-1})^{\imath_{n-1}-\jmath_{n-1}})y_1^{\imath_1}x_1^{\jmath_1}...y_{n-1}^{\imath_{n-1}}x_{n-1}^{\jmath_{n-1}}$$

s'il existe $1 \le k \le n-1$ tel que $i_k \ne j_k$ on a:

$$P_\nu x_n^\imath = \frac{ad_{P_\nu x_n^{\imath-1}}(x_n)}{1 - [(\lambda_{1,n}q_1)^{\imath_1-\jmath_1}...(\lambda_{n-1,n}q_{n-1})^{\imath_{n-1}-\jmath_{n-1}}]}.$$

D'où, modulo des dérivations de la forme:

$$\frac{ad_{P_\nu x_n^{\imath-1}}}{1 - [(\lambda_{1,n}q_1)^{\imath_1-\jmath_1}...(\lambda_{n-1,n}q_{n-1})^{\imath_{n-1}-\jmath_{n-1}}]}$$

les monômes $b_\imath x_n^\imath$ dans l'expression de $D(x_n)$ sont tels que $b_\imath = \sum_\nu \alpha_\nu y_1^{\imath_1}x_1^{\imath_1}y_2^{\imath_2}x_2^{\imath_2}...y_{n-1}^{\imath_{n-1}}x_{n-1}^{\imath_{n-1}}$, ce qui entraîne que $b_\imath$ centralise z_n, x_n et y_n.

2-6   Lemme

L'expression de $D(x_n)$ et $D(z_n)$ peut être simplifier sous la forme:

$$D(x_n) = ax_n \quad et \quad D(z_n) = \alpha + \sum_{\imath\ge1}\beta_\imath x_n^\imath + \sum_{\imath\ge1}\lambda_\imath y_n x_n^\imath,$$

avec $a,\alpha,\beta_\imath,\lambda_\imath, \in A_{n-1}^{\bar q,\Lambda}$, *qui centralisent* z_n,y_n *et* x_n, *pour tout* $\imath \ge 1$.

Preuve. Si on prend: $D(z_n) = \alpha + \sum_{\imath\ge1}\beta_\imath x_n^\imath + \sum_{\imath\ge1}\gamma_\imath y_n^\imath + \sum_{\imath,\jmath\ge1}\xi_{\imath,\jmath}y_n^\imath x_n^\jmath$, avec $\alpha,\beta_\imath,\gamma_\imath,\xi_{\imath,\jmath} \in A_{n-1}^{q,\Lambda}$ pour tout $i,j \in \mathbb{N}^*$. Appliquons D à l'équation $x_nz_n = q_nz_nx_n$, on aura:

$$D(x_nz_n) = D(x_n)z_n + x_nD(z_n) = az_n + \sum_{\imath\ge1}q_n^\imath b_\imath z_nx_n^\imath + \sum_{\imath\ge1}c_\imath y_n^\imath z_n + \alpha^{(2)}x_n +$$

$$\sum_{\imath\ge1}\beta_\imath^{(2)}x_n^{\imath+1} + \sum_{\imath\ge1}\gamma_\imath^{(2)}(q_n^\imath y_n^\imath x_n + (i)_{q_n}z_{n-1}y_n^{\imath-1}) + \sum_{\imath,\jmath\ge1}\xi_{\imath,\jmath}^{(2)}(q_n^\imath y_n^\imath x_n + (\imath)_{q_n}z_{n-1}y_n^{\imath-1})x_n^\jmath.$$

En vertu de l'identité $z_n = z_{n-1} + (q_n-1)y_nx_n$, on déduit que:

$$D(x_nz_n) = (a+\gamma_1^{(2)})z_{n-1} + (\alpha^{(2)} + q_nb_1z_{n-1} + \xi_{1,1}^{(2)}z_{n-1})x_n +$$

$$\sum_{\imath\ge1}(\beta_\imath^{(2)} + q_n^{\imath+1}b_{\imath+1}z_{n-1} + \xi_{1,\imath+1}^{(2)}z_{n-1})x_n^{\imath+1} + ((q_n-1)a + q_n\gamma_1^{(2)} + (2)_{q_n}\xi_{2,1}^{(2)}z_{n-1})y_nx_n +$$

$$\sum_{\imath\ge1}(c_\imath z_{n-1} + (\imath+1)_{q_n}\gamma_{\imath+1}^{(2)}z_{n-1})y_n^\imath + \sum_{\imath\ge1}((q_n-1)c_\imath + q_n^{\imath+1}\gamma_{\imath+1}^{(2)} + (\imath+2)_{q_n}\xi_{\imath+2,1}^{(2)}z_{n-1})y_n^{\imath+1}x_n +$$

$$\sum_{\imath\ge1}(q_n^\imath(q_n-1)b_\imath + q_n\xi_{1,\imath}^{(2)} + (2)_{q_n}\xi_{2,\imath+1}^{(2)}z_{n-1})y_nx_n^{\imath+1} + \sum_{\imath,\jmath\ge1}(q_n^{\imath+1}\xi_{\imath+1,\jmath}^{(2)} + (i+2)_{q_n}\xi_{\imath+2,\jmath+1}^{(2)}z_{n-1})y_n^{\imath+1}x_n^{\jmath+1}.$$

D'autre part:

$$D(z_nx_n) = a^{(n)}z_{n-1} + (\alpha + b_1z_{n-1})x_n + \sum_{\imath\ge1}(\beta_\imath + b_{\imath+1}z_{n-1})x_n^{\imath+1} +$$

$$((q_n - 1)a^{(n)} + \gamma_1)y_n x_n + \sum_{i \geq 1}(q_n^i(q_n - 1)c_i^{(n)} + \gamma_{i+1})y_n^{i+1}x_n + \sum_{i \geq 1}(q_n^i c_i^{(n)} z_{n-1})y_n^i +$$

$$\sum_{i \geq 1}((q_n - 1)b_i + \xi_{1,i})y_n x_n^{i+1} + \sum_{i,j \geq 1}\xi_{i+1,j}y_n^{i+1}x_n^{i+1}.$$

En comparant les coefficients de $D(x_n z_n)$ et ceux de $q_n D(z_n x_n)$, on déduit:

1) $q_n^{i+1}\xi_{i+1,j}^{(2)} + (i+2)_{q_n}\xi_{i+2,j+1}^{(2)}z_{n-1} = q_n\xi_{i+1,j}.$

Supposons que les $\xi_{i,j}$ ne sont pas tous nuls pour $i \geq 2$ et $j \geq 1$, soit m maximal tel que $\xi_{m,j} \neq 0$ pour certains $j \geq 1$. Remplaçant i par $m - 1$ dans l'équation précédente on obtient:

$$q_n^{m-1}\xi_{m,j}^{(2)} = \xi_{m,j}.$$

ce qui donne, en vertu du lemme 2.4. que $\xi_{m,j} = 0$. ce qui est absurde. D'où $\xi_{i,j} = 0$ pour tout $i \geq 2$ et $j \geq 1$.

2) $q_n^i(q_n - 1)b_i + q_n\xi_{1,i}^{(2)} = q_n(q_n - 1)b_i + q_n\xi_{1,i} \implies (q_n - 1)(q_n^{i-1} - 1)b_i = \xi_{1,i} - \xi_{1,i}^{(2)}$. D'où $\xi_{1,i} - \xi_{1,i}^{(2)}$ centralise z_n, x_n et y_n et par suite $\xi_{1,i}$ centralise x_n et y_n ce qui signifie $(q_n^{i-1} - 1)b_i = 0$ pour tout $i \geq 1$ c'est à dire $b_i = 0$ pour tout $i \geq 2$.

3) $(q_n - 1)c_i + q_n^{i+1}\gamma_{i+1}^{(2)} = q_n^{i+1}(q_n - 1)c_i^{(n)} + q_n\gamma_{i+1}$
$$\implies q_n(q_n^i\gamma_{i+1}^{(2)} - \gamma_{i+1}) = (q_n - 1)(q_n^{i+1}c_i^{(n)} - c_i).$$

4) $(c_i + (i+1)_{q_n}\gamma_{i+1}^{(2)})z_{n-1} = q_n^{i+1}c_i^{(n)}z_{n-1} \implies q_n^{i+1}c_i^{(n)} - c_i = (i+1)_{q_n}\gamma_{i+1}^{(2)}.$

En combinant 3) et 4) on déduit que pour tout $i \geq 1$. on a:
$$q_n(q_n^i\gamma_{i+1}^{(2)} - \gamma_{i+1}) = (q_n - 1)(i+1)_{q_n}\gamma_{i+1}^{(2)} \implies q_n(q_n^i\gamma_{i+1}^{(2)} - \gamma_{i+1}) = (q_n^{i+1} - 1)\gamma_{i+1}^{(2)} \implies \gamma_{i+1} = 0$$
et par suite $q_n^{i+1}c_i^{(n)} = c_i \implies c_i = 0$, pour tout $i \geq 1$.

5) $(q_n - 1)a + q_n\gamma_1^{(2)} = q_n(q_n - 1)a^{(n)} + q_n\gamma_1. \implies (q_n - 1)(a - q_n a^{(n)}) = q_n(\gamma_1 - \gamma_1^{(2)}).$

6) $\beta_i^{(2)} + \xi_{1,i+1}z_{n-1} = q_n\beta_i \implies q_n\beta_i - \beta_i^{(2)} = \xi_{1,i+1}z_{n-1}$. D'où β_i centralise x_n, y_n et z_n et par suite $(q_n - 1)\beta_i = \xi_{1,i+1}z_{n-1}$. pour tout $i \geq 1$.

7) $\alpha^{(2)} + q_n b_1 z_{n-1} + \xi_{1,1}z_{n-1} = q_n(\alpha + b_1 z_{n-1}) \implies \alpha^{(2)} - q_n\alpha = -\xi_{1,1}z_{n-1}$, d'où α centralise $z_n. y_n$ et x_n de plus $(q_n - 1)\alpha = \xi_{1,1}z_{n-1}$.

8) $(a + \gamma_1^{(2)})z_{n-1} = q_n a^{(n)}z_{n-1} \implies \gamma_1^{(2)} = q_n a^{(n)} - a$, en substituant dans l'équation 5) on obtient: $(1 - q_n)\gamma_1^{(2)} = q_n(\gamma_1 - \gamma_1^{(2)}) \implies \gamma_1^{(2)} = q_n\gamma_1 \implies \gamma_1 = 0$ et $a = 0$.

On déduit de ce qui précède que $D(x_n)$ et $D(z_n)$ sont de la forme:
$$D(x_n) = b_1 x_n, \quad D(z_n) = \alpha + \sum_{i \geq 1}\beta_i x_n^i + \sum_{i \geq 1}\xi_{1,i}y_n x_n^i.$$

avec $b_1, \alpha, \beta_i, \xi_{1,i} \in A_{n-1}^{\bar{q}\Lambda}$, centralisant z_n, y_n et x_n, pour tout $i \geq 1$.

2-7 Lemme

Modulo une dérivation intérieure on peut écrire

$$D(x_n) = ax_n, \quad D(z_n) = \alpha + \lambda y_n x_n + \sum_{i \geq 1} \beta_i x_n^i$$

avec $a, \alpha, \lambda, \beta_i \in \Lambda_{n-1}^{\bar{q}, \Lambda}$. *qui centralisent* z_n, y_n *et* x_n, *pour tout* $i \geq 1$.

Preuve. Avec les notations du lemme 2.6 on a:

$$D(z_n) = \alpha + \lambda y_n x_n + \sum_{i \geq 1} \beta_i x_n^i + \sum_{i \geq 1} \gamma_i y_n x_n^{i+1}.$$

avec $\alpha, \lambda = \lambda_1, \beta_i, \gamma_i = \lambda_{i+1}, \in A_{n-1}^{q, \Lambda}$. centralisant z_n, y_n et x_n. Soit $P_i = \gamma_i x_n^i$ on a: $ad_{P_i}(x_n) = 0$ et $ad_{P_i}(z_n) = P_i z_n - z_n P_i = \gamma_i x_n^i z_n - z_n \gamma_i x_n^i = \gamma_i (q_n^i z_n x_n^i - z_n x_n^i) = (q_n^i - 1)\gamma_i(z_{n-1} + (q_n - 1)y_n x_n)x_n^i = (q_n^i - 1)\gamma_i z_{n-1} x_n^i + (q_n^i - 1)(q_n - 1)\gamma_i y_n x_n^{i+1}$, ce qui implique que

$$\gamma_i y_n x_n^{i+1} = \frac{ad_{P_i}(z_n)}{(q_n^i - 1)(q_n - 1)} - \frac{\gamma_i z_{n-1}}{q_n - 1} x_n^i.$$

Modulo la dérivation intérieure $\delta = \sum_{i \geq 1} \frac{ad_{P_i}}{(q_n^i - 1)(q_n - 1)}$ on aura:
$D(z_n) = \alpha + \lambda y_n x_n + \sum_{i \geq 1}(\beta_i - \frac{\gamma_i z_{n-1}}{q_n - 1})x_n^i$, ceci termine la preuve du lemme.

2-8 Lemme

Les expressions de $D(x_n), D(y_n)$ *et* $D(z_n)$ *sont de la forme:*
$D(x_n) = ax_n$
$D(z_n) = \alpha + \lambda y_n x_n$
$D(y_n) = b + cy_n + \sum_{i \geq 1} b_i x_n^i$, *avec* $\alpha, \lambda, a, b, c, b_i \in A_{n-1}^{q, \Lambda}$. *centralisant* z_n, y_n, x_n *tels que* $(q_n - 1)(c + a) = \lambda$ *et* $(a + c)z_{n-1} = \alpha$.

Preuve. Soit $D(y_n) = b + \sum_{i \geq 1} b_i x_n^i + \sum_{i \geq 1} c_i y_n^i + \sum_{i,j \geq 1} d_{i,j} y_n^i x_n^j$. on vérifie facilement que:

$$D(x_n y_n) = (a + c_1^{(2)})z_{n-1} + (b^{(2)} + d_{1,1}^{(2)} z_{n-1})x_n + (q_n a + q_n c_1^{(2)} + (2)_{q_n} d_{2,1}^{(2)} z_{n-1})y_n x_n +$$

$$\sum_{i \geq 1}(b_i^{(2)} + d_{1,i+1}^{(2)} z_{n-1})x_n^{i+1} + \sum_{i \geq 1}(i+1)_{q_n} c_{i+1}^{(2)} z_{n-1} y_n^i + \sum_{i \geq 1}(q_n^{i+1} c_{i+1}^{(2)} + (i+2)_{q_n} d_{i+2,1}^{(2)} z_{n-1})y_n^{i+1} x_n +$$

$$\sum_{i,j \geq 1}(q_n^j d_{i,j}^{(2)} + (i+1)_{q_n} d_{i+1,j+1}^{(2)} z_{n-1})y_n^i x_n^{j+1}.$$

De même on a:

$$D(y_n x_n) = bx_n + (a + c_1)y_n x_n + \sum_{i \geq 1} b_i x_n^{i+1} + \sum_{i \geq 1} c_{i+1} y_n^{i+1} x_n + \sum_{i,j \geq 1} d_{i,j} y_n^i x_n^{j+1}.$$

D'où:

$$D(z_n) = (a + c_1^{(2)})z_{n-1} + (b^{(2)} + d_{1,1}^{(2)}z_{n-1} - b)x_n + ((q_n - 1)a + q_n c_1^{(2)} - c_1 + (2)_{q_n} d_{2,1}^{(2)} z_{n-1})y_n x_n +$$

$$\sum_{i \geq 1}(b_i^{(2)} + d_{1,i+1}^{(2)}z_{n-1} - b_i)x_n^{i+1} + \sum_{i \geq 1}(i+1)_{q_n} c_{i+1}^{(2)} z_{n-1} y_n^i +$$

$$\sum_{i \geq 1}(q_n^{i+1} c_{i+1}^{(2)} + (i+2)_{q_n} d_{i+2,1}^{(2)} z_{n-1} - c_{i+1})y_n^{i+1} x_n + \sum_{i,j \geq 1}(q_n^i d_{i,j}^{(2)} + (i+1)_{q_n} d_{i+1,j+1}^{(2)} z_{n-1} - d_{i,j})y_n^i x_n^{j+1}.$$

En comparant avec l'expression de $D(z_n)$ du lemme 2.7, on déduit:

1) $q_n^i d_{i,j}^{(2)} + (i+1)_{q_n} d_{i+1,j+1}^{(2)} - d_{i,j} = 0$. soit m maximal tel que $d_{m,j} \neq 0$ pour certains $j \geq 1$. Si on remplace i par m dans l'équation précédente on obtient $q_n^m d_{m,j}^{(2)} - d_{m,j} = 0$. celà signifie que $d_{m,j} = 0$, ce qui est absurde, d'où $d_{i,j} = 0$ pour tout $i,j \geq 1$.

2) $q_n^{i+1} c_{i+1}^{(2)} - c_{i+1} = 0 \implies c_{i+1} = 0$. pour tout $i \geq 1$.

3) $b_i^{(2)} - b_i = \beta_{i+1}$. implique que b_i centralise x_n, y_n et z_n on en déduit que $\beta_{i+1} = 0$ pour tout $i \geq 1$.

4) $(q_n - 1)a + q_n c_1^{(2)} - c_1 = \lambda \implies \lambda - (q_n - 1)a = q_n c_1^{(2)} - c_1$, d'où c_1 centralise x_n, y_n, z_n et $\lambda = (q_n - 1)(c_1 + a)$.

5) $b^{(2)} - b = \beta_1$. entraine que b centralise x_n, y_n, z_n et par suite $\beta_1 = 0$.

6) $a = (a + c_1)z_{n-1}$. Ceci achève la preuve du lemme.

2-9   Lemme

Les expressions de $D(x_n)$, $D(y_n)$ et $D(z_n)$ sont de la forme: $D(x_n) = ax_n$, $D(z_n) = \beta z_n$ et $D(y_n) = cy_n$. avec $a, \beta, c \in A_{n-1}^{q,\Lambda}$. centralisant z_n, y_n, x_n tels que $\beta = a + c$.

Preuve. Prenant $D(y_n)$ et $D(z_n)$ comme dans le lemme 2.8, on a:

$$D(y_n z_n) = bz_{n-1} + (a + cz_{n-1})y_n + (q_n - 1)by_n x_n +$$

$$(\lambda + (q_n - 1)c)y_n^2 x_n + \sum_{i \geq 1} q_n^i b_i z_{n-1} x_n^i + \sum_{i \geq 1}(q_n - 1)q_n^i b_i y_n x_n^{i+1}.$$

D'autre part:

$$D(z_n y_n) = bz_{n-1} + (a + q_n cz_{n-1} + \lambda z_{n-1})y_n + (q_n - 1)by_n x_n +$$

$$q_n(\lambda + (q_n - 1)c)y_n^2 x_n + \sum_{i \geq 1} b_i z_{n-1} x_n^i + \sum_{i \geq 1}(q_n - 1)b_i y_n x_n^{i+1}.$$

De l'équation $q_n y_n z_n = z_n y_n$ on obtient:

1) $(q_n - 1)q_n^{i+1}b_i = (q_n - 1)b_i \implies b_i = 0$, pour tout $i \geq 1$.

2) $(q_n - 1)b = q_n(q_n - 1)b \implies b = 0$.

3) $\alpha + q_n c z_{n-1} + \lambda z_{n-1} = q_n \alpha + q_n c z_{n-1} \implies (q_n - 1)\alpha = \lambda z_{n-1}$.

Finalement on a $D(x_n) = a x_n$, $D(y_n) = c y_n$ et

$$(q_n - 1)D(z_n) = (q_n - 1)\alpha + (q_n - 1)\lambda y_n x_n = \lambda(z_{n-1} + (q_n - 1)y_n x_n) = \lambda z_n.$$

et d'après le lemme 2.8 on a $a + c = \beta$ avec $\beta = \frac{\lambda}{q_n - 1}$.

Remarque: On a:

$$D(z_{n-1}) = D(z_n) - (q_n - 1)D(y_n x_n) =$$

$$\beta z_n - (q_n - 1)(c y_n x_n + a y_n x_n) = \beta(z_n - (q_n - 1)y_n x_n) = \beta z_{n-1},$$

ce qui implique $D(z_{n-1}) = \beta z_{n-1}$. De même $D(z_{n-2}) = \beta z_{n-2}, \ldots, D(z_1) = \beta z_1$.

Preuve du théorème . Jusqu'à maintenant, on a montré que si D est une dérivation de $A_n^{q,\Lambda}$, modulo une dérivation intérieure, il existe $a, b, c \in A_{n-1}^{q,\Lambda}$ centralisant z_n, y_n et x_n tels que $a = b + c$ avec:

$i) D(z_r) = a z_r$, pour tout $1 \leq r \leq n$; $ii) D(y_n) = b y_n$; $iii) D(x_n) = c x_n$.

Soit $1 \leq r < n$ et posons: $D(y_r) = \alpha + \sum_{i \geq 1} \beta_i y_n^i + \sum_{i \geq 1} \gamma_i x_n^i + \sum_{i,j \geq 1} \delta_{i,j} y_n^i x_n^j$. avec $\alpha, \beta_i, \gamma_i, \xi_{i,j} \in A_{n-1}^{q,\Lambda}$ pour tout $i, j \in N^*$. Appliquons D à l'équation $q_r y_r z_n = z_n y_r$. on aura:

$$D(z_n y_r) = D(z_n)y_r + z_n D(y_r) = a z_n y_r + z_n(\alpha + \sum_{i \geq 1} \beta_i y_n^i + \sum_{i \geq 1} \gamma_i x_n^i + \sum_{i,j \geq 1} \delta_{i,j} y_n^i x_n^j) =$$

$$[(q_r a y_r + \alpha^{(n)}) + \sum_{i \geq 1} q_n^i \beta_i^{(n)} y_n^i + \sum_{i \geq 1} q_n^{-i} \gamma_i^{(n)} x_n^i + \sum_{i,j \geq 1} q_n^{i-j} \delta_{i,j}^{(n)} y_n^i x_n^j] z_n.$$

De même on a:

$$D(y_r z_n) = [(y_r a + \alpha) + \sum_{i \geq 1} \beta_i y_n^i + \sum_{i \geq 1} \gamma_i x_n^i + \sum_{i,j \geq 1} \delta_{i,j} y_n^i x_n^j] z_n.$$

En comparant les coefficients de $D(z_n y_r)$ et $q_r D(y_r z_n)$, on déduit:

1) $q_r a y_r + \alpha^{(n)} = q_r y_r a + q_r \alpha \implies a y_r - y_r a = \alpha - q_r^{-1}\alpha^{(n)}$.

2) $q_n^i \beta_i^{(n)} = q_r \beta_i$. pour tout $i \geq 1$. Soit $\alpha_\nu y_1^{i_1} x_1^{j_1} \ldots y_{n-1}^{i_{n-1}} x_{n-1}^{j_{n-1}}$ un monôme dans l'expression de β_i, on a:

$$q_n^i \alpha_\nu (y_1^{i_1} x_1^{j_1} \ldots y_{n-1}^{i_{n-1}} x_{n-1}^{j_{n-1}})^{(n)} = q_r y_1^{i_1} x_1^{j_1} \ldots y_{n-1}^{i_{n-1}} x_{n-1}^{j_{n-1}}$$

implique que:

$$\alpha_\nu(q_n^i q_1^{i_1-j_1}...q_{r-1}^{i_{r-1}-j_{r-1}} q_r^{i_r-j_r-1} q_{r+1}^{i_{r+1}-j_{r+1}}...q_{n-1}^{i_{n-1}-j_{n-1}} - 1)y_1^{i_1}x_1^{j_1}...y_{n-1}^{i_{n-1}}x_{n-1}^{j_{n-1}} = 0$$

D'où $\alpha_\nu = 0$ et par suite $\beta_i = 0$, pour tout $i \geq 1$.

3) $q_r\gamma_i = q_n^{-1}\gamma_i^{(n)} \Longrightarrow \gamma_i = 0$ pour tout $i \geq 1$.

4) $q_i\delta_{i,j} = q_n^{-1}\delta_{i,j}^{(n)}$, comme dans 2) si $\alpha_\nu y_1^{i_1}x_1^{j_1}...y_{n-1}^{i_{n-1}}x_{n-1}^{j_{n-1}}$ est un monôme dans l'expression de $\delta_{i,j}$ on a:

$$\alpha_\nu(q_n^{i-j}q_1^{i_1-j_1}...q_{r-1}^{i_{r-1}-j_{r-1}} q_r^{i_r-j_r-1} q_{r+1}^{i_{r+1}-j_{r+1}}....q_{n-1}^{i_{n-1}-j_{n-1}} - 1)y_1^{i_1}x_1^{j_1}...y_{n-1}^{i_{n-1}}x_{n-1}^{j_{n-1}} = 0$$

Si $\alpha_\nu \neq 0$ on a: $i = j, i_1 = j_1, ... i_{r-1} = j_{r-1}. i_r = j_r + 1, i_{r+1} = j_{r+1},..., i_{n-1} = j_{n-1}$ et si on pose $\delta_i = \delta_{i,j}$ on déduit que $D(y_r)$ est de la forme:

$$D(y_r) = \alpha + \sum_{i \geq 1} \delta_i y_n^i x_n^i.$$

avec $\alpha, \delta_i(i \geq 1) \in A_{n-1}^{\bar{q},\Lambda}$ tels que $ay_r - y_r a = \alpha - q_r^{-1}\alpha^{(n)}$ et δ_i s'écrit comme combinaison linéaire d'éléments de la forme:

$$y_1^{i_1}x_1^{i_1}y_2^{i_2}x_2^{i_2}...y_{r-1}^{i_{r-1}}x_{r-1}^{i_{r-1}}y_r^{i_r+1}x_r^{i_r}y_{r+1}^{i_{r+1}}x_{r+1}^{i_{r+1}}...y_{n-1}^{i_{n-1}}x_{n-1}^{i_{n-1}}.$$

Appliquons maintenant D à l'équation $x_n y_r = \lambda_{r,n} q_r y_r x_n$, on obtient:

$$D(x_n y_r) = (\alpha^{(2)} + \lambda_{r\,n}q_r cy_r + \delta_1^{(2)}z_{n-1})x_n + \sum_{i \geq 1}(q_n^i \delta_i^{(2)} + (i+1)_{q_n}\delta_{i+1}^{(2)}z_{n-1})y_n^i x_n^{i+1}.$$

$$D(y_r x_n) = (\alpha + y_r c)x_n + \sum_{i \geq 1} \delta_i y_n^i x_n^{i+1}.$$

On en déduit que:

1) $\alpha^{(2)} + \lambda_{r,n}q_i cy_r + \delta_1^{(2)}z_{n-1} = \lambda_{r,n}q_r(\alpha + y_r c)$

$$\Longrightarrow cy_r - y_r c = \alpha - \lambda_{r,n}^{-1}q_r^{-1}\alpha^{(2)} - \lambda_{r,n}^{-1}q_r^{-1}\delta_1^{(2)}z_{n-1}.$$

2) $q_n^i \delta_i^{(2)} + (i+1)_{q_n}\delta_{i+1}^{(2)}z_{n-1} = \lambda_{r,n}q_r \delta_i$, pour tout $i \geq 1$.

De l'expression de δ_i donné ci-dessus, on déduit que:

$$x_n \delta_i = \lambda_{r,n}q_r \delta_i x_n \Longrightarrow \delta_i^{(2)} = \lambda_{r\,n}q_r \delta_i.$$

En remplaçant dans l'équation 2) précédente on trouve que pour tout $i \geq 1$:

$$q_n^i \delta_i^{(2)} + (i+1)_{q_n}\delta_{i+1}^{(2)}z_{n-1} = \delta_i^{(2)} \Longrightarrow (1 - q_n^i)\delta_i^{(2)} = (i+1)_{q_n}\delta_{i+1}^{(2)}z_{n-1},$$

ce qui implique que $\delta_i = 0$ pour tout $i \geq 1$ et que $D(y_r) = \alpha \in A_{n-1}^{q,\Lambda}$, pour tout $1 \leq r \leq n - 1$ avec:

$$\begin{cases} ay_r - y_r a = \alpha - q_r^{-1}\alpha^{(n)}. \\ cy_r - y_r c = \alpha - \lambda_{r,n}^{-1}q_r^{-1}\alpha^{(2)}. \end{cases}$$

De même on montre que. pour tout $1 \leq r < n$. on a $D(x_r) = \beta \in A_{n-1}^{q,\Lambda}$, avec:

$$\begin{cases} ax_r - x_r a = \beta - q_r \beta^{(n)}. \\ bx_r - x_r b = \beta - \lambda_{r,n}^{-1} \beta^{(1)}. \end{cases}$$

On a montré que, modulo une dérivation intérieure, D induit une dérivation sur $A_{n-1}^{\tilde{q},\Lambda}$ et par hypothèse de récurrence, il existe $k_r \in K$ tel que $\alpha = -k_r y_r, \beta = k_r x_r$, ce qui implique que:

1) $D(z_r) = az_r = 0$, ce qui entraine que $a = 0$ et par suite $b = -c$.

2) $\alpha^{(2)} = \lambda_{r,n} q_r \alpha$, ce qui signifie que c commute avec y_r.

3) $\beta^{(1)} = \lambda_{r,n} \beta$, donc b commute avec x_r.

D'où $b = -c$ commute avec x_n, y_n et x_r, y_r, pour tout $1 \leq r \leq n - 1$, c 'est à dire $b, c \in K = Z(A_n^{q,\Lambda})$. le centre de $A_n^{q,\Lambda}$; En effet:

Soit $a = \alpha + \sum_{i \geq 1} \beta_i x_n^i + \sum_{i \geq 1} \gamma_i y_n^i + \sum_{i,j \geq 1} \delta_{i,j} y_n^i x_n^j \in Z(A_n^{\tilde{q},\Lambda})$, avec $\alpha, \beta_i, \gamma_i, \delta_{i,j} \in A_{n-1}^{\tilde{q},\Lambda}$, pour tout $i, j \geq 1$. L'équation $z_n a = az_n$ implique que a est de la forme $a = \mu + \sum_{i \geq 1} \nu_i y_n^i x_n^i$. avec $\mu, \nu_i (i \geq 1)$, commute avec z_n, y_n, x_n respectivement. L'identité $x_n a = ax_n$, implique que $\nu_i = 0$ pour tout $i \geq 1$, ce qui signifie que $a \in Z(A_{n-1}^{q,\Lambda})$ et par récurrence on déduit le résultat. Ce qui achève la preuve du théorème.

References

[1] J.ALEV ET F.DUMAS. *Rigidité des plongements des quotients primitifs minimaux de $U_q(sl(2))$ dans l'algèbre quantique de Weyl-Hayashi.* Nagoya.Math.J, Vol 143, 119 - 146, (1996).

[2] J.ALEV ET F.DUMAS. *Sur le corps des fractions de certaines algèbres quantiques.* J.Algebra. Vol, 170. No. 1 (15). 1994.

[3] J.ALEV ET F.DUMAS. *Automorphismes de certains complétés du corps de Weyl quantique.* (Preprint).

[4] L.BEN YAKOUB ET M.P.MALLIAVIN. *Caractérisation des dérivations intérieures de l'algèbre de Weyl et de l'algèbre de Heisenberg quantiques.*Comm. Algebra, 24(10). 3131 - 3148. (1996).

[5] K.R.GOODEARL. *Prime Ideal in Skew Polynomial Rings and Quantized Weyl Algebras* : J.Algebra : 150. 324 - 377 (1992).

[6] G. MALTSINIOTIS. *Calcul différentiel quantique.*Groupe de travail, Université de Paris VII. 1992.

[7] L. RIGAL. *Spectre de l'algèbre de Weyl quantique.* Beiträge für Geometrie and Algebra. 37. Vol 1. 119 - 148, (1996).

Tensor Hochschild Homology and Cohomology

Claude Cibils

Département de Mathématiques,
Université de Montpellier 2,
F-34095 Montpellier cedex 5.

cibils@math.univ-montp2.fr

Abstract

We consider the non commutative setting given by a ring A, an A-bimodule M and T the corresponding tensor algebra. We prove that the Hochschild cohomology of quotients T/I by positive ideals coincides with the homology of A whenever the quiver of M has no oriented cycles.

If the quiver is an arrow (i.e. T is a triangular two by two matrix algebra) the Hochschild cohomology belongs to a long exact sequence. For other quivers, a spectral sequence converging to the Hochschild cohomology will be described in a forthcoming paper.

1991 Mathematics Subject Classification: 16E40, 15A72.

1 Introduction

The purpose of this paper is two-fold: in one direction we prove that the Hochschild homology of quotients of a tensor algebra T over a ring A coincides with the Hochschild homology of A, provided a certain quiver has no oriented cycles. This result has its origin in [4], see also [5] (Proposition 2.3) and [8].

In the other direction, we study the Hochschild *cohomology* of T when the quiver is only an arrow, which corresponds to the case of a triangular two by two algebra of matrices: the Hochschild cohomology of T can be compared with the Hochschild cohomology of A combined with the extensions groups of the corner two-sided ideal of T, through a long exact sequence. The connecting homomorphism is provided by a cup-product associated to the arrow of the quiver.

Actually we use a canonical free resolution of a bimodule M, which provides a complex of cochains computing the Ext functor between M and a bimodule X. This complex is of interest by itself, when the two rings coincide and M is the ring, it specializes to the Hochschild complex which compute the Hochschild cohomology of the ring with coefficients in X.

The homology results enlarges considerations of J.-L. Loday in [9] and the cohomology computations generalizes the one-point extensions considered by D. Happel, see [8]. B. Bendiffalah and D. Guin ([1, 2]) have obtained partial results in the cohomological direction for a triangular algebra.

When the quiver is not just an arrow, there is a spectral sequence converging to the Hochschild cohomology of the tensor algebra, but the techniques are more intricate. The spectral sequence is obtained through a filtration given by the trajectories of the quiver that we define in 3.5. In a forthcoming paper we will describe it, as well as the differentials at the first level which are obtained through a cup product with an element provided by the sum of all the arrows of the quiver.

The first section of the present paper provides an introduction to the subject, we notice relations to known facts and we set apart trivial cases. In the last sections we give the complete statements and proofs of the results.

Acknowledgements : I thank J.-M. Oudom for suggesting the right way of proving the exactness of the resolution of Proposition 4.1.

2   Tensor algebras

In the following we recall the definition of a triangular two by two ring and some well known results. Triangular rings will appear to be special instances of tensor rings; their module categories are morphism categories.

Let A_1 and A_2 be rings, M be an $A_2 - A_1$ bimodule. The *triangular ring T* is the ring $\begin{pmatrix} A_1 & 0 \\ M & A_2 \end{pmatrix}$, namely the set of triples (a_1, a_2, m) in $A_1 \times A_2 \times M$ which multiplies as matrices do, using the bimodule structure of M and the products of A_1 and A_2.

The sub-ring $A_1 \times A_2$ has two central orthogonal idempotents: $e_1 = (1, 0, 0)$ and $e_2 = (0, 1, 0)$. The proof of the next trivial Lemma is based on the existence of these elements.

Lemma 2.1 *The category of left T-modules is equivalent to the category which objects are triples (U_1, U_2, φ) where U_1 is an A_1-module, U_2 is an A_2-module and $\varphi : M \otimes_{A_1} U_1 \longrightarrow U_2$ is a morphism of A_2-modules. Morphisms of this category are couples consisting of an A_1 and an A_2-morphism $(f_1 : U_1 \longrightarrow U_1',$ $f_2 : U_2 \longrightarrow U_2')$ such that $f_2\varphi = \varphi'(1 \otimes f_1)$.*

Proof. A T-module X provides the triple $(e_1 X, e_2 X, \varphi)$ where φ is given by the action of $\begin{pmatrix} 0 & 0 \\ M & 0 \end{pmatrix}$ on X. Conversely a triple (U_1, U_2, φ) gives an $A_1 \times A_2$-

module structure on $U_1 \oplus U_2$. The action of M is determined by φ. All the verifications are straightforward.

Example 2.2 Let $g : A_2 \longrightarrow A_1$ be a ring morphism. The ring A_1 is an A_2 module, where A_2 acts on A_1 via g. The module category of the corresponding triangular algebra is equivalent to the category which objects are morphisms of A_2-modules $\varphi : U_1 \longrightarrow U_2$, where U_1 is an A_1-module considered as an A_2-module via g, and U_2 is an A_2-module.

Example 2.3 Let k be a field, A_1 be a K-algebra and M be a right module. Since M is actually a k-A_1-bimodule the triangular construction can be performed, the resulting algebra is called a "one-point extension" (see for instance [8]).

Definition 2.4 *Let A be a ring and M an A-bimodule. The tensor ring $T_A(M) = A \oplus M \oplus (M \otimes_A M) \oplus (M \otimes_A M \otimes_A M) \oplus \ldots$ has a product given by the natural A-bimodule map*

$$M^{\otimes_A i} \otimes_A M^{\otimes_A j} \longrightarrow M^{\otimes_A i+j}.$$

The next result recalls that the triangular two by two ring $T = \begin{pmatrix} A_1 & 0 \\ M & A_2 \end{pmatrix}$ is a tensor ring. Consider $A = A_1 \times A_2$ and let $\overline{M}$ be the A-bimodule given by M with actions of A_2 on the left and of A_1 on the right extended by zero.

Lemma 2.5 *The rings T and $T_A(M)$ are isomorphic.*

Proof. The key property is that $\overline{M} \otimes_A \overline{M} = 0$. Indeed, $m \otimes m' = me_1 \otimes m' = m \otimes e_1 m' = m \otimes 0 = 0$. The map $\begin{pmatrix} a_1 & 0 \\ m & a_2 \end{pmatrix} \mapsto (a_1, a_2) + m$ is a bijective, and is clearly a ring homomorphism.

Generalizations are immediate, at least in the following direction: let A_1, A_2, A_3 be rings and M_1, M_2 be $A_2 - A_1$ and $A_3 - A_2$ bimodules respectively. The ring

$$\begin{pmatrix} A_1 & 0 & 0 \\ M_1 & A_2 & 0 \\ M_2 \otimes_{A_2} M_1 & M_2 & A_3 \end{pmatrix}$$

is isomorphic to the tensor algebra $T_{A_1 \times A_2 \times A_3}(\overline{M_1 \oplus M_2})$, where $\overline{M_1 \oplus M_2}$ is $M_1 \oplus M_2$ with actions extended by zero. Notice that

$$\left(\overline{M_1 \oplus M_2}\right)^{\otimes_A 3} = 0.$$

Objects of the module category are sequences $(U_1, U_2, U_3, \varphi_1, \varphi_2)$, where U_i is an A_i-module and $\varphi_i : M_i \otimes_{A_i} U_i \to U_{i+1}$ is an A_{i+1}-morphism.

Recall that a quiver Q is a finite oriented graph, more precisely Q is given by two finite sets Q_0 (vertices) and Q_1 (arrows), and two maps s and t from Q_1 to Q_0 assigning the source and the target vertices to each arrow. We define next the quiver of a bimodule; it will provide an appropriate setting for results concerning Hochschild cohomology and homology.

Definition 2.6 *Let A be a ring. A central idempotent system Q_0 of A is a finite set of non zero orthogonal central idempotents of A such that $\sum_{s \in Q_0} s = 1$. The quiver of an A-bimodule M with respect to Q_0 is an oriented graph with set of vertices Q_0 and one arrow from the vertex s to the vertex t if and only if $tMs \neq 0$. Notice that we do not require that the idempotents are primitive: $\{1\}$ is the trivial central idempotent system.*

Remark 2.7 The idea of considering the quiver of a bimodule belongs to P. Gabriel, he used the quiver of $\mathrm{rad}\Lambda/\mathrm{rad}^2\Lambda$ for a finite-dimensional basic and split k-algebra Λ where $\mathrm{rad}\Lambda$ denotes the Jacobson radical of Λ, see for instance [6, 7]. This combinatorial object is at the origin of the developments of the representation theory of non semi-simple algebras during the past decades.

Example 2.8 Let T be a triangular two by two ring $\begin{pmatrix} A_1 & 0 \\ M & A_2 \end{pmatrix}$. The quiver of $\overline{M}$ with respect to the central idempotent system $Q_0 = \{e_1, e_2\}$ of $A_1 \times A_2$ is an arrow.

Example 2.9 If a bimodule is non zero, the quiver corresponding to the trivial central idempotent system $Q_0 = \{1\}$ is a loop.

Proposition 2.10 *Let A be a ring and $Q_0 = \{e_1, e_2\}$ be a central idempotent system of A. Let M be an A-bimodule such that its quiver is an arrow from e_1 to e_2. Then the tensor ring $T_A(M)$ is isomorphic to $\begin{pmatrix} Ae_1 & 0 \\ M & Ae_2 \end{pmatrix}$.*

Proof. Since $\{e_1, e_2\}$ is an idempotent system, there is a ring decomposition $A = Ae_1 \times Ae_2$. Moreover $e_1 M e_2 = e_1 M e_1 = e_2 M e_2 = 0$, consequently $M = e_2 M e_1$. We infer that $M \otimes_A M = 0$; this enables to obtain the isomorphism.

Remark 2.11 The next consideration shows that besides the trivial case of a loop, the quiver of a bimodule depends on the choice of the central idempotent system. The 3×3 triangular matrix algebra considered above provides the quiver

$. \to . \to .$ for the natural choice of the central idempotent system $\{e_1, e_2, e_3\}$, while $\{e_1, e_2 + e_3\}$ provides the quiver with two vertices, one arrow between them and a loop at the target vertex.

Tensor algebras over separable algebras are almost trivial on a homological point of view, as the following known result shows.

Proposition 2.12 *Let k be a field and let A be a separable k-algebra. Let M be an A-bimodule and $T = T_A(M)$ be the corresponding tensor algebra.*
The projective dimension of T as a T-bimodule is at most one.

Proof. First we will show that for a non necessarily separable k-algebra A there is an exact sequence of T-bimodules

$$0 \longrightarrow T \otimes_A M \otimes_A T \xrightarrow{\; g \;} T \otimes_A T \xrightarrow{\; f \;} T \longrightarrow 0$$

where $f(t_1 \otimes t_2) = t_1 t_2$ and $g(t_1 \otimes m \otimes t_2) = t_1 m \otimes t_2 - t_1 \otimes m t_2$.

Consider $V = M^{\otimes_A n} = T^n$ the homogeneous component of T, and notice that $(T \otimes_A T)^n = V \oplus \ldots \oplus V = nV$ since

$$(T \otimes_A T)^n = (T^n \otimes_A A) \oplus (T^{n-1} \otimes_A T^1) \oplus \ldots \oplus (A \otimes_A T^n).$$

Similarly, $(T \otimes_A M \otimes_A T)^n = (n-1)V$, since

$$(T \otimes_A M \otimes_A T)^n = (T^{n-1} \otimes_A M \otimes_A A) \oplus (T^{n-2} \otimes_A M \otimes_A T^1) \oplus \ldots \oplus (A \otimes_A M \otimes_A T^{n-1}).$$

Through these canonical isomorphisms, it is easy to prove that the above sequence becomes a direct sum of sequences

$$0 \longrightarrow (n-1)V \xrightarrow{\; g_n \;} nV \xrightarrow{\; f_n \;} V \longrightarrow 0$$

where

$$g_n(v_1, \ldots, v_{n-1}) = (v_1, v_2 - v_1, \ldots, v_{n-1} - v_{n-2}, -v_{n-1})$$

and

$$f_n(v_1, \ldots, v_n) = v_1 + \ldots + v_n.$$

Finally let A be a separable k-algebra. By definition A is projective as an A-bimodule. Consequently any A-bimodule is projective, hence $T \otimes_A M \otimes_A T$ is a projective T-bimodule, as well as $T \otimes_A T = T \otimes_A A \otimes_A T$. We infer that the above exact sequence is a projective resolution of the T-bimodule T.

Example 2.13 Let Q be a quiver. A path of Q is a finite sequence of concatenated arrows, a trivial one is just a vertex. The path algebra kQ is the vector space which basis the set of paths, the multiplication of two paths is given by its concatenation if it can be done and 0 otherwise; the sum of all the vertices is the unit element of kQ. Actually it is easy to see that kQ is a tensor algebra: consider the separable sub-algebra kQ_0 and the kQ_0-bimodule kQ_1. Then $kQ = T_{kQ_0}(kQ_1)$ since we have an isomorphism of kQ_0-bimodules

$$kQ_i \otimes_{kQ_0} kQ_j \longrightarrow kQ_{i+j}$$

where Q_i is the set of paths having i arrows and kQ_i the corresponding vector space equipped with its natural kQ_0-bimodule structure.

3 Homology

Let k be a field and let Λ be a k-algebra; the Hochschild cohomology and homology vector spaces with coefficients in a Λ-bimodule X are given by

$$H^i(\Lambda, X) = \mathrm{Ext}^i_{\Lambda \otimes_k \Lambda^{op}}(\Lambda, X)$$

$$H_i(\Lambda, X) = \mathrm{Tor}_i^{\Lambda \otimes_k \Lambda^{op}}(X, \Lambda).$$

Recall that Λ-bimodules and left modules over the enveloping algebra $\Lambda \otimes_k \Lambda^{op}$ coincide since Λ^{op} is the underlying vector space of Λ equipped with the reverse product.

We consider now a k-algebra A, an A-bimodule M and $T = T_A(M)$ the tensor algebra. A two-sided ideal I of T is called *positive* if it is contained in the two-sided ideal of T generated by M. When I is positive, A is a sub-algebra of T/I.

Notice that a T/I-bimodule X is also an A-bimodule by restriction. If Q_0 is a central idempotent system of A, let $X = X_\Delta \oplus X_\perp$ be the decomposition of X given by

$$X_\Delta = \bigoplus_{u \in Q_0} uXu$$

$$X_\perp = \bigoplus_{\substack{u,v \in Q_0 \\ u \neq v}} vXu$$

Theorem 3.1 *Let A be a k-algebra, M be an A-bimodule and I a positive two-sided ideal of T. Let X be a T/I-bimodule and assume there exist a system Q_0 of central idempotent system of A such that the quiver of $M \oplus X_\perp$ has no oriented cycles.*

Then

$$H_*(T/I, X) = H_*(A, X_\Delta).$$

Before proving the Theorem, we infer some consequences and links to known results.

Corollary 3.2 *Let T/I be a k-algebra as above, and assume there exist a central idempotent system of A such that the quiver of M has no oriented cycles.*
    Then

$$H_*(T/I, T/I) = H_*(A, A).$$

Proof of the Corollary. We consider the T/I-bimodule of coefficients T/I and we need to show two properties in order to apply the Theorem 3.1:
    1) the quiver of $M \oplus (T/I)_\perp$ has no oriented cycles
    2) $(T/I)_\Delta = A$.
    First notice that the quiver of a direct sum is an appropriate union of quivers: the set of vertices do not change, and there is an arrow from s to t if there is already one at least for one of the quivers of the factors.
    Secondly we remark that since I is positive, $T/I = A \oplus (T^{>0}/I)$, where $T^{>0} = M \oplus (M \otimes_A M) \oplus \ldots$. Moreover $(T/I)_\perp \subset (T^{>0}/I)$ since $A = \bigoplus_{s \in Q_0} As = \bigoplus_{s \in Q_0} sAs$.
    Next we examine the quiver of $T^{>0}/I$ which is obtained from the quiver of $T^{>0}$ by deleting some arrows. We will show that the quiver of $M \otimes_A M$ has no oriented cycles, the same property will hold for any A-tensor power of M by induction. Using the central idempotent system Q_0, we obtain a decomposition

$$M \otimes_A M = \bigoplus_{u_1, u_2, u_3 \in Q_0} (u_3 M u_2) \otimes_{A u_2} (u_2 M u_1).$$

Hence the quiver of $M \otimes_A M$ is obtained by deleting some arrows from the square quiver of M that we define now: its set of vertices remains the same, and the set of arrows is the set of the original arrows added with a new arrow from a vertex s to a vertex t for each length two path from s to t in the original quiver. The square quiver of a quiver without oriented cycles still have this property, as well as quivers obtained by deleting some arrows.
    These considerations shows that the existence of an oriented cycle in the quiver of $M \oplus (T/I)_\perp$ would imply the existence of a (perhaps longer) oriented cycle in the quiver of M.
    Concerning the second property, we have to show that $u (T^{>0}/I) u = 0$ for each $u \in Q_0$; this is equivalent to the fact that the quiver of $T^{>0}/I$ has no loops, and we already know that it has not even oriented cycles.

Example 3.3 If the quiver is just an arrow, we know by Proposition 2.10 that

$$T_A(M) = \begin{pmatrix} A_1 & 0 \\ M & A_2 \end{pmatrix}.$$

The Corollary reduces in this case to a result presented in [9].

In order to prove Theorem 3.1, we will use a projective resolution of T obtained through the separable sub-algebra $E = \times_{u \in Q_0} ku$ of A provided by the central idempotent system Q_0.

Proposition 3.4 *Let Λ be a k-algebra and E a separable sub-algebra. The following is a resolution of Λ by projective Λ-bimodules :*

$$\ldots \longrightarrow \Lambda^{\otimes_E m} \xrightarrow{d} \Lambda^{\otimes_E m-1} \longrightarrow \ldots \longrightarrow \Lambda \otimes_E \Lambda \longrightarrow \Lambda \longrightarrow 0$$

where

$$d(x_1 \otimes \ldots \otimes x_n) = \sum_{i=1}^{n-1} (-1)^i (x_1 \otimes \ldots \otimes x_i x_{i+1} \otimes \ldots \otimes x_n).$$

Notice that tensor products are considered over E and that the differentials are well defined; since E is separable, any E-bimodule Z is projective, hence $\Lambda \otimes_E Z \otimes_E \Lambda$ is a projective Λ-bimodule.

Moreover $t : \Lambda^{\otimes_E n-1} \longrightarrow \Lambda^{\otimes_E n}$ defined by $t(x) = 1 \otimes x$ is an homotopy contraction, i.e. $dt + td = 1$; consequently $\mathrm{Ker} d = \mathrm{Im} d$.

For a tensor algebra $\Lambda = T = T_A(M)$, there is a decomposition of the previous resolution. For this purpose we define below the trajectories of a quiver.

Definition 3.5 The set $TR_n(Q)$ of n-*trajectories* of a quiver Q is the set of sequences $(\gamma_n, \ldots, \gamma_1)$ of n concatenated paths of the quiver, with no restriction on the length (i.e. number of arrows) of each γ_i. In particular 0-length paths (i.e. vertices) are allowed. Concatenated paths means that for each $i \geq 2$ we have $s(\gamma_i) = t(\gamma_{i-1})$ where the source s (resp. target t) of a path is the source (resp. target) of its first (resp. last) arrow. If u is a vertex, $s(u) = t(u) = u$.

We set some notation. Let A be a k-algebra, Q_0 a central idempotent system of A and let M be an A-bimodule with quiver Q with respect to Q_0.

For each vertex u we set $M_u = Au$.
Let a be an arrow; then $M_a = t(a)Ms(a)$.

Let $\gamma = a_p \ldots a_1$ be a path of the quiver; then

$$M_\gamma = M_{a_p} \otimes_{As(a_p)} \cdots \otimes_{At(a_1)} M_{a_1}.$$

Let $\tau = (\gamma_n, \ldots, \gamma_1)$ be a trajectory; then

$$M_\tau = M_{\gamma_n} \otimes_k \ldots \otimes_k M_{\gamma_1}.$$

Proposition 3.6 *Let $T = T_A(M)$ be the tensor algebra of an A-bimodule M, and let Q be the quiver of M with respect to a central idempotent system Q_0 of A. Let $E = \times_{u \in Q_0} ku \subset A$. Then*

$$T^{\otimes_E n} = \bigoplus_{\tau \in TR_n(Q)} M_\tau.$$

Proof. First notice that the homogeneous component $M^{\otimes_A p}$ of T decomposes as follows

$$M^{\otimes_A p} = \bigoplus_{\gamma \in Q_p} M_\gamma$$

where Q_p is the set of paths of length p of the quiver Q. Indeed, consider the decomposition of M as an A-bimodule

$$M = \bigoplus_{u,v \in Q_0} vMu$$

and

$$M \otimes_A M = \bigoplus_{u,v,u',v' \in Q_0} (v'Mu') \otimes_A (vMu).$$

The terms with $u' \neq v$ are 0, since u' and v are orthogonal idempotents belonging to A. Moreover

$$v'Mv \otimes_A vMu = v'Mv \otimes_{Av} vMu$$

since the components of A different from Av acts as zero on the right of $v'Mv$ and on the left of vMu.

These considerations shows that

$$M \otimes_A M = \bigoplus_{\gamma \in Q_2} M_\gamma.$$

Secondly notice that

$$T \otimes_E T = \bigoplus M_{\gamma'} \otimes_E M_\gamma$$

where γ' and γ are paths of Q of any length, – recall that $M_u = Au$ if u is a vertex –. Finally we have that

$$M_{\gamma'} \otimes_E M_\gamma = 0 \quad \text{if} \quad t(\gamma) \neq s(\gamma')$$

and

$$M_{\gamma'} \otimes_E M_\gamma = M_{\gamma'} \otimes_k M_\gamma \quad \text{if} \quad t(\gamma) = s(\gamma').$$

We infer

$$T \otimes_E T = \bigoplus_{\tau \in TR_2(Q)} M_\tau.$$

Corollary 3.7 *Considering the same setting and notation as before, let u and v be vertices, and let $v[TR_n(Q)]u$ be the set of n-trajectories beginning at u and ending at v. Then*

$$v\left(T^{\otimes_E n}\right)u \neq 0 \iff v[TR_n(Q)]u \neq \emptyset.$$

We record now the following useful result for quotient algebras by positive ideals.

Proposition 3.8 *Let A be a k-algebra, and let M be an A-bimodule having a quiver Q with respect to a central idempotent system. Let I be a positive ideal of $T = T_A(M)$ and let u and v be two vertices of Q. Then*

$$v\left((T/I)^{\otimes_E n}\right)u \neq 0 \Rightarrow v[TR_n(Q)]u \neq \emptyset.$$

Proof. There is a surjective morphism

$$T^{\otimes_E n} \longrightarrow (T/I)^{\otimes_E n}$$

which provides surjections

$$v\left(T^{\otimes_E n}\right)u \longrightarrow v[(T/I)^{\otimes_E n}]u$$

for each couple (v, u) of idempotents belonging to the system Q_0.

Proof of Theorem 3.1. In order to compute the Hochschild homology, we consider the projective resolution of Proposition 3.4 for the algebra T/I, and we tensor it by the bimodule X over the enveloping algebra of T/I .
For $n \geq 2$ we obtain

$$(T/I)^{\otimes_E n} \bigotimes_{T/I-T/I} X = (T/I)^{\otimes_E n-2} \bigotimes_{E-E} X$$

and in degree 2

$$(T/I)^{\otimes_E 2} \bigotimes_{T/I - T/I} X = X/\langle ux - xu \mid u \in Q_0 \rangle.$$

Actually we have for $n \geq 1$

$$(T/I)^{\otimes_E n} \bigotimes_{E-E} X = \bigoplus_{u \in Q_0} (Au)^{\otimes_k n} \otimes_k uXu$$

and

$$X/\langle ux - xu \rangle = \bigoplus_{u \in Q_0} uXu = X_\Delta.$$

We will show that these assertions follows directly from the hypothesis of the Theorem, namely that the quiver of $M \oplus X_\perp$ has no oriented cycles.

Indeed, first there is a decomposition which holds independently of the hypothesis:

$$(T/I)^{\otimes_E n} = \bigoplus_{u,v \in Q_0} v(T/I)^{\otimes_E n}u.$$

We know that if a piece of the above direct sum is non zero, then there is trajectory from u to v, and then a path from u to v. Notice that the path can be trivial, in that case $u = v$.

It is useful to recall that if Y and Z are bimodules over a ring Λ (i.e. left $\Lambda \otimes \Lambda^{op}$-modules), the tensor product $Y \otimes_{\Lambda - \Lambda} Z$ (i.e. $Y \otimes_{\Lambda \otimes \Lambda^{op}} Z$) affords the relations $y\lambda \otimes z = y \otimes \lambda z$ and $\lambda y \otimes z = y \otimes z\lambda$ for $y \in Y$, $z \in Z$ and $\lambda \in \Lambda$. Consequently

$$\begin{aligned} (T/I)^{\otimes_E n} \otimes_{E-E} X &= \bigoplus_{u,v,u',v' \in Q_0} v[(T/I)^{\otimes_E n}]u \otimes_k v'Xu' \\ &= \bigoplus_{u,v \in Q_0} v[(T/I)^{\otimes_E n}]u \otimes_k uXv. \end{aligned}$$

A non-zero component with $u \neq v$ in this direct sum provides two non-zero vector spaces $v[(T/I)^{\otimes_E n}]u$ and uXv. This gives an oriented cycle in the quiver of $M \oplus X_\perp$, contradicting the hypothesis. Hence the only possibly non-zero components of the direct sum corresponds to $u = v$. Moreover $u(T^{\otimes_E n})u = Au^{\otimes_k n}$ since we know that

$$u(T^{\otimes_E n})u = \bigoplus_{\tau \in (TR^n(Q))u} M_\tau$$

and the trivial trajectory $(u, u, \ldots, u)$ is the only element in $uTR^n(Q)u$ (recall that Q has no oriented cycles). Then

$$u((T/I)^{\otimes_E n})u = Au^{\otimes_k n}.$$

An easier consideration shows the equality

$$X/\langle ux - xu \rangle = X_\Delta.$$

Actually these decompositions provide a decomposition of the complex of chains as a direct sum of sub-complexes indexed by u; each of them coincides precisely with the Hochschild complex of chains computing the Hochschild homology, (see for instance [3]).

4   Hochschild cohomology

In order to compute the Hochschild cohomology of a triangular algebra

$$\begin{pmatrix} A_1 & 0 \\ M & A_2 \end{pmatrix}$$

we introduce a standard complex which has cohomology groups isomorphic to $\mathrm{Ext}^*_{A_2 - A_1}(M, Y)$, if M and Y are $A_2 - A_1$-bimodules. Recall that an $A_2 - A_1$-bimodule is an $A_2 \otimes_k A_1^{op}$-left module and that $\mathrm{Ext}^*_{A_2 - A_1}$ is $\mathrm{Ext}^*_{A_2 \otimes A_1^{op}}$.

In the following we drop the the tensor product sign $\otimes_k$ between vector spaces, and we replace the sign $\otimes$ between vectors by a comma.

Let A_1 and A_2 be k-algebras, and let M and Y be $A_2 - A_1$-bimodules. Let

$$M^{i,n} = A_2 A_2 \ldots A_2 M A_1 \ldots A_1$$

where the number of vector spaces in this tensor product is n and M is in the ith position from the left.

We consider the following complex of cochains

$$0 \to \mathrm{Hom}_k(M, Y) \xrightarrow{\partial} \mathrm{Hom}_k(M^{1,2}, Y) \xrightarrow{\partial} \ldots \xrightarrow{\partial} \mathrm{Hom}_k(\bigoplus_1^n M^{i,n}, Y) \xrightarrow{\partial} \ldots$$

with coboundaries analogous to the Hochschild ones:

$$\partial : \mathrm{Hom}_k(M^{i,n}, Y) \longrightarrow \mathrm{Hom}_k(M^{i,n+1}, Y) \oplus \mathrm{Hom}_k(M^{i+1,n+1}, Y)$$

is given by $\partial f = \partial_1 f + \partial_2 f$ where

$$\begin{aligned}
\partial_1 f(b_1, b_2, \ldots, \ & b_{i-1}, m, a_{i+2}, \ldots, a_{n+1}) = \\
& (-1)^i f(b_1, b_2, \ldots, b_{i-1}, m a_{i+2}, \ldots, a_{n+1}) \\
& + \Sigma (-1)^j f(b_1, b_2, \ldots, b_{i-1}, m, a_{i+2}, \ldots, a_j a_{j+1}, \ldots, a_{n+1}) \\
& + (-1)^{n+1} f(b_1, b_2, \ldots, b_{i-1}, m, a_{i+2}, \ldots, a_n) a_{n+1}.
\end{aligned}$$

and

$$\partial_2 f(b_1, b_2, \ldots, \quad b_i, m, a_{i+1}, \ldots, a_{n+1}) =$$
$$b_1 f(b_2, \ldots, b_i, m, a_{i+1}, \ldots, a_{n+1})$$
$$+ \Sigma(-1)^j f(b_1, b_2, \ldots, b_j b_{j+1}, \ldots, b_i, m, a_{i+1}, \ldots, a_{n+1})$$
$$+ (-1)^i f(b_1, b_2, \ldots, b_i, m, a_{i+1}, \ldots, a_{n+1})$$

Proposition 4.1 *The cohomology of the preceding complex of cochains is* $\mathrm{Ext}^*_{A_2 - A_1}(M, Y)$.

Remark 4.2 : Notice that for $f \in \mathrm{Hom}_k(M, Y)$ we have
$(\partial_1 f)(m, a) = f(ma) - f(m)a$ and
$(\partial_2 f)(b, m) = bf(m) - f(bm)$.
Consequently $\ker \partial = \mathrm{Hom}_{A_2 - A_1}(M, Y)$.

Proof of Proposition 4.1. Consider the free $A_2 - A_1$-bimodule $\overline{M} = A_2 M A_1$ and the following complex

$$\ldots \xrightarrow{d} \bigoplus_{i=1}^{n} \overline{M}^{n,i} \xrightarrow{d} \ldots \xrightarrow{d} A_2 \overline{M} \oplus \overline{M} A_1 \xrightarrow{d} \overline{M} \xrightarrow{d} M \longrightarrow 0$$

where $\overline{M}^{n,i}$ is defined as before.
    The differential $d : \overline{M}^{n,i} \longrightarrow \overline{M}^{n-1, i-1} \oplus \overline{M}^{n-1, i}$ is given by

$$d(b_1, b_2, \ldots, b_{i-1}, \overline{m}, a_{i+1}, \ldots, a_n) =$$
$$b_1 b_2, \ldots, b_{i-1}, \overline{m}, a_{i+1}, \ldots, a_n$$
$$+ \Sigma_i^{i-2}(-1)^j b_1, \ldots, b_j b_{j+1}, \ldots, b_{i-1}, \overline{m}, a_{i+1}, \ldots, a_n$$
$$+ (-1)^{i-1} b_1, \ldots, b_{i-2}, b_{i-1} \overline{m}, a_{i+1}, \ldots, a_n$$
$$+ (-1)^i b_1, \ldots, b_{i-1}, \overline{m} a_{i+1}, \ldots, a_n$$
$$+ \Sigma(-1)^k b_1, \ldots, b_{i-1}, \overline{m}, a_{i+1}, \ldots, a_k a_{k+1}, \ldots, a_n.$$

In order to verify that we obtain a resolution of the $A_2 - A_1$-bimodule M, notice that there is an evident double complex which total complex is the one above. Each column and each row are acyclic, using the standard homotopy contraction obtained by inserting a one.
    In order to finish the proof, we apply the functor $\mathrm{Hom}_{A_2 - A_1}(-, M)$ to the above resolution. Recall that

$$\mathrm{Hom}_{A_2 - A_1}(A_2 Z A_1, M) = \mathrm{Hom}_k(Z, M),$$

and notice that through this identification, the dual of the differential becomes the announced coboundary ∂.

We compute now the Hochschild cohomology of a triangular algebra. From the resolution of Proposition 3.4 of a k-algebra Λ with a given separable sub-algebra E, we have that the Hochschild cohomology vector spaces $H^*(\Lambda, X)$ with coefficients in a Λ-bimodule X are given by the cohomology of the following complex of cochains:

$$\mathcal{H} : 0 \to X^E \xrightarrow{\delta} \mathrm{Hom}_{E-E}(\Lambda, X) \xrightarrow{\delta} \mathrm{Hom}_{E-E}(\Lambda_{\otimes_E}\Lambda, X) \xrightarrow{\delta} \ldots$$

where $X^E = \{x \in X \mid ex = xe \; \forall e \in E\}$ and

$$\partial f(x_1 \otimes \ldots \otimes x_n) = \begin{aligned}[t] & x_1 f(x_2 \otimes \ldots \otimes x_n) \\ & + \Sigma(-1)^i f(x_1 \otimes \ldots \otimes x_i x_{i+1} \otimes \ldots \otimes x_n) \\ & + (-1)^{n-1} f(x_1 \otimes \ldots \otimes x_{n-1}) x_n. \end{aligned}$$

We restrict ourselves to the case where Λ is a tensor algebra $T = T_A(M)$ such that there exists in A a central idempotent system of order two $\{e_1, e_2\}$ verifying

$$e_1 M_1 e_1 = e_2 M_2 e_2 = e_1 M e_2 = 0.$$

We know from Proposition 2.10 that $T = \begin{pmatrix} A_1 & 0 \\ M & A_2 \end{pmatrix}$. Moreover $E = \begin{pmatrix} k & 0 \\ 0 & k \end{pmatrix}$ for $e_1 = \begin{pmatrix} 1 & 0 \\ 0 & 0 \end{pmatrix}$ and $e_2 = \begin{pmatrix} 0 & 0 \\ 0 & 1 \end{pmatrix}$.

Lemma 4.3 *For T as above, we have*

$$T^{\otimes_E n} = A_1^{\otimes_E n} \oplus A_2^{\otimes_E n} \oplus \bigoplus_{i=1}^{n} M^{i,n}$$

where $M^{i,n} = A_2 \otimes \ldots \otimes A_2 \otimes M \otimes A_1 \otimes \ldots \otimes A_1$ ($M^{i,n}$ has n tensor vector spaces and M is at place i).

Proof. Notice that

$$0 = A_1 \otimes_E A_2 = A_2 \otimes_E A_1 = M \otimes_E A_2 = A_1 \otimes_E M.$$

Hence the "missing" terms in the decomposition of $T^{\otimes_E n}$ are actually zero.

Remark 4.4 We have proved at the preceding section a more general result, namely

$$T^{\otimes_E n} = \bigoplus_{\tau \in TR^n Q} M_\tau.$$

Since the quiver is now only an arrow a, a trajectory of length n is either trivial staying in one vertex or it goes through the arrow, namely the trajectory is

$$(e_2, e_2, \ldots, e_2, a, e_1, e_1, \ldots, e_1).$$

This agrees to the decomposition of the preceding Lemma.

Theorem 4.5 *Let T be as above, let X be a T-bimodule, $_2X_1 = e_2 X e_1$ and $X_0 = e_1 X e_1 \oplus e_2 X e_2$. There is a long exact sequence*

$$
\begin{array}{lcccl}
& 0 & \longrightarrow & H^0(T, X) & \longrightarrow & H^0(A, X_0) & \longrightarrow \\
\longrightarrow & \operatorname{Hom}_{A_2-A_1}(M, {}_2X_1) & \longrightarrow & H^1(T, X) & \longrightarrow & H^1(A, X_0) & \longrightarrow \\
\longrightarrow & \operatorname{Ext}^1_{A_2-A_1}(M, {}_2X_1) & \longrightarrow & H^2(T, X) & \longrightarrow & H^2(A, X_0) & \longrightarrow
\end{array}
$$

$$\cdots$$

$$
\longrightarrow \quad \operatorname{Ext}^n_{A_2-A_1}(M, {}_2X_1) \quad \longrightarrow \quad H^{n+1}(T, X) \quad \longrightarrow \quad H^{n+1}(A, X_0) \quad \longrightarrow
$$

$$\cdots$$

Remark 4.6 If $A_2 = k$ and M is a right A_1-module the triangular algebra T is a one-point extension. The above long exact sequence has been obtained by D. Happel [8] in this case.

Proof of Theorem 4.5. Notice that the following vector spaces provides a sub-complex of cochains of $\mathcal{H}$:

$$
\mathcal{M}^n = \operatorname{Hom}_{E-E}\left(\bigoplus_{i=1}^{n} M^{i,n}, X\right) = \operatorname{Hom}_k\left(\bigoplus M^{i,n}, e_2 X e_1\right).
$$

By contrast

$$
\operatorname{Hom}_{E-E}\left(A_j^{\otimes_E n}, X\right) = \operatorname{Hom}_k(A_j^{\otimes_E n}, e_j X e_j)
$$

for $j = 1$ or 2 do not give in general a sub-complex of $\mathcal{H}$. However we obtain an exact sequence

$$
0 \longrightarrow \mathcal{M} \longrightarrow \mathcal{H} \longrightarrow \mathcal{H}/\mathcal{M} \longrightarrow 0
$$

of complexes of cochains and it is easy to make an identification of $\mathcal{H}/\mathcal{M}$ with the usual complex of cochains computing the Hochschild cohomology of A with coefficients in $X_0 = e_1 X e_1 \oplus e_2 X e_2$.

In order to finish the proof of the Theorem, just notice that the sub-complex $\mathcal{M}^*$ is precisely the complex of cochains that we have considered in the Proposition above.

Corollary 4.7 *Assume M is a projective $A_2 - A_1$-bimodule, and let T be the triangular algebra $\begin{pmatrix} A_2 & 0 \\ M & A_1 \end{pmatrix}$. Then*

$$
H^*(T, X) \cong H^*(A, X_0)
$$

for $* \geq 2$, where X is a T-bimodule, $A = A_1 \times A_2$ and $X_0 = e_1 X e_1 + e_2 X e_2$.

Finally, we give an interpretation of the connecting homomorphism

$$\delta : H^n(A, X_0) \longrightarrow \text{Ext}^{n+1}_{A_2 - A_1}(M, {}_2X_1)$$

using an analogous of the usual cup-product.

Proposition 4.8 *Let* $T = \begin{pmatrix} A_1 & 0 \\ M & A_2 \end{pmatrix}$ *be the triangular algebra, where* M *is an* $A_2 - A_1$-*bimodule. A* T-*bimodule* X *is given by four* $A_i - A_j$-*bimodules* ${}_iX_j$ *for all* $i, j \in \{1, 2\}$ *and four maps*

$$\begin{array}{rcccl} a: & M & \otimes_{A_1} & {}_1X_1 & \longrightarrow {}_2X_1 \\ b: & M & \otimes_{A_1} & {}_1X_1 & \longrightarrow {}_2X_2 \\ c: & {}_1X_2 & \otimes_{A_2} & M & \longrightarrow {}_1X_1 \\ d: & {}_2X_2 & \otimes_{A_2} & M & \longrightarrow {}_2X_1 \end{array}$$

of the corresponding bimodule structures, verifying

$$d(b \otimes 1) = a(1 \otimes c)$$

Proof : The decomposition

$$X = e_1 X e_1 \oplus e_2 X e_2 \oplus e_1 X e_2 \oplus e_2 X e_1$$

and the actions of M on X provides the morphisms a, b, c and d. The commutation formula reflects the bimodule requirement

$$(tx)t' = t(xt')$$

for $t, t' \in T$ and $x \in X$.

Consider the i-cochain vector space of the complex computing the Hochschild cohomology, $f \in \text{Hom}_k(A_1^{\otimes n}, {}_1X_1)$ and $g \in \text{Hom}(A_2^{\otimes n}, X_2)$. Let also h be an element of $\text{End}_{A_2 - A_1}(M)$.

Notice that

$$h \smile f : M \otimes A_1^{\otimes n} \xrightarrow{h \otimes f} M \otimes {}_1X_1 \xrightarrow{a} {}_2X_1$$

and

$$g \smile 1_M : A_2^{\otimes n} \otimes M \xrightarrow{g \otimes h} {}_2X_2 \otimes M \xrightarrow{d} {}_2X_1$$

gives cochains of the complex computing $\text{Ext}^{n+1}_{A_2 - A_1}(M, {}_2X_1)$; actually this provides cup-products

$$\text{End}_{A_2 - A_1}(M) \otimes H^n(A_1, {}_1X_1) \longrightarrow \text{Ext}^{n+1}_{A_2 - A_1}(M, {}_2X_1)$$

$$H^n(A_2, {}_2X_2) \otimes \text{End}_{A_2 - A_1}(M) \longrightarrow \text{Ext}^{n+1}_{A_2 - A_1}(M, {}_2X_1).$$

Proposition 4.9 *The connecting homomorphism*

$$\partial : H^n(A_1, {}_1X_1) \oplus H^n(A_2, {}_2X_2) \longrightarrow \mathrm{Ext}^n_{A_2 - A_1}(M, {}_2X_1)$$

of the long exact sequence of the Theorem above is given by the formula

$$\partial(f + g) \ = \ 1_M \smile f \ + \ (-1)^{n+1}g \smile 1_M.$$

References

[1] Bendiffalah, B., Guin, D.: Catégorie de morphismes et cohomologie de Hochschild. Prépublication 1997/08, Université de Montpellier 2 (1997).

[2] Bendiffalah, B., Guin, D.: Cohomologie des morphismes, to appear in Comm. in Algebra (1998).

[3] Cartan, H. Eilenberg, S.: Homological algebra. Princeton University Press, 1956.

[4] Cibils, C.: Hochschild homology of an algebra whose quiver has no oriented cycles. Representation Theory I - Finite Dimensional Algebras, Ottawa 1984, Springer Lect. Notes Math., **1177**, 55–59, (1986).

[5] Cibils, C.: Rigidity of truncated quiver algebras, Adv. in Math **79**, 18–42, (1990).

[6] Gabriel, P.: Finite representation type is open. Representations of algebras, Ottawa 1974, Lect. Notes Math. **488**, 132–155 (1975).

[7] Gabriel P.: Indecomposable representations, II, Symposia Matematica **2** (Instituto Nazionale di Alta Matematica, Roma), 81–104 (1973).

[8] Happel, D.: Hochschild cohomology of finite-dimensional algebras. Séminaire d'algèbre Paul Dubreuil et Marie-Paule Malliavin, Lect. Notes Math. **1404**, 108–126 (1989).

[9] Loday, J.-L.: Cyclic homology. Grundlehren der mathematischen Wissenschaften, Second edition 1998 (513 p.)

Generalized Path Algebras

Flávio Ulhoa Coelho

Departamento de Matemática -IME
Universidade de São Paulo
CP 66281 São Paulo - SP CEP 05315-970 Brazil

Shao-Xue Liu

Beijing Normal University
Beijing 100875, PRC

Dedicated to Prof. Helmut Lenzing on his 60th Birthday

The description of basic algebras of finite dimension over algebraically closed fields as quotients of path algebras has been very useful in the representation theory, not only in providing a wide range of examples but mainly because of the description of the finitely generated modules over the given algebras (see, for instance [1, 6]).

The purpose of these notes is to look at a generalization of the concept of path algebras. Instead of assigning to each vertex of a given quiver the base field K, we shall assign a K-algebra. With this, we lose the uniqueness of the quiver associated to a given algebra but we hope to get a better insight in some properties of the ring structure, for instance primeness and noetherianness. Some related problems can be found in [2, 3, 4, 5].

These notes are organized as follows. In section 1, after giving the definition of the generalized path algebras we prove some preliminary results on them. Section 2 is devoted to characterising generalized path algebras which are prime and noetherian, while in the last section we look at the so-called isomorphism problem for generalized path algebras.

Part of this work was completed when the second named author was visiting the University of São Paulo. He would like to take this opportunity to express his thanks to the department of mathematics of USP, especially Professor H. Merklen, for hospitality, and also for the financial support of NSFC (Grant No: 19331013). The first named author acknowledges the support given by CNPq. The authors thank the referee for the comments on a preliminary version of this work.

1   Preliminaries

This section is devoted to define generalized path algebras and to establish some preliminary results on them. Along this paper, K will denote a fixed field. For an algebra, we mean an associative K-algebra.

A *quiver* Δ is given by two sets Δ_0 and Δ_1 together with two maps $s, e\colon \Delta_1 \longrightarrow \Delta_0$. The elements of Δ_0 are called *vertices*, while the elements of Δ_1 are called *arrows*. For an arrow $\alpha \in \Delta_1$, the vertex $s(\alpha)$ is the *start vertex of* α and the vertex $e(\alpha)$ is the *end vertex of* α, and we draw $s(\alpha) \xrightarrow{\alpha} e(\alpha)$. A *path in* Δ is $(a|\alpha_1 \cdots \alpha_n|b)$, where $\alpha_i \in \Delta_1$, for $i = 1, \cdots, n$, and $s(\alpha_1) = a$, $e(\alpha_i) = s(\alpha_{i+1})$ for $i = 1, \cdots, n-1$, and $e(\alpha_n) = b$. The *length of a path* is the number of arrows in it. to each arrow α we can assign an edge $\overline{\alpha}$ where the orientation is forgotten. A *walk* between two vertices a and b is given by $(a|\overline{\alpha_1} \cdots \overline{\alpha_n}|b)$, where $a \in \{s(\alpha_1), e(\alpha_1)\}$, $b \in \{s(\alpha_n), e(\alpha_n)\}$, and for each $i = 1, \cdots, n-1$, $\{s(\alpha_i), e(\alpha_i)\} \cap \{s(\alpha_{i+1}), e(\alpha_{i+1})\} \neq \emptyset$. A quiver is said to be *connected* if for each pair of vertices a and b, there exists a walk between them.

Let $\Delta = (\Delta_0, \Delta_1)$ be a quiver and $\mathcal{A} = \{A_i : i \in \Delta_0\}$ be a family of

K-algebras A_i with identity, indexed by the vertices of Δ. Unless otherwise stated, we shall indicated the identity of A_i as e_i, for $i \in \Delta_0$. The elements of $\cup_{i \in \Delta_0} A_i$ are called the $\mathcal{A}$-*paths of length zero*, and for each $n \geq 1$, an $\mathcal{A}$-*path of length n* is given by $a_1 \beta_1 a_2 \beta_2 \cdots a_n \beta_n a_{n+1}$, where $(s(\beta_1)|\beta_1 \cdots \beta_n|e(\beta_n))$ is a path in Δ of length n, for each $i = 1, \cdots, n$, $a_i \in A_{s(\beta_i)}$, and $a_{n+1} \in A_{e(\beta_n)}$. Consider now the quotient R of the K-vector space with basis the set of all $\mathcal{A}$-paths of Δ by the subspace generated by all the elements of the form

$$(a_1 \beta_1 \cdots \beta_{j-1}(a_j^1 + \cdots + a_j^m)\beta_j a_{j+1} \cdots \beta_n a_{n+1}) - \sum_{l=1}^{m} a_1(\beta_1 \cdots \beta_{j-1} a_j^l \beta_j \cdots \beta_n a_{n+1})$$

where $(s(\beta_1)|\beta_1 \cdots \beta_n|e(\beta_n))$ is a path in Δ of length n, for each $i = 1, \cdots, n$, $a_i \in A_{s(\beta_i)}$, $a_{n+1} \in A_{e(\beta_n)}$, and $a_j^l \in A_{s(\beta_j)}$ for $l = 1, \cdots, m$. Define now in R the following multiplication. Given two elements $[a_1 \beta_1 \cdots \beta_n a_{n+1}]$ and $[b_1 \gamma_1 \cdots \gamma_m b_{m+1}]$, we define

$$[a_1 \beta_1 \cdots \beta_n a_{n+1}] \cdot [b_1 \gamma_1 \cdots \gamma_m b_{m+1}]$$

to be equal to $[a_1 \beta_1 \cdots \beta_n(a_{n+1}b_1)\gamma_1 \cdots \gamma_m b_{m+1}]$, if a_{n+1} and b_1 belong to the same A_i and equal to 0 otherwise. It is easy to check that the above multiplication in R is well-defined and gives to R an structure of K-algebra. Also, R has identity if and only if Δ_0 is finite. The algebra R defined above is called the $\mathcal{A}$-*path algebra of Δ* and we denote it by $R = K(\Delta, \mathcal{A})$.

Remark 1.1 (i) Observe that if $A_i = K$ for each $i \in \Delta_0$, then the algebra $K(\Delta, \mathcal{A})$ defined above is the usual path algebra $K\Delta$ of Δ.

(ii) Any K-algebra R with identity can be *realized* as an $\mathcal{A}$-path algebra $K(\Delta, \mathcal{A})$ by just taking Δ as the quiver consisting of a unique vertex and $\mathcal{A} = \{R\}$. Also, it is not difficult to see that a *realization* of a K-algebra as $\mathcal{A}$-path algebra is not necessarily unique. We shall discuss the problem of uniqueness in section 3 below.

(iii) Let $R = K(\Delta, \mathcal{A})$. We leave to the reader the verification that $\dim_K R <$

∞ if and only if $\dim_K A_i < \infty$ for each $i \in \Delta_0$, and Δ is a finite quiver without oriented cycles.

We can give an alternative definition for an $\mathcal{A}$-path algebra $R = K(\Delta, \mathcal{A})$ as follows. Let ${}_i M_j$ be the free A_i-A_j-bimodule with free generators given by the arrows from i to j. If $A = \bigoplus_{i \in \Delta_0} A_i$, then ${}_i M_j$ is also an A-A-bimodule by defining $A_k \cdot_i M_j = 0$ if $k \neq i$ and ${}_i M_j \cdot A_k = 0$ if $k \neq j$. Let $M = \bigoplus_{i \longrightarrow j} {}_i M_j$, which is clearly an A-A-bimodule. We leave to the reader to check that R is isomorphic to the algebra

$$A \oplus M \oplus (M \otimes_K M) \oplus (M \otimes_K M \otimes_K M) \oplus \cdots$$

with multiplication given by the tensor product.

Examples 1.2 Let Δ be the quiver

$$\begin{array}{ccccc} \bullet & \longleftarrow & \bullet & \longleftarrow & \bullet \\ 1 & & 2 & & 3 \end{array}$$

and let $\mathcal{A} = \{K, K[x], K\}$. Observe that the algebra $R = K(\Delta, \mathcal{A})$ is isomorphic to the (usual) path algebra $K\Delta'$, where Δ' is the quiver

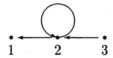

$$\begin{array}{ccccc} \bullet & \longleftarrow & \bullet & \longleftarrow & \bullet \\ 1 & & 2 & & 3 \end{array}$$

$\square$

We finish this section characterizing the Jacobson radical $J(R)$ of an $\mathcal{A}$-path algebra $R = K(\Delta, \mathcal{A})$. We shall need the following definition.

Definition. Let Δ be a quiver and $\mathcal{A}$ be a family of K-algebras $\{A_i\}_{i \in \Delta_0}$.
(i) We say that a path $(i|\beta_1 \cdots \beta_n|j)$, $n \geq 1$, in Δ is *regular* provided it is not a subpath of an oriented cycle in Δ.

(ii) Let $(i|\beta_1 \cdots \beta_n|j)$, $n \geq 1$, be a regular path in Δ. Then, any $\mathcal{A}$-path $a\beta_1 \cdots \beta_n b$, with $a \in A_i$ and $b \in A_j$ is called *regular*. Also, if there are no oriented cycles passing through a vertex i, then the elements of $J(A_i)$ are called *regular $\mathcal{A}$-paths (of length zero)*.

Proposition 1.3 Let $R = k(\Delta, \mathcal{A})$ be the $\mathcal{A}$-path algebra of Δ. Then the Jacobson radical $J(R)$ of R is just the K-subspace of R generated by all regular paths of R.

Proof: Denote by M the K-subspace of R generated by all regular paths of R. Also, for $X \subset R$, denote by (X) the (bilateral) ideal generated by X. Let $u \in R$ be a regular path of length greater than zero. Clearly, $(u)^2 = 0$ and so $(u) \subset J(R)$. Let now i be a vertex at which there are no oriented cycles and consider $U_i = \sum_u (u)$, where u runs over all regular paths of length greater than zero in R, starting or ending at i.

Observe that $(J(A_i)) = J(A_i) + U_i$. On the other hand, U_i is a quasi-regular ideal of R and since $(J(A_i) + U_i)/U_i \cong J(A_i)$ is also quasi-regular, we infer that $(J(A_i))$ is quasi-regular. Hence $(J(A_i)) \subset J(R)$, and this shows that $M \subset J(R)$.

Suppose that $M \neq J(R)$. Then there exists a nonzero element $x = \sum_i p_i \in J(R)$, where each p_i is not regular. Clearly, there are l, m such that $0 \neq e_l x e_m \in J(R)$. So, without loss of generality we can assume that $e_l p_i e_m = p_i$, for each i. By hypothesis, the p_i's are subpaths of oriented cycles, and so there exists a path $q = e_m q e_l$ from m to l. Multiplying x by q, we still get a sum of nonzero paths in $J(R)$ which are clearly oriented cycles. Therefore, we can assume that each p_i is an oriented cycle at the vertex l. Moreover, multiplying x on both sides by a path $(e_l|\gamma \cdots \delta|e_l)$, we may finally assume that $0 \neq x = \sum_i p_i$, where all p_i's have the form $(e_l|\gamma \cdots \delta|e_l)$. Observe that if z is a path in R with $e_l z \neq 0$, then $p_i z \neq 0$ and so x is a quasi-regular element in R. Hence, there exists $y \in R$ such that $x + y + xy = 0$ $(*)$. Multiplying $(*)$ on both side by e_l, we may suppose $e_l y e_l = y$. But now $(*)$ leads to a contradiction because $xy \neq 0$ and has a bigger length than $x + y$. Therefore, $M = J(R)$, as required. $\square$

The next result is an easy consequence of the above proposition.

Corollary 1.4 Let $R = K(\Delta, \mathcal{A})$ be the $\mathcal{A}$-path algebra of Δ. Then, R is semiprimitive if and only if Δ has no regular paths, and $J(A_i) = 0$ whenever i is an isolated vertex.

2   Criteria for primeness and noetherianness

Along this section, let us assume that Δ is a quiver, $\mathcal{A} = \{A_i : i \in \Delta_0\}$ is a family of K-algebras A_i indexed by the vertices of Δ and with identity e_i, and $R = K(\Delta, \mathcal{A})$ is the $\mathcal{A}$-path algebra of Δ. We shall give now criteria for R to be prime and to be right and left noetherian in terms of the quiver Δ and the family $\mathcal{A}$. We start with a definition.

Definition. We say that a quiver Δ is *oriented connected* provided for any pair of distinct vertices i and j, there exists a path in Δ from i to j.

Clearly, a connected quiver Δ is oriented connected if and only if any path of Δ is a subpath of an oriented cycle.

Theorem 2.1 With the above notations and assuming that Δ is connected and has at least 2 vertices, the following are equivalent:

(a) R is prime;

(b) Δ is oriented connected;

(c) No $\mathcal{A}$-path is regular.

Proof: $(a) \Rightarrow (b)$. Suppose Δ is not oriented connected. So, there are distinct vertices i and j with no paths from i to j. Hence, there is no $\mathcal{A}$-path in R from i to j, which implies that $e_i R e_j = 0$, a contradiction.

$(b) \Rightarrow (a)$. Let I and J be two nonzero ideals of R. Observe that there are vertices $i, j, l, m \in \Delta_0$ such that $e_i I e_j \neq 0$ and $e_l J e_m \neq 0$. Assuming now that Δ is oriented connected we infer that there exists a path $\gamma \in \Delta$ from j to l. Therefore, $e_i I e_j \gamma e_l J e_m \neq 0$, and so R is prime as required.

$(b) \Rightarrow (c)$. It is enough to show that any path in Δ is non-regular and each vertex belongs to an oriented cycle. Let $(i|\alpha_1 \cdots \alpha_n|j)$ be a path in Δ with $i \neq j$. Since Δ is oriented connected, then there exists a path $(j|\beta_1 \cdots \beta_m|i)$ and so $(i|\alpha_1 \cdots \alpha_n \beta_1 \cdots \beta_m|i)$ is a cycle. Since $|\Delta_0| \geq 2$, the same argument can be used to show that each vertex belongs to an oriented cycle.

$(c) \Rightarrow (b)$. Since Δ is connected, given two distinct vertices i and j, there always exists a walk between them. Denote by $l = l(i, j)$ the length of a shortest walk between i and j. We shall prove the statement by induction on $l \geq 1$. Suppose that $l = 1$, that is, that there exists either an arrow $\alpha: i \longrightarrow j$ or an arrow $\beta: j \longrightarrow i$. There is nothing to show in the former case, so let us consider the later. By hypothesis, there exists an oriented cycle $(j|\beta \beta_1 \cdots \beta_m|j)$ and so $(i|\beta_1 \cdots \beta_m|j)$ is the required path from i to j. Suppose now $l > 1$. So there exists a walk $(i|\overline{\gamma_1} \cdots \overline{\gamma_l}|j)$ of length l between i and j. If $s(\gamma_1) = i$, then there exists a walk $(e(\gamma_1)|\overline{\gamma_2} \cdots \overline{\gamma_l}|j)$ of length $l-1$ and so, by induction, there exists a path from $e(\gamma_1)$ to j which composed on the left with γ_1 gives a path from i to j. If now $e(\gamma_1) = i$, then the same argument used above, gives a path from i to $s(\gamma_1)$. The induction hypothesis yields a path from $s(\gamma_1)$ to j, leading to the required path from i to j. $\square$

We shall now prove a criterion for R to be right noetherian.

Theorem 2.2 With the above notations, the algebra R is right noetherian if and only if the following conditions hold:

(i) Δ is finite.

(ii) There exists only one arrow starting at each vertex lying in an oriented cycle of Δ.

(iii) The algebra A_i is one dimensional for each vertex i lying in an oriented cycle.

(iv) If j is not a sink vertex, then A_j is finite dimensional, and if j is a sink vertex, then the algebra A_j is right noetherian.

Proof: Let us assume first that R is right noetherian. Clearly, Δ is finite, and so we have (i).

Suppose now that Δ has a cycle γ starting and ending at a vertex i and that there is an arrow $\alpha: i \longrightarrow j$ which does not belong to γ. To get a contradiction, just consider the following infinite ascending chain of distinct right ideals

$$(\gamma\alpha) \subset (\gamma\alpha, \gamma^2\alpha) \subset (\gamma\alpha, \gamma^2\alpha, \gamma^3\alpha) \subset \cdots$$

This proves (ii).

Let now γ be an oriented cycle in Δ passing through a vertex i, and consider the arrows α and β such that $s(\alpha) = i$ and $e(\beta) = i$. If A_i is not one-dimensional, then there exist two elements K-linearly independent a and b in A_i. Consider $x = e_i \alpha \cdots \beta a$ and $y = e_i \alpha \cdots \beta b$. Then,

$$(xy) \subset (xy, x^2y) \subset (xy, x^2y, x^3y) \subset \cdots$$

is an infinite ascending chain of distinct right ideals, which proves (iii).

If j is a sink vertex, then every right ideal of A_j is also a right ideal of R, so A_j must be right noetherian. If j is not a sink vertex, then there exists an arrow $\alpha: j \longrightarrow l$ starting at j. If now A_j has infinite dimension, let $\{a_i : i \in N\}$ be a linearly independent infinite set of elements of A_j. Then,

$$(a_1\alpha) \subset (a_1\alpha, a_2\alpha) \subset (a_1\alpha, a_2\alpha, a_3\alpha) \subset \cdots$$

gives a contradiction to the fact that R is right noetherian.

So, (i) - (iv) are necessary for R to be right noetherian.

Let us now assume that (i) - (iv) hold and prove that R is right noetherian. To do so, let

$$I_1 \subset I_2 \subset \cdots \subset I_n \subset \cdots$$

be an ascending chain of right ideals of R. Since Δ_0 is finite, say $|\Delta_0| = m$, then $\sum_{\iota=1}^{m} e_\iota$ gives the identity of R. So it is enough to show that

$$e_\iota I_1 e_\jmath \subset e_\iota I_2 e_\jmath \subset \cdots \subset e_\iota I_n e_\jmath \cdots \qquad (*)$$

stops after a finite number of steps for each pair (i,j). Clearly, for each n, $e_\iota I_n e_\jmath \subset e_\iota R e_\jmath$. Let γ be a path in $e_\iota R e_\jmath$. First assume that γ does not belong to an oriented cycle. Therefore, using (ii), γ has no subpaths which are oriented cycles. If furthermore, j is not a sink vertex, then by hypotehsis $e_\iota R e_\jmath$ is a finite dimensional K-space and so $(*)$ stops after a finite number of steps. In case j is a sink vertex, then $e_\iota R e_\jmath$ is a finitely generated right $A_\jmath$-module. Since, by hypothesis, $A_\jmath$ is right noetherian, we infer using the Hilbert basis theorem, that the chain $(*)$ should stop.

Suppose now j belongs to an oriented cycle. By (ii), there exists a unique basic cycle θ at j and the end vertices of paths starting at j should belong to θ. In this case, $e_\iota R e_\jmath$ is a finite generated right $K[\theta]$-module. Using now the well-known fact that a finitely generated right R-module over a right noetherian ring R is noetherian, we infer that the chain $(*)$ should stop. $\square$

It is not difficult to dualize the above result to get the following characterization of left noetherian $\mathcal{A}$-path algebras.

Theorem 2.3 With the above notations, the algebra R is left noetherian if and only if the following conditions hold:

(i) Δ is finite.

(ii) There exists only one arrow ending at each vertex lying in an oriented cycle of Δ.

(iii) The algebra A_i is one dimensional for each vertex i lying in an oriented cycle.

(iv) If j is not a source vertex, then A_j is finite dimensional, and if j is a source vertex, then the algebra A_j is left noetherian.

As a consequence, we get the following result.

Corollary 2.4 Let $R = K(\Delta, \mathcal{A})$ be an $\mathcal{A}$-path algebra. Then R is noetherian if and only if Δ is finite and for each connected component Γ of Δ one of the following facts happens:

(a) $\Gamma = \{i\}$ is an isolated vertex and A_i is noetherian;

(b) Γ has no oriented cycles and A_i is finite dimensional for each vertex i;

(c) Γ is a basic cycle and A_i is one-dimensional for each vertex i in Γ (so that the corresponding ring direct summand of R is of type $K\tilde{A}_n$ with $\tilde{A}_n$ linearly ordered).

3 The isomorphism problem

In this section, we shall study the so-called *isomorphism problem* for the generalized path algebras, as defined in the first section. In other words, let Δ and Γ be two quivers, and $\mathcal{A} = \{A_i : i \in \Delta_0\}$ and $\mathcal{B} = \{B_j : j \in \Gamma_0\}$ be two families of K-algebras with identities indexed by the vertices of Δ and Γ, respectively. The isomorphism problem can be posed as follows: if $\Phi: K(\Delta, \mathcal{A}) \longrightarrow K(\Gamma, \mathcal{B})$ is an algebra isomorphism, at which conditions do there exist a quiver isomorphism $\psi: \Delta \longrightarrow \Gamma$ and an isomorphism $A_a \xrightarrow{\cong} B_{\psi(a)}$, for each $a \in \Delta_0$?

For the classical case of path algebras, that is, when $A_i \cong K$ for each $i \in \Delta_0$, and $B_j \cong K$ for each $j \in \Gamma_0$, it is well-known that an isomorphism of algebras $K\Delta \xrightarrow{\cong} K\Gamma$ induces an isomorphism of quivers $\Delta \xrightarrow{\cong} \Gamma$. On the other hand, any K-algebra can be realized as a generalized path algebra of a quiver consisting of a unique vertex. We shall look in the sequel at some situations at which the isomorphism theorem holds. We keep the notation from the previous paragraph.

From now on, let us assume that Δ and Γ are finite quivers with no oriented cycles. Assume, furthermore, that, for each $a \in \Delta_0$, the identity e_a of A_a is the unique nonzero idempotent of A_a, and, for each $b \in \Gamma_0$, the identity f_b of B_b is the unique nonzero idempotent of B_b. Denote $R = K(\Delta, \mathcal{A})$ and $S = K(\Gamma, \mathcal{B})$ and assume that $\Phi : R \longrightarrow S$ is an algebra isomorphism. For convenience, we shall consider both R and S as $\mathbf{Z}^+$-graded algebras as follows: $R = R_0 + R_1 + \cdots$ and $S = S_0 + S_1 + \cdots$, where $R_0 = \bigoplus_{a \in \Delta_0} A_a$, $R_i = (\mathcal{A}$-paths of length i in Δ), $S_0 = \bigoplus_{b \in \Gamma_0} B_b$, $S_i = (\mathcal{B}$-paths of length i in Γ). Assume, furthermore, that $|\Delta_0| = n$, $|\Gamma_0| = m$. By our hypothesis, we infer that $\{e_1, \cdots, e_n\}$ and $\{f_1, \cdots, f_m\}$ are complete systems of primitive and pairwise orthogonal idempotents for R and S, respectively. In particular, $\sum_{i \in \Delta_0} e_i = 1_R$ and $\sum_{i \in \Gamma_0} f_i = 1_S$ are the identities of R and S, respectively.

Observe that due to the imposed conditions on the algebras of $\mathcal{A}$, if Δ has no arrows, then R_0 has exactly $2^n - 1$ distinct idempotents. It should be also clear that if Δ has an arrow $\alpha : a \longrightarrow b$ which is not a loop, then $e_a + x_a \alpha x_b$ is also an idempotent for each $x_a \in A_a$ and $x_b \in A_b$. Actually, if e is an idempotent in R, then $e = r_0 + r_1 + \cdots$, where r_0 is an idempotent in R_0, that is, r_0 is a sum of some e_a's. Since Φ is an isomorphism, we have $1_S = \Phi(1_R) = \Phi(e_1) + \cdots + \Phi(e_n)$, and so $\Phi(e_i) = f_{\sigma(i)} + s_{\sigma(i)}$, where $s_{\sigma(i)} \in S_1 + S_2 + \cdots$. Since $\{f_j : j \in \Gamma_0\}$ is a complete set of primitive and pairwise orthogonal idempotents, and rearranging the terms if necessary, we infer that $\Phi(e_i) = f_i + s_{(i)}$, with $s_{(i)} \in S_1 + S_2 + \cdots$, and so $m = n$. In particular, Φ induces an one-to-one correspondence between Δ_0 and Γ_0.

Lemma 3.1 Let e be an idempotent of R and $f = e + r_{10} + r_{01} + r_{00}$, where $r_{10} + r_{01} + r_{00} \in R_1 + R_2 + \cdots$, $e \cdot r_{00} = 0 = r_{00} \cdot e$, $e \cdot r_{10} = r_{10}$ and $r_{01} \cdot e = r_{01}$. Then

(a) f is an idempotent if and only if $r_{00} = r_{01} \cdot r_{10}$ and $r_{00} \cdot r_{00} = r_{10} \cdot r_{00} = r_{00} \cdot r_{01} = 0$.

(b) Suppose f is an idempotent, then f is primitive if and only if e is primitive.

Proof: (a) Observe that since Δ has no oriented cycles, then $r_{10} \cdot r_{01} = 0$. Therefore,

$$(f)^2 = (e + r_{10} + r_{01} + r_{00})^2 = e + r_{10} + r_{10} \cdot r_{00} + r_{01} + r_{01} \cdot r_{10} + r_{00} \cdot r_{01} + r_{00} \cdot r_{00}$$

If f is an idempotent, then $r_{10} = r_{10} + r_{10} \cdot r_{00}$, $r_{01} = r_{01} + r_{00} \cdot r_{01}$, and $r_{00} = r_{00} \cdot r_{00} + r_{01}\dot{r}_{10}$ (∗). Clearly now, $r_{10} \cdot r_{00} = 0 = r_{00} \cdot r_{01}$ and, by (∗), we get $r_{00} \cdot r_{00} = (r_{00} \cdot r_{00} + r_{01} + r_{10}) \cdot r_{00} = r_{00} \cdot r_{00} \cdot r_{00}$. Iterating this procedure, we get for each n, that $(r_{00})^2 = (r_{00})^3 = \cdots = (r_{00})^n$. Therefore, by length, we infer that $(r_{00})^2 = 0$ and so $r_{00} = r_{01} \cdot r_{10}$. The converse is clear.

(b) Suppose $e = e_2 + e_3$, where e_2 and e_3 are orthogonal idempotents. Writing $r_{10} = r_{20} + r_{30}$ with $e_2 \cdot r_{20} = r_{20}$, $e_3 \cdot r_{30} = r_{30}$ and $r_{01} = r_{02} + r_{03}$ with $r_{02} \cdot e_2 = r_{02}$, $r_{03} \cdot e_3 = r_{03}$, we will have $f = (e_2 + e_3) + (r_{20} + r_{30}) + (r_{02} + r_{03}) + (r_{02} + r_{03}) \cdot (r_{20} + r_{30}) = (e_2 + r_{20} + r_{02} + r_{02} \cdot r_{20}) + (e_3 + r_{30} + r_{03} + r_{03} \cdot r_{30})$, and so e is primitive if and only if f is primitive. $\square$

Corollary 3.2 The ideal generated by $\Phi(e_i)S\Phi(e_i)$ equals the one generated by f_iSf_i. In particular, $(B_i) = (\Phi(e_i)S\Phi(e_i))$.

We shall now state and prove our results concerning the isomorphism problem. Keeping the above notation, we have just seen that $|\Delta_0| = |\Gamma_0|$, and

$\Phi(A_i) = B_i$, for each i.

Theorem 3.3 Suppose neither Δ nor Γ have multiple arrows. Then the isomorphism $\Phi\colon K(\Delta, \mathcal{A}) \longrightarrow K(\Gamma, \mathcal{B})$ induces an isomorphism $\Psi\colon \Delta \longrightarrow \Gamma$ of quivers such that for each $a \in \Delta_0$, $A_a \cong B_{\Psi(a)}$.

Proof: We shall prove it by induction on $|\Delta_0|$. If $|\Delta_0| = 1$, then $\Delta_1 = \emptyset = \Gamma_1$, because neither Δ nor Γ have oriented cycles. Also, by the above $\Phi(A_a) = B_a$, where a is the unique vertex of Δ (identifying Δ_0 with Γ_0). Suppose now that $|\Delta_0| = n > 1$. Let I_i be the ideal of R generated by A_i, and J_i the ideal of S generated by B_i. Since $\Phi(A_i) = B_i$, we have that $\Phi(I_i) = J_i$ and so Φ induces an isomorphism $\Phi'\colon R/I_i \longrightarrow S/J_i$. Writing $\hat{R}_i = R/I_i$ and $\hat{S}_i = S/J_i$, we have clearly that $\hat{R}_i = K(\Delta \setminus \{i\}, \mathcal{A} \setminus \{A_i\})$ and $\hat{S}_i = K(\Gamma \setminus \{i\}, \mathcal{B} \setminus \{B_i\})$. Using the induction hypothesis, we have that $\Delta \setminus \{i\}$ is isomorphic to $\Gamma \setminus \{i\}$, and $A_j \cong B_j$, for each $j \neq i$. Since this can be done for each i, and using the fact that neither Δ nor Γ have multiple arrows, we infer that the quivers Δ and Γ are isomorphic and, $A_j \cong B_j$, for each j as required. $\square$

For the next result we shall have the same hypothesis on the families $\mathcal{A}$ and $\mathcal{B}$ but Δ and Γ are arbitrary finite quivers.

Theorem 3.4 Assume furthermore that the algebras A_i, $i \in \Delta_0$, B_j, $j \in \Gamma_0$, are finite dimensional. Then the isomorphism $\Phi\colon K(\Delta, \mathcal{A}) \longrightarrow K(\Gamma, \mathcal{B})$ induces an isomorphism $\Psi\colon \Delta \longrightarrow \Gamma$ of quivers such that for each $a \in \Delta_0$, $A_a \cong B_{\Psi(a)}$.

Proof: By counting dimensions over K, one can prove the result for $|\Delta_0| = |\Gamma_0| = 2$. By induction, using the same argument as before, we get the general case. $\square$

References

[1] M. Auslander, I. Reiten, S. O. Smalø, *Representation theory of artin algebras*, Cambridge University Press (1995).

[2] S. X. Liu, Y. L. Luo, J. Xiao, *Isomorphism of path algebras*, Bull. Beijing Normal Univ. **3** (1986) 13-20.

[3] S. X. Liu, *Geometric properties of directed graphs and algebraic properties of their path algebras*, Acta Math. Sinica **31** (1998), 483-487.

[4] S. X. Liu, *Isomorphism problem for tensor algebras over vulued graphs*, Science in China (series A) **34:3** (1991), 268-272.

[5] S. X. Liu, *Tensor product of path algebras and direct product of directed graphs*, Chinese J. of Contemporary Math. **13:2** (1992) 111-120.

[6] C. M. Ringel, *Tame algebras and integral quadratic forms*, Lecture Notes in Math. **1099** (1984).

Representations and Corepresentations of Hopf Algebras

Miriam Cohen[*†‡]
Department of Mathematics and Computer Sciences
Ben-Gurion University of the Negev
Beer-Sheva 84105, Israel
mia@cs.bgu.ac.il

Abstract

We explain some basic ideas connected with Hopf algebras, their representations and corepresentations, with special emphasis on H-commutative algebras. We focus on some selected topics in our work that exhibit the role played by ring, module and category theories in the study of Hopf algebras.

Introduction

One of the "hot" topics in ring theory during the 70s was, so called, Non-commutative Galois theory. Specifically, if A is an algebra over a field k, G a finite group of automorphisms of A and $A^G = \{a \in A \mid g(a) = a,$ for all $g \in G\}$, then the study concerned connections between the ideal structures of A and A^G. Much of the information is encoded in the skew group ring $A * G$ [M1] and in the connection:

$$A^G \subset A \subset A * G.$$

Another situation with a similar flavor is:

---

[*]The author was supported by the Israel Science forndation funded by the Isreal Academy of Sciences and Humanities.

[†]The author was on Sabattical at the Mathematics Institute of Fudan University, China while this paper was written. She wishes to thank Prof. Xu Yonghua, Prof. Wu Quanshui and Prof. Zhu Shenglin for fruitful discussions and an outstanding hospitality.

[‡]She also wishes to express appreciation to the organizers of the Euroconference on Interactions between ring theory and representation theory of algebras held at the University of Murcia, Spain for a very successful conference and for the opportunity to present the ideas in this paper.

A is an algebra over a field k, L is a Lie algebra of derivations of A and $A^L = \{a \in A \mid l(a) = 0,$ for all $l \in L\}$. Here again $A^L \subset A$ but the analogue of $A * G$ is not obvious.

A third set-up of the same nature is that of group-graded algebras [CR, NO]. Specifically, let A be an algebra over a field k, G a finite group and $A = \sum_{g \in G} \oplus A_g$ where each A_g is a linear subspace of A and $A_g A_h \subset A_{gh}$. The analogue of A^G and A^L is A_1 (where 1 is the identity element of G). Here it is even less obvious what should be the analogue of $A * G$.

Some of us working on these topics were puzzled by the question:

Why do these three setups "feel" the same?

It turned out that **the secret is: Hopf algebras**.

This is when my interest in Hopf algebra began [CM]. Although quite cumbersome for a beginner, as part of the structure lacks algebraic intuition studying them is very rewarding. Hopf algebras unify not only these setups but phenomena ranging from a variety of areas in mathematics to physics, computer sciences, chemistry and even genetics.

In what follows we explain how Hopf algebras unify these setups. We survey basic definitions, examples and selected topics in the theory related to our work. Specifically, we consider H-commutativity in Yetter-Drinfeld categories with special emphasis on triangular Hopf algebras (Section 5, 6). We consider H-quotient algebras (Section 7) and apply them to H-commutative algebras (Section 8). Finally in Section 9 we give an easy proof using basic concepts from ring theory of the fact that the character ring of a semisimple and cosemisimple Hopf algebra over an algebraically closed field is semisimple. This applies to semisimple Hopf algebra over k which also have characteristic 0. The fact about the character ring is an essential step in the proof [Z] of Kaplansky's Conjecture about Hopf algebra of prime dimension.

1   The Definition and examples of Hopf algebras

Throughout , let k denote a field, and $\otimes = \otimes_k$. Basic references are [M3, S].

Definition *A Hopf algebra H over a field k is an algebra with unit 1_H endowed with a coalgebra structure: comultiplication $\Delta : H \to H \otimes H$, counit $\varepsilon : H \to k$ and antipode $S : H \to H$.*

The maps satisfy the following:

(1) Δ satisfies coassociativity: $(\Delta \otimes \mathrm{id})\Delta(h) = (\mathrm{id} \otimes \Delta)\Delta(h)$. That is, after "opening up" h for the first time one can either "open up" the left or the right tensorands. Coassociativity means that the result is the same. Symbolically we write $(\Delta \otimes \mathrm{id})\Delta(h) = \sum h_1 \otimes h_2 \otimes h_3$.

(2) ε satisfies: $h = \sum \varepsilon(h_1)h_2 = \sum h_1\varepsilon(h_2)$, for all $h \in H$.

   (if (C, Δ, ε) satisfy (1) and (2) then C is called a coalgebra). Next have the properties that relate to the algebra structure and the antipode.

(3) Δ and ε arc algebra maps.

(4) $\sum S(h_1)h_2 = \sum h_1 S(h_2) = \varepsilon(h)1_H$, for all $h \in H$.

Comultiplication is in a sense going the "opposite" way of multiplication. When multiplying one takes a pair of elements and gets a single element, while on comultiplying one takes a single element which "opens up" to a sum of pairs. We use the "sigma notation" introduced by Sweedler and Heyneman: $\Delta(h) = \sum h_1 \otimes h_2$ (where the subscripts 1 and 2 arc symbolic and do not indicate particular elements of H).

Examples *(1) The most obvious example is $H = kG$, the group algebra of a group G. Here $\Delta(g) = g \otimes g$, $\varepsilon(g) = 1$ and $S(g) = g^{-1}$, for all $g \in G$. Let us check coassociativity. Indeed*

$$(\Delta \otimes \mathrm{id})\Delta(g) = (g \otimes g) \otimes g = g \otimes (g \otimes g) = (\mathrm{id} \otimes \Delta)\Delta(g).$$

(2) $H = U(L)$ the enveloping algebra of a Lie algebra L. Here $\Delta(l) = l \otimes 1 + 1 \otimes l$, $\varepsilon(l) = 0$ and $S(l) = -l$ for all $l \in L$.

(3) If G is a finite group then $H = (kG)^* = \mathrm{Hom}_k(kG, k)$ is a Hopf algebra. If $\{g\}_{g \in G}$, $\{P_g\}_{g \in G}$ are dual basis for kG and $(kG)^*$ respectively, then $\Delta(P_g) = \sum_{h \in G} P_{gh^{-1}} \otimes P_h$, $\varepsilon(P_g) = \delta_{1g}$ and $S(P_g) = P_{g^{-1}}$.

In examples (1) and (2) Δ is "symmetric". That is: $\Delta = \tau \circ \Delta$, where τ is the flip map, $\tau : a \otimes b \longmapsto b \otimes a$. Explicitly, $\sum h_1 \otimes h_2 = \sum h_2 \otimes h_1$ for all $h \in H$. Such Hopf algebras are called *cocommutative*. Example (3) is easily seen to be cocommutative only if G is an abelian group. It is however a commutative algebra.

Are all Hopf algebras either commutative or cocommutative? the answer is negative. Here is a counter-example of least dimension.

Example *Sweedler's 4-dimensional Hopf algebra $H_4 = k < 1, g, x, gx \mid g^2 = 1. x^2 = 0, xg = -gx >$ with $\Delta(g) = g \otimes g$, $\Delta(x) = x \otimes 1 + g \otimes x$, $\varepsilon(g) = 1$, $\varepsilon(x) = 0$, $S(g) = g$ and $S(x) = -x$. It is obviously non-commutative, and it is non-cocommutative since $\Delta(x) \neq (\tau \circ \Delta)(x)$.*

In fact, the class of Hopf algebras termed quantum groups, introduced independently by [D]. [J] and [W] are sometimes called non-commutative, non-cocommutative Hopf algebra (see Section 6).

2 Representations and actions of H

Let R be an algebra over k and $_R\mathrm{Mod}$, the category of left R-modules (where the morphisms are $R - module$ maps). Usually not much can be said about R-modules which are also k-algebras. This changes dramatically if $R = H$ is a Hopf algebra, for here the dual structure can be naturally applied, as seen in property (2) of the following.

Definition *Let H be a Hopf algebra over k and A an algebra over k. Then A is called a left H-module algebra if*

    1. $A \in {}_H\mathrm{Mod}$

    (2) $h \cdot (ab) = \sum (h_1 \cdot a)(h_2 \cdot b)$, all $h \in H$, $a, b \in A$.

We also say: H acts on A.

For any $M \in {}_H\mathrm{Mod}$, set $M^H = \{m \in M \mid h \cdot m = \varepsilon(h)m, \text{ for all } h \in H\}$. If H acts on A then A^H is a subalgebra of A. We define the "Smash product" of A and H. It is $A \otimes H$ as a vector space. We write $a\#h$ for $a \otimes h$ and multiply according to:

$$(a\#h)(b\#g) = \sum a(h_1 \cdot b)\#h_2 g, \text{ for all } a, b \in A, \ h, g \in H.$$

We identify A with $A\#1$ and have:

$$A^H \subset A \subset A\#H.$$

Returning to the setups in the introduction, it is now evident that all are H-module algebras with $H = kG$, $U(L)$ and $(kG)^*$ respectively, see [B, CM]. We can now fill in the gap. The analogues of the skew group-ring are $A\#H$ for the appropriate H.

There is a categorical way to view H-module algebras. The category ${}_H\mathrm{Mod}$ has additional features that originate from the coalgebra structure. It is closed under $\otimes$ and Hom. Specifically, while for a general ring R and $M, N \in {}_R\mathrm{Mod}$ there exists no canonical structure of a left R-module on $M \otimes N$, the structure for $R = H$, a Hopf algebra, is given by:

$$h \cdot (m \otimes n) = \sum (h_1 \cdot m) \otimes (h_2 \cdot n)$$

for all $h \in H$, $m \in M$, $n \in N$. Moreover, if $f \in \mathrm{Hom}_R(M, N)$, $h \in H$ and $m \in M$ then defining $(h \cdot f)(m) = \sum h_1 \cdot f(Sh_2 \cdot m)$ endows $\mathrm{Hom}_k(M, N)$ with a left H-module structure. Thus $({}_H\mathrm{Mod}, \otimes, k)$ is a monoidal category.

Now, the categorical way to view H-module algebras is: A is an algebra, the multiplication of which is a morphism in the category.

3   Corepresentations and coactions of H

Let M be a vector spaces over k. Then M is a left corepresentation of H, or, a left H-comodule if there exists a map $\rho_M : M \to H \otimes M$ such that the following diagrams commute:

$$
\begin{array}{ccc}
M & \xrightarrow{\rho_M} & H \otimes M \\
\rho_M \downarrow & & \downarrow \Delta \otimes \mathrm{id} \\
H \otimes M & \xrightarrow{\mathrm{id} \otimes \rho_M} & H \otimes H \otimes M
\end{array}
\qquad
\begin{array}{ccc}
M & \xrightarrow{\rho_M} & H \otimes M \\
& \searrow{\mathrm{id} \otimes 1} & \downarrow \mathrm{id} \otimes \varepsilon \\
& & M \otimes k
\end{array}
$$

There is also a sigma notation for left H-comodules. We write

$$\rho_M(m) = \sum m_{-1} \otimes m_0 \in H \otimes M.$$

If M and N are left H-comodules then a map $f : M \to N$ is called a comodule map if the following diagram commutes:

$$
\begin{array}{ccc}
M & \xrightarrow{f} & N \\
\rho_M \downarrow & & \downarrow \rho_N \\
H \otimes M & \xrightarrow{\mathrm{id} \otimes f} & H \otimes N
\end{array}
$$

The category whose objects are left H-comodule and where morphisms are H-comodule maps is denoted by $^H\mathrm{Com}$. It is also closed under $\otimes$ as follows:

If $M, N \in {}^H\mathrm{Com}$ then

$$\rho_{M \otimes N} \;:\; M \otimes N \to H \otimes M \otimes N \quad \text{is given by}$$
$$m \otimes n \longmapsto \sum m_{-1} n_{-1} \otimes m_0 \otimes n_0. \text{ for all } m \in M, n \in N.$$

In this setting, $({}^H\mathrm{Com}, \otimes, k)$ forms a monoidal category.

If $M = A$ is an algebra over k then it is called a left H-comodule algebra if

(1) $(A, \rho_A) \in {}^H\mathrm{Com}$

(2) $\rho_A(ab) = \sum a_{-1} b_{-1} \otimes a_0 b_0$ and $\rho_A(1_A) = 1_H \otimes 1_A$.

We also say, H left acts on A.

This can again be stated categorically, namely the multiplication of A is a morphism in the category. For each $M \in {}^H\mathrm{Com}$ define the coinvariants $M^{\mathrm{Co}\,H} = \{m \in M \mid \rho(m) = 1 \otimes m\}$.

Example of an H-comodule algebra *Let G be a group (not necessarily finite) and $A = \sum_{g \in G} \oplus A_g$ be a G-graded algebra. Let $H = kG$. Then A is an H-comodule algebra via: $\rho_A(a) = g \otimes a$ for each $a \in A_g$. Extend this definition linearly to A. Note that $A_1 = A^{\mathrm{Co}\,H}$.*

Thus if G is a finite group, and A is a G $-$ graded algebra then A is both a $(kG)^*$-module algebra (as seen in Section 2) and a kG-comodule algebra.

4 Algebras which are both H-module algebras and H-comodule algebras

Example 1 *H itself is an H-module algebra via the adjoint action:*

$$h \cdot x = \sum h_1 x S h_2. \quad \text{all } x, h \in H.$$

It is an H-comodule algebra via $\rho_H = \Delta$. We write this structure as (H, ad, Δ).

Example 2 *Commutative superalgebras*

Assume Char $k \neq 2$. Let $Z_2 = \{1, g\}$ be the cyclic group of order 2. Let $A = A_1 \oplus A_g$ be a Z_2-graded algebra, such an algebra is called a superalgebra. It is a commutative superalgebra if $ab = ba$ for all $a \in A_1$ and $b \in A_1$ or $b \in A_g$. If both $a, b \in A_g$ then $ab = -ba$ (the elements of A_1 and A_g are called homogeneous elements). A is a kZ_2-module algebra as follows. For $a \in A_1$, and $b \in A_g$ define $g \cdot a = a$ and $g \cdot b = -b$ and extend linearly. As we have seen in Section 3 A is a kZ_2-comodule algebra by $\rho(a) = 1 \otimes a$ and $\rho(b) = g \otimes b$. For later references (Section 6) note that

$$\rho(x) = \frac{1}{2}[1 \otimes (1 \cdot x) + g \otimes (1 \cdot x) + 1 \otimes (g \cdot x) - g \otimes (g \cdot x)]$$

for $x = a$ or b.

This way of viewing commutative superalgebras appears in [Maj].

A more complicated example is

Example 3 *The quantum plane* $A = \mathbb{C}_q < x, y \mid xy = q^{-1}yx$, *where q is an nth root of $1 >$. Let $H = k(Z_n \times Z_n)$, where Z_n is the cyclic group of order n. For $(s, t) \in Z_n \times Z_n$ define*

$$(s, t) \cdot x = q^{-t}x \text{ and } (s, t) \cdot y = q^t y$$

extend this to $\mathbb{C}_q < x, y >$. This gives an action of H on A.

The coaction is given by:

$$\rho(a) = \frac{1}{n^2} \sum_{i,j,s,t=0}^{n-1} q^{sj-ti}(i,j) \otimes (s,t) \cdot a$$

for all $a \in A$.

This way of viewing the quantum plane appears in [CW2, p. 2991].

5 *H*-commutativity

Both the commutative superalgebra and the quantum plane have the same flavor. The generators commute up to multiplication by some scalar, while the first example does not have any such visible property. There is however a categorical way to unify them. They are all H-commutative (or sometimes called quantum-commutative) algebras. This notion introduced in [CW1] was hinted at in [Ma], and appeared in some form in [H] for $H = kG$ or $(kG)^*$.

Specifically, let us first express usual commutativity in a fancy way. Denote by m the multiplication $m : A \otimes A \to A$. Then A is a *commutative* algebra if

$$m \circ \tau = m,$$

where τ is the flip map mentioned in Section 1.

An H-module H-comodule algebra gives rise to another natural flip map which we denote by

$$\Psi : A \otimes A \to A \otimes A \text{ via } a \otimes b \longmapsto \sum (a_{-1} \cdot b) \otimes a_0.$$

Now by analogy, A is called H-*commutative* if

$$m \circ \Psi = m.$$

All the examples considered in Section 4 are H-commutative. Let us prove it for Example 1. We must show that for all $x, y \in H$

$$\sum (x_{-1} \cdot y)x_0 = xy.$$

In this case the notation $\sum x_{-1} \otimes x_0 = \Delta(x) = \sum x_1 \otimes x_2$. And indeed,

$$\begin{aligned}
\sum (x_{-1} \cdot y)x_0 &= \sum (x_1 \cdot y)x_2 \\
&= \sum x_1 y S(x_2)x_3 \\
&= \sum x_1 y \varepsilon(x_2) \qquad \text{(by property (4) of the antipode)} \\
&= \left(\sum x_1 \varepsilon(x_2)\right) y \\
&= xy. \qquad \text{(by property (2) of } \varepsilon)
\end{aligned}$$

It is straight forward to check H-commutativity of the other two examples.

The natural setting of these examples is the so called Yetter-Drinfeld category. ${}^H_H \mathcal{YD}$ see [Y. RT]. An object in this category is a left H-module and left H-comodule M which satisfies the compatibility condition:

$$\rho_M(h \cdot m) = \sum h_1 m_{-1} S h_3 \otimes h_2 \cdot m_0, \text{ for all } h \in H, m \in M.$$

The morphisms in this category are maps which are both H-module and H-comodule maps. This category is closed under $\otimes$ and for each $M, N \in {}^H_H \mathcal{YD}$ there is a flip map which is a morphism in the category given by

$$\begin{aligned}
\Psi_{M,N} : \quad & M \otimes N \to N \otimes M \qquad \text{via} \\
& m \otimes n \longmapsto \sum (m_{-1} \cdot n) \otimes m_0.
\end{aligned}$$

Note that the previously mentioned flip Ψ is $\Psi_{A,A}$. Examples 1-3 in Section 4 are algebras in this category. In fact, example 1 can be easily checked and examples 2 and 3 belong to the category as explained in the next section.

6   (Quasi)triangular Hopf algebras (quantum groups) and ${}^H_H \mathcal{YD}$

Definition [D] *A quasitriangular Hopf algebra is a pair (H, R) where H is a Hopf algebra over k and $R = \sum R^1 \otimes R^2 \in H \otimes H$ is invertible, such that the following holds (with $r = R$)*

(1) $\sum \Delta(R^1) \otimes R^2 = \sum R^1 \otimes r^1 \otimes R^2 r^2$

(2) $\sum R^1 \otimes \Delta(R^2) = \sum R^1 r^1 \otimes r^2 \otimes R^2$

(3) $(\tau \circ \Delta)(h) = R\Delta(h)R^{-1}$, *all* $h \in H$.

(4) *If* $R^{-1} = \tau \circ R$ *then* (H, R) *is called a triangular Hopf algebra.*

If (H, R) is quasitriangular then any $M \in {}_H\mathrm{mod}$ belongs to ${}^H\mathrm{Com}$ via:

$$\rho_M^R(m) = \sum SR^1 \otimes R^2 \cdot m, \text{ all } m \in M.$$

In fact with this coaction $M \in {}_H^H\mathcal{YD}$. The flip map is explicitly given by $\Psi_{M,M}(m \otimes n) = \sum (SR^1 \cdot n) \otimes (R^2 \cdot m)$. Or if (H, R) is triangular then

$$\Psi_{M,N}(m \otimes n) = \sum R^2 \cdot n \otimes R^1 \cdot m,$$

all $m, n \in M$ (for then $R^{-1} = \sum SR^1 \otimes R^2 = \sum R^2 \otimes R^1$). The Hopf algebra $k\mathbb{Z}_2$ and $k(\mathbb{Z}_n \times \mathbb{Z}_n)$ are quasitriangular Hopf algebra with

$$R = \frac{1}{2}(1 \otimes 1 + 1 \otimes g + g \otimes 1 - g \otimes g) \text{ and } R = \frac{1}{n^2} \sum_{i,j,s,t=0}^{n-1} q^{sj-it}(s,t) \otimes (i,j)$$

respectively. It is now evident that the comodule structure of the commutative superalgebra and the quantum plane are just the ones induced from R, as above. and so both belong to ${}_H^H\mathcal{YD}$.

Thus the collection of all $(M, \cdot, \rho_M^R)$ is a subcategory of ${}_H^H\mathcal{YD}$. An important element in the structure theory of triangular Hopf algebras is $u = \sum (SR^2)R^1$. If (H, R) is also a semisimple algebra over an algebraically closed field of characteristic 0 then $u^2 = 1$ and u is a central element. Thus u acts on every irreducible representation of H as 1 or -1. This is the categorical dimension of the representation [EG1]. When (H, R) is triangular this subcategory is "symmetric" [Mac, p. 180]. That is $\Psi_{M,N}\Psi_{N,M} = \mathrm{id}$, and it is a so called rigid braided tensor category.

Since $\Psi_{M,M}$ is a symmetry there exists a well defined action of the symmetric group S_n on the n-th fold tensor product $M^{\otimes n}$ [Ma],given by

$$(i, i+1) \cdot_\Psi (M_1 \otimes \cdots \otimes M_i \otimes M_{i+1} \otimes \cdots \otimes M_n) = M_1 \otimes \cdots \otimes \Psi_{M_i,M_{i+1}}(M_i \otimes M_{i+1}) \otimes \cdots \otimes M_n$$

where $M_i = M$, all i. Then extend the action to $\sigma \in S_n$ by representing σ as a product of elementary transpositions.

This is different from the usual way to view $M^{\otimes n}$ as a representation of S_n. For the usual way uses the flip τ rather than Ψ.

How are these two representations of S_n related?

Theorem [CWZ, Theorem 2.11] *Let* (H, R) *be a triangular Hopf algebra over* k. *Let* M *be an m-dimensional left H-module so that* u *acts on* M *as the identity, then*

(1) $\chi_\Psi(\sigma) = \chi(\sigma)$, for all $\sigma \in S_n$, all n. (where $\chi(\sigma) = \text{Trace}\,(\sigma)$ via the representation induced by τ and $\chi_\Psi(\sigma) = \text{Trace}\,(\sigma)$ via the representation induced by Ψ).

(2) If, moreover Char $k = 0$ or Char $k > m$ then $M^{\times n}$ via τ and $M^{\otimes n}$ via Ψ are isomorphic representations of kS_n, for all n.

As a consequence of this theorem one can form Ψ-exterior products which have the same well known dimensions as τ-exterior products, and get certain determinant functions:

Theorem [CWZ. Theorem 2.19] *Let $(H.R)$, M and k be as in the previous theorem. Let A be an H-commutative H-module algebra. Then there exists a function*

$$\det : \text{End}_{A\#H}(A \otimes M) \to A^H$$

(here $A^H \subset Z(A)$) so that $\det(\text{id}) = 1$ and $\det(S \circ T) = (\det S)(\det T)$ for all $S.T \in \text{End}_{A\#H}(A \otimes M)$.

Using this there is a partial Cayley-Hamilton theorem which is used to prove

Theorem [CWZ, Theorem 4.7] *If (H, R) is a triangular semisimple Hopf algebra over a field k of characteristic 0 and A an H-commutative H-module algebra, then*

1. *A is integral over A^H.*

2. *A is a PI ring.*

3. *If A is also k-affine then*

    (a) *A^H is k-affine (this is an analogue of Noether's theorem for group actions)*

    (b) *A is a finitely generated left (and right) A^H-module.*

    (c) *A is a left and right Noetherian PI ring.*

The proof of these theorems involves methods from group representations and ring theory. They are adaptation of [FS] to this more complex situation.

7 H-Quotient rings

A fundamental construction in ring theory is that of a right Martindale quotient ring of a ring R with respect to the filter $\mathcal{F}$ of ideals with zero left and right annihilators. Denote this ring by $Q^r_{\mathcal{F}}(R)$, see [M1, Chapter 3]. There also exists a symmetric ring of quotients $Q(R)$ see [P, Chapter 3]. These constructions can be adapted to our setup of Hmodule algebras.

Theorem [C] *let H be a Hopf algebra with a bijective antipode. Let A be an H-module algebra and let $\mathcal{F}_H$ be the filter of H-stable ideals of A with 0 left and right annihilators. Then*

(1) *One can repeat the constructions of $Q^r_{\mathcal{F}}(R)$ replacing $\mathcal{F}$ by $\mathcal{F}_H$ and obtain the algebra $Q^r_H(R)$ that contains A.*

(2) *$Q^r_H(R)$ is an H-module algebra, with an H-action extending the action on A. Specifically.*

If $f : I_\Lambda \to A$, for $I \in \mathcal{F}_H$, determine an element of $Q^r_H(R)$. Then define

$$(h \cdot f)(a) = \sum h_1 \cdot f(Sh_2 \cdot a), \text{ for all } a \in I, h \in H.$$

When does an H-action extend to the symmetric Martindale ring of quotients $Q(A)$?

Theorem [M2] *Let H be a pointed Hopf algebra (i.e all its simple subcoalgebra are 1-dimensional) and let A be an H-module algebra. Then the H action on A extends to an action on $Q(A)$.*

8 H-Quotient algebras of H-commutative algebras

The right Martindale ring of quotients was used by [K] in the study of non-commutative Galois theory. He introduced so called X-inner automorphism. An automorphism σ of a ring R is X-inner if there exists $0 \neq q \in Q^r_{\mathcal{F}}(R)$ so that

$$aq = q\sigma(a) \qquad \text{for all } a \in R.$$

Let $o_\sigma = \{q \in Q \mid aq = q\sigma(a), \text{ for all } a \in R\}$.

Theorem [K] *If R is a semiprime ring, and $\sigma \in \mathrm{Aut}\,(R)$ then $Q(R)_\sigma = Cx_\sigma$, for some x_σ, where C is the center of $Q^r_{\mathcal{F}}(R)$.*

A possible source for such automorphisms is semiinvariants of H-commutative H-module algebras. (see Section 6).

For simplicity of exposition let us assume that H is finite dimensional.

Definition [BCM] *Let A be an H-module algebra and $\sigma \in H^*$ so that $\Delta(\sigma) = \sigma \otimes \sigma$ (σ is so called group-like element). Define the set of σ-semiinvariants by:*

$$A_\sigma = \{a \in A \mid h \cdot a = \, <\sigma, h> a. \text{ for all } h \in H\}.$$

These are "weight spaces" with respect to H-actions. If σ is a group-like element of H^* then we can define an automorphism $\hat{\sigma} : A \to A$ by: $\hat{\sigma}(a) = \sum <\sigma, a_{-1}> a_0$. Denote for short $Q = Q^r_H(A)$. We assume in what follows that A is an H-semiprime algebra. This means that A has no non-trivial nilpotent H-stable ideals.

Lemma [CKW] *Let $A \in {}^H_H \mathcal{YD}$ be an H-commutative H-semiprime H-module algebra. If $q \in Q_\sigma$ then $aq = q\hat{\sigma}(a)$ for all $a \in A$.*

Though similar, this does not necessarily mean that $\hat{\sigma}$ is X-inner for Q may not equal the full Martindale ring of quotients. If equality holds (e.g if A is a simple algebra), then indeed these $\hat{\sigma}$ are X-inner, though they may be trivial.

We prove results similar in flavour to the ones known for X-inner automorphisms.

Theorem [CKW] *Let H, A, Q and σ be as in the previous lemma. Then*

(1) $Q_\sigma = Q^H x_\sigma$, for some x_σ. Here $Q^H \subset$ center of Q.

(2) Q is a non-singular, injective Q^H-module and Q^H is self injective.

(3) When H^* is pointed then part (1) of this theorem is a first step in a filtration of A. Using this we show that:

Q is generated over Q^H by at most $\dim H$ elements.

9 Semisimple Hopf algebras

The best understood Hopf algebras are semisimple, though their classification is far from completed. They are necessarily finite-dimensional [So, 3.1]. Some of the best classification results so far are [Z] and [EG2] who proved respectively that if Char $k = 0$ and k is algebraically closed then if $\dim H = p$ or $\dim H = pq$, p and q distinct primes then H is trivial (that is, $H = kG$ or $(kG)^*$).

Many of the techniques used in the proofs come from group representations. The first step in the proof in [Z] depends heavily on the fact that a certain bilinear form is positive definite. this cannot be used in positive characteristic.

We end this paper by giving a short proof of this step that depends only on facts from ring theory and is applicable to any characteristic. First, some background, see [Z, So].

Let H be a semisimple Hopf algebra over an algebraically closed field. Choose a system of non-isomorphic left H-modules $V_1, \ldots, V_n$, let $l_i : H \to \text{End}(V_i)$ denote the corresponding representation, and let the character χ_i defined by:

$$\chi_i(h) = \text{Trace}(l_i(h)), \text{ all } h \in H.$$

Then the character ring of H, $Ch(H)$ is the linear span of $\{\chi_i\}_{i=1}^n$. It is a subalgebra of H^*. Moreover, $S^2 = \text{id}$ on $Ch(H)$ and $S(\chi_i) = \chi_j$. We denote this j by i^*.

Let $0 \neq t \in H$ so that $ht = th = \varepsilon(h)t$, for all $h \in H$, and $\varepsilon(t) = 1$. Such an element (called an integral for H) exists for semisimple Hopf algebras [S]. Moreover, just as for group representations, $\{\chi_i\}_{i=1}^n$ satisfy the orthogonality relations:

$$< \chi_i \chi_j, t > = \delta_{i^* j},$$

a version of [L, Theorem 3.a].

Finally, by [R, Theorem 3.a], if H^* is semisimple then

$$< pq, t > \ = \ < qS^2(p), t > \qquad \text{for all } p, q \in H^*.$$

Since $S^2(p) = p$ for all $p \in Ch(H)$, we obtain:

$$< pq, t > \ = \ < qp, t >, \qquad \text{for all } p \in Ch(H), \ q \in H^*. \tag{1}$$

We are ready to prove:

Theorem [CZ] *If H and H^* are semisimple Hopf algebras over an algebraically closed field k, then $Ch(H)$ is a semisimple algebra.*

Proof. Assume $p \in Ch(H)$ and $pCh(H)p = 0$. To prove semisimplicity we must show that $p = 0$. We have,

$$0 = \ < pCh(H)pH^*, t > \ = \ < Ch(H)pH^*p, t > \quad \text{(by (1))}. \tag{2}$$

Since H^* is semisimple, it is a von Neumann regular ring and so $p \in pH^*p$. In particular (2) implies that

$$< Ch(H)p, t > \ = 0.$$

The orthogonality relations imply that $p = 0$. □

Note that if k is of characteristic 0 and algebraically closed and H is semisimple then H^* is semisimple [LR]. So the above theorem applies. We also remark that the result in [CZ] is more general, but is beyond the scope of this paper.

References

[B] G. Bergman, *Everybody knows what a Hopf algebra is*, **Contemporary Math.** **43**(1985), 25--48.

[BCM] R.J. Blattner, M. Cohen and S. Montgomery, *Crossed products and inner actions of Hopf algebra*, **Trans. AMS 298**(1986), 671--711.

[C] M. Cohen, *Smash products, inner actions and quotient rings*, **Pacific J. Math.** **125**(1986), 46--65.

[CM] M. Cohen and S. Montgomery, *Group-graded rings, Smash products and group actions*, **Trans. AMS 282**(1984). 237--258.

[CR] M. Cohen and L. Rowen, *Group graded rings*, **Comm. in Algebra 11**(1983), 1253--1270.

[CW1] M. Cohen and S. Westreich, *From supersymmetry to quantum commutativity*, **J. Algebra 168**(1994). 1--27.

[CW2] M. Cohen and S. Westreich, *Central invariants of H-module algebras*, **Comm. in Algebra 21**(8)(1993), 2859 -2883.

[CWZ] M. Cohen, S. Westreich and S. Zhu. *Determinants, integrality and Noether's theorem for quantum commutative algebras*, **Israel J. Math. 96**(1996), 185—222.

[CKW] M. Cohen, A. Koryukin and S. Westreich, *On generalized invariants of injective non-singular module algebras over their invariants*, **J. Algebra** (accepted).

[CZ] M. Cohen and S. Zhu. *Invariants of the adjoint coaction and Yetter-Drinfeld categories*, submitted.

[D] V.G. Drinfeld, *Quantum groups*, **Proc. Int. Cong. Math. Berkeley, 1**(1986), 789- -820.

[EG1] P. Etingof and S. Gelaki, *Some properties of finite-dimensional semisimple Hopf algebras*, **Math. Research Letters 5**(1998), 191—197.

[EG2] P. Etingof and S. Gelaki, *Semisimple Hopf algebras of dimension pq are trivial*, **J. Algebra 210**(1998), 664—669.

[FS] W. Ferrer-Santos, *Finite generation of the invariants of finite-dimensional Hopf algebras*, **J. Algebra 165**(1994). 543 549.

[H] S. Haran, *An invitation to dyslectic geometry*, **J. Algebra 155**(1993), 455 -481.

[J] M. Jimbo, *A q-difference analogue of $U(g)$ and the Yang-Baxter equation*, **Lett. Mat. Phy. 10**(1985), 63 -69.

[K] V.K. Kharchenko, *Fixed elements under a finite group acting on a semiprime ring*, **Algebra and Logic 14**(3)(1975), 328 -344.

[L] R.G. Larson, *Characters of Hopf algebras*, **J. Algebra 17**(1971), 352—368.

[LR] R.G. Larson and D.E. Radford, *Finite dimensional cosemisimple Hopf algebras in characteristic 0 are semisimple*, **J. Algebra 117**(1988), 267—289.

[Mac] S. MacLane, "Categories for the working Mathematicians", Graduate texts in Math. 5, Springer-Verlag, Berlin, 1971.

[Maj] S. Majid, *Examples of braided groups and braided matrices*. **J. Math. Phys. 32**(1991), 3246—3253.

[Ma] Y. Manin, "Quantum groups and non-commutative geometry", University of Montreal Lectures, 1988.

[M1] S. Montgomery, "Fixed rings of finite automorphism groups of associative rings", Lecture Notes in Math. 818. Springer. Berlin, 1980.

[M2]   S. Montgomery, *Biinvertible actions of Hopf algebras*, **Israel J. Math. 83**(1993),
        45—72.

[M3]   S. Montgomery, "Hopf algebras and their actions on rings", CBMS Regional
        Conference Series in Math. No. 82, Amer. Math. Soc. Providence RI, 1993.

[NO]   C. Nastasescu and F. von Oystaeyen, "Graded ring theory", North-Holland, Am-
        sterdam, 1982.

[P]   D.S. Passman, "Infinite crossed products", Academic Press, N. Y., 1989.

[R]   D.E. Radford, *The trace function and Hopf algebras*, **J. Algebra 163**(1994),
        583--622:

[RT]   D.E. Radford and J. Towber, *Yetter-Drinfeld categories associated to an arbitrary
        bialgebra*, **J. Pure Appl. Algebra** (1993), 259 −279.

[So]   Y. Sommerhauser. *On Kaplansky's fifth conjecture.* **J. Algebra 204**(1998), 202-
        224.

[S]   M.E. Sweedler. "Hopf algebras", Benjamin, N. Y. 1969.

[W]   S.L. Woronowicz, *Twisted SU(2) group. An example of non-commutative differ-
        ential calculus*, Publ. RIMS, Kyoto Univ. **23(1987)**, 117--181.

[Y]   D.N. Yetter, *Quantum groups and representations of monoidal categories*, **Math.
        Proc. Cambridge Phil. Soc. 108**(1990), 261- 290.

[Z]   Y. Zhu, *Hopf algebra of prime dimension*, **International Math. Research No-
        tices, No. 1**(1994), 53—59.

Cotilting Bimodules and Their Dualities

RICCARDO COLPI

DIPARTIMENTO DI MATEMATICA PURA E APPLICATA, UNIVERSITÀ DI
PADOVA, VIA BELZONI 7, 35131 PADOVA, ITALY
E-mail address: COLPI@MATH.UNIPD.IT

ABSTRACT The Brenner and Butler theorem allows us to read tilting theory as a
far reaching generalization of Morita theory, where the equivalence concerns torsion
and torsion-free classes rather than the whole module categories.

Here we present some steps for a dual theory. First we introduce the notion
of a cotilting bimodule $_SU_R$ as a dual of a tilting bimodule. Then we show that
this theory generalizes Morita dualities: namely, $_SU_R$ cogenerates torsion theories
in Mod-R and S-Mod, and it defines four functors realizing dualities between nice
subcategories of torsion and torsion-free modules, respectively.

We denote by R and S two arbitrary associative rings with unit, and by Mod-R
and S-Mod the category of all unitary right R- and left S-modules, respectively. All
the classes of modules that we introduce are to be considered as full subcategories
of modules closed under isomorphisms, and all the functors are additive functors.
Given a module U, we denote by Cogen(U) the class of all modules *cogenerated* by
U, that is all the modules M such that there exists an exact sequence $0 \to M \to U^\alpha$,
for some cardinal α. We denote by $\mathrm{Rej}_U(-)$ the *reject* radical, defined by the
position $\mathrm{Rej}_U(M) = \cap\{\mathrm{Ker}(f) \mid f \in \mathrm{Hom}_R(M, U)\}$, i.e., the least submodule M_0
of M such that M/M_0 belongs to Cogen(U). The torsion theories that we consider
are generally non-hereditary. For further notation, we refer to [AF], [St] and [We].

1. Definition. A *cotilting module* U_R is a right R-module which satisfies the condition Cogen(U_R) = Ker $\mathrm{Ext}^1_R(-, U_R)$ or, equivalently, the conditions (see [CpDeTo,
Proposition 1.7])

    i) $\mathrm{inj\,dim}(U_R) \leq 1$,
    ii) $\mathrm{Ext}^1_R(U_R^\alpha, U_R) = 0$ for any cardinal α,
    iii) Ker $\mathrm{Hom}_R(-, U_R) \cap$ Ker $\mathrm{Ext}^1_R(-, U_R) = 0$.

Roughly speaking, we can say that a cotilting module is a module U_R which is
Ext-injective exactly in the class of all modules cogenerated by U_R. Thus cotilting modules naturally generalize injective cogenerators. Some results on cotilting
modules and the duality between them and tilting modules over arbitrary rings are
studied in [CpToTr].

Now, Morita dualities are induced by faithfully balanced injective cogenerators.
Therefore we are naturally lead to consider, more generally, dualities associated to
faithfully balanced bimodules $_SU_R$, which are cotilting modules on both sides. For
short, we call such a bimodule $_SU_R$ a *cotilting bimodule*.

The ideas and techniques that we present in the study of these dualities draw
inspiration from the Colby's papers [Cb1] and [Cb2], to which we are in debt. The
main differences between our and Colby's setting is that we are not assuming the

further hypothesis that the class of reflexive modules is closed under submodules, and generally the rings does not satisfies chain conditions.

Given any bimodule $_SU_R$, we have four functors $\Delta, \Gamma \colon \text{Mod-}R \to S\text{-Mod}$ and $\Delta, \Gamma \colon S\text{-Mod} \to \text{Mod-}R$, defined as follows:

$$\Delta = \text{Hom}_?(-, {}_SU_R) \qquad \text{and} \qquad \Gamma = \text{Ext}^1_?(-, {}_SU_R)$$

where $? = R$ or S.

For any right R-module (respectively, left S-module) M, the *evaluation morphism* δ_M is defined by

$$\delta_M \colon M \to \Delta^2(M), \qquad x \mapsto [\xi \mapsto \xi(x)], \quad \text{for all } \xi \in \Delta(M).$$

2. Lemma. *Let $_SU_R$ be a cotilting bimodule. Then*

   a) $(\text{Ker}\,\Delta, \text{Ker}\,\Gamma)$ *is a torsion theory in* Mod-R *(respectively, in S-Mod), associated to the radical* $\text{Rej}_U(-) = \text{Ker}(\delta_-)$;
   b) δ_M *is monic if and only if* $M \in \text{Ker}\,\Gamma$;
   c) *the canonical projection* $M \twoheadrightarrow M/\text{Rej}_U(M)$ *induces a natural isomorphism* $\Delta(M) \cong \Delta(M/\text{Rej}_U(M))$, *and the canonical inclusion* $\text{Rej}_U(M) \hookrightarrow M$ *induces a natural isomorphism* $\Gamma(\text{Rej}_U(M)) \cong \Gamma(M)$;
   d) $\text{Im}\,\Delta \subseteq \text{Ker}\,\Gamma$.

Proof. a). Clearly $\text{Cogen}(U_R) = \text{Ker}\,\Gamma$ is a torsion-free class, thus $M \in \text{Mod-}R$ is a torsion module if and only if $\Delta(M) = 0$, and it is torsion-free if and only if $\text{Rej}_U(M) = 0$. The same argument works for $_SU$.

   b). It follows directly from a).

   c). From the exact sequence

$$0 \to \text{Rej}_U(M) \to M \to M/\text{Rej}_U(M) \to 0$$

we get the long exact sequence

$$0 \to \Delta(M/\text{Rej}_U(M)) \xrightarrow{\cong} \Delta(M) \to \Delta(\text{Rej}_U(M)) = 0 \to$$
$$\to 0 = \Gamma(M/\text{Rej}_U(M)) \to \Gamma(M) \xrightarrow{\cong} \Gamma(\text{Rej}_U(M)) \to 0.$$

   d). For any module M, we get $\Delta(M) \le U^M$, therefore $\Delta(M) \in \text{Cogen}(U) = \text{Ker}\,\Gamma$. $\square$

From the previous condition a), it is clear that the torsion theories in Mod-R and in S-Mod cogenerated by the cotilting bimodule $_SU_R$ are both trivial (i.e., every module is torsion-free) if and only if $_SU_R$ is an injective cogenerator on both sides.

On the other hand, condition c) shows that any module M, for which $M \cong \Delta^2(M)$ canonically, is necessarily torsion-free. Similarly, if there exists any canonical isomorphism $M \cong \Gamma^2(M)$, then M must be a torsion module.

Our next goal is to find representative subclasses of $\text{Ker}\,\Gamma$ and of $\text{Ker}\,\Delta$ whose members are, respectively, Δ-reflexive and Γ-reflexive.

3. Notation. Given a cotilting bimodule $_SU_R$, the following classes of right R-modules (respectively, left S-modules) are defined:

    i) $\mathcal{Y} = \{M \mid \delta_M \text{ is an isomorphism}\} \subseteq \operatorname{Ker}\Gamma$,
    ii) $\mathcal{C} = \{M \mid M \cong L/K, \text{ for some } L, K \in \mathcal{Y}\}$,
    iii) $\mathcal{X} = \mathcal{C} \cap \operatorname{Ker}\Delta \subseteq \operatorname{Ker}\Delta$.

Our next aim is to show that $\mathcal{Y}$ has nice properties. In particular $\mathcal{Y}$, and hence $\mathcal{C}$, are quite large classes. Moreover, it will turn out (see Theorem 6) that $\mathcal{X}$ consists of all the modules in $\mathcal{C}$ which are Γ-reflexive.

4. Lemma. *Let $_SU_R$ be a cotilting bimodule, and let $M \in$ Mod-R (respectively, $M \in S$-Mod). Then:*

    a) *$M \in \operatorname{Ker}\Gamma$ and $\Delta(M)$ is finitely generated if and only if there exists an exact sequence $0 \to M \to U^n \to C \to 0$ with $C \in \operatorname{Ker}\Gamma$;*
    b) *if $M \in \operatorname{Ker}\Gamma$ and $\varphi: L \hookrightarrow M$ is a monomorphism such that $\Delta(\varphi)$ is an isomorphism, then φ is an isomorphism;*
    c) *$\Delta(M) \in \mathcal{Y}$ if and only if δ_M is an epimorphism;*
    d) *let $0 \to K \to L \xrightarrow{\varphi} M \to 0$ be an exact sequence:*
        i) *if $L \in \mathcal{Y}$ (and so $K \in \operatorname{Ker}\Gamma$), then $\Delta(M), \operatorname{Coker}(\Delta(\varphi)) \in \mathcal{Y}$,*
        ii) *if $K \in \mathcal{Y}$ and $L \in \operatorname{Ker}\Gamma$, then $\Gamma(M) \in \operatorname{Ker}\Delta$ and $\Gamma^3(M) \cong \Gamma(M)$ naturally.*

Proof. a). Let $M \in \operatorname{Ker}\Gamma$ such that $\Delta(M)$ is finitely generated, say $\Delta(M) = \langle f_1, \ldots, f_n \rangle$. Then there is an exact sequence of the form

(ex) $$0 \to M \xrightarrow{f} U^n \to C \to 0,$$

where $f = (f_1, \ldots, f_n)$ and $C = \operatorname{Coker} f$. From that, we obtain the exact sequence $0 \to \Delta(C) \to \Delta(U^n) \xrightarrow{\Delta(f)} \Delta(M) \to \Gamma(C) \to 0$, where $\Delta(f)$ is epic by construction. Therefore $\Gamma(C) = 0$. Conversely, suppose that $M \in$ Mod-R is such that there exists an exact sequence (ex), with $C \in \operatorname{Ker}\Gamma$. Then $M \in \operatorname{Cogen}(U_R) = \operatorname{Ker}\Gamma$, and from (ex) we derive the exact sequence $0 \to \Delta(C) \to S^n \to \Delta(M) \to 0$, proving that $\Delta(M)$ is finitely generated.

    b). From the exact sequence $0 \to L \xrightarrow{\varphi} M \to N \to 0$ with $\Delta(\varphi)$ iso we get the exact sequence $0 \to \Delta(N) \to \Delta(M) \xrightarrow{\cong} \Delta(L) \to \Gamma(N) \to 0$. It follows that $\Delta(N) = 0 = \Gamma(N)$, i.e., $N = 0$.

    c). From the commutative diagram

$$
\begin{array}{ccc}
\Delta(M) & \xrightarrow{\delta_{\Delta(M)}} & \Delta^3(M) \\
\| & & \downarrow{\scriptstyle \Delta(\delta_M)} \\
\Delta(M) & =\!\!=\!\!= & \Delta(M)
\end{array}
$$

we see that δ_M epic $\Rightarrow \Delta(\delta_M)$ monic $\Rightarrow \Delta(\delta_M)$ iso $\Rightarrow \delta_{\Delta(M)}$ iso, i.e., $\Delta(M)$ is Δ-reflexive. On the other hand, if $\Delta(M)$ is Δ-reflexive we get that $\Delta(\delta_M)$ is an isomorphism. Hence, from the short exact sequences

$$0 \to \mathrm{Rej}_U(M) \to M \xrightarrow{\alpha} \mathrm{Im}(\delta_M) \to 0$$

$$0 \to \mathrm{Im}(\delta_M) \xrightarrow{\beta} \Delta^2(M) \to C \to 0$$

with $\beta\alpha = \delta_M$, we get that both $\Delta(\delta_M) = \Delta(\alpha)\Delta(\beta)$ and $\Delta(\alpha)$ are isomorphisms, so that $\Delta(\beta)$ is an isomorphism too. From b) we conclude that β is an isomorphism, i.e., δ_M is epic.

d). Given the exact sequence $0 \to K \xrightarrow{k} L \xrightarrow{\varphi} M \to 0$, we get the exact sequence $0 \to \Delta(M) \xrightarrow{\Delta(\varphi)} \Delta(L) \xrightarrow{\Delta(k)} \Delta(K) \to \Gamma(M) \to 0$ that can be split in the two short exact sequences

(ex1) $$0 \to \Delta(M) \xrightarrow{\Delta(\varphi)} \Delta(L) \xrightarrow{\alpha} C \to 0$$

and

(ex2) $$0 \to C \xrightarrow{\beta} \Delta(K) \to \Gamma(M) \to 0,$$

where $\beta\alpha = \Delta(k)$ and $C = \mathrm{Coker}(\Delta(\varphi)) \in \mathrm{Ker}\,\Gamma$, because $C \hookrightarrow \Delta(K) \in \mathrm{Ker}\,\Gamma$.

Now, let us first suppose that $L \in \mathcal{Y}$. From (ex1) and using Lemma 2 d), we get the commutative diagram

$$
\begin{array}{ccccccccc}
0 & \longrightarrow & \Delta(M) & \longrightarrow & \Delta(L) & \longrightarrow & C & \longrightarrow & 0 \\
 & & \downarrow{\scriptstyle \delta_{\Delta(M)}} & & \cong\downarrow{\scriptstyle \delta_{\Delta(L)}} & & \downarrow{\scriptstyle \delta_C} & & \\
0 & \longrightarrow & \Delta^3(M) & \longrightarrow & \Delta^3(L) & \longrightarrow & \Delta^2(C) & \longrightarrow & 0
\end{array}
$$

where $\delta_{\Delta(L)}$ is an isomorphism, as L — hence $\Delta(L)$ — is Δ-reflexive, and δ_C is monic, as $C \in \mathrm{Ker}\,\Gamma$. Therefore both δ_C and $\delta_{\Delta(M)}$ are isomorphisms.

Next, let us suppose that $K \in \mathcal{Y}$ and $L \in \mathrm{Ker}\,\Gamma$. Thus δ_K is an isomorphism and δ_L is a monomorphism. This implies that $\Delta^2(k) = \Delta(\alpha)\Delta(\beta)$ is monic too, so that $\Delta(\beta)$ is a monomorphism. On the other hand, from (ex2) we get the exact sequence

$$0 \to \Delta\Gamma(M) \to \Delta^2(K) \xrightarrow{\Delta(\beta)} \Delta(C) \to \Gamma^2(M) \to 0,$$

from which we conclude that $\Delta\Gamma(M) = 0$.

Finally, in order to prove that in this case there is a natural isomorphism $\Gamma^3(M) \cong \Gamma(M)$, thanks to Lemma 2 c) we can assume, without loss of generality, that $M \in \mathrm{Ker}\,\Delta$. Thus we have the exact sequence $0 \to \Delta(L) \to \Delta(K) \to \Gamma(M) \to 0$, where $\Delta(L) \in \mathrm{Ker}\,\Gamma$, $\Delta(K) \in \mathcal{Y}$ and $\Gamma(M) \in \mathrm{Ker}\,\Delta$. Therefore we get the commutative diagrams with exact rows and columns

$$
\begin{array}{ccccccccc}
& & 0 & & 0 & & 0 & & \\
& & \downarrow & & \downarrow & & \downarrow & & \\
0 & \longrightarrow & K & \longrightarrow & L & \longrightarrow & M & \longrightarrow & 0 \\
& & \cong \downarrow {\scriptstyle \delta_K} & & \downarrow {\scriptstyle \delta_L} & & \downarrow {\scriptstyle f} & & \\
0 & \longrightarrow & \Delta^2(K) & \longrightarrow & \Delta^2(L) & \longrightarrow & \Gamma^2(M) & \longrightarrow & 0 \\
& & \downarrow & & \downarrow & & \downarrow {\scriptstyle g} & & \\
& & 0 & \longrightarrow & N & = & N & \longrightarrow & 0 \\
& & & & \downarrow & & \downarrow & & \\
& & & & 0 & & 0 & &
\end{array}
$$

and

$$
\begin{array}{ccccccccc}
& 0 & & & & & & & \\
& \uparrow & & & & & & & \\
& \Gamma(M) & & & & & & & \\
& \uparrow {\scriptstyle \Gamma(f)} & & & & & & & \\
& \Gamma^3(M) & & 0 & & & & & \\
& \uparrow & & \uparrow & & & & & \\
& \Gamma(N) & = & \Gamma(N) & & 0 & & & \\
& \uparrow & & \uparrow & & \uparrow & & & \\
& 0 & \longrightarrow & \Delta(L) & \longrightarrow & \Delta(K) & \longrightarrow & \Gamma(M) & \longrightarrow & 0 \\
& \uparrow & & \oplus \uparrow {\scriptstyle \Delta(\delta_L)} & & \cong \uparrow {\scriptstyle \Delta(\delta_K)} & & \uparrow {\scriptstyle \Gamma(f)} & \\
0 \longrightarrow \Delta\Gamma^2(M) & \longrightarrow & \Delta^3(L) & \longrightarrow & \Delta^3(K) & \longrightarrow & \Gamma^3(M) & \longrightarrow & 0 \\
\uparrow {\scriptstyle \Delta(g)} & & \uparrow & & \uparrow & & & \\
0 \longrightarrow \Delta(N) & = & \Delta(N) & \longrightarrow & 0 & & & \\
\uparrow & & \uparrow & & & & & \\
0 & & 0 & & & & &
\end{array}
$$

where $\Delta(\delta_L)$ is a split epimorphism, because of the commuting diagram

$$
\begin{array}{ccc}
\Delta(L) & \xrightarrow{\;\delta_{\Delta(L)}\;} & \Delta^3(L) \\
\| & & \downarrow {\scriptstyle \Delta(\delta_L)} \\
\Delta(L) & = & \Delta(L)
\end{array}
$$

Hence, looking at the second column, we get $\Gamma(N) = 0$, so that from the first column we see that $\Gamma(f)$ is an isomorphism. The naturality of $\Gamma(f)$ follows from the commutativity of the two central exact rows. $\square$

5. Proposition. *Let $_SU_R$ be a cotilting bimodule, and let $M \in$ Mod-R (respectively, $M \in S$-Mod). Then:*

    a) *if $0 \to L \to M \to N \to 0$ is an exact sequence in $\operatorname{Ker}\Gamma$, then $M \in \mathcal{Y}$ if and only if both L and N belong to $\mathcal{Y}$;*

    b) *$M \in \mathcal{Y}$ if and only if $M \in \operatorname{Ker}\Gamma$ and $\Delta(M) \in \mathcal{Y}$;*

    c) *if $M \in \operatorname{Ker}\Gamma$ and either M or $\Delta(M)$ is finitely generated, then $M \in \mathcal{Y}$;*

    d) *$\mathcal{C}$ contains all the finitely presented modules.*

Proof. a). Let $0 \to L \to M \to N \to 0$ be an exact sequence in $\operatorname{Ker}\Gamma$. As $\operatorname{Im}\Delta \subseteq \operatorname{Ker}\Gamma$ by Lemma 2 d), we get the commutative diagram with exact rows

$$
\begin{array}{ccccccccc}
0 & \longrightarrow & L & \longrightarrow & M & \longrightarrow & N & \longrightarrow & 0 \\
& & \downarrow{\scriptstyle \delta_L} & & \downarrow{\scriptstyle \delta_M} & & \downarrow{\scriptstyle \delta_N} & & \\
0 & \longrightarrow & \Delta^2(L) & \longrightarrow & \Delta^2(M) & \longrightarrow & \Delta^2(N) & \longrightarrow & 0
\end{array}
$$

where $\delta_L, \delta_M, \delta_N$ are monic, because of Lemma 2 a). It is easy to see that δ_M is an isomorphism if and only if both δ_L and δ_N are isomorphisms.

    b). It follows from Lemma 4 c).

    c). First of all, we note that R_R and U_R (respectively, $_SS$ and $_SU$) are Δ-reflexive. Therefore all their finite direct sums belong to $\mathcal{Y}$. Now let $M \in \operatorname{Ker}\Gamma$. On the one hand, if M is finitely generated, then there exists an exact sequence, in $\operatorname{Ker}\Gamma$, of the form $0 \to K \to R^n \to M \to 0$, the case in S-Mod being analogous. On the other hand, if $\Delta(M)$ is finitely generated, then by Lemma 4 a) there is an exact sequence, in $\operatorname{Ker}\Gamma$, of the form $0 \to M \to U^n \to C \to 0$. In both cases, as R^n and U^n, respectively, belong to $\mathcal{Y}$, using a) we conclude that $M \in \mathcal{Y}$ too.

    d). Any finitely presented module M is of the form $M \cong L/K$, with L, K finitely generated and L projective. Thus both L and K belong to $\operatorname{Ker}\Gamma$, and so to $\mathcal{Y}$, because of c). This shows that $M \in \mathcal{C}$. $\square$

We are now ready to prove a "Cotilting Theorem":

6. Theorem. *Let $_SU_R$ be a cotilting bimodule and let $M \in \mathcal{C}$. Then:*

    a) *$\Delta(M) \in \mathcal{Y}$ and $\Gamma(M) \in \mathcal{X}$;*

    b) *There is a natural morphism $\gamma_M : \Gamma^2(M) \to M$ such that the canonical sequence*

$$
0 \to \Gamma^2(M) \xrightarrow{\gamma_M} M \xrightarrow{\delta_M} \Delta^2(M) \to 0
$$

       is exact, with $\operatorname{Im}(\gamma_M) = \operatorname{Rej}_U(M)$;

    c) *$(\mathcal{X}, \mathcal{Y})$ is a torsion theory in $\mathcal{C}$, $\Delta{\restriction}\,\mathcal{X} = 0 = \Gamma{\restriction}\,\mathcal{Y}$, $\Delta{\restriction}\,\mathcal{Y}$ and $\Gamma{\restriction}\,\mathcal{X}$ are exact functors, defining dualities $\mathcal{Y} \underset{\Delta}{\overset{\Delta}{\rightleftarrows}} \mathcal{Y}$ and $\mathcal{X} \underset{\Gamma}{\overset{\Gamma}{\rightleftarrows}} \mathcal{X}$.*

Proof. Let $M \in \mathcal{C}$ and let $0 \to K \xrightarrow{k} L \xrightarrow{\varphi} M \to 0$ be exact, with $K, L \in \mathcal{Y}$.

First, we note that from Lemma 4 c) we get both $\Delta(M) \in \mathcal{Y}$ and $\Gamma(M) \in \operatorname{Ker} \Delta$. Moreover the exact sequence (ex2) in the proof of Lemma 4 d) shows that $\Gamma(M) \in \mathcal{C}$. Hence $\Gamma(M) \in \mathcal{X}$. This proves a).

Next, using the same notation as in the proof of Lemma 4 d), we obtain the commutative diagrams with exact rows

$$
\begin{array}{ccccccccc}
0 & \longrightarrow & K & \xrightarrow{k} & L & \xrightarrow{\varphi} & M & \longrightarrow & 0 \\
& & \downarrow{\scriptstyle \Delta(\beta) \circ \delta_K} & & \cong \downarrow{\scriptstyle \delta_L} & & \downarrow{\scriptstyle \delta_M} & & \\
0 & \longrightarrow & \Delta(C) & \xrightarrow{\Delta(\alpha)} & \Delta^2(L) & \xrightarrow{\Delta^2(\varphi)} & \Delta^2(M) & \longrightarrow & 0
\end{array}
$$

and

$$
0 \to \Delta^2(K) \xrightarrow{\Delta(\beta)} \Delta(C) \to \Gamma^2(M) \to 0.
$$

Now, by the Snake Lemma [Pi, 11.3], we have $\operatorname{Ker}(\delta_M) \cong \operatorname{Coker}(\Delta(\beta) \circ \delta_K)$, canonically. Moreover, since δ_K is an iso, $\operatorname{Coker}(\Delta(\beta)) \cong \operatorname{Coker}(\Delta(\beta) \circ \delta_K)$, so that we get the canonical isomorphisms and inclusion

$$
\Gamma^2(M) \cong \operatorname{Coker}(\Delta(\beta)) \cong \operatorname{Ker}(\delta_M) \hookrightarrow M,
$$

which show that $\gamma_M \colon \Gamma^2(M) \to M$ is well defined, injective, canonical and, finally, $\operatorname{Im}(\gamma_M) = \operatorname{Ker}(\delta_M) = \operatorname{Rej}_U(M)$ by construction. This proves b).

In order to prove c), we first note that from a) and b) it is immediately seen that $(\mathcal{X}, \mathcal{Y})$ is a torsion theory in $\mathcal{C}$, with associated radical $\operatorname{Rej}_U(-)$. Moreover $\Delta \upharpoonright \mathcal{X} = 0$ because $\mathcal{X} \subseteq \operatorname{Ker} \Delta$ and, similarly, $\Gamma \upharpoonright \mathcal{Y} = 0$ because $\mathcal{Y} \subseteq \operatorname{Ker} \Gamma$. Next, $\Delta(\mathcal{Y}) \subseteq \mathcal{Y}$ and $\Gamma(\mathcal{X}) \subseteq \mathcal{X}$ because of a), and for any $M \in \mathcal{Y}$ we get $M \cong \Delta^2(M)$ by definition. Similarly, if $M \in \mathcal{X} \subseteq \operatorname{Ker} \Delta$, then $M = \operatorname{Rej}_U(M)$, therefore γ_M is an isomorphism because of b), so that $M \cong \Gamma^2(M)$. This proves that Δ and Γ induce the stated dualities. Finally, $\Delta \upharpoonright \mathcal{Y}$ and $\Gamma \upharpoonright \mathcal{X}$ are exact functors, as $\Gamma \upharpoonright \mathcal{Y} = 0 = \Delta \upharpoonright \mathcal{X}$ and $\operatorname{inj} \dim(U) \le 1$. $\square$

The following remark justifies the definition of the class $\mathcal{C}$ given in 3 ii), showing that $\mathcal{C}$ is, in a certain sense, the largest class for which the previous theorem holds:

7. Remark. *Let $_S U_R$ be a cotilting bimodule, and let M be any factor of a Δ-reflexive module. Then M satisfies the conditions a) and b) of Theorem 6 if and only if $M \in \mathcal{C}$.*

Proof. Let M be a factor of the Δ-reflexive module L. Then we get the commutative

exact diagram

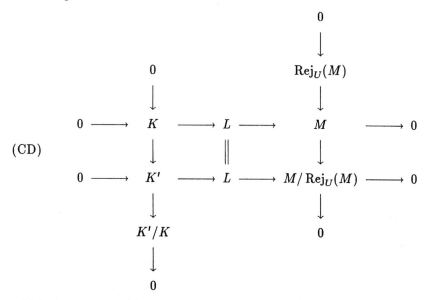

which shows that $K'/K \cong \mathrm{Rej}_U(M)$ and, thanks to Proposition 5 a), that K' and $M/\mathrm{Rej}_U(M)$ are Δ-reflexive too.

First, from the second exact row of (CD) we obtain the exact sequence

$$0 \to \Delta(M) \cong \Delta(M/\mathrm{Rej}_U(M)) \to \Delta(L) \to \Delta(K') \to 0$$

which shows that $\Delta(M)$ is Δ-reflexive and that the sequence $0 \to \Delta^2(K') \to \Delta^2(L) \to \Delta^2(M) \to 0$ is exact, so that

$$\boxed{\Delta^2(M) \text{ is } \Delta\text{-reflexive and } \Delta^2(M) \cong M/\mathrm{Rej}_U(M).}$$

Next, from the first exact column of (CD) we obtain the exact sequence $0 \to \Delta(K') \to \Delta(K) \to \Gamma(\mathrm{Rej}_U(M)) \cong \Gamma(M) \to 0$ with $\Delta(K')$ Δ-reflexive, which shows that

$$\boxed{\Gamma^2(M) \in \mathrm{Ker}\,\Delta \text{ and } \Gamma^2(M) \text{ is } \Gamma\text{-reflexive.}}$$

thanks to Lemma 4 d).

Finally, we have the commutative exact diagram

$$
\begin{array}{ccccccccc}
 & & & & & & & & 0 \\
 & & & & & & & & \uparrow \\
0 & \longrightarrow & \Delta\Gamma(M) & \longrightarrow & \Delta^2(K) & \longrightarrow & \Delta^2(K') & \longrightarrow & \Gamma^2(M) & \longrightarrow & 0 \\
 & & & & \uparrow{\scriptstyle \delta_K} & & \cong\uparrow{\scriptstyle \Delta_{K'}} & & \uparrow{\scriptstyle \varphi} \\
 & & 0 & \longrightarrow & K & \longrightarrow & K' & \longrightarrow & \mathrm{Rej}_U(M) & \longrightarrow & 0 \\
 & & & & \uparrow \\
 & & & & 0
\end{array}
$$

which defines a natural epimorphism $\varphi\colon \mathrm{Rej}_U(M) \twoheadrightarrow \Gamma^2(M)$ and shows that

$$\boxed{\Gamma(M) \in \operatorname{Ker}\Delta \text{ and } \Gamma^2(M) \cong \mathrm{Rej}_U(M) \text{ naturally} \Rightarrow K \text{ is } \Delta\text{-reflexive, i.e., } M \in \mathcal{C}.}$$

□

From Proposition 5 a), we immediately see that a factor of a Δ-reflexive module is Δ-reflexive if and only if it is torsion-free. Now, using the previous result, we are able to give a criterion for a submodule (respectively, an extension) of a Δ-reflexive module to be Δ-reflexive.

8. Proposition. *Let $_SU_R$ be a cotilting bimodule, and let $0 \to K \to L \to M \to 0$ be an exact sequence. Then:*

  a) *if $L \in \mathcal{Y}$ (and so $K \in \operatorname{Ker}\Gamma$), then $K \in \mathcal{Y}$ if and only if $M \in \mathcal{C}$;*
  b) *if $K \in \mathcal{Y}$ and $L \in \operatorname{Ker}\Gamma$, then $L \in \mathcal{Y}$ if and only if $M \in \mathcal{C}$.*

Proof. One implication both in a) and in b) follows trivially by definition of $\mathcal{C}$. Conversely, let us assume that $M \in \mathcal{C}$ and $L \in \operatorname{Ker}\Gamma$. Then $\Delta(M) \in \mathcal{Y}$ and $\Gamma(M) \in \mathcal{X}$ because of Theorem 6 a), and we get the two exact sequences

(ex1) $$0 \to \Delta(M) \to \Delta(L) \to C \to 0$$
(ex2) $$0 \to C \to \Delta(K) \to \Gamma(M) \to 0.$$

From (ex 2) we see that $C \in \operatorname{Ker}\Gamma$, and we get the short exact sequence $0 \to \Delta^2(K) \to \Delta(C) \to \Gamma^2(M) \to 0$ and the commutative diagram with exact rows

(CD)
$$
\begin{array}{ccccccccc}
0 & \longrightarrow & \Delta^2(C) & \longrightarrow & \Delta^3(K) & \longrightarrow & \Gamma^3(M) & \longrightarrow & 0 \\
 & & \uparrow{\scriptstyle\delta_C} & & \uparrow{\scriptstyle\delta_{\Delta(K)}} & & \cong\downarrow{\scriptstyle\gamma_{\Gamma(M)}} & & \\
0 & \longrightarrow & C & \longrightarrow & \Delta(K) & \longrightarrow & \Gamma(M) & \longrightarrow & 0.
\end{array}
$$

From the one hand, if we assume that $L \in \mathcal{Y}$, then $\Delta(L) \in \mathcal{Y}$ too, and from (ex1) we get $C \in \mathcal{Y}$ because of Proposition 5 a). Therefore δ_C is an isomorphism, and from (CD) we get that $\Delta(K) \in \mathcal{Y}$. As $K \in \operatorname{Ker}\Gamma$, by Proposition 5 b) we conclude that $K \in \mathcal{Y}$.

On the other hand, if we assume that $K \in \mathcal{Y}$, then $\Delta(K) \in \mathcal{Y}$ too, and from (CD) we get $C \in \mathcal{Y}$. From (ex1) we have $\Delta(L) \in \mathcal{Y}$, because of Proposition 5 a). As $L \in \operatorname{Ker}\Gamma$, by Proposition 5 b) we conclude that $L \in \mathcal{Y}$. □

9. Corollary. *Let $_SU_R$ be a cotilting bimodule, and let $0 \to L \to M \to N \to 0$ be an exact sequence with $M \in \mathcal{C}$. Then $L \in \mathcal{C}$ if and only if $N \in \mathcal{C}$.*

Proof. Let $0 \to Y_1 \to Y_2 \to M \to 0$ be an exact sequence, with $Y_1, Y_2 \in \mathcal{Y}$. From the commutative diagram with exact rows and columns

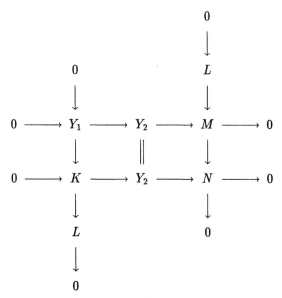

we see, using Proposition 8, that $L \in \mathcal{C}$ if and only if $K \in \mathcal{Y}$ if and only if $N \in \mathcal{C}$.   □

Of course, one of the main question concerns the characterization of the classes $\mathcal{C}$, $\mathcal{Y}$ and $\mathcal{X}$ involved in the Cotilting Theorem. By definition, the main role is played by $\mathcal{Y}$, i.e., the class of all Δ-reflexive modules.

In case of a Morita duality, $\mathcal{X} = 0$, and $\mathcal{C} = \mathcal{Y}$ consists of all the *linearly compact* modules, i.e., the modules M such that for any inverse system $\{M \xrightarrow{p_\lambda} M_\lambda \mid \lambda \in \Lambda\}$ of epimorphisms, the morphism

$$\varprojlim p_\lambda : M \to \varprojlim M_\lambda$$

is epic too. Similarly, we have the following

10. Proposition. *Given a cotilting bimodule ${}_SU_R$, let $M \in \operatorname{Ker}\Gamma$. Let us consider the following assertions:*

    i) *For any inverse system of morphisms $\{M \xrightarrow{p_\lambda} M_\lambda \mid \lambda \in \Lambda\}$ with $M_\lambda \in \operatorname{Ker}\Gamma$ and $\operatorname{Coker}(p_\lambda) \in \operatorname{Ker}\Delta$, we have $\operatorname{Coker}(\varprojlim p_\lambda) \in \operatorname{Ker}\Delta$;*

    ii) *M is Δ-reflexive.*

Then i) ⇒ ii).

Proof. Let $\{N_\lambda \xrightarrow{i_\lambda} \Delta(M) \mid \lambda \in \Lambda\}$ be the direct system given by the finitely generated submodules N_λ of $\Delta(M)$ and the corresponding natural inclusions. Then $\varinjlim i_\lambda$ is an isomorphism, and each N_λ is Δ-reflexive, because of Proposition 5 c). Let $p_\lambda = \Delta(i_\lambda)\delta_M$. Then we have the inverse system $\{M \xrightarrow{p_\lambda} \Delta(N_\lambda) \mid \lambda \in \Lambda\}$ in $\operatorname{Ker}\Gamma$.

First, we show that each $C_\lambda = \operatorname{Coker}(p_\lambda)$ is in $\operatorname{Ker}\Delta$. Indeed, on one hand from the exact sequence $M \xrightarrow{p_\lambda} \Delta(N_\lambda) \xrightarrow{c_\lambda} C_\lambda \to 0$ we derive the exact sequence

$$(1) \qquad\qquad 0 \to \Delta(C_\lambda) \xrightarrow{\Delta(c_\lambda)} \Delta^2(N_\lambda) \xrightarrow{\Delta(p_\lambda)} \Delta(M).$$

On the other hand, the commutative diagram

(CD)

$$
\begin{array}{ccccccc}
0 & \longrightarrow & N_\lambda & \xrightarrow{\imath_\lambda} & \Delta(M) & =\!=\!= & \Delta(M) \\
& & \cong \downarrow{\scriptstyle\delta_{N_\lambda}} & & \downarrow{\scriptstyle\delta_{\Delta(M)}} & & \| \\
& & \Delta^2(N_\lambda) & \xrightarrow{\Delta^2(\imath_\lambda)} & \Delta^3(M) & \xrightarrow{\Delta(\delta_M)} & \Delta(M)
\end{array}
$$

shows that $\Delta(p_\lambda) = \Delta(\delta_M)\Delta^2(i_\lambda)$ is a monomorphism. Combining this with (1), we get $\Delta(C_\lambda) = 0$, so that hypothesis i) applies, producing

(2)
$$\mathrm{Coker}(\varprojlim p_\lambda) \in \mathrm{Ker}\,\Delta.$$

Now, from the canonical isomorphisms $\varprojlim p_\lambda \cong \varprojlim \Delta(i_\lambda)\delta_M \cong \Delta(\varinjlim i_\lambda)\delta_M$, we get

(3)
$$\mathrm{Coker}(\varprojlim p_\lambda) \cong \mathrm{Coker}(\delta_M).$$

Combining (2) with (3), we see that $\Delta(\delta_M)$ is a monomorphism. Therefore, from (CD) it follows that $\delta_{\Delta(M)}$ is an isomorphism, i.e., $\Delta(M)$ is Δ-reflexive. By Proposition 5 b) we conclude that M is Δ-reflexive. $\square$

11. Remark. *We do not know if, in Proposition 10, ii) $\Rightarrow$ i) holds true.*

If $_S U_R$ is a cotilting bimodule inducing the functors Δ and Γ with $\mathcal{T} = \mathrm{Ker}\,\Delta$ and $\mathcal{F} = \mathrm{Ker}\,\Gamma$, and $\mathcal{Y}$ is the class of Δ-reflexive modules in Mod-R (respectively, S-Mod), then the following conditions hold:

12. **Duality conditions.** *$(\mathcal{T}, \mathcal{F})$ is a torsion theory in Mod-R (respectively, S-Mod) and $\mathcal{Y}$ is a subclass of $\mathcal{F}$, satisfying the following conditions:*

  i) *$\mathcal{F}$ contains every projective module;*
  ii) *$\mathcal{Y}$ contains all the finitely generated modules of $\mathcal{F}$;*
  iii) *if $0 \to L \to M \to N \to 0$ is an exact sequence with $M \in \mathcal{Y}$ and $N \in \mathcal{F}$, then $N \in \mathcal{Y}$, i.e., $\mathcal{Y}$ is closed under factors in $\mathcal{F}$;*
  iv) *there exists a duality $\mathcal{Y} \rightleftarrows \mathcal{Y}$.*

Proof. Condition i) is clear, as $\mathcal{F} = \mathrm{Ker}\,\mathrm{Ext}^1_?(-, U)$. Conditions ii) and iii) are proved in Proposition 5. Finally, condition iv) is part of Theorem 6 c). $\square$

Conversely, a natural question is whether the previous Duality conditions characterize cotilting bimodules, at least when R_R and $_S S$ are noetherian. Trlifaj in [Tr] has proved that the answer to this question is negative, even for $R = S = \mathbb{Z}$, considering the usual torsion theory in Mod-$\mathbb{Z}$.

Nevertheless, Duality conditions 12 are always realized by a bimodule which has a cotilting behaviour, with respect to the finitely presented modules, by means of the following

13. Proposition. *Under the notation and hypotheses of the Duality conditions 12, the following facts hold:*

  a) *there exists a unique, up to isomorphism, faithfully balanced bimodule $_S U_R$ such that the duality in iv) is naturally isomorphic to $\Delta = \mathrm{Hom}_?(-, U)$; in particular $\mathrm{Cogen}(U) \subseteq \mathcal{F}$;*

    b) $\operatorname{Ext}^1_?(M, U) = 0$ for any finitely generated $M \in \mathcal{F}$;

    c) for any finitely presented module M, we have: $M \in \operatorname{Cogen}(U) \Leftrightarrow M \in \mathcal{F}$
$\Leftrightarrow M \in \operatorname{Ker}\operatorname{Ext}^1_?(-, U)$;

    d) if R_R (respectively, $_SS$) is noetherian, then $\operatorname{inj\,dim}_R(U) \le 1$ (respectively, $\operatorname{inj\,dim}_S(U) \le 1$).

Proof. a). The regular modules R_R and $_SS$ are in $\mathcal{Y}$, because of 12 i). It follows easily that the bimodule $_SU_R$, dual of R_R, is canonically isomorphic to the dual of $_SS$, and it represents the duality by means of the functors $\Delta = \operatorname{Hom}_R(-, U)$ and $\Delta = \operatorname{Hom}_S(-, U)$. Finally, $U \in \mathcal{Y} \subseteq \mathcal{F}$, so that $\operatorname{Cogen}(U) \subseteq \mathcal{F}$.

    b). Let $M \in \mathcal{F}$ and let $0 \to K \xrightarrow{k} R^n \xrightarrow{\varphi} M \to 0$ be exact in Mod-R (the case in S-Mod being analogous). From that, we get the exact sequence $0 \to \Delta(M) \xrightarrow{\Delta(\varphi)} \Delta(R^n) \xrightarrow{\Delta(k)} \Delta(K) \to \Gamma(M) \to 0$ that can be split in the two short exact sequences

(ex1) $\qquad\qquad\qquad 0 \to \Delta(M) \xrightarrow{\Delta(\varphi)} \Delta(R^n) \xrightarrow{\alpha} C \to 0$

and

(ex2) $\qquad\qquad\qquad 0 \to C \xrightarrow{\beta} \Delta(K) \to \Gamma(M) \to 0,$

where $\beta \circ \alpha = \Delta(k)$. Now R^n and M are in $\mathcal{Y}$, because of 12 i) and ii), and so $\Delta(R^n)$ is in $\mathcal{Y}$ too, because of 12 iv). On the other hand $C \hookrightarrow \Delta(K) \in \operatorname{Cogen}(U) \subseteq \mathcal{F}$, so that $C \in \mathcal{F}$. Therefore 12 iii) applies to (ex1), providing $C \in \mathcal{Y}$. From that we have the commutative diagram with exact rows

$$
\begin{array}{ccccccccc}
0 & \longrightarrow & K & \xrightarrow{\ k\ } & R^n & \xrightarrow{\ \varphi\ } & M & \longrightarrow & 0 \\
& & \Big\downarrow{\scriptstyle \Delta(\beta)\circ\delta_K} & & {\scriptstyle\cong}\Big\downarrow{\scriptstyle \delta_{R^n}} & & {\scriptstyle\cong}\Big\downarrow{\scriptstyle \delta_M} & & \\
0 & \longrightarrow & \Delta(C) & \xrightarrow{\Delta(\alpha)} & \Delta^2(R^n) & \xrightarrow{\Delta^2(\varphi)} & \Delta^2(M) & &
\end{array}
$$

from which we see that $\Delta(\beta)\circ\delta_K$ is an isomorphism, thus $K \cong \Delta(C) \in \mathcal{Y}$, because of 12 iv). Therefore δ_K is iso, so that $\Delta(\beta)$ is iso too. Then $\Delta^2(\beta)$ is an isomorphism, and both C and $\Delta(K)$ being in $\mathcal{Y}$, it follows that β is an isomorphism too. From (ex2) we conclude that $\Gamma(M) = 0$.

    c). Let M be a finitely presented right R-module. First, if $M \in \operatorname{Cogen}(U)$, then $M \in \mathcal{F}$ because of a). Next, if $M \in \mathcal{F}$, then $M \in \operatorname{Ker}\operatorname{Ext}^1_R(-, U)$ because of b). Finally, let $M \in \operatorname{Ker}\operatorname{Ext}^1_R(-, U)$ and let $0 \to K \to R^n \to M \to 0$ be exact, with K finitely generated. Then we get the exact sequence $0 \to \Delta(M) \to \Delta(R^n) \to \Delta(K) \to \Gamma(M) = 0$, and the commutative diagram with exact rows

$$
\begin{array}{ccccccccc}
0 & \longrightarrow & K & \longrightarrow & R^n & \longrightarrow & M & \longrightarrow & 0 \\
& & {\scriptstyle\cong}\Big\downarrow{\scriptstyle \delta_K} & & {\scriptstyle\cong}\Big\downarrow{\scriptstyle \delta_{R^n}} & & \Big\downarrow{\scriptstyle \delta_M} & & \\
0 & \longrightarrow & \Delta^2(K) & \longrightarrow & \Delta^2(R^n) & \longrightarrow & \Delta^2(M) & &
\end{array}
$$

where δ_K and δ_{R^n} are isomorphisms, because of 12 ii). Therefore δ_M is a monomorphism, i.e., $M \in \operatorname{Cogen}(U)$.

    d). It is sufficient to prove that $\operatorname{Ext}^2_R(M, U) = 0$ for every finitely generated module M. Let $0 \to K \to R^n \to M \to 0$ be exact. Now $K \in \mathcal{F}$, because of 12 i), and K is finitely generated, as R_R is noetherian. From b) we get the exact sequence $0 = \operatorname{Ext}^1_R(K, U) \to \operatorname{Ext}^2_R(M, U) \to \operatorname{Ext}^2_R(R^n, U) = 0$, so that $\operatorname{Ext}^2_R(M, U) = 0$. $\qquad\square$

REFERENCES

[AF]   F. D. Anderson and K. R. Fuller, *Rings and Categories of Modules*, (2nd edition), Springer, New York, 1992.

[Cb1]   R. R. Colby, *A generalization of Morita duality and the tilting theorem*, Comm. Algebra **17 (7)** (1989), 1709–1722.

[Cb2]   R. R. Colby, *A cotilting theorem for rings*, Methods in Module Theory, M. Dekker, New York, 1993, pp. 33–37.

[CpDeTo] R. Colpi, G. D'Este and A. Tonolo, *Quasi-tilting modules and counter equivalences*, J. Algebra **191** (1997), 461–494.

[CpToTr] R. Colpi, A. Tonolo and J. Trlifaj, *Partial cotilting modules and the lattices induced by them*, Comm. Algebra **25** (1997), 3225–3237.

[Pi]   R. S. Pierce, *Associative Algebras*, Springer-Verlag GTM 88, Berlin, Heidelberg, New York, 1982.

[St]   B. Stenström, *Rings of Quotients*, Springer-Verlag, Berlin, Heidelberg, New York, 1975.

[Tr]   J. Trlifaj, *Remarks on duality conditions, cotilting, and quasi-duality modules* (Preprint 1997).

[We]   C. A. Weibel, *An introduction to homological algebra*, vol. 38, Cambridge Studies in Advanced Mathematics, 1994.

Finding Splitting Elements and Maximal Tori in Matrix Algebras

Willem A. de Graaf* Gábor Ivanyos†

*School of Mathematical and Computational Sciences, University of St Andrews, North Haugh, St Andrews, Fife, KY16 9SS, Scotland E-mail: wdg@dcs.st-and.ac.uk.
†Computer and Automation Institute, Hungarian Academy of Sciences, Lágymányosi u. 11., H-1111 Budapest, Hungary E-mail: Gabor.Ivanyos@sztaki.hu. Research supported by FKFP Grant 0612/1997, OTKA Grant 016503, and NWO-OTKA Grant 048.011.022.

Abstract

Roughly speaking, a splitting element of a matrix algebra is an element with a maximal number of eigenvalues. Using randomization, splitting elements can be found very easily and have proved to be extremely useful in computational representation theory. Efficient deterministic algorithms for finding splitting elements have been known only for special cases. In this paper we present a deterministic polynomial time method for constructing splitting elements and closely related subalgebras – maximal tori – which works over a wide range of ground fields.

1 Introduction

Let K be a field and let $\overline{K}$ denote the algebraic closure of K. A splitting element of a matrix algebra $A \leq M_n(K)$ is an element $a \in A$ with a maximal number of eigenvalues among the elements of $\overline{K} \otimes_K A \leq M_n(\overline{K})$. Being objects in "sufficiently general position", splitting elements have proved to be very useful in computing the structure of matrix algebras. The notion (in special cases) was introduced by W. M. Eberly [E1, E2, E3]. He observed that if K is sufficiently large then a random element of A has a good chance to being a splitting element. Below we give a partial list of applications.

[E2] Assume that K is an algebraic number field or a sufficiently large finite field and A is a commutative and semisimple algebra over K. Then a splitting element $a \in A$ is nothing else than a generator of A as a K-algebra; therefore it can be used to construct an isomorphism $A \cong K[x]/(f(x))$ where $f(x)$ is the a minimal polynomial of a. An immediate consequence is a randomized

95

polynomial time reduction from finding the simple components of algebras over K to factoring univariate polynomials over K.

[BR] Suppose that K is an algebraic number field, A is a central simple algebra over K, and a is a splitting element of A. Let α be any eigenvalue of a and $K' = K[\alpha]$. Then K' is a splitting field of A (this justifies the terminology "splitting element") and an explicit isomorphism $K' \otimes_K A \cong M_m(K')$ $(m = \sqrt{\dim_K A})$ can be constructed in polynomial time. The result is a randomized polynomial time algorithm for finding the complex irreducible representations of A.

[E3] Suppose that K is a real number field, A is a central simple algebra over K, and a is a splitting element of A. Let α be an imaginary eigenvalue of a, $\overline{\alpha}$ its complex conjugate, and $K' = K[\alpha + \overline{\alpha}, \alpha\overline{\alpha}]$. Then a minimal left ideal of $K' \otimes A$ can be constructed in polynomial time. The consequence is a randomized polynomial time algorithm for computing the local indices of simple algebras over number fields at the infinite primes. Complemented with the method of [IR] for the local indices at the finite primes, this gives an algorithm for calculating the global index.

[I] Recently, splitting elements appeared to be useful in computing the radical of algebras (in particular over fields of prime characteristic.)

The only point where randomization is used in the algorithms listed above is finding a splitting element. Hence these methods can be derandomized as soon as we have an efficient deterministic algorithm for exhibiting splitting elements. The first result in this direction is due to Rónyai [Ró]. Using the theory of orders, he showed how to find a splitting element in a simple algebra over a number field deterministically. In [GIR] methods for the closely related task of constructing Cartan subalgebras of certain Lie algebras are presented. Unfortunately, the cases efficiently treated in [GIR] do not include Lie algebras of associative algebras over certain important infinite fields of positive characteristic (e.g., global function fields).

In this paper we present a method which works over arbitrary sufficiently large ground fields, including infinite fields of positive characteristic. The algorithm performs a polynomial number of arithmetic steps. Over ground fields admitting efficient linear algebra (e.g., finite fields, number fields, function fields of bounded transcendence degree) a modified version of the algorithm runs in time which is polynomial in the size of the input.

The paper is structured as follows. The remainder of this section is devoted to a brief description of the computational models we work with. In Section 2 we summarize some known facts about splitting elements and subalgebras generated by semisimple splitting elements (the maximal tori), as well as algorithms for some subtasks. A general method for finding a maximal torus is described in Section 3. The algorithm is based on Lemma 2 which provides us with a test for maximality of a torus. In Section 4 we show how to modify the algorithm in order to keep the sizes of the intermediate objects small. This results in a polynomial time algorithm over ground fields admitting efficient linear algebra. We have implemented the algorithm in Version 4 of the GAP programming system [GAP97]. A comparison of the running time of the present algorithm to two other methods (the randomized algorithm and an algorithm using Cartan subalgebras, based on the methods of [GIR]) is given in Section 5.

We assume an encoding for the elements of the ground field K. The matrix algebra $A \leq M_n(K)$ is assumed to be given by an array of n by n matrices over K which form a K-linear basis of A. One usual way to measure the cost of an algorithm is the *uniform cost* which is the number of arithmetic operations (including zero tests) performed in the worst case, measured in the total number of field elements representing the input. The basic subtasks of linear algebra (performing the usual matrix operations, computing characteristic and minimal polynomials, solving systems of linear equations) used as subroutines of our algorithms have solutions of polynomial uniform cost (c.f. [BP]).

A more realistic measure is the *Boolean cost* where the total number of "bit operations" is taken into account. The Boolean cost of an algorithm is compared to the total number of bits representing the input. Working in this model it is natural to require that the costs of the field operations of K are bounded by a polynomial of the total size of the operands. By a *polynomial time* algorithm we mean an algorithm the Boolean cost of which is bounded by a polynomial in the size of the input. The Boolean cost of an algorithm of polynomial uniform cost is also polynomial provided that the size of the intermediate objects computed by the algorithm remains below a bound which is a polynomial in the input size.

We require that the principal tasks of linear algebra can be accomplished in polynomial time. Also, in order to exclude pathological encodings it is natural to assume that for every integer k, $\min\{|K|, k\}$ distinct elements of K can be listed in time $k^{O(1)}$ (in the Boolean sense). We refer to fields satisfying these requirements as fields admitting *efficient linear algebra*. Finite fields, algebraic number fields, or function fields over these which have transcendence degree bounded by a constant (using an appropriate, "standard" encoding for the elements of K) admit efficient linear algebra [F, IRSz]. We note that the standard encoding for a finite field K represents elements with $\theta(\log |K|)$ bits and hence an algorithm of polynomial uniform cost has automatically polynomial Boolean cost as well.

2 Preliminaries

We fix some notation used throughout the paper. K is always a field, $\overline{K}$ denotes its algebraic closure and $A \leq M_n(K)$ is a matrix algebra over K containing the identity matrix. For a matrix $a \in M_n(K)$ we denote by $d(a)$ the number of eigenvalues of a (in $\overline{K}$). Obviously, $d(a)$ is the degree of the squarefree part of the characteristic (or minimal) polynomial of a over $\overline{K}$. By $K[a]$ we denote the matrix algebra over K generated by a and the identity matrix. Then $\dim_K K[a]$ is the degree of the minimal polynomial of a. Let

$$a^* = \begin{cases} a_s & \text{if char } K = 0 \text{ or char } K > n, \\ a^{p^{\lceil \log_p n \rceil}} & \text{if } 0 < \text{char } K = p < n. \end{cases}$$

Here a_s stands for the semisimple part from the Jordan decomposition of a. It is immediate that $d(a) = d(a^*) = \dim_K K[a^*]$. If $0 < \text{char } K < n$ then a^* can be computed in a obvious way. In the remaining cases first we compute the squarefree part of the characteristic polynomial of a and use the Newton-Hensel iteration for computing a_s. The detailed description as well as an analysis of the algorithm is given in [BBCIL] or [Be]. It is shown that the method requires $n^{O(1)}$ arithmetical operations, and the Boolean cost is also a polynomial of the input size provided that systems of linear equations over K can be solved in polynomial time.

By a *torus* over K or a K-torus we mean a finite dimensional commutative algebra with identity which is separable over K. Recall that a finite dimensional commutative K-algebra T is separable over K if T is a direct sum of separable extension fields of K (c.f. [P], Section 10.7). A commutative matrix algebra $T \leq M_n(K)$ containing the identity matrix is a torus if and only if every element of T is a semisimple matrix. Also, the matrices in T can be simultaneously diagonalized over $\overline{K}$: there exists a matrix $u \in GL_n(\overline{K})$ such that $u^{-1}au$ is a diagonal matrix for every $a \in T$. Obviously $d(a) \leq \dim_K T$ and equality holds if and only if $T = K[a]$. If K is sufficiently large then equality can be reached: the idea of the usual proof of Steinitz's theorem about simple extensions of fields (c.f. [Ba], Theorem 3.6.1.) is applicable to this situation as well. It means that $a \in T$ is a splitting element of T if and only if a is a generator of T.

The proof mentioned above suggests the following simple method for finding a splitting element of T. Assume that $b_1, \ldots, b_s$ form a system of algebra generators of T and Ω is a subset of K such that $|\Omega| > (\dim_K T)(\dim_K T - 1)$. Assume that we have an element $a \in T$ such that $d(a) < \dim_K T$. Then we can choose $b \in \{b_1, \ldots, b_s\} \setminus K[a]$ and replace a with an element c of the form $c = a + \beta b$ ($\beta \in \Omega$) such that $d(c) > d(a)$. In fact, the set $\{a + \beta b | \beta \in \Omega\}$ contains at least one generator of the subalgebra $K[a, b]$. In the worst case the algorithm requires ·calculation of $d(c)$ for $O((\dim_K T)^3)$ elements $c \in \{\alpha_1 b_1 + \ldots + \alpha_s b_s | \alpha_1, \ldots, \alpha_s \in \Omega\}$.

We remark that if K is a finite field such that $\dim_K T \leq |K|$ but $|K| \leq (\dim_K T)(\dim_K T - 1)$ then by [FR] the Wedderburn decomposition of T can be computed in time $(\dim_K T)^{O(1)}$ and the decomposition can be used to find a splitting element of T efficiently.

Now let A be a finite dimensional (associative) algebra over K with identity element 1_A. A K-torus $T \leq A$ in the algebra A is called a maximal K-torus of A if there exists no torus $T' \leq A$ such that $T < T'$. Below we summarize some known facts about maximal tori. By the Wedderburn-Malcev principal theorem (c.f. [P], Section 11.6), T is a maximal torus if and only if $T + Rad(A)$ is a maximal torus of $A/Rad(A)$. It follows that we can calculate the dimension of a maximal torus of A once we know the dimension of maximal tori of the simple components of $A/Rad(A)$.

Assume that A is a simple algebra over K with center Z and let K' be the separable closure of K in Z. It is obvious that $K' = T \cap Z$ for every maximal torus T in A, and a commutative subalgebra T of A containing K' is a torus over K if and only if T is a torus over K'. Let $T \leq A$ be a torus containing K'. Then TZ is a commutative semisimple subalgebra and hence $\dim_{K'} T = \dim_Z ZT \leq \sqrt{\dim_Z A}$. On the other hand, there exists a Z-torus U of A such that $\dim_Z(U) = \sqrt{\dim_Z A}$ (in [P], Section 13.5, this is proved for division algebras but the result can be extended to simple algebras in an obvious way). The sum of the separable closures of K' in the simple components of U is a K'-torus of U of dimension $\sqrt{\dim_Z A}$.

We obtained that the dimension of a maximal torus in an arbitrary finite dimensional K-algebra is independent of the particular choice of the torus. An easy consequence is that T is a maximal torus of A if and only if $\overline{K} \otimes_K T$ is a maximal torus of $\overline{K} \otimes_K A$.

For an arbitrary element $a \in A$ the set $T(a)$ of the elements of $K[a]$ which are separable over K form a torus. (The element b is called separable over K iff the minimal polynomial of b over K is a separable polynomial.) We remark that if $a \in M_n(K)$ is a matrix then $T(a) = K[a^*]$ and $\dim_K T(a) = d(a)$. Since generators of tori over sufficiently large fields (in particular, over $\overline{K}$) do exist we obtain the following.

Observation 1 *Let $A \leq M_n(K)$ be a matrix algebra containing the identity matrix. Then the matrix $a \in A$ is a splitting element of A if and only if $T(a)$ is a maximal torus of A.*

The discussion of this section implies that we can efficiently find a splitting element in $A \leq M_n(K)$ provided that we have an efficient algorithm for finding a maximal torus and $|K| \geq n$.

3   The main algorithm

Every torus containing a fixed torus T is obviously a subtorus of the centralizer $C_A(T)$ of T. Also, if T is not a maximal torus then there exists a semisimple matrix $a \in C_A(T) \setminus T$ and the subalgebra generated by T and a is a torus bigger than T. The following lemma provides us with an efficient way for finding such a matrix a.

Lemma 2 *Let A be a finite dimensional algebra over K, let $T \leq A$ be a torus in A, and let $S \subseteq A$ be a subset of A such that $T + TS = C_A(T)$ (e.g., S is a basis of $C_A(T)$). Then T is a maximal torus of A if and only if $T(b) \leq T$ and $T(b + c) \leq T$ for every $b, c \in S$.*

Proof. The "only if part" of the statement is obvious. To prove the reverse implication assume that $C_A(T) = T + TS$ and $T(b), T(b + c) \leq T$ for every $b, c \in S$. These conditions remain true if we extend the ground field to $\overline{K}$ and it is sufficient to prove that $\overline{K} \otimes_K T$ is a maximal $\overline{K}$-torus in $\overline{K} \otimes_K A$. Hence we can and will assume that K is algebraically closed. For every $b \in S$ we consider the Jordan decomposition $b = b_s + b_n$ of b. Since b_s generates $T(b)$ our assumption is equivalent to that b is a sum of an element of T and a nilpotent element $u(b) = b_n$. Similarly, $b + c$ is the sum of an element of T and a nilpotent element $u(b + c)$. We set $S' = \{u(b) | b \in S\}$. Then $C_A(T) = T + TS'$.

Let $\phi : C_A(T) \rightarrow M_m(K)$ be an arbitrary irreducible matrix representation of $C_A(T)$. Then $\phi(C_A(T)) = M_m(K)$ and $\phi(T) \leq Z(M_m(K)) = KI$, whence $\phi(T) = KI$ (here, I stands for the appropriate identity matrix). For every $u \in S'$, $\phi(u)$ is a nilpotent matrix and the linear span of $\phi(S') \cup \{I\}$ is $M_m(K)$. This is impossible if m is not a unit in K because in that case the identity matrix I as well as every nilpotent matrix has zero trace. Hence m is a unit in K and therefore the only scalar matrix of trace 0 is zero. For every $u, v \in \phi(S')$ the sum $u + v$ is required to be the sum of a scalar matrix and a nilpotent matrix. By comparing traces it is immediate that $u + v$ must be is nilpotent.

Assume that $m > 1$. For an arbitrary matrix $a \in M_m(K)$ we consider the characteristic polynomial $\det(a + tI) = t^m + c_1(a)t^{m-1} + c_2(a)t^{m-2} +$ terms of lower degree. Note that $c_1(a) = Tr(a)$. Amitsur's formulas [A] for the coefficients of the characteristic polynomial of a sum tell us that

$$c_2(u + v) = c_2(u) + c_2(v) + c_1(u) + c_1(v) - c_1(uv).$$

Since the matrices $u, v, u + v$ are nilpotent we have $c_2(u + v) = c_2(u) = c_2(v) = c_1(u) = c_1(v) = 0$ and hence $c_1(uv) = 0$ for every $u, v \in \phi(S')$. We obtained that $Tr(u) = Tr(uv) = 0$ for every $u, v \in \phi(S')$. By (bi-)linearity, these equalities extend to the whole linear span U of $\phi(S')$. On the other hand, from $M_m(K) = KI + U$ we infer that U must coincide with the set $sl_m(K)$ of matrices of trace zero. This is a contradiction because if $m > 1$ then $sl_m(K)$ does contain matrices u, v with $Tr(uv) \neq 0$.

What we have seen so far is that every irreducible representation ϕ of $C_A(T)$ is linear and hence $\phi(S') = (0)$. This means that $S' \subseteq Rad(C_A(T))$ therefore $TS \subseteq Rad(C_A(T))$ and $C_A(T) = T + Rad(C_A(T))$. From this we infer that T is a maximal torus of $C_A(T)$. Hence it is immediate that T is a maximal torus of A as well. $\square$

As a consequence we obtain the following.

Theorem 3 *Assume that the matrix algebra $A \leq M_n(K)$ containing the identity matrix is given by a set of matrices which form a K-basis of A. Then a basis of a maximal torus T of A can be computed by an algorithm using $n^{O(1)}$ arithmetical operations in K.*

Proof. We start with (a basis of) an initial torus T, say the trivial one generated by the identity and iterate the following step until T is proved to be maximal. We compute a basis S of the centralizer $C_A(T)$ of T by solving the system of linear equations $xa - ax = 0$ where a runs over the basis of T. For every pair $b, c \in S \cup \{0\}$ we compute $(b + c)^*$ using the method described in Section 2. Then we test whether $(b + c)^* \in T$. Again, this can be done by solving a system of linear equations. If $(b + c)^* \notin T$ for some pair b, c then we calculate a basis of the bigger torus generated by $T \cup \{b + c\}^*$ and proceed with it in place of T. Otherwise we can terminate as T is a maximal torus by the lemma. Obviously, the algorithm terminates in $O(n)$ rounds. The systems of linear equations involved are all of dimension $O(n^2)$ therefore they can be solved using $n^{O(1)}$ operations. $\square$

Unfortunately we cannot prove that the Boolean cost of this algorithm (over an infinite field K with efficient linear algebra) is polynomial. The point is that we are unable to give a polynomial bound on the sizes of the tori calculated in the course of the algorithm because of the iterative centralizer computation. However, as size problems over finite fields do not occur we immediately obtain the following.

Corollary 4 *Assume that K is a finite field. Then a maximal torus of A can be found in polynomial time. Furthermore, if $|K| \geq n$, then a splitting element of A can be computed in polynomial time.*

4   Reduction of sizes

In this section we describe a modification of the algorithm which runs in polynomial time over infinite ground fields with efficient linear algebra. The approach is analogous to the reduction technique of [GIR]. The idea is based on the fact that $d(a)$ can be described by polynomial functions on $M_n(K)$.

We observe that if char $K = 0$ or char $K > n$ then for an arbitrary matrix $a \in M_n(K)$ we have $d(a) = \dim_K(K[a]/Rad(K[a]))$ and

$$Rad(K[a]) = \{b \in K[a] | Tr(a^i b) = 0 \ \ (i = 0, 1, \ldots, n - 1)\}$$

(c.f. [D]). From this it is immediate that $d(a)$ is the rank of the Bezoutian $Tr(a^i a^j)_{i,j=0}^{n-1}$.

Unfortunately this does not work if $0 < \text{char } K = p \leq n$. In that case we have to apply the cruder approach based on $d(a) = \dim_K K[a^*]$. We obtain that $d(a) < k$ for some $1 < k \leq n$ if and only if the matrices $I = a^{*0}, a^{*1}, \ldots, a^{*k-1}$ are linearly dependent. Recall that $a^* = a^{p^{\lceil \log_p n \rceil}}$.

Lemma 5 *Set $N = n(n - 1)$ if $0 < \text{char } K \leq n$, and $N = n(n - 1)^3/2$ otherwise. Assume that $|K| > N$ and let Ω be a subset of $K \setminus \{0\}$ such that $|\Omega| > N$. Let $a, b \in M_n(K)$ two matrices. Then there exists an element $\omega \in \Omega$ such that $d(a + \omega b) \geq d(a)$.*

Proof. Assume that $d(a) = k$. If char $K = 0$ or char $K > n$ then let $\mathcal{F} \subseteq K[t]$ be the set of the determinants of the k by k minors of the matrix $Tr((a + tb)^i(a + tb)^j))_{i,j=0}^{n-1} \in M_n(K[t])$. The degrees of the polynomials in $\mathcal{F}$ are bounded by $2\sum_{j=n-k}^{n-1} j = k(2n - k - 1) \le n(n - 1) = N$. We have $d(a + \omega b) < k$ if and only if ω is a common zero of $\mathcal{F}$.

If $0 < \text{char } K = p \le n$ then set $q = p^{\lceil \log_p n \rceil}$ and let $\mathcal{F} \subseteq K[t]$ consist of the determinants of the k by k minors of the matrix the rows of which are $(a + tb)^{q^0}, (a + tb)^{q^1} \ldots, (a + tb)^{q^{(k-1)}}$, each considered as a row vector of length n^2. Now the degree bound is $p^l k(k - 1)/2 \le (n - 1)^2 k(k - 1)/2 \le n(n - 1)^3/2 = N$. Again, $d(a + \omega b) < k$ if and only ω is a common zero of $\mathcal{F}$.

In both cases, $d(a) = k$ therefore there exists at least one polynomial $f(t) \in \mathcal{F}$ such that $f(0) \ne 0$. In particular $f(t)$ is not the zero polynomial. Since $\deg f(t) \le N$, it is impossible that $f(t)$ vanishes on the set Ω of cardinality greater then N. $\square$

We are ready to outline the modified algorithm. First we choose a set $\Omega \subseteq K$ of cardinality $N + 1$ consisting of elements of "small" size. (In practice, if char $K = 0$ then we can take $\Omega = \{0, \ldots, N\}$. If K is a function field over a finite field, we take $N + 1$ polynomials of smallest degree.) Assume that A is given by the basis $b_1, \ldots, b_r$. Initially we can take $a = I$. We compute a basis S of the centralizer $C_A(a^*) = C_A(T(a))$ and use the test of Lemma 2. We either terminate with the conclusion that $T(a)$ is a maximal torus (or, equivalently a is a splitting element) or find an element b of the form $b = (b' + c')$ $(b', c' \in S \cup \{0\})$ such that $b^* \notin T(a)$. Then we use the method described in Section 2 to find an element $a' \in a^* + \Omega b^*$ such that $d(a') > d(a)$. We express a' in terms of the basis $b_1, \ldots, b_r$: $a' = \alpha_1 b_1 + \ldots + \alpha_r b_r$. Then, for the indices $i = 1$ in increasing order we do the following. If $\alpha_i \notin \Omega$ we replace a' with an element $a' + (\omega - \alpha_i)b_i$ $(\omega \in \Omega)$ such that $d(a' + (\omega - \alpha_i)b_i) \ge d(a)$. By Lemma 5, such an element exists and obviously can be found using $O(N) = n^{O(1)}$ trials. We achieve that the new a' is of the form $\omega_1 b_1 + \ldots + \omega_r b_r$ $(\omega_i \in \Omega)$ and $d(a') > d(a)$. We proceed with a' in place of a. The "small" coordinates guarantee that the sizes of matrices we compute with in the course of the algorithm remain below a bound which is polynomial in the size of the input. We have proved the following.

Theorem 6 *Assume that K is an infinite field with efficient linear algebra. Then a splitting element (and a maximal torus) of A can be computed in polynomial time.*

5   Evaluation

In this section we compare the method described in Section 3 to two other methods. The first of these is the randomized method mentioned briefly in the introduction. Let Ω be a sufficiently large subset of the ground field K. Let $\{a_1, \ldots, a_m\}$ be the basis of the algebra A that is input to the algorithm. Choose randomly and uniformly m elements α_i from Ω $(1 \leq i \leq m)$. Set $a = \sum_i \alpha_i a_i$. Then with large probability a^* generates a maximal torus of A. Note that this algorithm is probabilistic of Monte Carlo type: the output is correct with a probability that can be made arbitrarily close to 1.

The second method uses the Lie algebra structure of A. Let A_L denote the vector space A equipped with the multiplication $[\,,\,]: A_L \times A_L \to A_L$ defined by $[a, b] = ab - ba$. Then it is well known that A_L is a Lie algebra. First we calculate a Cartan subalgebra H of L (see [GIR]), then by Theorem 4.5.17 of [W], H is the centralizer of a maximal torus T of A. Then we calculate the centre $C(H)$ of H (see [BKS]). Then $T \subset C(H)$ and since $C(H)$ is a space of commuting matrices, we have that T is spanned by a^* where a runs through a basis of $C(H)$.

We have tested the methods on the algebras $M_n(\mathbf{Q})$ for various values of n. The results are displayed in Table 1.

n	$\dim(A)$	Deterministic	Cartan	Random
8	64	8	13	3
9	81	9	26	7
10	100	16	50	14
11	121	24	93	29
12	144	31	160	58

Table 1: Running times (in seconds) for the calculation of a maximal torus of the associative algebra of all $n \times n$ matrices over $\mathbf{Q}$. "Deterministic" denotes the method described in Section 3, "Cartan" is the method using Cartan subalgebras and "Random" refers to the randomized method.

One can observe that the deterministic method performs quite well. The randomized method is very fast for small examples, but the running times of this method increase rapidly with the dimension. This is caused by the fact that a random element a will be a dense matrix (i.e., a matrix with many nonzero entries). This makes the calculation of a^* and of a basis of the torus generated by a^* very time consuming. The deterministic method, on the contrary, works mainly with sparse matrices (containing many zeros). This reduces the time needed for the linear algebra computations. Finally it is also apparent from the table that calculating a Cartan subalgebra of A_L is rather expensive compared to the other methods.

Acknowledgement

The authors are grateful to Lajos Rónyai for his useful remarks and suggestions.

References

[A]   S. A. Amitsur, On the characteristic polynomial of a sum of matrices, *Linear and Multilinear Algebra 8 (1980) 177–182.*

[Ba]   J. R. Bastida, Field Extensions and Galois Theory, in *Rota, G-C. (ed), Encyclopedia of Mathematics and Its Applications, Vol. 22. Cambridge University Press and Addison-Wesley, 1984.*

[BBCIL] L. Babai, R. Beals, J.-Y. Cai, G. Ivanyos, E. M. Luks, Multiplicative equations over commuting matrices, *Proc. 7th ACM-SIAM Symp. on Discrete Algorithms, (1996), 498-507.*

[Be]   R. Beals, Algorithms for matrix groups and the Tits alternative, *Proc. 36th IEEE FOCS, (1995), 593-602.*

[BKS] Beck, R.E., Kolman, B. and Stewart, I.N. Computing the structure of a Lie algebra. in: Beck, R.E. and Kolman, B. eds., *Computers in Non-associative Rings and Algebras*, Academic Press, New York, 167–188 (1977).

[BP]   D. Bini, V. Pan, *Polynomial and matrix computations, Vol 1 (Fundamental algorithms) (Birkhäuser, Basel, 1994).*

[BR]   L. Babai, L. Rónyai, Computing irreducible representations of finite groups, *Mathematics of Computation 55, 192 (1990), 705-722.*

[D]   L. E. Dickson, Algebras and their arithmetics, University of Chicago, Chicago, 1923.

[E1]   W. M. Eberly, Computations for algebras and group representations, *Ph. D. Thesis, Dept. of Computer Science, University of Toronto, 1989.*

[E2]   W. M. Eberly, Decomposition of algebras over finite fields and number fields, *Computational Complexity 1 (1991) 179-206.*

[E3]   W. M. Eberly, Decompositions of algebras over **R** and **C**, *Computational Complexity 1, (1991), 207-230.*

[F]   M. A. Frumkin; Polynomial time algorithms in the theory of linear diophantine equations, *Proc. of FCT'76, Lecture Notes in Computer Science 56, Springer-Verlag, 1976, 386-392.*

[FR]   K. Friedl, L. Rónyai, Polynomial time solution of some problems in computational algebra, *Proc. 17th ACM STOC, (1985), 153-162.*

[GAP97] The GAP Group, Lehrstuhl D für Mathematik, RWTH Aachen, Germany and School of Mathematical and Computational Sciences, U. St. Andrews, Scotland. *GAP - Groups, Algorithms, and Programming, Version 4,* 1997.

[GIR] W. A. de Graaf, G. Ivanyos, L. Rónyai, Computing Cartan subalgebras of Lie algebras, *Applicable Algebra in Engineering, Communication and Computing 7, (1996), 71-90.*

[I] G. Ivanyos, Finding the radical of matrix algebras using Fitting decompositions, *Manuscript, 1998.*

[IR] G. Ivanyos, L. Rónyai, Finding maximal orders in semisimple algebras over **Q**, *Computational Complexity 3, (1993), 245-261.*

[IRSz] G. Ivanyos, L. Rónyai, Á. Szántó, Decomposition of algebras over $F_q(X_1, \ldots, X_m)$, *Applicable Algebra in Engineering, Communication and Computing 5 (1994) 71-90.*

[P] R. S. Pierce, Associative algebras *Springer-Verlag, Berlin, 1982.*

[Ró] L. Rónyai, A deterministic method for computing splitting elements in semisimple algebras over Q, *Journal of Algorithms 16, (1994), 24-32.*

for verification

[W] D. J. Winter, Abstract Lie Algebras, *M.I.T. Press, Cambridge, Mass., 1972.*

Cohen–Macaulay Modules and Vector Bundles

YURI A. DROZD

DEPARTMENT OF MECHANICS AND MATHEMATICS, KIEV TARAS SHEVCHENKO
UNIVERSITY, 252033 KIEV, UKRAINE
E-mail address: drozd@uni-alg.kiev.ua

1. GENERALITIES

We refer to [33] or [4] for standard facts concerning the commutative
algebra.

Throughout this paper **R** will denote a commutative noetherian, local complete ring having no nonzero nilpotents, $\mathfrak{m}$ its maximal ideal and
$\mathbf{k} = \mathbf{R}/\mathfrak{m}$ its residue field. All rings are supposed to be **R**-algebras; $\otimes$
and Hom denote, respectively, $\otimes_{\mathbf{R}}$ and $\mathrm{Hom}_{\mathbf{R}}$. 'Category' also means
an **R**-category, i.e., such that all morphism sets are **R**-modules and
the multiplication is **R**-bilinear. Respectively, all functors are supposed **R**-linear. A *module* over a category $\mathcal{A}$ is, by definition, a functor
$\mathcal{A} \to \mathbf{R}$-Mod, the category of (left) **R**-modules. A category $\mathcal{A}$ is said
to be *fully additive* if it is additive and each idempotent in $\mathcal{A}$ splits,
i.e., corresponds to a direct decomposition of the object. For each cat-
egory $\mathcal{A}$, denote add $\mathcal{A}$ its *additive hull*, i.e., the smallest fully additive
category containing $\mathcal{A}$. Remind that such a category always exists and
is equivalent to the category $\mathcal{A}$-pro of finitely generated projective left
$\mathcal{A}$-modules. In particular, this is the case when $\mathcal{A}$ is just an **R**-algebra.

Let M be an **R**-module. Remind that a sequence $(a_1, a_2, \ldots, a_n)$ of
elements of $\mathfrak{m}$ is called an *M-sequence* if, for each $i = 1, \ldots, n$, the element a_i is not a zero divisor on the factor-module $M/(a_1, \ldots, a_{i-1})M$.
A finitely generated **R**-module M is said to be a *maximal Cohen–
Macaulay* **R**-*module* if there is an M-sequence containing $d = \mathrm{Kr.\,dim}\,\mathbf{R}$
elements. Later on we omit the word 'maximal' and simply say *Cohen–
Macaulay module*. In particular, if **R** is itself a Cohen–Macaulay **R**-
module, one says that **R** is a Cohen–Macaulay ring. An **R**-algebra **A**
(not necessary commutative) is said to be a *Cohen–Macaulay* **R**-*algebra*
(or simply Cohen–Macaulay algebra, if **R** is fixed) if it is Cohen–
Macaulay as an **R**-module. Given a Cohen–Macaulay **R**-algebra **A**,
we call an **A**-module M a *Cohen–Macaulay* **A**-*module* provided it is
(maximal) Cohen–Macaulay as an **R**-module. If we suppose that **A** is

This work was performed in collaboration with G.-M. Greuel and under partial
support of DFG and CRDF, Grant UM1-327.

faithful as an **R**-module, these notions do not depend on the choice of **R**; in particular, **A** is a Cohen–Macaulay algebra over **R** if and only if it is Cohen–Macaulay over its centre. For each Cohen–Macaulay algebra **A** denote by $\mathsf{Cm}(\mathbf{A})$ the category of all Cohen–Macaulay **A**-modules.

Denote by $\widetilde{\mathbf{R}}$ the full ring of fractions of **R** and consider, for every **R**-module M, the natural map $M \to \widetilde{M} = \widetilde{\mathbf{R}} \otimes M$. It is an embedding if and only if the module M is *torsion free*, i.e., such that $au \neq 0$ for each nonzero element u of the module and each non-zero-divisor $a \in R$. In particular, this is the case when the module M is Cohen-Macaulay. We always identify a torsion free module M with its image in $\widetilde{M}$. Then for any set P of prime ideals from **R**, denote by M_P the submodule $\left\{ u \in \widetilde{M} \mid i_{\mathfrak{p}}(u) \in M_{\mathfrak{p}} \text{ for all } \mathfrak{p} \in P \right\}$ of $\widetilde{M}$, where $i_{\mathfrak{p}}$ denotes the natural homomorphism $\widetilde{M} \to \widetilde{M_{\mathfrak{p}}}$.

If **A** is a Cohen–Macaulay **R**-algebra, $\widetilde{\mathbf{A}}$ is an artinian $\widetilde{\mathbf{R}}$-algebra. Call an *over-ring* of **A** any other Cohen–Macaulay **R**-algebra **A**′ such that $\mathbf{A} \subseteq \mathbf{A}' \subset \widetilde{\mathbf{A}}$. In this case $\mathsf{Cm}(\mathbf{A}')$ is a full subcategory in $\mathsf{Cm}(\mathbf{A})$. If **A**′ is isomorphic to an over-ring of **A**, one says that **A**′ *dominates* **A**. We usually consider only *semi-prime* algebras **A**, i.e., those having no nilpotent ideals. As we suppose **A** being **R**-faithful, then **R** contains no nilpotents. In this case the algebra $\widetilde{\mathbf{A}}$ is semi-simple. Let $\left\{ \widetilde{\mathbf{A}}_i \mid 1 \leq i \leq s \right\}$ be its simple components and let S_i be the simple $\widetilde{\mathbf{A}}_i$-module. If M is a Cohen–Macaulay $\widetilde{\mathbf{A}}$-module, then $\widetilde{M}$ is an $\widetilde{\mathbf{A}}$-module, hence $\widetilde{M} \simeq \bigoplus_{i=1}^{s} r_i S_i$ for some multiplicities r_i. Call the vector $\mathbf{r}(M) = (r_1, r_2, \ldots, r_s)$ the *rank-vector* of the module M and denote by $\mathsf{Cm}_{\mathbf{r}}(\mathbf{A})$ the set of all Cohen–Macaulay **A**-modules having the prescribed rank-vector **r**. We also denote by $\mathsf{Cmi}(\mathbf{A})$ the set of isomorphism classes of indecomposable Cohen–Macaulay **A**-modules and by $\mathsf{Cmi}_{\mathbf{r}}(\mathbf{A})$ its subset consisting of those having the rank-vector **r**.

We are mostly interested in the cases when $d = 1$ or $d = 2$. It is well-known (and easy to see) that if $\mathrm{Kr.\,dim}\,R = 1$, **R** is always a Cohen–Macaulay ring and Cohen–Macaulay modules are just (finitely generated) torsion free ones. On the other hand, each normal ring **R** of Krull dimension 2 is Cohen–Macaulay and in this case an **R**-module M is Cohen–Macaulay if and only if it is *reflexive*, i.e., the natural homomorphism $M \to M^{**}$ is an isomorphism, where $M^* = \mathrm{Hom}(M, \mathbf{R})$. An equivalent condition is that M is torsion free and, moreover, coincides with $M_{\mathfrak{P}}$ where $\mathfrak{P}$ is the set of all prime ideals of height 1 in **R**. Remark that the last condition automatically holds for torsion-free modules if $d = 1$. This enable us to construct a left adjoint functor to the embedding $\mathrm{Cm}(\mathbf{A}) \to \mathrm{Cm}(\mathbf{A}')$, where $\mathbf{A}'$ is an over-ring of $\mathbf{A}$. Namely, this functor maps a Cohen–Macaulay $\mathbf{A}$-module M to the "Cohen–Macaulayzation" of $\mathbf{A}'M$, i.e., to $(\mathbf{A}'M)_{\mathfrak{P}}$ (in Krull dimension 1 the latter module is just $\mathbf{A}'M$ itself).

To define representation types of Cohen–Macaulay algebras we need some technical preparation, especially for cases when there is no ground field (for instance, when $\mathbf{R} = \mathbb{Z}_p$, the classical case of "p-adic representations"). Call a *correspondence* between two categories, $\mathcal{A}$ and $\mathcal{B}$, a pair of functors $F = (F_{\mathcal{A}}, F_{\mathcal{B}})$, where $F_{\mathcal{A}} : \mathcal{C} \to \mathcal{A}$, $F_{\mathcal{B}} : \mathcal{C} \to \mathcal{B}$. We write $F : \mathcal{A} \leftrightarrow \mathcal{B}$. For an object $x \in \mathrm{Ob}\,\mathcal{A}$, denote $Fx = \{\, y \in \mathrm{Ob}\,\mathcal{B} \,|\, (\exists z \in \mathrm{Ob}\,\mathcal{B})(F_{\mathcal{A}}z = x,\, F_{\mathcal{B}}z = y) \,\}$. In the same way Fy is defined for $y \in \mathrm{Ob}\,\mathcal{B}$.

We say that such a correspondence:

- *reflects isomorphisms in $\mathcal{A}$* if, given any $x, x' \in \mathrm{Ob}\,\mathcal{A}$, $y \in Fx$ and $y' \in Fx'$, $y \simeq y'$ implies $x \simeq x'$;
- *preserves indecomposability in $\mathcal{A}$* if, given any indecomposable $x \in \mathrm{Ob}\,\mathcal{A}$, all objects from Fx are also indecomposable.
- is *dense in $\mathcal{A}$* if each $x \in \mathrm{Ob}\,\mathcal{A}$ is isomorphic to an object from Fy for some $y \in \mathrm{Ob}\,\mathcal{B}$.

Quite in the same way, the "$\mathcal{B}$-analogues" of all these notions are defined. It is evident that whenever F is dense in $\mathcal{B}$ and reflects isomorphisms in $\mathcal{A}$, it also preserves indecomposability in $\mathcal{A}$.

If a correspondence reflects isomorphisms and preserves indecomposability in $\mathcal{A}$, we call it a *representation embedding* of $\mathcal{A}$ into $\mathcal{B}$. If it is dense and reflects isomorphisms both in $\mathcal{A}$ and in $\mathcal{B}$ (hence, preserves indecomposability both in $\mathcal{A}$ and in $\mathcal{B}$), call it a *representation equivalence* between $\mathcal{A}$ and $\mathcal{B}$.

Certainly, each functor $F : \mathcal{A} \to \mathcal{B}$ defines a correspondence (Id, F) between $\mathcal{A}$ and $\mathcal{B}$, which is automatically dense in $\mathcal{A}$, preserves indecomposability and reflects isomorphisms in $\mathcal{B}$. So, in this case, we can omit the names of categories, e.g., say that F reflects isomorphisms if this correspondence reflects isomorphisms in $\mathcal{A}$, etc.

Remark 1.1. In what follows, we often consider $\mathbf{A}$-$\mathbf{B}$-bimodules U, where $\mathbf{A}$ and $\mathbf{B}$ are two $\mathbf{R}$-algebras. Usually it does not mean that it is an *algebra-bimodule* (over $\mathbf{R}$), i.e., that the elements of $\mathbf{R}$ act in the same way on the right and on the left. Moreover, in a lot of important cases (e.g., in the "arithmetic" case, when $\mathbf{R}$ is the ring of p-adic integers) our bimodules are almost never of the latter type. Nevertheless, we need and always suppose satisfied the following condition:

$\mathbf{m}U = U\mathbf{m}$, *where* $\mathbf{m}$ *is the maximal ideal of* $\mathbf{R}$, *and* $ru \equiv ur$ (mod $\mathbf{m}U$) *for all elements* $r \in \mathbf{R}$ *and* $u \in U$.

Let $\mathbf{S}$ be an $\mathbf{R}$-algebra. Denote by $\mathsf{Cm}(\mathbf{A}, \mathbf{S})$ the full subcategory of the category of $\mathbf{S}$-$\mathbf{A}$-bimodules consisting of all bimodules $_{\mathbf{A}}F_{\mathbf{S}}$ satisfying the following conditions:

1. F is finitely generated (as bimodule).
2. $F_{\mathbf{S}}$ is flat.
3. There is an F-sequence $(a_1, a_2, \ldots, a_d)$ in $\mathbf{R}$ (where $d = \mathrm{Kr.\,dim\,}\mathbf{R}$) such that all $\mathbf{S}$-modules $F/(a_1, a_2, \ldots, a_m)F$ $(1 \leq m \leq d)$ are flat.

In this case, for every $\mathbf{S}$-module L which is of finite length over $\mathbf{R}$, the $\mathbf{A}$-module $F \otimes_{\mathbf{S}} L$ is a Cohen–Macaulay $\mathbf{A}$-module. Hence, we get a functor $F \otimes : \mathbf{S}\text{-mod} \to \mathsf{Cm}(\mathbf{A})$, where $\mathbf{S}$-mod denotes the category of (left) $\mathbf{S}$-modules which are of finite length over $\mathbf{R}$.

Put $\overline{\mathbf{S}} = \mathbf{S}/\mathfrak{m}\mathbf{S}$ and $\overline{M} = M/\mathfrak{m}M$ for each $\mathbf{S}$-module M. The functors $^{-} : \mathbf{S}\text{-mod} \to \overline{\mathbf{S}}\text{-mod}$ and $F \otimes$ define a correspondence between $\overline{\mathbf{S}}$-mod and $\mathsf{Cm}(\mathbf{A})$. We denote this correspondence by $\hat{F}$. Call the bimodule F *strict* if this correspondence reflects isomorphisms and preserves indecomposability in $\overline{\mathbf{S}}$-mod, i.e., is a representation embedding of $\overline{\mathbf{S}}$-mod into $\mathsf{Cm}(\mathbf{A})$. If $\mathbf{S}$ is indeed a $\mathbf{k}$-algebra, this notion coincides with that of 'strict bimodule' defined in [15].

Let now $\mathbf{F} = \{F_i\}$ be a set of bimodules, $F_i \in \mathsf{Cm}(\mathbf{A}, \mathbf{S}_i)$. We say that this set is *parametrizing* if, given any rank-vector $\mathbf{r}$, almost all (i.e., all but a finite number, up to isomorphism) indecomposable modules from $\mathsf{Cmi}_{\mathbf{r}}(\mathbf{A})$ are isomorphic to those from $\hat{F}_i X$ for some i and some $X \in \overline{\mathbf{S}}_i$-mod.

Define a *1-parametre family* of Cohen–Macaulay $\mathbf{A}$-modules as a bimodule $F \in \mathsf{Cm}(\mathbf{A}, \mathbf{S})$ such that:

- $\mathbf{S}$ is a commutative finitely generated flat $\mathbf{R}$-algebra and $\overline{\mathbf{S}}$ is a smooth $\mathbf{k}$-algebra of Krull dimension 1.
- The correspondence $\hat{F}$ is a representation embedding of $\overline{\mathbf{S}}$-mod into $\mathsf{Cm}(\mathbf{A})$.

The *rank-vector* $\mathbf{r}(F)$ of a 1-parametre family F is, by definition, that of F/FI, where I is a maximal ideal of $\mathbf{S}$ (it is easy to see that it does not depend on the choice of I). Given any set $\mathbf{F}$ of 1-parametre families, denote by $\nu(\mathbf{r}, \mathbf{F})$ the number of bimodules from $\mathbf{F}$ having the rank vector $\mathbf{r}$. Call such a set $\mathbf{F}$

- *locally finite* if $\nu(\mathbf{r}, \mathbf{F}) < \infty$ for all $\mathbf{r}$.
- *bounded* if there is a constant N such that $\nu(\mathbf{r}, \mathbf{F}) < N$ for all $\mathbf{r}$.

Now we are ready to define *Cohen–Macaulay types* of Cohen–Macaulay algebras.

Definition 1.2. *Let $\mathbf{A}$ be a Cohen–Macaulay algebra. Call it:*

1. Cohen–Macaulay finite *if $\mathsf{Cmi}(\mathbf{A})$ is finite.*

2. Cohen–Macaulay tame *if there is a non-empty locally finite para-metrizing set of 1-parametre families of Cohen–Macaulay* **A**-*modules. In this case, call* **A**:
    - bounded *if there is a bounded parametrizing set of 1-parametre families of Cohen–Macaulay* **A**-*modules;*
    - unbounded *otherwise.*
3. Cohen–Macaulay wild *if, for every finitely generated* **R**-*algebra* **S**, *there is a strict bimodule* $F \in \mathsf{Cm}(\mathbf{A}, \mathbf{S})$.

Remind the following useful result (cf. [23, 12]).

Proposition 1.3. *The following properties are equivalent:*

1. **A** *is wild.*
2. *There is a strict bimodule* $F \in \mathsf{Cm}(\mathbf{A}, \mathbf{R}\langle x, y \rangle)$ *(free algebra in 2 generators).*
3. *There is a strict bimodule* $F \in \mathsf{Cm}(\mathbf{A}, \mathbf{R}[x, y])$.
4. *There is a strict bimodule* $F \in \mathsf{Cm}(\mathbf{A}, \mathbf{R}[[x, y]])$.
5. *There is a strict bimodule* $F \in \mathsf{Cm}(\mathbf{A}, \mathbf{R}[x, y]/(x^2, y^3, xy^2))$.

Proof. $1 \Rightarrow 2 \Rightarrow 3 \Rightarrow 4 \Rightarrow 5$ is evident. So, we only have to prove $5 \Rightarrow 1$. Let $\Phi = \mathbf{R}[x, y](x^2, y^3, xy^2)$, $F \in \mathsf{Cm}(\mathbf{A}, \Phi)$ be a strict bimodule and $\mathbf{S} = \mathbf{R}\langle a_1, a_2, \ldots, a_m \rangle$ be a finitely generated **R**-algebra. Consider the Φ-**S**-bimodule M defined as follows. As right **S**-module, M is free of rank $9n$; elements of **R** commute with elements of M, while the action of x and y is given, respectively, by the following two matrices X and Y:

$$X = \begin{pmatrix} 0 & 0 & I_{4n} \\ 0 & 0 & 0 \\ 0 & 0 & 0 \end{pmatrix}, Y = \begin{pmatrix} Y_1 & 0 & Y_2 \\ 0 & 0 & Y_3 \\ 0 & 0 & Y_1 \end{pmatrix},$$

where

$$Y_1 = \begin{pmatrix} 0 & 0 & 0 & I_n \\ 0 & 0 & 0 & 0 \\ 0 & 0 & 0 & 0 \\ 0 & 0 & 0 & 0 \end{pmatrix}, Y_2 = \begin{pmatrix} 0 & 0 & 0 & 0 \\ I_n & 0 & 0 & 0 \\ 0 & I_n & 0 & 0 \\ 0 & J & D & 0 \end{pmatrix}, Y_3 = \begin{pmatrix} 0 & 0 & I_n & 0 \end{pmatrix}.$$

Here I_m denotes, as usually, the $m \times m$ identity matrix, J is the nilpotent Jordan block of size $n \times n$ and $D = \mathrm{diag}(a_1, a_2, \ldots, a_n)$. A straitforward calculation, like in [12], shows that the functor $M\otimes$ reflects isomorphisms and preserves indecomposability in **S**-mod. Hence, $F \otimes_\Phi M$ is a strict bimodule from $\mathsf{Cm}(\mathbf{A}, \mathbf{S})$. $\square$

2. One-dimensional case: technique

In this section and in the next one, we consider the case of Krull dimension 1. Remind relations existing in this case between Cohen–Macaulay modules and "bimodule problems," first used (in other terms), probably, by Jacobinski [28] and then developed systematically in [24, 34, 15]. First of all, remind the definitions related to bimodule problems themselves [11, 14]. Just as in [14], we only consider *bipartite bimodules*. It is also more convenient here to consider bimodules over categories further than over rings.

Suppose given two categories, $\mathcal{A}$ and $\mathcal{B}$, and an $\mathcal{A}$-$\mathcal{B}$-bimodule U, i.e., a biadditive functor $\mathsf{U} : \mathcal{B}^\circ \times \mathcal{A} \to \mathrm{Ab}$. We can always prolong it to the additive hulls $\mathrm{add}\,\mathcal{A}$ and $\mathrm{add}\,\mathcal{B}$. Consider the category of elements of U, whose objects are indeed the elements of $\mathsf{U}(B, A)$ ($B \in \mathrm{add}\,\mathcal{B}$ and $A \in \mathrm{add}\,\mathcal{A}$) and a morphism from $u \in \mathsf{U}(B, A)$ to $u' \in \mathsf{U}(B', A')$ is, by definition, a pair of morphisms $\beta : B \to B'$, $\alpha : A \to A'$ such that $u'\beta = \alpha u$. Here (and later on) we write αu instead of $\mathsf{U}(\mathrm{id}_B, \alpha)u$, etc. In this way we obtain the category $\mathrm{El}(\mathsf{U})$. (In [11] the elements of U were called 'U-matrices.')

Suppose now that U is indeed an *algebra-bimodule* over $\mathbf{R}$, i.e., $ru = ur$ for all $r \in \mathbf{R}$, $u \in \mathsf{U}$. For any $\mathbf{R}$-algebra $\mathbf{S}$, consider the $\mathcal{B}\otimes\mathbf{S}$-$\mathcal{A}\otimes\mathbf{S}$-bimodule $\mathsf{U}^{\mathbf{S}} = \mathsf{U}\otimes\mathbf{S}$. Given any element u of $\mathsf{U}^{\mathbf{S}}$, $u \in \mathsf{U}(B, A)$, and any *representation* of $\mathbf{S}$ over $\mathbf{R}$, i.e., an $\mathbf{S}$-module L finitely generated and projective over $\mathbf{R}$, one can get an element $u \otimes L \in \mathsf{U}(B \otimes_{\mathbf{S}} L, A \otimes_{\mathbf{S}} L)$ of U using the natural isomorphisms

$$A \otimes_{\Lambda\otimes\mathbf{S}} (\mathsf{U} \otimes S) \;\simeq\; A \otimes_\Lambda \mathsf{U}\,;$$
$$(A \otimes_\Lambda \mathsf{U}) \otimes_{\mathbf{S}} L \;\simeq\; (A \otimes_{\mathbf{S}} L) \otimes_\Lambda \mathsf{U}\,.$$

Hence, one can define the *matrix type* (or simply *type*) of a bimodule U, just in the same way as it has been done above for the Cohen–Macaulay type of Cohen–Macaulay algebras.

Let $\mathbf{A}$ be a Cohen–Macaulay algebra, $\mathbf{B} \supset \mathbf{A}$ its over-ring and $J \subset \mathbf{A}$ a $\mathbf{B}$-ideal such that $\mathbf{A}/J$ is artinian (it always exists, e.g., the conductor of $\mathbf{B}$ in $\mathbf{A}$). Denote $\overline{M} = M/JM$ for every $\mathbf{A}$-module M. Given any Cohen–Macaulay $\mathbf{A}$-module, we have inclusions $JM = JBM \subseteq M \subseteq BM$ or $\overline{M} \subseteq \overline{BM}$. Consider the category $\mathcal{I}_J$ whose objects are pairs (B, A), where B is a Cohen–Macaulay $\mathbf{B}$-module and A is a $\overline{\mathbf{A}}$-submodule in $\overline{B}$, while a morphism $f : (B, A) \to (B', A')$ is, by definition, a homomorphism $f : B \to B'$ such that $\overline{f}(A) \subseteq A'$. We have a natural functor $F_J : \mathrm{Cm}(\mathbf{A}) \to \mathcal{I}_J$, $F_J M = (\mathbf{B}, \overline{M})$. Let $\mathcal{I}_J^g$ be the full subcategory of $\mathcal{I}_J$ consisting of such pairs (B, A) that A generates $\overline{B}$ as $\overline{\mathbf{B}}$-module. The following assertion is then evident:

Proposition 2.1. F_J *induces an equivalence of the categories* $\mathsf{Cm}(\mathbf{A})$ *and* $\mathcal{I}_J^g$.

The category $\mathcal{I}_J$ has an advantage that it is closely related to one of the elements of a bimodule. Namely, consider the $(\overline{\mathbf{A}}\text{-pro})\text{-}\mathsf{Cm}(\mathbf{B})$-bimodule U such that $\mathsf{U}(P, B) = \mathrm{Hom}_{\overline{\mathbf{A}}}(P, \overline{B})$. For each $u : P \to \overline{B}$, $\mathrm{Im}\,u$ is an $\overline{\mathbf{A}}$-submodule in $\overline{B}$; it gives us the functor $\mathrm{Im} : \mathrm{El}(\mathsf{U}) \to \mathcal{I}_J$. Denote by $\mathrm{El}_0(\mathsf{U})$ the full subcategory of $\mathrm{El}(\mathsf{U})$ consisting of all homomorphisms $u : P \to \overline{B}$ such that $\ker u \subset \mathrm{rad}\,P$ and by $\mathrm{El}_0^g(\mathsf{U})$ the full subcategory of $\mathrm{El}_0(\mathsf{U})$ consisting of such u that $\mathrm{Im}\,u$ generates $\overline{B}$ as $\overline{\mathbf{B}}$-module. Then the following statement holds (the proof being again quite evident).

Proposition 2.2. *The functor* Im *defines a representation equivalence between* $\mathrm{El}_0(\mathsf{U})$ *and* $\mathcal{I}_J$ *as well as between* $\mathrm{El}_0^g(\mathsf{U})$ *and* $\mathcal{I}_J^g$.

Indeed, there is not a big difference between El and El_0, as any element of $\mathrm{El}(\mathsf{U})$ is a direct sum of an element of $\mathrm{El}_0(\mathsf{U})$ and a "trivial" one from $\mathsf{U}(P, 0)$ for some P. The indecomposable trivial elements are in one-to-one correspondence with the indecomposable projective $\overline{\mathbf{A}}$-modules, so there is only finitely many of them. In any case, they do not imply the matrix type. On the other hand, the difference between El_0 and El_0^g can be very big indeed. There are "natural" examples (even arising from Cohen–Macaulay modules) where El_0^g is of finite type, while El is wild. This is the case, for instance, when one consider the plane curve singularities of type E (cf. below).

Remark that the bimodule U is not faithful. On the contrary, consider the ideal J^* of the category $\mathsf{Cm}(\mathbf{B})$ such that $J^*(B, B') = \{f : B \to B' \mid \mathrm{Im}\,f \subseteq JB'\}$. Then $J^*\mathsf{U} = 0$, hence U is "indeed" a $(\overline{\mathbf{A}}\text{-pro})\text{-}\mathsf{Cm}(\mathbf{B})/J^*$-bimodule. In the category $\mathsf{Cm}(\mathbf{B})/J^*$ all Hom-sets are finite length modules over $\mathbf{R}/(\mathbf{R} \cap J)$; in particular, in "geometric" case, when $\mathbf{R}$ is a $\mathbf{k}$-algebra, they are finite dimensional over $\mathbf{k}$. Hence, we can then apply to $\mathrm{El}(\mathsf{U})$ all usual technique of "matrix problems."

The most usual choice for the ideal J is the (Jacobson) radical $\mathrm{rad}\,\mathbf{A}$. Then $\left\{ a \in \widetilde{\mathbf{A}} \mid aJ \cup Ja \in J \right\}$ is a proper over-ring of $\mathbf{A}$ provided $\mathbf{A}$ is not hereditary [18] and it is often chosen for $\mathbf{B}$. Moreover, in this case $J \supseteq \mathfrak{m}\mathbf{B}$, hence, the categories $\overline{\mathbf{A}}$-pro and $\mathsf{Cm}(\mathbf{B})/J^*$ are again finite dimensional categories over the filed $\mathbf{k}$ and U is a finite dimensional bimodule. In, particular, it allows to apply some elementary geometric observations, like "calculation of parametres" (i.e., dimensions of varieties). Sometimes it gives immediate results of the following kind (cf. [10, 14]).

Proposition 2.3. *Let P be an indecomposable $\mathbf{A}$-module, $l = \text{length}_{\widetilde{\mathbf{A}}}(\widetilde{L})$. If $l > 3$, $\mathbf{A}$ is of infinite Cohen–Macaulay type; if $l > 4$, $\mathbf{A}$ is Cohen–Macaulay wild.*

3. ONE-DIMENSIONAL CASE: COHEN–MACAULAY -TYPE

In this section we suppose that $\text{Kr. dim } \mathbf{R} = 1$ and $\mathbf{A}$ is commutative. Denote by $\mathbf{A}_0$ the unique maximal over-ring of $\mathbf{A}$, or, the same, its integral closure in $\widetilde{\mathbf{A}}$ (it is known [13] that this integral closure is indeed finitely generated as $\mathbf{A}$-module, hence, is an over-ring of $\mathbf{A}$). In this case there is a rather complete knowledge of the Cohen–Macaulay type of $\mathbf{A}$. First, the following result (cf. [21, 28]) gives a criterion for $\mathbf{A}$ to be Cohen–Macaulay finite. We denote by $\mu_{\mathbf{A}}(M)$ the minimal number of generators of an $\mathbf{A}$-module M.

Theorem 3.1. *Denote J the radical of $\mathbf{A}$, $\mathbf{A}_1 = J\mathbf{A}_0 + \mathbf{A}$. $\mathbf{A}$ is Cohen–Macaulay finite if and only if the following conditions hold:*

1. $\mu_{\mathbf{A}}(\mathbf{A}_0) \leq 3$.
2. $\mu_{\mathbf{A}}(\mathbf{A}_1) \leq 2$.

In geometric situation, when $\mathbf{R} = \mathbf{k}[[t]]$, a power series ring over an algebraically closed field $\mathbf{k}$, this criterion was reformulated in [25] in terms related to the theory of singularities. Indeed, in this case $\mathbf{A}$ can be considered as the comletion of a local ring of a point of an algebraic curve over $\mathbf{k}$. Call $\mathbf{A}$ a *plane singularity* if its maximal ideal has 2 generators (then indeed $\mathbf{A}$ is a completion of a point of a plane curve). In this case, $\mathbf{A} \simeq \mathbf{k}[[x,y]]/(f)$ for some power series $f(x,y)$. For the sake of simplicity, suppose that $\text{char } \mathbf{k} \neq 2$. The following plane singularities (given by the corresponding series $f(x,y)$) are called *simple* (or *0-modal*) as they have only finitely many non-isomorphic deformations (cf. [2]) (if $\text{char } \mathbf{k} = 2$, this list should be changed a little):

type A_n	$x^{n+1} + y^2$
type D_n	$x^{n-1} + xy^2$
type E_6	$x^4 + y^3$
type E_7	$x^3 y + y^3$
type E_8	$x^5 + y^3$

The names of these types correspond to their relations to simple Lie algebras, though we have no possibility to explain them here. The following result is a consequence of Theorem 3.1 (especially in the form of Jacobinski, who further gave not "over-ring conditions" but a list of Cohen–Macaulay finite rings.[1]

Corollary 3.2. *In geometric situation, **A** is Cohen–Macaulay finite if and only if it dominates one of the simple plane singularities.*

There is another list of the so called *unimodal* plane singularities (cf. [2]). It includes an infinite family ("type T_{pq}") given by the polynomials $f(x, y) = x^p + y^q + \lambda x^2 y^2$ with $1/p + 1/q \leq 1/2$ and 14 "special" singularities (we are not including their list here). In [16] the following result was proved.

Theorem 3.3. *Let* $\mathbf{R} = \mathbf{k}[[t]]$, *where* $\mathbf{k}$ *is an algebraically closed field,* $\mathbf{A}$ *be a Cohen–Macaulay algebra of infinite type. Then* $\mathbf{A}$ *is Cohen–Macaulay tame if and only if it dominates a plane singularity of type* T_{pq}. *Moreover, it is bounded if and only if it dominates* T_{44} *or* T_{36} *(the so called "parabolic" plane singularities, when* $1/p + 1/q = 1/2$).

The representation type of the singularities of types T_{36} and T_{44} was established by Dieterich [8, 9].

The proof in [16] was also based on some over-ring conditions. Namely, let J be the radical of $\mathbf{A}$, J_0 that of $\mathbf{A}_0$, $\mathbf{A}' = J_0 + \mathbf{A}$ (the maximal *local* over-ring of $\mathbf{A}$) and $\mathbf{A}'' = J\mathbf{A}' + \mathbf{A}$. Consider the following conditions:

Conditions 3.4.   1. $\mu_{\mathbf{A}}(\mathbf{A}_0) \leq 4$ *and* $J_0^2 \subseteq J\mathbf{A}_0$.
   2. $\mu_{\mathbf{A}}(\mathbf{A}') \leq 3$.
   3. *If* $\mu_{\mathbf{A}}(\mathbf{A}_0) = \mu_{\mathbf{A}}(\mathbf{A}') = 3$, *then* $\mu_{\mathbf{A}}(\mathbf{A}'') = 2$.

The following result was proved in [16].

Proposition 3.5. *If* $\mathbf{A}$ *is Cohen–Macaulay infinite but not Cohen–Macaulay wild, it satisfies Conditions 3.4.*

Indeed, the proof in [16] do not use the "geometric" specifics, so this result is also true in general case. In geometric situation, Conditions 3.4 imply that $\mathbf{A}$ dominates a plane singularity of type T_{pq}. To prove that all singularities of type T_{pq} are tame indeed, they used the technique of the deformation theory. Namely, they considered other singularities, called P_{pq}. Namely, $P_{pq} = \mathbf{k}[[x, y, z]]/(xy, x^p + y^q + z^2)$. Though P_{pq} is no more plane, it is a complete intersection, hence, Gorenstein. It is well-known then (cf. [6, 18]) that each indecomposable Cohen–Macaulay $\mathbf{A}$-module is either $\mathbf{A}$ itself or a $\mathbf{B}$-module, where $\mathbf{B}$ is the minimal over-ring of $\mathbf{A}$ (which is unique in Gorenstein case). In particular, $\mathbf{A}$ and $\mathbf{B}$ are of the same Cohen–Macaulay type. If $\mathbf{A} = P_{pq}$, the radical

---

[1]Though Jacobinski considered only "arithmetic case," when $\mathbf{R} = \mathbf{Z}_p$, his calculations are valid in general situation as well.

J of $\mathbf{B}$ is decomposable and its endomorphism ring $\mathbf{C}$ is a Bassian ring in the sense of [20, 18], i.e., all its over-rings are Gorenstein. Hence, all indecomposable Cohen–Macaulay $\mathbf{C}$-modules are direct summands of over-rings of $\mathbf{C}$ and it is easy to apply the procedure of the previous section to the ring $\mathbf{B}$, its over-ring $\mathbf{C}$ and the ideal J. It results in a bimodule problem known as a "bunch of semi-chains" (cf. [7]), which is tame. Therefore, $\mathbf{B}$ and $\mathbf{A}$ are Cohen–Macaulay tame.

Now one can easily check that the singularity T_{pq} is a deformation of P_{pq}. Hence, one can use a result of Knörrer [31] which implies that a deformation of a Cohen–Macaulay tame singularity is always either Cohen–Macaulay tame or Cohen–Macaulay finite. Therefore, the singularities T_{pq} and all their over-rings are Cohen–Macaulay tame too.

In non-geometric situation, one has analogues of the rings T_{pq} and P_{pq}. It is again true that Conditions 3.4 imply that $\mathbf{A}$ dominates a ring of type T_{pq} and that all rings of type P_{pq} are Cohen–Macaulay tame (the proofs are almost the same as in [16]). Nevertheless, we do not have any analogue of the Knörrer semi-continuity theorem, so cannot proceed as above.

By the way, the above criterion of Cohen–Macaulay finiteness can be generalized to non-commutative *local* Cohen–Macaulay algebras. To do it, one only has to replace the unique maximal over-ring of a commutative algebra by the *intersection* of all maximal over-rings of a non-commutative one. It gives the following result [19].

Theorem 3.6. *Let $\mathbf{A}$ be a local (not necessarily commutative) Cohen–Macaulay algebra. Denote by $\mathbf{A}_0$ the intersection of all its maximal over-rings, $J = \operatorname{rad} \mathbf{A}$, $\mathbf{A}_1 = J\mathbf{A}_0 + \mathbf{A}$. Then $\mathbf{A}$ is Cohen–Macaulay finite if and only if the following conditions hold:*

1. $\mathbf{A}_0$ *is hereditary.*
2. $\mu_{\mathbf{A}}(\mathbf{A}_0) \leq 3$.
3. $\mu_{\mathbf{A}}(\mathbf{A}_1) \leq 2$.

Of course, in the non-commutative case we cannot restrict ourselves by local rings (as in commutative one). So the problem of finding a general criterion of Cohen–Macaulay finiteness still remains open.

Remind also the following result, which shows the place of other unimodal and bimodal singularities given in [2] in the theory of Cohen–Macaulay modules (cf. [36, 17]).

Theorem 3.7. *In the geometric situation,* **A** *has not more than one-parametre families of ideals if and only if it dominates a simple, a unimodal or a bimodal plane singularity.*[2]

4. TWO-DIMENSIONAL CASE

Consider now the case when $\mathrm{Kr.\,dim}\,\mathbf{R} = 1$. Here we restrict ourselves by commutative algebras and even by *geometric* situation, when $\mathbf{R} = \mathbf{k}[[x, y]]$ for an algebraically closed field $\mathbf{k}$. In other words, we consider the *surface singularities*, i.e., the completions of local rings of points of algebraic surfaces. It is not too much known about Cohen–Macaulay type of **A** in this situation. The only complete result is the following criterion of Cohen–Macaulay finiteness (cf. [22, 5]):

Theorem 4.1. *A commutative two-dimensional Cohen–Macaulay algebra* **A** *is Cohen–Macaulay finite if and only if it is a* quotient *singularity, i.e.* $\mathbf{A} \simeq \mathbf{k}[[x, y]]^G$, *the ring of invariants of a finite subgroup* $G \subset \mathbf{GL}(2, \mathbf{k})$.

An important technics for calculating Cohen–Macaulay modules was developed by Kahn [29]. Namely, suppose that **A** is *normal* (integrally closed in $\widetilde{\mathbf{A}}$). Equivalent condition is that **A** is an *isolated singularity*, i.e., for each non-maximal prime ideal $\mathfrak{p}$, the localization $\mathbf{A}_{\mathfrak{p}}$ is a regular ring. Put $X = \mathrm{spec}\,\mathbf{A}$. Then one can consider its *resolution of singularity* [1], which is, by definition, a birational proper morphism of schemes $\nu : \widetilde{X} \to X$, where $\widetilde{X}$ is a smooth scheme, ν is an isomorphism outside $C = \nu^{-1}(\mathfrak{m})$, while C is a projective curve over $\mathbf{k}$ (the *exceptional curve*). Usually, one consider a *minimal resolution*, i.e., such that it cannot be factored through another one. The result of Kahn can be stated as follows:

Theorem 4.2. *There is an effective divisor Z on $\widetilde{X}$ (called the reduction cycle) with the support C such that isomorphism classes of Cohen–Macaulay* **A**-*modules are in one-to-one correspondence with isomorphism classes of vector bundles $\mathcal{B}$ over Z satisfying the following conditions:*

1. *$\mathcal{B}$ is generically generated (i.e., generated on a dense open subset) by its global sections.*
2. *There is a vector bundle $\mathcal{B}^{\#}$ over $2Z$ such that its restriction on Z coincides with $\mathcal{B}$ and the mapping $\mathrm{H}^0(C, \mathcal{B}(Z)) \to \mathrm{H}^1(C, \mathcal{B})$ defined by the exact sequence $0 \to \mathcal{B} \to \mathcal{B}^{\#} \to \mathcal{B}(Z) \to 0$ is injective.*

---

[2]In [36, 17] unimodal and bimodal singularities are called, following Wall, *strictly unimodal*.

Here one identify an effective divisor Z on $\widetilde{X}$ with the closed sub-scheme defined by the sheaf of ideals $\mathcal{O}_{\widetilde{X}}(Z)$. Then, in particular, Z can be considered as a closed subscheme of $2Z$. Remark that the divisor Z is constructed in [29] effectively if one knows effectively the curve C and the intersection multiplicities of its components on $\widetilde{X}$.

Hence, to consider Cohen–Macaulay **A**-modules, one has to deal with vector bundles on the projective curve Z (maybe, non-reduced).

Remark that all quotient singularities are know to be *ratioanl*, i.e., such that $H^1(\widetilde{X}, \mathcal{O}_{\widetilde{X}}) = 0$. In this case the conditions on the vector bundle $\mathcal{B}$ in Kahn's theorem becomes much simpler, namely:

1. $\mathcal{B}$ is generated by global sections.
2. $H^0(C, \mathcal{B}(Z)) = 0$.

These conditions are very restrictive indeed as the first one requires a lot of global sections for $\mathcal{B}$ while the second one prohibits global sections for its fixed shift. Maybe, this is the main reason which makes possible for some rational singularities to be Cohen–Macaulay finite.

The reduction cycle Z is related to the so called *fundamental cycle* Z_0. Remind that Z_0 is the smallest effective divisor such that $Z.C_i \leq 0$ for each component C_i of the exceptional curve C. One knows [29] that always $Z \geq Z_0$ and it can really happen that $Z > Z_0$ even for rational singularities.

5. VECTOR BUNDLES ON PROJECTIVE CURVES

In this section we consider projective curves over an algebraically closed field **k**. The signs $\otimes$ and Hom without subscripts here mean $\otimes_{\mathbf{k}}$ and $\mathrm{Hom}_{\mathbf{k}}$. Respectively, for any **k**-algebra **S**, **S**-mod denote the category of **S**-modules which are finite dimensional over **k**. For an algebraic curve C over **k**, denote by $\mathsf{Vb}(C)$ the category of vector bundles over C. It is convenient to identify vector bundles with locally free coherent sheaves of $\mathcal{O}_C$-modules and we always do it. We usually suppose our curves to be connected and *reduced* (with no nilpotents in the structure sheaf) but often *reducible*.

Concerning the classification of vector bundles over projective curves, only two complete results have been known till now. Namely, they are the Grothendieck classification of vector bundles over the projective line $\mathbb{P}^1$ [26] and the Atiyah classification of vector bundles over elliptic curves [3]. To be complete, remind them. We use the book [27] for standard references in algebraic geometry.

First of all, as all spaces $\mathrm{Hom}(\mathcal{F}, \mathcal{G})$ are finite dimensional for coherent sheaves over any projective variety, they always decompose uniquely into direct sums of indecomposables. Hence, one is interested in finding out the indecomposable vector bundles. If the curve C is fixed, we denote by $\mathcal{O}$ its structure sheaf (or the corresponding trivial one-dimensional bundle).

Theorem 5.1. *Any indecomposable vector bundle over $\mathbb{P}^1$ is isomorphic to one of the shifts $\mathcal{O}(n)$ of the structure sheaf.*

Theorem 5.2. *Let C be an elliptic curve. Then, for each r and d, there is a vector bundle $\mathcal{F}(r,d)$ on $C \times \mathrm{Pic}^{\circ}(C)$ such that the set of its fibres $\{\, \mathcal{F}(r,d)(p) \,|\, p \in \mathrm{Pic}^{\circ}(C) \,\}$ is a full set of representatives of the isomorphism classes of indecomposable vector bundles over C of rank r and degree d.*

Remark that, given any algebraic family of vector bundles over a projective curve C, in other words, a vector bundle $\mathcal{M}$ over $C \times X$, where X is a connected algebraic variety, the functions $\mathrm{rk}\,\mathcal{M}(x)$ and $\deg \mathcal{M}(x)$ $(x \in X)$ are always constant. Moreover, if the curve is not irreducible, $C = \cup_{i=1}^{s} C_i$, where C_i are its components, the ranks and degrees of the restrictions of a vector bundle onto these components are also constant in any algebraic family. Put $\mathbf{r}(\mathcal{F}) = (r_1, r_2, \ldots, r_s)$, where $r_i = \mathrm{rk}\,\mathcal{F}|_{C_i}$, and $\mathbf{d}(\mathcal{F}) = \mathbf{d}$, where $d_i = \deg \mathcal{F}|_{C_i}$, and call these vectors, respectively, the *rank vector* and the *degree vector* of the vector bundle $\mathcal{F}$. If we are going to define *vector bundle types* of projective curves, we should consider both rank and degree vectors as "discrete" parametres. Moreover, there is a "big" discrete group of shifts acting on the category of vector bundles. Namely, the function $\mathbf{d}$ defines an epimorphism $\mathbf{d} : \mathrm{Pic}\,C \to \mathbb{Z}^s$. Consider some section $\omega : \mathbb{Z} \to \mathrm{Pic}\,C$ of this homomorphsm. For instance, we can choose one smooth point a_i at each of its components C_i and put $\omega\,(d_1, d_2, \ldots, d_s) = \mathcal{O}_C(\sum_i d_i a_i)$. Denote by $\mathcal{O}(\mathbf{d})$ the line bundle $\omega(\mathbf{d})$ and put $\mathcal{F}(\mathbf{d}) = \mathcal{F} \otimes_{\mathcal{O}} \mathcal{O}(\mathbf{d})$. Thus the group $\mathbb{Z}^s$ becomes a group of automorphisms of the category $\mathsf{Vb}(C)$. If we are going to classify the vector bundles, it is enough to classify them up to the action of this group. Hence, one should include this action in the definitions of vector bundle finite and vector bundle tame curves. On the other hand, we can define *vector bundle wild* curves just as we have defined above Cohen–Macaulay wild algebras.

Definitions 5.3.   1. *Let $\mathbf{S}$ be a $\mathbf{k}$-algebra. Call a family of vector bundles over C based on $\mathbf{S}$ any locally free coherent sheaf $\mathcal{F}$ (of finite rank) of $\mathbf{S} \otimes \mathcal{O}_C$-modules.*
   Certainly, if $\mathbf{S}$ is the coordinate ring of an affine variety X, this notion is equivalent to that of a family of vector bundles over C with the base X. For any family $\mathcal{F}$ of vector bundles over C based

on S and any S-module L finite-dimensional over **k**, $\mathcal{F} \otimes_S L$ is a vector bundle over C. Hence, we get a functor $\mathcal{F}\otimes :$ S-mod $\to$ Vb(C).

2. *Call a family $\mathcal{F}$ of vector bundles over C based on S* strict *if the functor $\mathcal{F}\otimes$ reflects isomorphisms and preserves indecomposability.*

3. *Call a curve C* vector bundle finite *if there is a finite set S of vector bundles over C such that each indecomposable vector bundle is isomorphic to $\mathcal{F}(\mathbf{d})$ for some $\mathcal{F} \in$ S and some $\mathbf{d} \in \mathbb{Z}^s$.*

4. *Call a curve C* VB-tame *if there is a non-empty set* F $= \{\mathcal{F}_i\}$ *of strict families of vector bundles over C, $\mathcal{F}_i$ based on S_i (remark that S_i may be different) satisfying the following conditions:*

   (a) *Each S_i is a commutative finitely generated smooth **k**-algebra of Krull dimension 1.*

   (b) *For each integer r and vector $\mathbf{d}$, the set* $F_{r,\mathbf{d}}$ *is finite, where* $F_{r,\mathbf{d}} = \{\mathcal{M} \in F \mid \mathrm{rk}(\mathcal{M}) = r, \deg \mathcal{B} = \mathbf{d}\}$.

   (c) *For each integer r and vector $\mathbf{d}_0$, all but a finite number of locally free indecomposable sheaves on C of rank r and vector-degree $\mathbf{d}_0$ are isomorphic to those of the form $\mathcal{M}_i(\mathbf{d}, N) = \mathcal{M}(\mathbf{d}) \otimes_{S_i} N$, for some $\mathcal{M}_i \in F$, $\mathbf{d} \in \mathbb{Z}^s$ and some finite dimensional S_i-module N.*

   In this case call F *a* parametrizing set *for vector bundles over C. Denote $\nu(r)$ the minimal number of sheaves in $F_{r,\mathbf{d}}$, where $\mathbf{d}$ runs through $\mathbb{Z}^s$. Then a VB-tame curve C is said to be:*
   - bounded *if the set* F *above can be chosen in such a way that $\nu(r) \leq m$ for some constant m and all ranks r;*
   - unbounded *otherwise.*

5. *Call the curve C* vector bundle wild *if, for every finitely generated **k**-algebra S, there is a strict family of vector bundles over C based on* S.

   (Just as in Proposition 1.3, one only need to find a strict family based on $\mathbf{k}\langle x, y \rangle$, or $\mathbf{k}[x,y]$, or $\mathbf{k}[[x,y]]$.)

From this point of view the projective line should be considered as vector bundle finite and elliptic curves as vector bundle tame. The folowing result shows that there are no other finite and tame curves among the smooth ones.

Proposition 5.4. *Any projective smooth curve of genus $g > 1$ is vector bundle wild.*

Proof. For any two points $a \neq b$ of C,

$$\mathrm{Hom}_{\mathcal{O}}(\mathcal{O}(a), \mathcal{O}(b)) \simeq \mathrm{H}^0(C, \mathcal{O}(b-a)) = 0.$$

On the other hand,

$$\text{Ext}^1_{\mathcal{O}}(\mathcal{O}(a), \mathcal{O}(b)) \simeq \text{H}^1(C, \mathcal{O}(b-a))$$

as $\mathcal{E}xt^1_{\mathcal{O}}(\mathcal{O}(a), \mathcal{O}(b)) = 0$. Using the Riemann–Roch theorem for the divisor $b - a$, we get

$$\dim \text{H}^1(C, \mathcal{O}(b-a)) = g - 1 \geq 1.$$

Choose 5 distinct points $a_1, \ldots, a_5$ and consider the class of locally free sheaves $\mathcal{A}$ admitting an exact sequence:

(1) $$0 \longrightarrow \mathcal{A}_1 \longrightarrow \mathcal{A} \longrightarrow \mathcal{A}_2 \longrightarrow 0,$$

where

$$\mathcal{A}_1 = r_1 \mathcal{O}(a_1) \oplus r_2 \mathcal{O}(a_2) \oplus r_3 \mathcal{O}(a_3)$$

and

$$\mathcal{A}_2 = r_4 \mathcal{O}(a_4) \oplus r_5 \mathcal{O}(a_5).$$

Let $\xi \in \text{Ext}_{\mathcal{O}}(\mathcal{A}_2, \mathcal{A}_1)$ be the element corresponding to the sequence (1). As there are no homomorphisms from the subsheaf to the factor-sheaf, one can easily check that two elements $\xi, \xi' \in \text{Ext}_{\mathcal{O}}(\mathcal{A}_2, \mathcal{A}_1)$ lead to isomorphic sheaves $\mathcal{A}$ and $\mathcal{A}'$ if and only if there are automorphisms $\alpha : \mathcal{A}_1 \xrightarrow{\sim} \mathcal{A}_1$ and $\beta : \mathcal{A}_2 \xrightarrow{\sim} \mathcal{A}_2$ such that $\alpha \xi = \xi \beta$ (we mean here the Yoneda multiplication). Choose some non-zero elements $\xi_{ij} \in \text{Ext}^1_{\mathcal{O}}(\mathcal{O}(a_j), \mathcal{O}(a_i))$. Put $\mathcal{S} = \mathcal{O} \otimes \mathbf{F}$, where $\mathbf{F} = \mathbf{k}\langle x, y \rangle$, the free $\mathbf{k}$-algebra in two generators, $\mathcal{S}(a) = \mathcal{S} \otimes_{\mathcal{O}} \mathcal{O}(a)$ for $a \in C$. Then $\text{Ext}^1_{\mathcal{S}}(\mathcal{S}(a), \mathcal{S}(b)) \simeq \text{Ext}^1_{\mathcal{O}}(\mathcal{O}(a), \mathcal{O}(b)) \otimes \mathbf{F}$. Consider the exact sequence of locally free $\mathcal{S}$-modules

$$0 \longrightarrow \mathcal{S}(a_1) \oplus \mathcal{S}(a_2) \oplus \mathcal{S}(a_3) \longrightarrow \mathcal{F} \longrightarrow \mathcal{S}(a_4) \oplus \mathcal{S}(a_5) \longrightarrow 0$$

corresponding to the element of the Ext-space given by the matrix

$$\begin{pmatrix} \xi_{14} & \xi_{15} \\ \xi_{24} & x\xi_{25} \\ \xi_{34} & y\xi_{35} \end{pmatrix}.$$

If N is any finite dimensional $\mathbf{F}$-module, then the locally free $\mathcal{O}$-module $\mathcal{F}(N) = \mathcal{F} \otimes_{\mathbf{F}} N$ corresponds to the element of Ext-space given by the matrix

$$\begin{pmatrix} \xi_{14} I & \xi_{15} I \\ \xi_{24} I & \xi_{25} X \\ \xi_{34} I & \xi_{35} Y \end{pmatrix}.$$

Here I denotes the identity matrix of size $\dim_{\mathbf{k}} N$, while X and Y are the matrices describing the action of x and y, respectively, in the module N. Now an easy straightforward calculation shows that $\mathcal{F}(N) \simeq \mathcal{F}(N')$ if and only if $N \simeq N'$ and if there is an idempotent endomorphism of $\mathcal{F}(N)$, there is also an idempotent endomorphism of N. $\qquad\square$

For a singular curve C, it is convenient to relate the classification of vector bundles to some bimodule problems. Namely, introduce first the following notations.

Notations 5.5. 1. $\nu : \widetilde{C} \to C$ is the normalization *of* C. (Remark that it can also be reducible or, the same, non-connected.)

2. $S = S(C)$ is the set of singular points *of* C and $\widetilde{S} = \nu^{-1}(S)$.

3. $\widetilde{\mathcal{O}} = \nu_*(\mathcal{O}_{\widetilde{C}})$. *We identify* $\mathcal{O}$ *with its natural image in* $\widetilde{\mathcal{O}}$.

4. $\mathcal{J}$ is the conductor *of* $\mathcal{O}$ *in* $\widetilde{\mathcal{O}}$, *i.e., the biggest sheaf of* $\widetilde{\mathcal{O}}$-*ideals contained in* $\mathcal{O}$.

5. $\Lambda = \mathcal{O}/\mathcal{J}$ *and* $\Gamma = \widetilde{\mathcal{O}}/\mathcal{J}$.

6. *For any vector bundle* $\mathcal{B}$, *put* $\widetilde{\mathcal{B}} = \widetilde{\mathcal{O}} \otimes_{\mathcal{O}} \mathcal{B}$ *(it is a vector bundle over* $\widetilde{C}$*) and* $\overline{\mathbf{B}} = \mathcal{B}/\mathcal{J}\mathcal{B}$. *We identify* $\mathcal{B}$ *with its image in* $\widetilde{\mathcal{F}}$. *In the same way, for a vector bundle* $\mathcal{A}$ *over* $\widetilde{C}$, *put* $\overline{\mathbf{A}} = \mathcal{A}/\mathcal{J}\mathcal{A}$.

Remark that Λ and Γ are sky-scraper sheaves, zero outside S and with finite dimensional fibres. Hence, we may (and will) identify them with finite dimensional **k**-algebras $\bigoplus_{x \in S} \Lambda_x$ and $\bigoplus_{x \in S} \Gamma_x$ respectively. Just in the same way we identify the sky-scraper sheaf $\overline{\mathbf{B}}$ with the Λ-module $\bigoplus_{x \in S} \overline{\mathbf{B}}_x$.

Denote by A the category of locally free (coherent) $\widetilde{\mathcal{O}}$-sheaves and by B the category of projective (finitely generated) Λ-modules. Define a B-A-bimodule U putting

$$\mathsf{U}(B, \mathcal{A}) = \mathrm{Hom}_\Lambda(B, \overline{\mathbf{A}}).$$

Each vector bundle $\mathcal{B}$ over C defines an element $u(\mathcal{B}) :\in \mathsf{U}(\overline{\mathbf{B}}, \widetilde{\mathcal{B}})$ induced by the imbedding $\mathcal{B} \to \widetilde{\mathcal{B}}$. This correspondence evidently defines a functor $\mathsf{Vb}(C) \to \mathrm{El}(\mathsf{U})$. Indeed, not all elemeth U are of this shape. Namely, introduce the following notion.

Definition 5.6. *An element* $u \in \mathsf{U}(B, \mathcal{A})$ *is said to be* correct *if the module B is free and u induces an isomorphism* $\Gamma \otimes_\Lambda B \simeq \overline{\mathbf{A}}$. *Denote by* $\mathrm{El}_c(\mathsf{U})$ *the full subcategory of* $\mathrm{El}(\mathsf{U})$ *consisting of all correct elements.*

It is clear that the element $u(\mathcal{B})$ is always correct and the following fact holds.

Proposition 5.7. *The correspondence* $\mathcal{B} \mapsto u(\mathcal{B})$ *defines a reprsentation equivalence of the categories* $\mathsf{Vb}(C)$ *and* $\mathrm{El}_c(\mathsf{U})$.

Proof. If $u \in \mathsf{U}(B, \mathcal{A})$ is correct, it induces an embedding $B \to \overline{\mathbf{A}}$. So we can consider from the very beginning B as a Λ-submodule in $\overline{\mathbf{A}}$ and

u as its embedding. Let $\mathcal{B}$ be the preimage of B in $\mathcal{A}$. Then $\mathcal{B}_x = \mathcal{A}_x$ for each point $x \notin S$. Moreover, as B generates $\overline{\mathbf{A}}$ as Γ-module, $\mathcal{B}_x$ generates $\mathcal{A}_x$ as $\mathcal{O}_x$-module for any point x. If $x \in S$, $\mathcal{B}_x/J\mathcal{B}_x \simeq B_x$. Hence, there is an epimorphism $r\mathcal{O}_x \to \mathcal{B}_x$, where $r = \operatorname{rk} B$. But $\operatorname{rk} B = \operatorname{rk}\mathcal{A}$, hence, this epimorphism is an isomorphism and $\mathcal{B}$ is a locally free sheaf of rank r. Moreover, the natural homomorphism $\widetilde{\mathcal{B}} \to \mathcal{A}$ induced by the embedding $\mathcal{B} \to \mathcal{A}$ is an isomorphism (as it is an isomorphism after localization at all points $x \in C$). It means that $u \simeq u(\mathcal{B})$, thus this functor is dense.

Suppose that $u(\mathcal{B}) \simeq u(\mathcal{B}')$, i.e., there are isomorphisms $f : \overline{\mathbf{B}} \to \overline{\mathbf{B}}'$ and $g : \widetilde{\mathcal{B}} \to \widetilde{\mathcal{B}}'$ such that $gu(\mathcal{B}) = u(\mathcal{B}')f$. Then $g(\mathcal{B})$ and $\mathcal{B}'$ coincide outside S and coincide modulo J, hence, coincide. It means that this functor reflects isomorphism, thus, it is a representation equivalence. $\square$

Remark that, as all Hom-spaces in both categories A and B are finite dimensional, U is indeed a (locally) finite dimensional bimodule of the sort usual in the representation theory of finite dimensional algebras.

Certainly, considering only correct elements may happen a rather tough restriciton. Nevertheless, we are also able to define *correct elements* in $\operatorname{El}(\mathbf{U}^{\mathbf{S}})$ for any k-algebra S (just by the same definition as above). Therefore, we can speak on the bimodule U being *correctly finite*, *correctly tame* or *correctly wild*. In particular, it is obvious now that if U is correctly wild, the curve C is vector bundle wild too.

Of course, if the curve $\widetilde{C}$ is vector bundle wild, so is C. Therefore, if C is not vector bundle wild, all components of $\widetilde{C}$ are either rational or elliptic curves. It happens that indeed the latter case cannot occur.

Proposition 5.8. *If the curve C is really singular and $\widetilde{C}$ has a non-rational component, then C is vector bundle wild.*

Proof. Let $\widetilde{C}_1, \widetilde{C}_2, \ldots, \widetilde{C}_s$ be the irreducible components of $\widetilde{C}$, $\mathcal{O}_k = \mathcal{O}_{\widetilde{C}_k}$. Suppose there is a component $\widetilde{C}_1$ which is elliptic. As C is singular and connected, there is a point $e \in C$ which lies on $\nu(\widetilde{C}_1)$ and such that $\mathcal{O}_e \neq \widetilde{\mathcal{O}}_e$. Consider the case when also $e \in \nu(\widetilde{C}_2)$ for another component $\widetilde{C}_2$ (other cases are even simpler to handle). Find 4 distinct points $a_1, \ldots, a_4$ on $\widetilde{C}_1 \setminus \widetilde{S}$ and a point $b \in \widetilde{C}_2 \setminus \widetilde{S}$. Consider the element u from $\mathrm{El}(\mathsf{U}, \mathbf{F})$, where $\mathbf{F} = \mathbf{k}\langle x, y \rangle$, defined as follows:

$$u \in \mathsf{U}(B, A), \quad \text{where:}$$

$$A = \bigoplus_{i=1}^{4} (\widetilde{\mathcal{O}}(a_\iota + ib) \otimes \mathbf{F}),$$

$$B = 4\mathcal{F} \otimes \mathbf{F}.$$

In this case $u : B \to A$ can be given by a set of 4×4 matrices u_{pk} with entries from $\mathcal{O}_{kp}/\mathcal{J} \otimes \mathbf{F}$, as such a matrix defines a homomorphism $B_p \to A_p$. Here p runs through $S(C)$ and $k = 1, 2, \ldots t$. We put all components equal identity matrices except of u_{e2} which is

$$\begin{pmatrix} x & y & 1 & 1 \\ 1 & 1 & 1 & 0 \\ 0 & 1 & 0 & 0 \\ 1 & 0 & 0 & 0 \end{pmatrix}.$$

Just as in the proof of Proposition 5.4, it is not difficult to verify that it is a strict element. Hence, C is VB-wild. $\square$

Hence, from now on we suppose all components of $\widetilde{C}$ being rational and identify each of them with the projective line. Then we call the curve C itself *rational*. In this case each vector bundle on $\widetilde{C}$ is isomorphic to $\mathcal{A}(\mathbf{d})$, where $\mathcal{A} = \sum_k \widetilde{\mathcal{O}}_{i_k}(m_k)$ with all $m_k \geq 0$, where $\widetilde{\mathcal{O}}_\iota$ denotes the restriction of $\widetilde{\mathcal{O}}$ onto C_ι. Of course, for any fixed vector rank and vector degree, there is only finitely many possibilities for such $\mathcal{A}$. It gives the following result.

Proposition 5.9. *If C is rational, it is vector bundle finite, tame or wild if and only if the corresponding bimodule U is correctly finite, tame or wild respectively.*

It is not difficult to determine the (correct) type of the bimodule U in the rational case. Remind that the category of vector bundles on $\mathbb{P}^1$ is equivalent to the additive hull of the category $\mathbb{L}$ defined as follows:

- $\mathrm{Ob}\,\mathbb{L} = \mathbb{Z}$,
- morphisms of $\mathbb{L}$ are generated by the set

$$\{x_n, y_n \mid n \in \mathbb{Z}\}, \quad \text{where } x_n, y_n \in \mathbb{L}(n, n+1),$$

  subject the relations:

$$x_n y_{n+1} = y_n x_{n+1} \quad \text{for all } n.$$

Indeed, n corresponds to the line bundle $\mathcal{O}(n)$, while x_n and y_n are generated by multiplication, respectively, by ξ and η (the homogeneous) coordinates on $\mathbb{P}^1$). Therefore, $\mathsf{Vb}(C) \simeq \mathbb{L}^s$.

Proposition 5.10. *In the following cases the curve C is vector bundle wild:*

1. *Γ is not semi-simple (i.e., contains non-zero nilpotents).*
2. *$\#(\nu^{-1}(a)) > 2$ for some point $a \in S$.*
3. *$\#(\widetilde{S} \cap \widetilde{C}_i) > 2$ for some component $\widetilde{C}_i$ of $\widetilde{C}$.*

Proof. All cases are treated similarly: we construct strict families based on $\mathbf{F} = \mathbf{k}[x, y]$. Checking their strictness is always a routine job very much alike the proof of Proposition 5.4 and we omit it.

1. Find a point $p \in \widetilde{S}$ such that Γ_p is not semi-simple. Then there is an element $\gamma \in \Gamma_p$ such that $I = \mathbf{k}\gamma$ is a minimal ideal in Γ, in particulaly, $\gamma^2 = 0$. As $\mathcal{J}$ is the conductor, Λ contains no ideal of Γ, hence, $I \cap \Lambda = 0$. Put

$$B = 2\Lambda, \ \mathcal{A} = \widetilde{\mathcal{O}}(n) \oplus \widetilde{\mathcal{O}}(n+1),$$

where n is any integer. An element of $\mathsf{U}(B, \mathcal{A})$ is given by a set of 2×2 matrices $\left\{ M_a \,|\, a \in \widetilde{S} \right\}$ defining the homomorphisms $2\Lambda_b \to \overline{\mathbf{A}}_a \simeq 2\Gamma_a$, where $b = \nu(a)$. The entries of these matrices are indeed elements of Γ_a. Then the following set gives a strict element of $\mathsf{U}^{\mathbf{F}}(B \otimes \mathbf{F}, \mathcal{A} \otimes \mathbf{F})$:

$$M_p = \begin{pmatrix} 1 & \gamma \\ x\gamma & 1 + y\gamma \end{pmatrix}, \quad M_a = \begin{pmatrix} 1 & 0 \\ 0 & 1 \end{pmatrix} \quad \text{for } a \neq p.$$

Moreover, It is evident that this element is correct as modulo radical all its components are identity matrices.

2. From now on suppose Γ and hence Λ semi-simple. Let p_1, p_2, p_3 be distinct points of $\widetilde{S}$ having the same image $b \in S$. Put $B = 4\Lambda$ and $\mathcal{A} = \bigoplus_{m=0}^{3} \widetilde{\mathcal{O}}(n+m)$ for some (arbitrary) $n \in \mathbb{Z}$. Remark that now $\mathrm{Hom}_\Lambda(\Lambda_b, \Gamma_a) \simeq \mathbf{k}$ for $b = \nu(a)$. Hence, the elements from $\mathsf{U}(B, \mathcal{A})$ can be considered as sets of 4×4 matrices $M = \left\{ M_a \,|\, a \in \widetilde{S} \right\}$ with entries in $\mathbf{k}$.

Now take M such that all its components are identity matrices except of the next two ones:

$$M_{p_1} = \begin{pmatrix} 0 & 0 & 0 & 1 \\ 0 & 0 & 1 & 0 \\ 0 & 1 & 0 & 0 \\ 1 & 0 & 0 & 0 \end{pmatrix},$$

$$M_{p_2} = \begin{pmatrix} 1 & 1 & x & y \\ 0 & 1 & 1 & 0 \\ 0 & 0 & 1 & 1 \\ 0 & 0 & 0 & 1 \end{pmatrix},$$

Then this element is again a strict correct element of $\mathsf{U}^{\mathbf{F}}$.

3. Suppose at last that some component $\widetilde{C}_i$ of $\widetilde{C}$ contains three distinct points p_1, p_2, p_3 from $\widetilde{S}$. In this case we put $B = \Lambda$ and $\mathcal{A} = \bigoplus_{k=0}^3 \widetilde{\mathcal{O}}(n + k)$ for some (arbitrary) n and take the set of matricese M, all of whose components are identity matrices except of the next two ones:

$$M_{p_1} = u_{\iota_0 \jmath_2} = \begin{pmatrix} 0 & 0 & 0 & 1 \\ 0 & 0 & 1 & 0 \\ 0 & 1 & 0 & 0 \\ 1 & 0 & 0 & 0 \end{pmatrix},$$

$$M_{p_2} = \begin{pmatrix} 0 & 0 & 0 & 1 \\ 0 & 0 & 1 & 1 \\ 0 & 1 & 1 & x \\ 1 & 1 & 0 & y \end{pmatrix}.$$

Again some easy straightforward calculation shows that u is a strict correct element.   □

Hence, if a rational curve C is not vector bundle wild, all its singular points are *simple nodes* (simple double-crossings) and each component of $\widetilde{S}$ contains at most two of them. As C is connected, its singular points $\{a_i\}$ and irreducible components C_j can be ordered in such a way that $C_i \cap C_{i+1} = \{a_i\}$ for all $i = 1, \ldots, s - 1$ and all other intersections of these components are empty, except, maybe, $C_s \cap C_1 = \{a_s\}$. In particular, $\#(S)$ is either $s - 1$ or s. In the former case, we say that C is *of type* A_s, while in the latter one we say that it is *of type* $\widetilde{A}_s$. In particular, if C is irreducible, it should be of type $\widetilde{A}_1$, i.e., having only one singular point, which is a simple node. It is also convenient to consider the projective line itself as the rational curve of type A_1 (one component, no singular points).

Proposition 5.11.   1. *Any rational curve of type* A_s *is vector bundle finite.*

2. *Any rational curve of type* $\widetilde{A}_s$ *is vector bundle tame unbounded.*

Proof. We prove the (more complicated) assertion *2* as the assertion *1* is proved by the same calculations. Moreover, these calculations show that the only indecomposable vector bundles over a curve of type A_s are indeed the shifts $\mathcal{O}(\mathbf{d})$ of the trivial line bundle.

So suppose C be of type $\widetilde{A}_s$, its singular points and irreducible components numbered as above. Then $\Lambda = \mathbf{k}^s$, indecomposable vector bundles over $\widetilde{C}$ are $\mathcal{O}_i(n)$, where $\mathcal{O}_i$ is the structure sheaf of the normalization of C_i and

$$\mathrm{Hom}_\Lambda(\mathbf{k}_j, \widetilde{\mathcal{O}}_i(n)) = \begin{cases} \mathbf{k} & \text{if } i = j \text{ or } j+1 \\ 0 & \text{otherwise.} \end{cases}$$

Here we denote by $\mathbf{k}_j$ the component of Λ corresponding to the point a_i and put $C_{s+1} = C_1$. For the sake of simplicity we suppose that the homogenious coordinates of the preimages of a_i on C_i and C_{i+1} are respectively $(1:0)$ and $(0:1)$. It can be done as any pair of points on $\mathbb{P}^1$ can be moved into any other pair by an automorphism of $\mathbb{P}^1$. Then the multiplication by ξ induces isomorphisms $\mathsf{U}(\mathbf{k}_i, \widetilde{\mathcal{O}}_i(n)) \to \mathsf{U}(\mathbf{k}_i, \widetilde{\mathcal{O}}_i(n+1))$ and zero mappings on $\mathsf{U}(\mathbf{k}_i, \widetilde{\mathcal{O}}_{i+1}(n))$, while the multiplication by η induces isomorphisms $\mathsf{U}(\mathbf{k}_i, \widetilde{\mathcal{O}}_{i+1}(n)) \to \mathsf{U}(\mathbf{k}_i, \widetilde{\mathcal{O}}_{i+1}(n+1))$ and zero mappings on $\mathsf{U}(\mathbf{k}_i, \widetilde{\mathcal{O}}_i(n))$. Therefore, the bimodule problem defined by the bimodule U is indeed a special case of *bunches of chains* considered in [7]. Hence, it is tame as well as the curve C is vector bundle tame. One can also easily see that it is unbounded. $\qquad\square$

As the result of our considerations, we obtain the following description of vector bundle types of projective curves.

Theorem 5.12. *Let C be a (reduced) projective curve. It is:*

1. *Vector bundle finite if C is rational of type A_s for some s.*
2. *Vector bundle tame bounded if C is a smooth elliptic curve.*
3. *Vector bundle tame unbounded if C is a rational curve of type $\widetilde{A}_s$ for some s.*
4. *Vector bundle wild otherwise.*

6. Minimal elliptic singularities

Theorem 5.12 together with the Kahn's result allows to determine Cohen–Macaulay types of a class of normal surface singularities called *minimal elliptic* singularities.

Definition 6.1. *A surface singularity $\mathbf{A}$ with a minimal resolution of singularity $\nu : \widetilde{X} \to X$ is said to be* minimal elliptic *if it is Gorenstein and $\dim \mathrm{H}^1(\widetilde{X}, \mathcal{O}_{\widetilde{X}}) = 1$.*

In what follows $\mathbf{A}$ denotes a minimal elliptic singularity, C denotes the exceptional curve of a minimal resolution of singularity of $\mathbf{A}$ and Z its reduction cycle, which coincide in this case with the *fundamental cycle* [29].

In this case the Kahn's theorem can be precised in the following way (cf. [29]).

Theorem 6.2. *Isomorphism classes of Cohen–Macaulay modules over a minimal elliptic singularity* **A** *are in one-to-one correspondence with the isomorphism classes of vector bundles* $\mathcal{F}$ *on the fundamental cycle* Z *satisfying the following conditions:*

1. $\mathcal{F} = \mathcal{B} \oplus n\mathcal{O}_Z$ *where* $n = \dim \mathrm{H}^0(C, \mathcal{B}(Z))$.
2. *The sheaf* $\mathcal{B}$ *is generically generated by global sections.*
3. $\mathrm{H}^1(C, \mathcal{B}(Z)) = 0$.

Moreover, this correspondence can be prolonged to the families of Cohen–Macaulay modules and vector bundles on Z.

Certainly, the conditions of this theorem are rather restrictive. However they do not imply Cohen–Macaulay type as one has the following simple result:

Proposition 6.3. *For every vector bundle* $\mathcal{B}$ *on* Z *there is a positive integer* m *such that* $\mathcal{B}(m)$ *satisfy the conditions 2 and 3 of Theorem 6.2.*

The proof is an immediate consequence of the Serre Theorem on cohomology of coherent sheaves over projective varieties.

Corollary 6.4. *The Cohen–Macaulay type of a minimal elliptic singularity coincide with the vector bundle type of the fundamental cycle of its minimal resolution of singularity.*

As no minimal elliptic singularity is Cohen–Macaulay finite, we get the following criterion.

Corollary 6.5. *A minimal elliptic singularity is Cohen–Macaulay tame if and only if the fundamental cycle of its minimal resolution of singularity is either an elliptic curve or a rational curve of type* $\widetilde{\mathrm{A}}_s$ *for some* s.

In particular, in both cases Z should coincide with the exceptional curve C. Fortunately, it is known (and easy to prove) that whenever C is an elliptic curve or a rational curve of type $\widetilde{\mathrm{A}}_s$, it always coincides with the fundamental cycle Z. In the former case the singularity is said to be *simple elliptic* (cf. [35]), while in the latter one it is called a *cusp singularity* (cf. [30]). Hence, our result can be reformulated as follows.

Theorem 6.6. *A minimal elliptic singularity is Cohen–Macaulay tame if and only if it is either a simple elliptic or a cusp singularity. In the former case it is tame bounded, while in the latter one it is tame unbounded.*

Taking into account the classification of unimodal complete intersection singularities given by Wall [37], we also get the following

Corollary 6.7. *A complete intersection singularity is tame if and only if it is of type* T_{pqr} *or* T_{pqrs}. *In particular, a hypersurface singularity is tame if and only if it is of type* T_{pqr}.

Remind that the singularities of type T_{pqr} ($p \geq q \geq r$ and $1/p + 1/q + 1/r \leq 1$) are $\mathbf{k}[[x, y, z]]/(x^p + y^q + z^r)$; those of type T_{pqrs} are $\mathbf{k}[[x, y, z, t]]/(xy + z^q + t^s, zt + x^p + y^r)$.

REFERENCES

[1] Abhyankar, S. S., *Resolution of singularities of arithmetical surfaces*, in: "Arithmetic Algebraic Geometry," Harper & Row, New York, 1965, 111–152.

[2] V. I. Arnold, A. N. Varchenko and S. M. Gusein-Zade, *Singularities of Differentiable Maps*, Vol. 1, Birkhäuser, 1985.

[3] Atiyah, M. F., *Vector bundles over an elliptic curve*, Proc. London Math. Soc. 7 (1957), 414–452.

[4] Atiyah, M. F., Macdonald, I. G., *Introduction to Commutative Algebra*, Addison-Wesley, Reading, 1969.

[5] Auslander M., *Rational singularities and almost split sequences*, Trans. Amer. Math. Soc., 293 (1986), 511–531.

[6] Bass, H., *On the ubiquity of Gorenstein rings*, Math. Z., 82 (1963), 8–28.

[7] Bondarenko, V. V., *Representations of bundles of semichained sets and their applications*, Algebra i Analiz 3, #5 (1991), 38–61 (English translation: St. Petersburg Math. J. 3 (1992), 973–996).

[8] Dieterich, E., *Solution of a non-domestic tame classification problem from integral representation theory of finite groups*, Mem. Amer. Math. Soc., 450 (1991).

[9] Dieterich, E., *Lattice categories over curve singularities with large conductor*, Preprint 92-069, SFB 343, Universität Bielefeld, 1992.

[10] Drozd, Y. A., *A generalization of a theorem of Dade*, Dopovidi Acad. Nauk Ukraine, 3 (1974), 204–207.

[11] Drozd, Y. A., *Matrix problems and categories of matrices*, Zapiski Nauchn. Semin. LOMI 28 (1972), 144–153.

[12] Drozd, Y. A., *Representations of commutative algebras*, Funkc. Anal. Prilozhen., 6:4 (1972), 41–43.

[13] Drozd, Y. A., *On existence of maximal orders*, Mat. Zametki, 37 (1985), 313–315.

[14] Drozd, Y. A., *Tame and wild matrix problems*, In: "Representations and Quadratic Forms," Inst. Math., Kiev, 1979, 39–74 (English translation: Amer. Math. Soc. Transl. (2) 128, 1986, 31–55).

[15] Drozd, Y. A., Greuel, G.-M., *Tame-wild dichotomy for Cohen-Macaulay modules*, Math. Ann. (1992), 387–394.

[16] Drozd, Y. A., Greuel, G.-M., *Cohen-Macaulay module type*, Compositio Math., 89 (1993), 315–338.

[17] Drozd, Y. A., Greuel, G.-M., *On Schappert characterization of strictly unimodal plane curve singularities*, to appear.

[18] Drozd, Y. A., Kirichenko, V. V., *On quasi-Bassian orders*, Izvestia Acad. Sci. USSR, 36 (1972), 328–370.

[19] Drozd, Y. A., Kirichenko, V. V., *Primary orders with a finite number of indecomposable representations*, Izvestia Acad. Sci. USSR, 37 (1973), 715–736.

[20] Drozd, Y. A., Kirichenko, V. V., Roiter, A. V., *On hereditary and Bassian orders*, Izvestia Acad. Sci. USSR, 31 (1967), 1415–1436.

[21] Drozd, Y. A., Roiter, A. V., *Commutative rings with a finite number of integral indecomposable representations*, Izvestia Acad. Sci. USSR, 31 (1967), 783–798.

[22] Esnault, H., *Reflixive modules on quotient surface singularities*, J. Reine Angew. Math., 362 (1985), 63–71.

[23] Gelfand, I. M., Ponomarev, V. A., *Remarks on the classification of a pair of commuting linear transformations in a finite dimensional space*, Funkc. Anal. Prilozhen. 3:4 (1969), 81-82.

[24] Green, E. L., Reiner, I., *Integral representations and diagrams*, Michigan Math. J., 25 (1978), 53–84.

[25] Greuel, G.-M., Knörrer, H., *Einfache Kurvesingularitäten und torsionfreie Moduln*, Math. Ann., 270 (1985), 417–425.

[26] Grothendieck, A. *Sur la classification des fibrés holomorphes sur la sphère de Riemann*, Amer. J. Math. 79 (1956), 121–138.

[27] Hartshorn, R., *Algebraic Geometry*, Springer-Verlag, Berlin, heidelberg, New York, 1977.

[28] Jacobinski, H., *Sur les ordres commutatifs avec un nombre fini de réseaux indécomposables*, Acta Math., 118 (1967), 1–31.

[29] C. Kahn, *Reflexive modules on minimally elliptic singularities*, Math. Ann. 285 (1989), 141–160.

[30] Karras, U., *Deformations of cusps singularities*, Proc. Symp. Pure Math. 30 (1), Amer. Math. Soc., 1977, 37–44.

[31] Knörrer H., Torsionfreie Moduln bei Deformation von Kurvensingularitäten, In: Greuel G.-M., Trautmann G. (ed.), Singularities, Representations of Algebras and Vector Bundles, Lambrecht 1985. Lecture Notes in Math., Vol. 1273, Springer, Berlin-Heidelberg-New York (1987), 150–155.

[32] Laufer, H. B., *On minimal elliptic singularities*, Amer. J. Math., 99 (1977), 1257–1295.

[33] Matsumura, H., *Commutative Ring Theory*, Cambridge University Press, Cambridge, 1986.

[34] Ringel, C. M. and Roggenkamp, K. W., *Diagrammatic methods in representation theory of orders*, J. Algebra, 60 (1979), 11–42.

[35] Saito, K, Einfach-elliptische Singularitäten, Invent. Math., 23 (1974), 289–325.

[36] Schappert A., A characterization of strict unimodal plane cure singularities, In: Greuel G.-M., Trautmann G. (ed.), Singularities, Representations of Algebras and Vector Bundles, Lambrecht 1985. Lecture Notes in Math., Vol. 1273, Springer, Berlin-Heidelberg-New York (1987), 168–177.

[37] C. T. C. Wall, *Classification of unimodal isolated singularities of complete intersections*, Proc. Symp. Pure Math. 40(2), Amer. Math. Soc., 1983, 625–640.

A Note on Filtrations and Valuations for Ordered Groups

Maha Helmy Ibrahim El Baroudy *
Ain Shams University, Dept. of Mathematics
Cairo, Egypt

Freddy van Oystaeyen,
University of Antwerp U.I.A., Dept. of Mathematics and
Computer Science, 2610 Antwerp, Belgium

0. Introduction

In [6] it is suggested that a theory for extensions of valuations may benefit from properties of certain associated filtrations. For discrete valuations this strategy has been used to discuss the extension of those valuations from the center to certain algebras of quantum type, cf. [9].

In this note we extend the applicability of the filtered methods to the case of totally ordered groups.

Section 1. contains the basic facts concerning filtrations by totally ordered groups, not necessarily abelian. Certain cases do allow, as in the $\mathbb{Z}$-filtered case, a reduction to graded ring theory both via the Rees ring and the associated graded ring. In particular there is a general version of the category equivalence between the category of Γ-filtered R-modules and Γ_+-torsion free graded modules over the Rees ring $\tilde{R}$ (Remark 1.3.2) etc ...

Whereas real trouble shows up when Noetherian-type conditions have to be considered, destroying the classical approach via Zariskian filtrations, all properties relating suitable filtrations to valuations reappear in the Γ-filtered case.

In section 2 we present the results concerning extension of Γ-valuations on fields to certain quantum-algebras over these fields, e.g. Weyl algebras cf. [8]....
The fact that the extension problem is easier for these algebras than for finite dimensional extensions of the center is due to the possibility to link the extension of valuations to the existence of good reductions for generators and relations defining the quantum algebras. These results follow from the results in section 1, characterising Γ-valuations as Γ-filtrations with a domain for the associated graded ring. We do not develop here a theory of good filtrations for arbitrary Γ, this seems to need a topological approach with respect to non-linear topologies, we aim to come back to these problems in forthcoming work.

* Supported by a research grant of the Egyptian government.

131

1   Filtration With Respect to an Ordered Group

We let Γ be a totally ordered group, let $\Gamma^+ = \Gamma_{>e}, \Gamma_+ = \Gamma_{\geq e}$. We look at a ring R and a family of additive subgroups $\{F_\gamma R, \gamma \in \Gamma\}$ satisfying :

(i) If $\gamma \leq \tau$ for γ, τ in Γ then $F_\gamma R \subset F_\tau R$

(ii) For $\gamma, \tau \in \Gamma$ we have $F_\gamma R F_\tau R \subset F_{\gamma+\tau} R$. (We write Γ additively)

(iii) $1 \in F_e R$, e is the neutral element in Γ

(iv) $\bigcup_{\gamma \in \Gamma} F_\gamma R = R$

We say that $FR := \{F_\gamma R, \gamma \in \Gamma\}$ is a **filtration of type Γ** on R. If $F_\gamma R = 0$ for $\gamma \not\in \Gamma_+$ then we say that FR is a **positive filtration (of type Γ)**.

For a filtration of type Γ we may define a Γ-graded associated graded ring $G_F(R) = \bigoplus_{\gamma \in \Gamma} G_F(R)_\gamma$, where $G_F(R)_\gamma = F_\gamma R / \sum_{\gamma' < \gamma} F_{\gamma'} R$.

The **Γ-graded Rees ring** $\widetilde{R} = \bigoplus_{\gamma \in \Gamma} F_\gamma R$ can be identified to the subring of the groupring $R\Gamma$ given as $\widetilde{R} = \sum_{\gamma \in \Gamma} F_\gamma R \gamma$. It is clear that $\Gamma_+ \subset \widetilde{R}$ and $\widetilde{R}\Gamma^+$ is graded ideal of $\widetilde{R}$. One easily verifies that $G_F(R) \cong \widetilde{R}/\widetilde{R}\Gamma^+$. Moreover, since elements of Γ commute with R in $R\Gamma$ it is clear that Γ^+ and Γ_+ are multiplicative systems in $\widetilde{R}$ consisting of homogenous elements.

1.1   Lemma

With notation as above, Γ_+ is an Ore set (left and right) of $\widetilde{R}$ (putting $1_{\widetilde{R}} = 1_R.e$).

Proof   Given $\tau \in \Gamma_+$ and $r_\gamma \gamma \in \widetilde{R}_\gamma$ then we see that $(\gamma\tau\gamma^{-1})(r_\gamma\gamma) = r_\gamma\gamma\tau$ yields the left Ore condition (other Ore conditions follow in a straight forward way). We have to establish these conditions with respect to nonhomogeneous $\sum r_\gamma \gamma \in \widetilde{R}$, where only finitely many r_γ are non zero. We can continue the proof by induction on the length of $\sum_{\gamma \in \Gamma} r_\gamma \gamma$. If only one term is nonzero, the first part establishes the claim. Suppose we have established the claim for elements of $\widetilde{R}$ having a decomposition of length strictly less than n. Consider $\sum_{i=1}^n r_{\gamma_i} \gamma_i \in \widetilde{R}, \tau \in \Gamma_+$; we may assume without loss of generality that $\gamma_1 \geq \gamma_i$ for $i = 1, \ldots, n$. Then $(\gamma, \tau\gamma_1^{-1})(\sum_{i=1}^n \gamma_i, \gamma_i) = r_{\gamma_1}\gamma_1\tau + \sum_{i=2}^n r_{\gamma_i}\gamma_1\tau\gamma_1^{-1}\gamma_i$.
The induction hypothesis yields the existence of some $\delta \in \Gamma_+$ such that $\delta \sum_{i=2}^n r_{\gamma_i}\gamma_1\tau\gamma_1^{-1}\gamma_i = r'\tau$ with $r' \in \widetilde{R}$.
Look at $\delta\gamma_1\tau\gamma_1^{-1} \in \Gamma_+$, then we obtain:

$$\delta(\gamma_1\tau\gamma_1^{-1}) \sum_{i=1}^n r_{\gamma_i}\gamma_i = r_{\gamma_1}\delta\gamma_1\tau + r'\tau.$$

Note that $r_{\gamma_1} \in F_\delta\gamma_1 R$ because $\gamma_1 \leq \delta\gamma$, hence $\delta(\gamma_1\tau\gamma_1^{-1}) \sum r_{\gamma_i}\gamma_i = r''\tau$ with $\tau'' \in \widetilde{R}$.

All other Ore conditions may be obtained either by left-right symmetry or in a straight forward way.

1.2 Corollary

$$(\Gamma_+)^{-1}\widetilde{R} = R\Gamma.$$

From hereon it is clear that all basic facts about $\mathbb{Z}$- filtrations, may be generalized to Γ-filtrations without extra problems, up to keeping in mind that Γ is not abelian i.e. respecting the order of operations in Γ. In fact enveloping algebras of finite dimensional Lie algebras over a field may be considered as $\mathbb{Z}^n$-filtered, the Weyl algebra. $A_n(\mathbb{C})$ may also be considered as $\mathbb{Z}^{zn}$-filtered.

1.3 Remarks

1. The change from $\mathbb{Z}$ to Γ may present serious problems on the level of Noetherian conditions, in particular $G_F(R)$ being gr-Noetherian is now not equivalent to being Noetherian. This of course does have an effect on the definition of Zariskian filtration, good filtration and other related notions. This is also clear from the fact that $R\Gamma$ need not be Noetherian even if R is a field (take Γ not of finite rank).

   Moreover $\widetilde{R}$ need not even be gr-Noetherian even where $G(R)$ and $R\Gamma$ are gr-Noetherian (hence R is Noetherian).

2. The total order on Γ allows to view Γ as a directed system so that for a filtration FM of type Γ we may indeed define :

$$\varprojlim_{\gamma \in \Gamma} M/F_\gamma M = \widehat{M}$$

and introduce completion as in the discrete case. Note however that expression in terms of (pseudo)-limits of sequences has to be modified , cf. [S]. As in the $\Gamma = \mathbb{Z}$-case we may obtain a category equivalence between R-filt and $\mathcal{F}_{-+} \subset \widetilde{R}$-gr, the full subcategory of graded Γ_+-torsion free $\widetilde{R}$-modules, via the functors $M \to \widetilde{M}$ and $\widetilde{M} \to \widetilde{M}/I\widetilde{M}$ where I is the ideal of $\widetilde{R}$ generated by $\{\gamma - 1, \gamma \in \Gamma_+\}$ or $\{\gamma - 1, \gamma \in \Gamma^+\}$). Note that the left ideal generated by these elements is an ideal because $r_\tau \tau(\gamma - 1) = (\tau\gamma\tau^{-1} - 1)\tau r_\tau$ for $\gamma \in \Gamma_+, r_\tau \in F_\tau R$, with $\tau\gamma\tau^{-1} \in \Gamma_+$ too !

We say that a Γ-filtration FR is Γ-**separated** if for every $x \in R$ there is a $\gamma \in \Gamma$ such that : $x \in F_\gamma R - \sum_{\gamma' < \gamma} F_{\gamma'} R$. For a Γ-separated filtration FR we may define a **principal symbol map** $\sigma : R \to G_F(R)$ by putting

$$\sigma(x) = x - mod \sum_{\gamma < \gamma} F_{\gamma'} R \text{if } x \in F_\gamma R - \sum_{\gamma' < \gamma} F_{\gamma'} R.$$

It is clear that in the Γ-separated case, $\gamma(x) = \deg \sigma(x) \in \Gamma$ is uniquely determined for every x; hence to a Γ-separated filtration FR we correspond a value function v_F by putting : $v_F : R \to \Gamma \cup \{\infty\}, v_F(0) = \infty, v_F(a) = - \deg \sigma(a)$ for $a \neq 0$. Observe that for discrete Γ (e.g. $\mathbb{Z}^n$) the Γ-separability of FR reduces to the usual condition $\cap_{\gamma \in \Gamma} F_\gamma R = 0$ expressing **separability** for FR..

1.4 Proposition

With notation as above :

1. The value-function of a Γ-separated filtration FR satisfies.

   a) $v_F(ab) \geq v_F(a) + v_F(b)$ for $a, b \in R$.

   b) $v_F(a + b) \geq \min\{v_F(a), v_F(b)\}$ for $a, b \in R$.

   c) $v_F(a) = \infty$ if and only if $a = 0$ (in the not necessarily separated case this can be generalized.

2. If $\sigma(a) \neq 0$ for $a \in R$, then $\sigma(a)$ is right regular in $G_F(R)$ if and only if $v_F(ab) = v_F(a) + v_F(b)$, for all $b \in R$.

   If $G_F(R)$ is a domain then in 1.a. above equality holds

3. If $G_F(R)$ is a domain, then R is a domain.

Proof : Straightforward.

1.5 Lemma

The following statements are equivalent for the Γ-separated filtration FR of type Γ :

1) $F_\gamma R F_\tau R = F_{\gamma+\tau} R$ for all $\gamma, \tau \in \Gamma$

2) $G_F(R)_\gamma G_F(R)_\tau = G_F(R)_{\gamma+\tau}$ for all $\gamma, \tau \in \Gamma$, i.e. $G_F(R)$ is strongly Γ-graded.

3) $G_F(R)_\gamma G_F(R)_{-\gamma} = G_F(R)_0$ for all $\gamma \in \Gamma$

4) $F_\gamma R F_{-\gamma} R = F_0 R$ for all $\gamma \in \Gamma$.

Proof Suppose 1). Then $\tilde{R}$ satisfies $\tilde{R}_\gamma \tilde{R}_{-\gamma} = \tilde{R}_0$ and hence $\tilde{R}$ is strongly graded (cf [NVO]). From the fact that $\tilde{R}\Gamma^+$ is a graded ideal and $\tilde{R}/\tilde{R}\Gamma+ \cong G_F(R)$ it follows that $G_F(R)$ is strongly graded, hence the equivalent statements 2) and 3) follow. Since 1) obviously implies 4) it suffices to establish the implication 2) $\Rightarrow$ 1) in order to finish the proof.

Since $\tilde{R}_\gamma$ maps onto $G_F(R)_\gamma$ modulo $\tilde{R}\Gamma_+$, statement 2) yields $F_\gamma R F_\tau R(\gamma+\tau) + \sum_{\delta < \gamma+\tau} \tilde{R}_\delta(-\delta + \gamma + \tau) = F_{\gamma+\tau} R(\gamma + \tau)$. Since $\tilde{R}_\delta(-\delta + \gamma + \tau) = F_\delta R(\gamma + \tau) \subset F_{\gamma+\tau} R(\gamma + \tau)$. It follows that $F_\gamma R F_\tau R = F_{\gamma+\tau} R$ for $\gamma, \tau \in \Gamma$.

1.6 Lemma

Suppose that A is a Γ-strongly graded ring

1) A is a domain if and only if A_0 is a domain.

2) If A_0 is a prime ring then A is a prime ring.

3) A is (left) gr-Noetherian if and only if A_0 is (left)Noetherian.

Proof

1. We only have to establish the claim that A is domain if A_0 is. Suppose $ab = 0$ for some nonzero $a, b \in A$ and let $a = a_{\gamma_1} + \ldots + a_{\gamma_d}, b = b_{\delta_1} + \ldots + b_{\delta_e}$, with $\gamma_1 \leq \ldots \leq \gamma_d$ and $\delta_1 \leq \ldots \leq \delta_e$ in Γ, be the homogenerous decompositions. From $ab = 0$ we then obtain $a_{\gamma_d} b_{\delta_e} = 0$ with nonzero $a_{\gamma_d}, b_{\delta_e}$.

   This yields $(A_{-\gamma_d} a_{\gamma_d})(b_{\delta_e} A_{-\delta_e}) = 0$ with $A_{-\gamma_d} a_{\gamma_d} \neq 0$ and $b_{\delta_e} A_{-\delta_e} \neq 0$, because if $A_{-\gamma_d} a_{\gamma_d} = 0$ then $A_{\gamma_d} A_{-\gamma_d} a_{\gamma_d} = 0$ would entail $a_{\gamma_d} = 0$ (similarly for the other claim). The former contradicts the assumption that A_0 is a domain.

2. If I and J are homogenous ideals of A such that $IJ = 0$ then $I = AI_0, J = AJ_0$ because A is strongly graded, hence $I_0 A J_0 = 0 = I_0 A_0 J_0$. Since A_0 is assumed to be prime I_0 or J_0 equals zero and therefore I or J equal zero. Consequently A is gr-prime. If now $aAb = 0$ with $a = a_{\gamma_1} + \ldots + a_{\gamma_d}, \gamma_1 \leq \ldots \leq \gamma_d \in \Gamma, b = b_{\delta_1} + \ldots + b_{\delta_e}, \delta_1 \leq \ldots \leq \delta_e \in \Gamma$, then clearly $a_{\gamma_d} y b_{\delta_e} = 0$ for every homogeneous $y \in A$ but then also $a_{\gamma_d} x b_{\delta_e} = 0$ for every $x \in A$. The gr-prime condition already established before now entails the contradiction : $a_{\gamma_d} = 0$ or $b_{\delta_e} = 0$.

3. Obvious because for every graded (left) ideal L of A we have $L = AL_0$.

1.7 Corollaries

1. For a Γ-separated filtration FR such that $G_\Gamma(R)$ is a domain it follows from proposition 1.3 that R is a domain and the value-function v_F satisfies the conditions of a valuation on elements of R (by 1.3 (2) and (1)).

2. If R as in 1 is also Artinian then it has to be a skewfield. In this case $G_F(R)$ is a gr-skewfield and $F_0 R$ is a Γ-valuation ring of R. If $\Gamma_s = \{\gamma \in \Gamma, G_F(R)_\gamma \neq 0\}$ then Γ_s is a normal subgroup of Γ such that $G_F(R)$ is Γ_s-strongly graded (note that for $\gamma \in \Gamma - \Gamma_s$ we have that $F_\gamma R = \sum_{\gamma' < \gamma} F_{\gamma'} R$).

Proof

1. All statements are evident.

2. An Artinian domain is a skewfield. If we look at $\sigma(a_\gamma) \in G_F(R)_\gamma$ with $a_\gamma \in F_\gamma R - \sum_{\gamma' < \gamma} F_{\gamma'} R$ then $a_\gamma^{-1} \in F_\tau R - \sum_{\tau' < \tau} F_{\tau'} R$ for some $\tau \in \Gamma$. Since σ is multiplicative as $G_F(R)$ is a domain, it follows that $\sigma(a_\gamma)\sigma(a_\gamma^{-1}) = 1$ and thus $\tau = \gamma^{-1}$. Consequently $G_F(R)$ is gr-field, hence of the form $\overline{\Delta} * \Gamma$, a crossed product over a skewfield $\overline{\Delta} = G_F(R)_0$. In view of 1, v_F is a valuation function on R and hence $F_0 R = \{r \in R, v_F(r) \geq 0\}$ is a valuation ring of R.

1.8 Remark

Usually we assume that $F_0 R \neq \sum_{\gamma < 0} F_\gamma R$, then in the foregoing $\Gamma_s = \Gamma$ and $G_F(R)$ is strongly Γ-graded. If we have a valuation ring Λ_v in a skewfield Δ then $F_\gamma \Delta = \{x \in \Delta, v(x) \geq -\gamma\}$ defines a Γ-separated filtration on Δ (using the sutjectivity of v) satisfying the properties listed in lemma 1.4. , where $\overline{\Delta}$ is the residue skewfield and $G_F(\Delta) = \overline{\Delta} * \Gamma$. The crossed product structure of $G_F(\Delta)$ derives directly from the structure of a strongly Γ-graded ring over a skewfield $\overline{\delta}$, however it can also be seen as follows :

For x and y in Δ we have $v(x) = v(y)$, if and only if $\Lambda_v x = \Lambda_v y$, consequently for every $\gamma \in \Gamma$ we have that $F_\gamma \Delta = \Lambda_v x + F_{<\gamma}\Delta$ for any $x \in F_\gamma \Delta - F_{<\gamma}\Delta$. Therefore $F_\gamma \Delta / F_{<\gamma}\Delta \cong \Lambda_v x / \Lambda_v x \cap F_{<\gamma}\Delta \cong (\Lambda_v / F_{<0}\Delta)x = \overline{\Delta}x$, i.e. $G_F(\Delta)_\gamma$ is one dimensional (left and right !) over $\overline{\Delta}$.

Note also that $\tilde{R}\Gamma_+$ is a completely prime gr-maximal ideal.

2   Extension of valuations

We fix notation as follows. Let K be a field with valuation ring $0_v \subset K$ and maximal ideal $w_v \subset 0_v$. Let A be a connected positively graded K-algebra, with A_1 a finite dimensional K-space, $A = K[A_1]$, writing $A_1 = \oplus_{i=1}^n Ka_i$. View A as a K-algebra given by generators and relations, say

$$0 \longrightarrow R \longrightarrow K < X_1, \ldots, X_n > \longrightarrow A \longrightarrow 0$$

where $K, < \underline{X} >$ is the free K-algebra on $\underline{X} = \{X_1, \ldots, X_n\}$ and π is defined by $\pi(X_i) = a_i, i = 1, \ldots, n$. The ideal of relations is homogenous. Restriction of π to $0_v < \underline{X} >$ defines a graded subring Λ of A with $\Lambda_0 = 0_v$, by :

$$0 \longrightarrow R \cap 0_v < \underline{X} > \longrightarrow 0_v < \underline{X} > \longrightarrow \Lambda \longrightarrow 0$$

Clearly π maps $w_v < \underline{X} >$ to $w_v \Lambda$ which is a graded ideal of Λ. Put $k_v = 0_v/w_v, \overline{\Lambda} = \Lambda/w_v\Lambda, \overline{R} = (R \cap 0_v < \underline{X} >) + w_v < \underline{X} > /w_v < \underline{X} >$. We have a commutative diagram with exact rows :

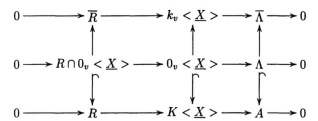

If R is generated by $P_1(\underline{X}), \ldots, P_d(\underline{X})$ say, then we may without loss of generality assume that each $P_i(\underline{X}) \in 0_v < \underline{X} > -w_v < \underline{X} >$. However it does not follow from this that $R \cap 0_v < \underline{X} >$ is generated (as a two-sided ideal) by $\{P_1(\underline{X}), \ldots, P_d(\underline{X})\}$, nor that $\overline{R}$ is generated by the $\overline{P}(\underline{X})$ obtained by reducing the coefficients of each one with respect to 0_v.

We say that R (or A) **reduces well at** 0_v or that Λ **defines a good reduction** if R is generated as an ideal by $\{\overline{P}_1(\underline{X}), \ldots, \overline{P}_d(\underline{X})\}$ (for this it is sufficient that $R \cap 0_v < \underline{X} >$ is generated by $\{P_1(\underline{X}), \ldots, P_d(\underline{X})\}$ as a two sided ideal of $0_v < \underline{X} >$).

We change notation of section 1 somewhat and write fK for the valuation filtration of K associated to the valuation v. We may define a Γ-filtration $fK < \underline{X} >$ by putting $f_\gamma K < \underline{X} > = (f_\gamma K) < \underline{X} >$ for $\gamma \in \Gamma$; this defines a strong filtration with $f_0 K < \underline{X} >$ equal to $0_v, \underline{X} >$. We say that a (left) ideal J of $0_v < \underline{X} >$ is v-**comaximal** if for all $\gamma \in \Gamma$ we have $J \cap (f_\gamma K) < \underline{X} > = (f_\gamma K)J$.

2.1 Lemma

Let L be the ideal of $O_v < \underline{X} >$ generated by $P_1(\underline{X}), \ldots, P_d(\underline{X})$. If L is v-comaximal then R reduces well at O_v.

Proof Since $fK < \underline{X} >$ is a strong filtration it follows that

$$R = K < \underline{X} > (f_0 K < \underline{X} > \cap R) = K < \underline{X} > (O_v < \underline{X} > \cap R).$$

By assumption we have $L^l K < \underline{X} >= R$, where L^l is the left ideal in $O_v < \underline{X} >$ generated by the $P_1(\underline{X}), \ldots, P_d(\underline{X})$. If $x \in f_0(L^l K < \underline{X} >)$ then there is a $\gamma \in \Gamma_+$ such that $x f_{-\gamma} K < \underline{X} >\subset L$ as well as $x f_{-\gamma} K < \underline{X} >\subset f_{-\gamma} K < \underline{X} >$ because x was in $f_0(L^l K < \underline{X} >\subset f_0 K < \underline{X} >$. Consequently, $x f_{-\gamma} K < \underline{X} >\subset L \cap f_{-\gamma} K < \underline{X} >= (f_{-\gamma} K) L$ by the v-comaximality of L. Therefore x is in L because $(f_{-\gamma} K) L f_\gamma K < \underline{X} >\subset O_v < \underline{X} > L O_v < \underline{X} >\subset L$.
Thus we obtain

$$L \subset R \cap O_v < \underline{X} >= f_0 R = f_0 (L^l k < \underline{X} >) \subset L$$

proving that $R \cap O_v < \underline{X} >= L$ and it is the ideal generated in $O_v < \underline{X} >$ by $\{P_1(\underline{X}), \ldots, P_d(\underline{X})\}$. Obviously ; $\overline{R} = (\overline{P}_1(\underline{X}), \ldots, \overline{P}_d(\underline{X}))$ follows from this.

2.2 Remark

In case the reduced relations $\{\overline{P}(\underline{X}), \ldots, \overline{P}_d(\underline{X})\}$ determine a simple k-algebra $k < \underline{X} > /(\ldots, \overline{P}_i(\underline{X}), \ldots)$ then Λ defines a good reduction, indeed $(R \cap O_v < \underline{X} >)$ mod $w_v < \underline{X} >$ must be equal to the maximal ideal $(\overline{P}_1(\underline{X}), \ldots, \overline{P}_d(\underline{X}))$ of $k < \underline{X} >$, i.e. $\overline{R} = (\overline{P}_1(\underline{X}), \ldots, \overline{P}_d(\underline{X}))$.

For example, if we look at the Weyl algebra $A_n(K)$ then the reduced relation defines the simple algebra $A_n(k)$, hence $A_n(O_v)$ defines a good reduction. A similar argument, up to minor modifications, can be applied to define that the quantized Weyl algebra defined by the relation $XY - qYX$ reduces well at every $O_v \subset K$ containing q in the constant field. Similar phenomena of good reduction exist for algebras having a standard basis, i.e. having a PBW theorem as in several filtered examples.

2.3 Lemma

Assume that Γ is discrete Let A be as before and assume moreover that A is gr-simple, then the filtration on A defined by $F_\gamma A = (f_\gamma K) \Lambda$ is separated and $G_F(A)$ is strongly graded. In this case if $\overline{\Lambda}$ is a domain, $G_F(A)$ is a domain too and Λ is a domain.

Proof Look at $I = \cap_{\gamma \in \Gamma_+} (f_{-\gamma} K) \Lambda$. Clearly $KI \subset I, IK \subset I$, hence $AI \subset I$ and $IA \subset I$ because $K\Lambda = A$. Then I is a graded ideal of A hence $I = 0$. That $G_F(A)$ is strongly graded follows from the obvious fact that the filtration FA is strong. If $\overline{\Lambda}$ is a domain then lemma 1.5. entails that $G_F(A)$ is a domain and from corollaries 1.6. (1) it follows that Λ is a domain too.

2.4   Remark

1. The undergraded version of the lemma holds for simple rings of the type $K < \underline{X} > /(P_1(\underline{X}), \ldots, P_d(\underline{X}))$. The proof is as above, forgetting the gradation of A and replacing "gr-simple" by "simple".

2. If A is a connected affine positively graded K-algebra, say $A = K \oplus A_1 \oplus A_2 \oplus \ldots$, where each A_i a finite dimensional K-space. We can then consider the graded O_v-algebra Λ in A and I as in the lemma. Since I is a graded ideal of A it now follows that $I \subset A_+$. Let I_d be the minimal degree part of I that is not zero. Then $w_v I_d = I_d$ follows and thus $I_d = 0$ because $I_d \subset \Lambda_d$ and the latter is a finitely generated O_v-module.

3. The proposition 2.4. of [9] is generally valid as long as the Γ-filtrations are being taken with respect to an abelian Γ. More precisely if we look at Γ-filtrations $F^I R, F^{II} R$ (exhaustive and separated) for an abelian torsionfree ordered group Γ, put $S^2 = G_{II}(R), S^1 = G_I(R)$. Now define :

$$f_\gamma^I S_\tau^2 = (F_\tau^{II} R_I \cap F_\gamma^I R)/(F_{<\tau}^{II} R \cap F_\gamma^I R),$$

$$f_\gamma^I S^2 = \oplus_\tau f_\gamma^I S_\tau^2 \text{ for all } \gamma \in \Gamma$$

Then $f^I S^2$ is an exhaustive and separated filtration on S^2. In a similar way we may induce $f^{II} S^1$, by interchanging S^1 and S^2, F^I and F^{II}. Again it follows that $G_{f^I}(G_{II}(R)) = G_{f^{II}}(G_I(R))$.

2.5   Theorem

(version of Theorem 2.7. of [9] for discrete Γ) Let $D_1(K)$ be the Weyl skewfield. Every valuation ring O_v of K extends to be a valuation ring Λ_v of $D_1(K)$ (with same value group Γ).

Proof In view of corollaries 1.6. (2) it suffices to construct a Γ-filtration, where Γ it the value group of v on K, on $D_1(K)$ that it extends the valuation filtration on K and the associated graded ring being a domain. Clearly we only have to construct such such a Γ-filtration on $A_1(K)$ because then it can always be extended to $D_1(K)$ in the obvious way. Remark 2.2. learns that $\Lambda = A_1(O_v)$ defines a good reduction at O_v and $\overline{\Lambda} \cong A_1(k_v)$ is a Weyl algebra over k hence a domain regardless of the value of char k_v. From Remark 2.4. (1) it follows that the associated graded ring corresponding to FA (defined by $F_\gamma A = (f_\gamma K)\Lambda$) is a domain, as desired.

2.6   Theorem

The statements of Lemma 2.3. (and Remark 2.4. (1)) remain valid for non-descrete Γ of one assumes that the $a_1, \ldots, a_n$ may be chosen to be a PBW-basis for A (i.e. elements of a have a unique expression as ordered polynomial expressions in the $a_1, \ldots, a_n$ with respect to some ordering on them). In particular Theorem 2.5. is true for non-discrete Γ.

Proof. We only have to establish that FA as defined in Lemma 2.3. is F separated; then the other arguments may still be applied. If FA is not separated then there is an $x \in A$ such that for every $\gamma \in \Gamma$ such that $x \in F_\gamma A$, there is a $\delta < \gamma$ in Γ such that $x \in F_\delta A$ too. This means that $x \in f_\gamma K \wedge$ entails $x \in f_\delta K \wedge$ for $\delta < \gamma$. Assume that the PBW-ordering on $\{a_1, -, a_n\}$ is exactly the one given by the indices used.

Then $x = \sum \xi_j \underline{a}^{\underline{j}} = \sum \eta_{\underline{i}} \underline{a}^{\underline{i}}$, in multi-index notation, where $\xi_j \in f_\gamma K, \eta_{\underline{i}} \in f_\delta K$. For some $c \in f_{-\gamma} K$ such that $c \xi_j \in m_v, c \eta_{\underline{i}} \in O_v$ and not all in m_v (possible since $\delta < \gamma$), we obtain : $\sum (c \xi_{\underline{i}} - c \eta_{\underline{i}}) \underline{a}^{\underline{i}} = 0$, where we have adepted a communion multi-index notation by adding zero coefficients $\xi_{\underline{i}}$ or $\eta_{\underline{i}}$ when necessary. Since $c \xi_{\underline{i}} \in m_v$ and not all $c \eta_{\underline{i}}$ an in m_v, this relation is nontrivial contradicting the PBW basis property for $\{a_1, \ldots, a_n\}$ in that ordering. Once Γ-separability is established, the general results preceding Lemma 2.3. apply and so all claims follow easily. $\square$ The extension problem for valuations of K-algebras to certain (graded) K-algebras defined by generators and relations has been reduced to checking "good reduction" for the relations and verification of the domain-property for the residue k_v-algebra. The examples obtained in the case of discrete valuations and $\mathbb{Z}$-filtrations also provide examples for more general obtained value group Γ (abelian because we start from an $O_v \subset K$ that is central in A).

Note Since Λ_v extends O_v, for $\lambda \in D_1(K)$ with $v(\lambda) = \alpha$ there is a $c \in K$ with $v(c) = \alpha^{-1}$ and thus $\lambda \in \Lambda_v^* c^{-1}$. Consequently if we look at $x_\sigma \in F_\sigma D_1(K) - F_{>\sigma} D_1(K)$ then $x_\sigma = \mu_0(\sigma) c_\sigma$ for some $c_\sigma \in K$ with $v(c_\sigma) = \sigma$ and $\mu_0(\sigma) \in \Lambda_v^*$. It follows that $G(D_1(K)) = D_1(k_v) \Gamma^t$, $t \in H^2(\Gamma, k_v^*)$ but no action of Γ on $D_1(k_v)$. In fact it is now also easily seen that $G(D_1(K)) \cong D_1(k_v) \Gamma$. Similar observations hold for extensions of central valuations with the same value group (then necessarily abelian).

3 References

1. M. J. Asensio, M. Van den Bergh, F. Van Oystaeyen, *A New Algebraic Approach to Microlocalization of Filtered Rings*, Trans. Amer. Math. Soc. 316, No. 2, 1989, 537-555.

2. M. Awami, F. Van Oystaeyen, *On Filtered Rings with Noetherian Associated Graded Rings*, in : Proceedings of Ring Theory Meeting, Granada, 1986, Springer-Verlag, Berlin and New York, 1987.

3. Bjork, J. E., *Rings of Differential Operators*, Math. Library, Vol. 21, North Holland, Amsterdam, 1979.

4. Endler, O. *Valuation Theory*, Springer Verlag, New York, 1972.

5. Fuchs, L., Salce, L., *Modules over Valuation Domains*, Marcel Dekker, New York, 1985.

6. Li Huishi, F. Van Oystaeyen, *Filtrations on Simple Artinian Rings*, J. Alg. 132, 361-376, 1990.

7. Li Huishi, F. Van Oystaeyen, *Zariskian Filtrations*, Comm. Alg. 17, 2945-2970, 1989.

8. Li Huishi, F. Van Oystaeyen, *Zariskian Filtrations*, Monograph, D. Reidel Publications, 1996.

9. H. Moawad, F. Van Oystaeyen, *Discrete Valuations Extend to Certain Algebras of Quantum Type*, Comm. Alg. 24, 8, 2551-2566, 1996.

10. C. Nastasescu, F. Van Oystaeyen, *Dimensions of Ring Theory*, D. Reidel Publishing Company, 1987.

11. C. Nastasescu, F. Van Oystaeyen, *Graded and Filtered Rings and Modules*, LNM 728, Springer Verlag, Berlin, 1979.

12. C. Nastasescu, F. Van Oystaeyen, *Graded Ring Theory*, Math. Library, 28, North Holland, Amsterdam, 1982.

13. O. F. Schilling, *The Theory of Valuations*, Mathematical Surveys, No. IV, 1950.

A Survey of Covers and Envelopes

Edgar E. Enochs
Department of Mathematics

University of Kentucky
Lexington, KY 40506-0027
USA
enochs@ms.uky.edu

Overtoun M.G. Jenda
Department of Discrete
and Statistical Sciences
Auburn University
Auburn, AL 36849-5307
USA
jendaov@mail.auburn.edu

1. Introduction

Covers and envelopes were defined in [12] and the same notions (but with the terminology right and left minimal approximations) were defined in [4]. In this article we will give some of the ideas we consider important, mention some open questions and point our directions taken by some recent research. The terminology of a cover comes from the notion of a simply connected covering space $U \to X$ of a connected topological manifold X and the terminology of an envelope comes from the notion of an injective envelope of a module.

2. Definitions and Remarks

If $\mathcal{F}$ is class of objects of a category $\mathcal{C}$, by an $\mathcal{F}$-precover of an object X of $\mathcal{C}$ we mean a morphism $\phi : F \to X$ with $F \in \mathcal{F}$ such that $\mathrm{Hom}\,(G, F) \to \mathrm{Hom}\,(G, X)$ is surjective for all $G \in \mathcal{F}$. If, moreover, any morphism $f : F \to F$ with $\phi \circ f = \phi$ is an automorphism of F, then $\phi : F \to X$ is called an $\mathcal{F}$-cover of X. Clearly $\mathcal{F}$-covers (if they exist) are unique up to isomorphism.

141

The dual notions are those of an $\mathcal{F}$-preenvelope and $\mathcal{F}$-envelope. When $\mathcal{F}$ is some class such as the class of flat modules in the category of left R-modules, an $\mathcal{F}$-cover is called a flat cover -- or just a cover if the class $\mathcal{F}$ is understood.

If $\mathcal{C}$ is an abelian category, an $\mathcal{F}$-precover $\phi : F \to X$ is said to be special if $\mathrm{Ext}^1(G, \ker(\phi)) = 0$ for all $G \in \mathcal{F}$. We note that if $\phi : F \to X$ is an epimorphism with $F \in \mathcal{F}$ where $\ker(\phi)$ satisfies this condition, then $\phi : F \to X$ is in fact a precover since $\mathrm{Hom}(G, F) \to \mathrm{Hom}(G, X) \to 0 = \mathrm{Ext}^1(G, \ker(\phi))$ is exact when $G \in \mathcal{F}$. This observation is very useful when trying to construct precovers. Special preenvelopes are defined dually.

3. Examples

The existence of injective envelopes is due to Eckmann and Schopf [7]. The dual notion of a projective cover was considered by various authors, but Bass in [5] showed that left perfect rings are precisely the rings for which every module has a projective cover. A. Gleason in [30] proved that projective covers exist in the category of topological spaces. Algebraic closures of fields are envelopes in the category of fields when our class is the class of algebraically closed fields. The notion of a total integral closure of a commutative ring R is an envelope which extends the notion of an algebraic closure ([10], [31], [6], [34]). However R has such an envelope if and only if R has no nonzero nilpotent elements.

Modules over integral domains always have torsion free covers [9]. By Pontrjagin duality this result implies that compact abelian groups have connected envelopes. However a locally compact abelian group G has a connected envelope if and only if $\text{Hom}(G, \mathbf{R})$ has finite dimension over $\mathbf{R}$ [19].

We propose these conjectures:

Every module has a flat cover

Nilpotent groups have torsion free covers (the notions of torsion free and torsion are well behaved for nilpotent groups, hence it seems possible that the main result of [9] carry over to this situation)

Every compact topological group G has a connected envelope $\phi : G \to C$ and $\pi_n(\phi)$ is an isomorphism for $n \geq 1$ (if we restrict ourselves to the case where the groups are also commutative, see [9] and [13] for proofs of the conjecture. Also see [19] for a related result)

4. Zorn's lemma for categories and applications

The result we give the name above is useful in proving that the existence of a precover (preenvelope) implies the existence of a cover (envelope).

Theorem. *If C is a locally small category and if every well-ordered inductive system in C admits a mapping to some object of C, then if C has an object X such that $\text{Hom}(Y, X) \neq \emptyset$ for all objects Y of C and such that all weak idempotents of X split, then X has a retract Z such that every endomorphism $Z \to Z$ is an isomorphism. (We note that*

$p:\ X \to X$ *is said to be a weak idempotent if for every object Y of*
$\mathcal{C}$, $\mathrm{Hom}\,(Y,f):\ \mathrm{Hom}\,(Y,X) \to \mathrm{Hom}\,(Y,X)$ *is such that the function*
$\mathrm{Im}\,\mathrm{Hom}(Y,f) \to \mathrm{Im}\,\mathrm{Hom}(Y,f)$ *induced by* $\mathrm{Hom}(Y,f)$ *is a bijection).*

We see that the Z of the conclusion of the theorem has the prop-
erties that $\mathrm{Hom}\,(Y,Z) \neq \emptyset$ for all objects Y of $\mathcal{C}$ and that every
endomorphism $g:\ Z \to Z$ is an automorphism. And clearly any
other object Z' of $\mathcal{C}$ having these properties is isomorphic to Z.

A sample of the various applications of this theorem is:

Corollary. *If a left R-module M has a flat precover, then it has a*
flat cover.

We let $\mathcal{C}$ be the category with objects the flat precovers $F \to M$
of M where morphisms are given by commutative diagrams

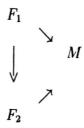

Then the hypotheses of the theorem are satisfied and the conclusion
implies the existence of a flat cover of M.

5. Sums of covers and local nilpotency

Let $\mathcal{F}$ be a class of left R-modules in the category of left R-
modules. Let $\phi:\ F \to M$ be an $\mathcal{F}$-cover such that $\psi:\ F \oplus F \oplus \cdots \to$
$M \oplus M \oplus \cdots$ is also an $\mathcal{F}$-cover (with $\psi = \phi \oplus \phi \oplus \cdots$). Let $g:\ F \to$
$\ker(\phi)$ be any linear map and let $f:\ F \oplus F \oplus \cdots \to F \oplus F \oplus \cdots$ be the
map $(x_1, x_2, x_3, \ldots) \mapsto (x_1, x_2 - g(x_1), x_3 - g(x_2), \ldots)$. Easily $\psi \circ f = \psi$

and so f must be an automorphism. So for any $x \in F$, $(x, 0, 0, 0, ...)$ must be in the image of f. If $f(x_1, x_2, x_3, ...) = (x, 0, 0, ...)$ then $x_1 = x$, $x_2 = g(x_1)$, $x_3 = g(x_2) = g^2(x)$ etc. But for large n, $x_n = 0$ and so $g^n(x) = 0$. So we say g is locally nilpotent on F. This result and a version for $\mathcal{F}$-envelopes gives strong information about $\mathcal{F}$-envelopes and and $\mathcal{F}$-covers.

For example, it implies that if $M \subset E$ is an injective envelope over a left noetherian ring R, any linear $f : E \to E$ with $f(E) \subset M$ is locally nilpotent on E. This result then says $E = \varinjlim \ker(f^n)$. It seems likely that under suitable hypothesis there is a dual result. This result would say that if $\phi : F \to M$ is an $\mathcal{F}$-cover and if $g : F \to F$ is such that $g(F) \subset \ker(\phi)$ then $F = \varprojlim \operatorname{coker}(g^n)$.

6. Wakamatsu lemmas

Let $\mathcal{C}$ be abelian category and let $\mathcal{F}$ be a class of objects of $\mathcal{C}$ which is closed under extensions. Then if $\phi : F \to X$ is an $\mathcal{F}$-cover and if we have a diagram

$$
\begin{array}{ccc}
A & \longrightarrow & B \\
\downarrow & & \downarrow \\
F & \longrightarrow & X
\end{array}
$$

where $A \to B$ is a monomorphism with $\operatorname{coker}(A \to B) \in \mathcal{F}$, then the diagram can be completed to a commutative diagram (with both triangles commuting (see [11] Proposition 2.2 for an argument in a particular case). This result implies what is now known as Wakamatsu's lemma ([36], Proposition 2.22).

Lemma. *In the situation above, $Ext^1(G, ker(\phi)) = 0$ for all $G \in \mathcal{F}$.*

Proof. In we consider any commutative diagram

$$
\begin{array}{ccccccc}
0 & \longrightarrow & \ker(\phi) & \longrightarrow & Y & \longrightarrow & G \longrightarrow 0 \\
  &   & \downarrow &   & \downarrow 0 & & \\
  &   & F & \longrightarrow & X & &
\end{array}
$$

with an exact top row and with $\ker(\phi) \to F$ the canonical map, we see that the map $Y \to F$ can be factored $Y \to \ker(\phi) \to F$. Hence the top exact sequence is split exact. This lemma has an obvious dual concerning envelopes.

Another variant of Wakamatsu's lemma concerns minimal generators. With the same hypothesis on $\mathcal{C}$ and $\mathcal{F}$, a short exact sequence $\xi : 0 \to X \to Y \to F \to 0$ with $F \in \mathcal{F}$ is said to be a minimal generator of $Ext^1(\mathcal{F}, X)$ if for any

$$
\xi' : 0 \to X \to Y' \to F' \to 0 \in Ext^1(F', X)
$$

with $F' \in \mathcal{F}$ there is a $g : F' \to F$ such that $Ext^1(g, X)(\xi) = \xi'$ and if any $g : F \to F$ with $Ext^1(g, X)(\xi) = \xi$ is an automorphism of F. Then if $\xi : 0 \to X \to Y \to F \to 0$ is such a minimal generator, $Ext^1(G, Y) = 0$ for all $G \in \mathcal{F}$. Then it follows that if we let $F^\perp$ consist of all objects Z with $Ext^1(G, Z) = 0$ for all $G \in \mathcal{F}$, $X \to Y$ is an $\mathcal{F}^\perp$-envelope. In fact this proceedure was developed to prove the existence of Gorenstein injective envelopes over Iwanaga-Gorenstein rings (see Theorem 6.1 of [29] and also section 10 below).

7. Xu's Theorem

In 1995, Jinzhong Xu published the best result to date concerning the existence of flat covers.

Theorem ([38]). *If R is commutative, noetherian and of finite Krull dimension, then all R-modules have flat covers.*

There are several interesting tools developed by Xu in his work. We give one sample. Let

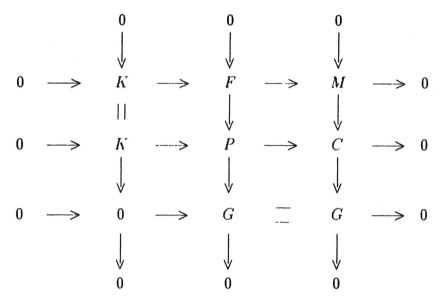

be a commutative diagram with exact rows and columns. If $F \to M$ is a flat cover (so an $\mathcal{F}$-cover with $\mathcal{F}$ the flat modules) and if $F \to P$ is an $\mathcal{F}^{\perp}$-envelope, then $P \to C$ is a flat cover and $M \to C$ is an $\mathcal{F}^{\perp}$-envelope. With proper restrictions on a class $\mathcal{F}$, this result has the obvious extensions. And then there is a dual result.

It seems that the next appropriate step in proving the existence of flat covers should be to remove the hypothesis that R have finite Krull dimension from Xu's theorem. The problem is in removing this hypothesis in Theorem 3.1 of [14]. We know of no progress in doing so.

The general conjecture is that modules over any ring have flat covers. If the general conjecture is to be settled, it seems likely that results with a set-theoretic flavor will be appealed to (eg. see Corollary 11 of [8] which gives such a result).

8. Flat Envelopes

It is not hard to prove that every left R-module has a flat preenvelope if and only if R is right coherent. Flat envelopes though are much rarer. For example, if R is a domain and every module has a flat envelope, then w.gl.dim $R \leq 2$ ([12], Theorem 6.1). Martínez Hernández improves this result by arguing that modules over a domain R have flat envelopes if and only if w.gl.dim $R \leq 2$ and R is coherent (Theorem 2.12 of [32]). Martínez Hernández and Asensio Mayor extend this result by showing that any R is such that left R-modules have flat envelopes that complete diagrams in a unique way if and only if w.gl.dim $R \leq 2$ and R is right coherent ([2], Proposition 2.1). In the same article they give a characterization of commutative R such that every R-module has a monomorphic flat envelope (Theorem 2.2) Saorín in ([35], Theorem 2.1) gives a deep structural characterization of these R. For results, concerning flat envelopes see [2], [12], [32], [33] and [35].

9. Balance and Relative Homological Algebra

If $\mathcal{G}$ is a class of objects that is preenveloping (i.e. every object has a $\mathcal{G}$-preenvelope) in an abelian category $\mathcal{C}$, then with every object X

of $\mathcal{C}$ we can associate a complex

$$0 \to X \to G^0 \to G^1 \to \cdots$$

by letting $X \to G^0$, $\text{coker}\,(X \to G^0) \to G^1$, $\text{coker}\,(G^0 \to G^1) \to G^2, \cdots$ all be $\mathcal{G}$-preenvelopes. The deleted complex $0 \to G^0 \to G^1 \to \cdots$ is unique up to homotopy of complexes and so can be used to compute derived functors of an additive functor $T : \mathcal{C} \to \mathcal{A}$ into another abelian category $\mathcal{C}$. If T is covariant we get the right derived functors and if T is contravariant we get the left derived functors.

The more interesting applications occur when we have a biadditive functor $T : \mathcal{C} \times \mathcal{D} \to \mathcal{E}$ of two variables (with $\mathcal{C}, \mathcal{D}$ and $\mathcal{E}$ all abelian) which is, say, contravariant in the first and covariant in the second variable. Then if there are classes $\mathcal{F}$ and $\mathcal{G}$ of objects of $\mathcal{C}$ and $\mathcal{D}$ respectively which are respectively precovering and preenveloping we say T is right balanced by the pair $(\mathcal{F}, \mathcal{G})$ if the functor $T(F, -)$ makes any complex $0 \to Y \to G^0 \to G^1 \to \cdots$ formed by taking $\mathcal{G}$-preenvelopes as above exact when $F \in \mathcal{F}$ and similarly $T(-, G)$ makes any complex $\cdots \to F_1 \to F_0 \to X \to 0$ formed by taking $\mathcal{F}$-precovers exact.

In this situation the right derived functors $(R^n T)(X, Y)$ can be computed using either of the two resolutions. This gives a tool which is particularly useful when comparing the global dimensions defined using the two classes $\mathcal{F}$ and $\mathcal{G}$. For examples see [21].

10. Gorenstein Homological Algebra

We call a ring R Iwanaga-Gorenstein if R is right and left noetherian and if R has finite self injective dimension on both sides. Over such rings there is an especially pleasant case of the balance mentioned in the last section.

Definition. A module G is said to be Gorenstein injective if there is an exact sequence

$$\cdots \to E^{-2} \to E^{-1} \to E^0 \to E^1 \to E^2 \to \cdots$$

of injective modules with $G = \ker(E^0 \to E^1)$ which is left exact by $\mathrm{Hom}\,(E,-)$ for any injective module E. Gorenstein projective modules are defined dually.

We have

Theorem ([22] and [29]). *If R is Iwanaga-Gorenstein then every module has a Gorenstein injective envelope and a Gorenstein projective precover. The functor $\mathrm{Hom}\,(-,-)$ is right balanced by $(\mathcal{C},\mathcal{G})$ where $\mathcal{C}$ is the class of Gorenstein projective modules and where $\mathcal{G}$ is the class of Gorenstein injective modules. If moreover inj. dim $_R R = n$, then for every module M there is an exact sequence $0 \to M \to G^0 \to E^1 \to \cdots \to E^n \to 0$ with G^0 Gorenstein injective and $E^1, ..., E^n$ injective. Likewise there is an exact sequence $0 \to P_n \to \cdots \to P_1 \to C_0 \to M \to 0$ with C_0 Gorenstein projective and $P_1, ..., P_n$ projective.*

If in addition R is commutative and local with residue field k and we let $M = k$ in the above we see that R resembles a ring of finite global dimension from another aspect. For then if we choose $0 \to k \to G^0 \to E^1 \to \cdots \to E^n \to 0$ as above but minimal, then $E^n \cong E(k)$ (the injective envelope of k). In fact it can be shown that

$E^i = E(k)^{\beta_{n-i}}$ for $i = 1, 2, ..., n$ where β_i is the i-th Betti number of the R-module k.

There is also a notion of a Gorenstein flat module.

Definition. A module F is said to be Gorenstein flat if there is an exact sequence

$$\cdots \to F^{-1} \to F^0 \to F^1 \to \cdots$$

of flat modules with $F = \ker(F^0 \to F^1)$ and such that for any injective right R-module E, $E \otimes -$ leaves the sequence exact.

The next result seems to justify calling these modules Gorenstein flat.

Proposition [19]. *If R is Iwanaga-Gorenstein, a module F is Gorenstein flat if and only if it is the inductive limit of Gorenstein projective modules.*

For graded versions of these notions see [1].

11. Orthogonal Classes and Torsion Theory Relative to Ext

In an abelian category $\mathcal{C}$ two objects X and Y are said to be orthogonal relative to Ext^1 if $\mathrm{Ext}^1(X, Y) = 0$. If $\mathcal{L}$ is a class of objects of $\mathcal{C}$ we let $\mathcal{L}^\perp$ consist of all objects Y such that $\mathrm{Ext}^1(X, Y) = 0$ for all $X \in \mathcal{L}$ and let $^\perp\mathcal{L}$ consist of all Y such that $\mathrm{Ext}^1(Y, X) = 0$ for all $X \in \mathcal{L}$. Then clearly $\mathcal{L} \subset {}^\perp(\mathcal{L}^\perp)$ and $\mathcal{L} \subset ({}^\perp(\mathcal{L}^\perp)^\perp$ for all classes $\mathcal{L}$. There are many important cases where equality holds and where $\mathcal{L}$ and $\mathcal{L}^\perp$ (or $^\perp\mathcal{L}$) have nice descriptions. For example if R is Iwanaga-Gorenstein and if $\mathcal{L}$ consist of all modules L with inj. $\dim L < \infty$ then

$\mathcal{L}^{\perp}$ is the class of Gorenstein injective modules and $^{\perp}\mathcal{L}$ is the class of Gorenstein projective modules. Furthermore $^{\perp}(\mathcal{L}^{\perp}) = \mathcal{L} = (^{\perp}\mathcal{L})^{\perp}$. In this situation the Gorenstein flat right modules are those F such that $\text{Tor}_1(F, L) = 0$ for all $L \in \mathcal{L}$. This suggests that the orthogonality notion may be usefully applied to other functors of two variables. Noting the similarity with the orthogonality relative to $\text{Hom}(-, -)$ which is considered in torsion theory, it is not hard to see how to define torsion theories relative to Ext^1. This program is carried out in [25]. There it is shown that there is a way to define both hereditary and perfect torsion theories relative to Ext^1 and that there are natural examples of each.

12. Auslander's Last Theorem

Before his death Auslander announced the beautiful result that if R is a commutative and local Gorenstein ring then every finitely generated module M has a maximal Cohen-Macaulay approximation $C \to M$ (with C finitely generated). In the language of this article this says M has a Gorenstein projective cover and that this cover is finitely generated (see [18]).

There are several possible ways of extending this result. Among them is a noncommutative version of the result.

Theorem ([26]. *If R is a complete local Gorenstein ring and G is a finite group then every finitely generated left RG-module has a Gorenstein projective cover which is also finitely generated.*

This result then allows one to create interesting modular representations of finite groups. For instance if we have a finite dimensional representation V of G over $Z/(p)$ (with p a prime), we can regard V as a $\hat{Z}_p G$-module. Then if $C \to V$ is a cover as above, C is a finitely generated free $\hat{Z}_p$-module. With $G = GL_n(Z/(p))$ and $V = Z/(p)^n$ as a natural G-module, the $\hat{Z}_p G$-module C gives a homomorphism $GL_n(Z/(p)) \to GL_m(\hat{Z}_p)$ where m is the rank of C over $\hat{Z}_p$. In [21] it is noted that this homomorphism can be regarded as giving generalized Teichmüller invariants.

13. Complexes and Covers and Envelopes

In the category of complexes of left R-modules over any ring R the class $\mathcal{E}$ of exact complexes play a special role. For example, they form one of the key ingredients in forming the derived category.

Theorem [28] and [15]. *Every complex of R-modules has an exact cover. Every complex of left R-modules has an exact envelope if and only if the ring R is left perfect.*

This result can be used to get the derived category without using fractions.

14. Galois Morphisms

Given a covering (or enveloping) class $\mathcal{F}$ in a category $\mathcal{C}$, we can consider properties of morphisms $f : X_1 \to X_2$ in $\mathcal{C}$ relative to $\mathcal{F}$. For example, we say f is covering if any map $F_1 \to F_2$ which makes the diagram

$$
\begin{array}{ccc}
F_1 & \cdots > & F_2 \\
\phi_1 \downarrow & & \phi_2 \downarrow \\
X_1 & \longrightarrow & X_2
\end{array}
$$

commutative (with $\phi_i : F_i \to X_1,$ $i = 1, 2$ covers) is an isomorphism. Examples suggest that in many cases any such $f : X_1 \to X_2$ is right minimal i.e. any $g : X_1 \to X_1$ with $f \circ g = f$ is an automorphism.

For an initial study of these notions see [17].

We propose the following conditions as a definition of a Galois morphism $f : X_1 \to X_2$ relative to a covering class $\mathcal{F}$.

1) $f : X_1 \to X_2$ is covering

2) $f : X_1 \to X_2$ is right minimal

3) if $\phi_1 : F \to X_1$ is a cover (so $\phi_2 = f \circ \phi_1 : F \to X_2$ is a cover by 1)), then for any $g : F \to F$ such that $\phi_2 \circ g = \phi_2$ there is an $h : X_1 \to X$, such that $\phi_1 \circ h = \phi$, and $f \circ h = f$.

4) if $f = f'' \circ f'$ with $f' : X_1 \to X'$ covering and f' not an isomorphism, then there is a $g : X_1 \to X_1$ not the identity such that $f' \circ g = f'$.

Given a Galois morphism relative to the covering class $\mathcal{F}$, $f : X_1 \to X_2$ we let the associated Galois group G consists of all morphism $g : X_1 \to X_1$ such that $f \circ g = f$. In some cases there is a natural topology to put on G such that the closed subgroup H of G are in one-to-one correspondence with the decompositions $X_1 \to X' \to X_2$ of f such that $X_1 \to X'$ is covering (see [16).

References

[1] M.J. Asensio, J.A. Lozez Ramos and B. Torrecillas, *Gorenstein gr-Injective and gr-Projective Modules*, to appear in Comm. Algebra.

[2] J. Asensio Mayor and J. Martínez Hernández, *On flat and projective envelopes*, J. Algebra **160** (1993), 434-440.

[3] M. Auslander, I. Reiten and S.O. Smalø, *Representation Theorey of Artin Algebras*, Cambridge Studies in Advanced Mathematics **36**, Cambridge University Press (1955).

[4] M. Auslander and S.O. Smalø, *Preprojective modules over artin algebras*, J. Algebra **66** (1980), 61-122.

[5] H. Bass, *Finitistic dimension and a homological generalization of semi-primary rings*, Trans. Amer. Math. Soc. **95** (1960), 466-488.

[6] W. Borho, *Wesentliche Erweiterungen Kommutative Rings*, Thesis, Hamburg, 1973.

[7] B. Eckmann and A. Schopf, *Über injektive Moduln*, Arch. Math. **4** (1953), 75-78.

[8] P. Eklof and J. Trlifaj, *How to make Ext vanish*, preprint.

[9] E. Enochs, *Torsion free covering modules*, Proc. Amer. Math. Soc. **14** (1963), 884-889.

[10] E. Enochs, *Totally integrally closed rings*, Proc. Amer. Math. Soc. **19** (1968), 701-706.

[11] E. Enochs, *Torsion free covering modules II*, Arch. Math. **22** (1971), 37-52.

[12] E. Enochs, *Injective and flat covers and resolvents*, Israel J. Math. **39** (1981), 189-209.

[13] E. Enochs, *Homotopy groups of connected envelopes of compact abelian groups*, to appear in Revue Romaine de Mathematique Pure et Appliques.

[14] E. Enochs, *Covers by flat modules and submodules of flat modules*, J. Pure Appl. Algebra **57** (1989), 33-38.

[15] E. Enochs and J.R. Garcia Rozas, *Exact envelopes of complexes*, to appear in Comm. Algebra.

[16] E. Enochs, J.R. Garcia Rozas, O. Jenda and L. Oyonarte, *Compact co-Galois groups*, to appear in Math. Proc. Camb. Phil. Soc.

[17] E. Enochs, J.R. Garcia Rozas and L. Oyonarte, *Galois morphisms*, preprint.

[18] E. Enochs, O. Jenda and J. Xu, *A generalization of Auslander's last theorem*, to appear in J. Algebra Represent. Theory.

[19] E. Enochs and W. Gerlach, *Connecting locally compact abelian groups*, Proc. Amer. Math. Soc. **89** (1983), 351-354.

[20] E. Enochs, J. Martínez Hernández and A. del Valle, *Coherent rings of finite weak global dimension*, Proc. Amer. Math. Soc. **126** (1998), 1611-1620.

[21] E. Enochs and O.M.G. Jenda, *Balanced functors applied to modules*, J. Algebra **92** (1985), 303-310.

[22] E. Enochs and O.M.G. Jenda, *Gorenstein injective and projective modules*, Math. Z. **220** (1995), 611-633.

[23] E. Enochs and O.M.G. Jenda, *Gorenstein balance of Hom and Tensor*, Tsukuba J. Math. **19** (1995), 1-13.

[24] E. Enochs, O.M.G. Jenda and B. Torrecillas, *Gorenstein flat modules*, Nanjing Daxue Xuebao Shuxue Bannian Kan **10** (1993), 1-9.

[25] E. Enochs, O.M.G. Jenda, B. Torrecillas and J. Xu, *Torsion theory relative to Ext*, preprint.

[26] E. Enochs, O.M.G. Jenda and J. Xu, *Lifting group representations to maximal Cohen-Macaulay representations*, J. Algebra **188** (1997), 58-68.

[27] E. Enochs, O.M.G. Jenda and J. Xu, *Zorn's lemma for categories*, preprint.

[28] E. Enochs, O.M.G. Jenda and J. Xu, *Orthogonality in the category of complexes*, Okayama J. Math. **38** (1996), 25-46.

[29] E. Enochs and O.M.G. Jenda and J. Xu, *Covers and envelopes over Gorenstein rings*, Tsukuba J. Math. **20** (1996), 487-503.

[30] A.M. Gleason, *Projective topological spaces*, Illinois J. Math. **2** (1958), 482-489.

[31] M. Hochster, *Totally integrally closed rings and extremal spaces*, Pac. J. Math. **32** (1970), 767-779.

[32] J. Martínez Hernández, *Relatively flat envelopes*, Comm. Algebra **14** (1986), 867-884.

[33] J. Martínez Hernández, M. Saorín and A. del Valle, *Noncommutative rings whose modules have essential flat envelopes*, J. Algebra **177** (1995), 434-450.

[34] R.M. Raphael, *Algebraic extensions of commutative regular rings*, Can. J. Math. **22** (1970), 1133-1155.

[35] M. Saorín, *The structure of commutative rings with monomorphic flat envelopes*, Comm. Algebra **23** (1995), 5383-5394.

[36] T. Wakamatsu, *Stable equivalence of self-injective algebras and a generalization of tilting modules*, J. Algebra **134** (1990), 298-325.

[37] J. Xu, *The existence of flat covers over noetherian rings of finite Krull dimensions*, Proc. Amer. Math. Soc. **123** (1995), 27-32.

[38] J. Xu, *Flat Covers of Modules*, Lecture Notes in Math. **1634**, Springer-Verlag, Berlin, Heidelberg, New York (1996).

On *D*-Gorenstein Modules

Edgar E. Enochs
Department of Mathematics
University of Kentucky
Lexington
KY 40506-0027 USA
enochs@ms.uky.edu

Overtoun M. G. Jenda
Department of Discrete
and Statistical Sciences
Auburn University
AL 36849-5307 USA
jendaov@mail.auburn.edu

Abstract

The aim of this paper is to introduce and study D-Gorenstein injective, projective, and flat modules which are a generalization of the usual Gorenstein injective, projective, and flat modules.

1 Introduction

R will denote a local Cohen-Macaulay ring of Krull dimension d admitting a dualizing module D.

We will let $\mathcal{G}_0(R)$ denote the class of R-modules M such that $\operatorname{Tor}_i(D, M) = \operatorname{Ext}^i(D, D \otimes M) = 0$ for all $i \geq 1$ and such that the natural map $M \to \operatorname{Hom}(D, D \otimes M)$ is an isomorphism, and $\mathcal{I}_0(R)$ denote the class of R-modules N such that $\operatorname{Ext}^i(D, N) = \operatorname{Tor}_i(D, \operatorname{Hom}(D, N)) = 0$ for all $i \geq 1$ and such that the natural map $D \otimes \operatorname{Hom}(D, N) \to N$ is an isomorphism. These classes were introduced and studied by Foxby [7].

The functor $D \otimes - : \mathcal{G}_0(R) \to \mathcal{I}_0(R)$ gives an equivalence between the two categories and so does $\operatorname{Hom}(D, -) : \mathcal{I}_0(R) \to \mathcal{G}_0(R)$. It is also easy to see that if $M_1, M_2 \in \mathcal{G}_0(R)$ then $\operatorname{Hom}(M_1, M_2) \cong \operatorname{Hom}(D \otimes M_1, D \otimes M_2)$, and if $M_1, M_2 \in \mathcal{I}_0(R)$ then $\operatorname{Hom}(M_1, M_2) \cong \operatorname{Hom}(\operatorname{Hom}(D, M_1), \operatorname{Hom}(D, M_2))$.

We now let $\mathcal{W}$ denote the class of R-modules W such that $W \cong D \otimes P$ for some projective R-module P, $\mathcal{V}$ denote the class of R-modules V such that $V \cong \operatorname{Hom}(D, E)$ for some injective R-module E, and $\mathcal{X}$ denote the class of all R-modules X such that $X \cong D \otimes F$ for some flat R-module F. We note from the above that $\mathcal{V} \subseteq \mathcal{G}_0(R)$ and $\mathcal{W}, \mathcal{X} \subseteq \mathcal{I}_0(R)$ since $\mathcal{I}_0(R)$ contains injective R-modules and $\mathcal{G}_0(R)$ contains flat R-modules.

An R-module M is said to be *D-Gorenstein injective* if there exists an exact sequence

$$\cdots \to V_1 \to V_0 \to V^0 \to V^1 \to \cdots$$

of modules in $\mathcal{V}$ such that $M = \operatorname{Ker}(V^0 \to V^1)$ and $\operatorname{Hom}(V, -)$ leaves the sequence exact whenever $V \in \mathcal{V}$. Similarly, M is said to be *D-Gorenstein projective* if there exists an exact sequence

$$\cdots \to W_1 \to W_0 \to W^0 \to W^1 \to \cdots$$

of modules in $\mathcal{W}$ such that $M = \operatorname{Ker}(W^0 \to W^1)$ and $\operatorname{Hom}(-, W)$ leaves the sequence exact whenever $W \in \mathcal{W}$, and M is said to be *D-Gorenstein flat* if there exists an exact sequence

$$\cdots \to F_1 \to F_0 \to F^0 \to F^1 \to \cdots$$

of modules in $\mathcal{X}$ such that $M = \operatorname{Ker}(F^0 \to F^1)$ and $V \otimes -$ leaves the sequence exact whenever $V \in \mathcal{V}$. We will study some basic properties of these modules in Section 2.

We note that each module in $\mathcal{V}, \mathcal{W}$, and $\mathcal{X}$ is D-Gorenstein injective, projective, and flat, respectively. Furthermore, if R is Gorenstein, then $D = R$ and so in this case $\mathcal{G}_0(R)$ and $\mathcal{I}_0(R)$ consist of all R-modules and $\mathcal{V}, \mathcal{W}, \mathcal{X}$ are simply the usual injective, projective, and flat R-modules respectively. Consequently, if M is D-Gorenstein flat, then there exists an exact sequence $\cdots \to F_1 \to F_0 \to F^0 \to F^1 \to \cdots$ of flat modules such that $M = \text{Ker}(F^0 \to F^1)$ and $E \otimes -$ leaves the sequence exact whenever E is injective. Thus in this case, D-Goresntein flat modules are simply the Gorenstein flat modules introduced and studied in Enochs-Jenda-Torrecillas [6]. This explains why we use the class $\mathcal{V}$ instead of $\mathcal{X}$ in the definition of D-Gorenstein flat modules. Similarly, D-Gorenstein injective and projective R-modules are Gorenstein injective and projective modules introduced in Enochs-Jenda [3] if R is Gorenstein.

In Section 3, we will show that $\mathcal{V}$ is a preenveloping and precovering class, that is, every R-module has a preenvelope and a precover. Note that preenvelopes and precovers are also known as left and right approximations respectively.

Since $\mathcal{V}$ is preenveloping, we can construct a complex $0 \to M \to V^0 \to V^1 \to \cdots$ of an R-module M where $M \to V^0$, $\text{Coker}(M \to V^0) \to V^1$, $\text{Coker}(V^{i-1} \to V^i) \to V^{i+1}$ for $i \geq 1$ are $\mathcal{V}$-preenvelopes. This complex is called a *right $\mathcal{V}$-resolution* of M. A *left $\mathcal{V}$-resolution* of M is defined dually using $\mathcal{V}$-precovers. We will say that an R-module M has *right $\mathcal{V}$-dimension* at most n, denoted *right $\mathcal{V}$-dim $M \leq n$*, if there is a right $\mathcal{V}$-resolution of M of the form $0 \to M \to V^0 \to V^1 \to \cdots \to V^n \to 0$. If n is the least, we set *right $\mathcal{V}$-dim $M = n$*.

The main aim of this paper is to show that there is an abundant supply of D-Gorenstein injective, projective, and flat R-modules. We will show in Section 2 that if M is Gorenstein injective, then $\text{Hom}(D, M)$ is D-Gorenstein injective (Lemma 2.1) and if M is Gorenstein projective (flat), then $D \otimes M$ is D-Gorenstein projective (flat) (Lemma 2.3). In Section 3, we will show that if $M \in \mathcal{G}_0(R)$, then every ith cosyzygy in the right $\mathcal{V}$-resolution of M is D-Gorenstein injective whenever $i \geq d$, and every ith syzygy in the left $\mathcal{V}$-resolution of M is also D-Gorenstein injective whenever $i \geq d - 1$ (Theorem 3.5). As a consequence, we get that $M \in \mathcal{G}_0(R)$ is D-Gorenstein injective if and only if $\text{Ext}^i(V, M) = 0$ for all $V \in \mathcal{V}$ and all $i \geq 1$.

2 D-Gorenstein injective, projective and flat modules

We start with the following result.

Lemma 2.1 M is a Gorenstein injective R-module if and only if $M \in \mathcal{I}_0(R)$ and $\text{Hom}(D, M)$ is D-Gorenstein injective.

Proof: If M is Gorenstein injective, then $M \in \mathcal{I}_0(R)$ by Enochs-Jenda-Xu [5, Proposition 1.4]. Furthermore, there exists an exact sequence $\cdots \to E_1 \to E_0 \to E^0 \to E^1 \to \cdots$ of injective R-modules such that $M = \text{Ker}(E^0 \to E^1)$ and $\text{Hom}(E, -)$ leaves the sequence exact whenever E is an injective R-module. But then $\text{Hom}(D, -)$ also leaves the sequence exact by Enochs-Jenda [3, Proposition 2.4] since *inj. dim* $D < \infty$. So we have an exact sequence $\cdots \to V_1 \to V_0 \to V^0 \to V^1 \to \cdots$ where $V_i = \text{Hom}(D, E_i)$ and $V^i = \text{Hom}(D, E^i)$. But then this sequence is left exact by $\text{Hom}(V, -)$ whenever $V \in \mathcal{V}$ since $\text{Hom}(V, V_i) \cong \text{Hom}(E, E_i)$ and $\text{Hom}(V, V^i) \cong \text{Hom}(E, E^i)$ where E is an injective R-module such that $V \cong \text{Hom}(D, E)$. Thus $\text{Hom}(D, M)$ is D-Gorenstein injective.

Conversely, suppose $\cdots \to V_1 \to V_0 \to V^0 \to V^1 \to \cdots$ is an exact sequence of R-modules in $\mathcal{V}$ with $\text{Hom}(D, M) = \text{Ker}(V^0 \to V^1)$ such that the sequence remains exact when $\text{Hom}(V, -)$ is applied to it whenever $V \in \mathcal{V}$. Then we get a complex $\cdots \to E_1 \to E_0 \to E^0 \to E^1 \to \cdots$ of injective modules by applying $D \otimes -$ to the exact sequence. But $M \in \mathcal{I}_0(R)$. So $\text{Hom}(D, M) \in \mathcal{G}_0(R)$. Thus $\text{Ker}(V^i \to V^{i+1}), \text{Ker}(V_{i+1} \to V_i) \in \mathcal{G}_0(R)$ for all $i \geq 0$. But $\text{Tor}_i(D, N) = 0$ for all $i \geq 1$ and all $N \in \mathcal{G}_0(R)$. So the complex $\cdots \to E_1 \to E_0 \to E^0 \to E^1 \to \cdots$ is exact and $M \cong D \otimes \text{Hom}(D, M) = \text{Ker}(E^0 \to E^1)$. It is now easy to see that $\text{Hom}(E, -)$ leaves this complex exact whenever E is injective. Thus M is a Gorenstein injective R-module.

$\square$

Proposition 2.2 *Let $M \in \mathcal{G}_0(R)$. Then M is D-Gorenstein injective if and only if $D \otimes M$ is Gorenstein injective.*

Proof: This easily follows from the lemma above since $M \in \mathcal{G}_0(R)$ implies that $M \cong \operatorname{Hom}(D, D \otimes M)$ and $D \otimes M \in \mathcal{I}_0(R)$. $\square$

Dual arguments give the following result and we include the proof here for completeness.

Lemma 2.3 *An R-module M is Gorenstein projective (flat) if and only if $M \in \mathcal{G}_0(R)$ and $D \otimes M$ is D-Gorenstein projective (flat).*

Proof: Suppose M is Gorenstein projective or Gorenstein flat. Then $M \in \mathcal{G}_0(R)$ by Enochs-Jenda-Xu [5]. If $\cdots \to P_1 \to P_0 \to P^0 \to P^1 \to \cdots$ is an exact sequence of projective modules such that $M = \operatorname{Ker}(P^0 \to P^1)$ and $\operatorname{Hom}(-, P)$ leaves the sequence exact, then $\operatorname{Hom}(-, V)$ also leaves the sequence exact whenever $V \in \mathcal{V}$ by induction since *proj. dim* $V < \infty$. Thus the complex $\cdots \to \operatorname{Hom}(D \otimes P^1, \mathbb{Q}/\mathbb{Z}) \to \operatorname{Hom}(D \otimes P^0, \mathbb{Q}/\mathbb{Z}) \to \operatorname{Hom}(D \otimes P_0, \mathbb{Q}/\mathbb{Z}) \to \cdots$ is exact and so we have an exact sequence $\cdots \to W_1 \to W_0 \to W^0 \to W^1 \to \cdots$ of modules in $\mathcal{W}$ where $W_i = D \otimes P_i$ and $W^i = D \otimes P^i$ for each i. But it is now easy to see that this complex remains exact when $\operatorname{Hom}(-, W)$ is applied to it whenever $W \in \mathcal{W}$. So $D \otimes M = \operatorname{Ker}(W^0 \to W^1)$ is D-Gorenstein projective.

Conversely, if $\cdots \to W_1 \to W_0 \to W^0 \to W^1 \to \cdots$ is as above, then applying $\operatorname{Hom}(D, -)$ to this exact sequence gives a complex $\cdots \to P_1 \to P_0 \to P^0 \to P^1 \cdots$ of projective R-modules. But $D \otimes M = \operatorname{Ker}(W^0 \to W^1) \in \mathcal{I}_0(R)$ since $M \in \mathcal{G}_0(R)$ by assumption. Furthermore $\mathcal{W} \subseteq \mathcal{I}_0(R)$. So $\operatorname{Ker}(W^i \to W^{i+1})$, $\operatorname{Ker}(W_{i+1} \to W_i) \in \mathcal{I}_0(R)$ for all $i \geq 0$. But

$\text{Ext}^1(D, N) = 0$ for all $N \in \mathcal{I}_0(R)$. So the complex $\cdots \to P_1 \to P_0 \to P^0 \to P^0 \to \cdots$ is exact and $M \cong \text{Hom}(D, D \otimes M) = \text{Ker}(P^0 \to P^1)$. Furthermore, if $W, W' \in \mathcal{W}$, then $W \cong \text{Hom}(D, P)$ and $W' \cong \text{Hom}(D, P')$ for some projectives P, P' and so $\text{Hom}(W, W') \cong \text{Hom}(P, P')$. Hence M is Gorenstein projective.

Now suppose M is Gorenstein flat. Then there exists an exact sequence $\cdots \to F_1 \to F_0 \to F^0 \to F^1 \to \cdots$ of flat R-modules such that $M = \text{Ker}(F^0 \to F^1)$ and $E \otimes -$ leaves the sequence exact whenever E is injective. But then by induction, $D \otimes -$ leaves the sequence exact since $inj.\ dim\ D < \infty$. So we have an exact sequence $\cdots \to X_1 \to X_0 \to X^0 \to X^1 \to \cdots$ of modules in $\mathcal{X}$. Now let $V \in \mathcal{V}$. Then $V \cong \text{Hom}(D, E)$ for some injective R-module E and so the complex $\cdots \to V \otimes X_1 \to V \otimes X_0 \to V \otimes X^0 \to V \otimes X^1 \to \cdots$ is equivalent to the complex $\cdots \to E \otimes F_1 \to E \otimes F_0 \to E \otimes F^0 \to E \otimes F^1 \to \cdots$ which is exact. Thus $D \otimes M = \text{Ker}(X^0 \to X^1)$ is D-Gorenstein flat.

Now suppose $M \in \mathcal{G}_0(R)$ and $D \otimes M$ is D-Gorenstein flat. Let $\cdots \to X_1 \to X_0 \to X^0 \to X^1 \to \cdots$ be an exact sequence of R-modules such that $D \otimes M = \text{Ker}(X^0 \to X^1)$ and $V \otimes -$ leaves the sequence exact whenever $V \in \mathcal{V}$. We apply $\text{Hom}(D, -)$ to this sequence to get an exact complex $\cdots \to F_1 \to F_0 \to F^0 \to F^1 \to \cdots$ of flat R-modules as in the case for D-Gorenstein projective above. But if E is injective and $X \cong D \otimes F$ with F flat, then $\text{Hom}(D, E) \otimes X \cong E \otimes F$. Thus it follows from the above that $E \otimes -$ leaves the exact sequence $\cdots \to F_1 \to F_0 \to F^0 \to F^1 \to \cdots$ exact. That is, M is Gorenstein flat. $\square$

Proposition 2.4 *Let $M \in \mathcal{I}_0(R)$. Then M is D-Gorenstein projective (flat) if and only if $\text{Hom}(D, M)$ is Gorenstein projective (flat).*

Proof: Since $M \in \mathcal{I}_0(R)$, we have that $\text{Hom}(D, M) \in \mathcal{G}_0(R)$ and $D \otimes \text{Hom}(D, M) \cong M$. So the result follows from the lemma above. $\square$

3 $\mathcal{V}$-resolutions

We start with the following.

Lemma 3.1 $\mathcal{V}$ *is preenveloping and precovering.*

Proof: Let M be an R-module and imbed $D \otimes M$ into an injective R-module E. Then the composition of the natural map $M \to \operatorname{Hom}(D, D \otimes M)$ and the inclusion $\operatorname{Hom}(D, D \otimes M) \subset \operatorname{Hom}(D, E)$ is a $\mathcal{V}$-preenvelope (see Enochs-Jenda-Xu [5, Proposition 1.5]).

Now let $\{E_k\}$ be a representative set of injective R-modules and set $V_k = \operatorname{Hom}(D, E_k)$. Then $\{V_k\}$ is a representative set of modules in $\mathcal{V}$. But it is easy to see that $\mathcal{V}$ is closed under direct sums. So $\oplus V_k^{(\operatorname{Hom}(V_k, M))} \to M$ is a $\mathcal{V}$-precover as in Enochs [1].   □

Remark 3.2 We note that if $M \in \mathcal{G}_0(R)$, then $\mathcal{V}$-preenvelopes are monomorphisms since $M \cong \operatorname{Hom}(D, D \otimes M)$ and hence right $\mathcal{V}$-resolutions of modules in $\mathcal{G}_0(R)$ are exact.

Lemma 3.3 $\operatorname{Ext}^i(V, V') = 0$ for all V, $V' \in \mathcal{V}$.

Proof:

$$\begin{aligned}
\operatorname{Ext}^i(V, V') &\cong \operatorname{Ext}^i(\operatorname{Hom}(D, E), \operatorname{Hom}(D, E')) \\
&\cong \operatorname{Hom}(\operatorname{Tor}_i(\operatorname{Hom}(D, E), D), E')
\end{aligned}$$

for some injective R-modules E, E'. But $E \in \mathcal{I}_0(R)$. So $\operatorname{Tor}_i(\operatorname{Hom}(D, E), D) = 0$ for all $i \geq 1$ and we are done.   □

Proposition 3.4 *Let M be an R-module. Then*

(1) If $0 \to M \to V^0 \to V^1 \to \cdots$ is an exact sequence with each $V^i \in \mathcal{V}$, then the complex $0 \to \operatorname{Hom}(V, M) \to \operatorname{Hom}(V, V^0) \to \operatorname{Hom}(V, V^1) \to \cdots$ is exact at $\operatorname{Hom}(V, V^i)$ for all $V \in \mathcal{V}$ and all $i \geq d$.

(2) If $\cdots \to V_1 \to V_0 \to M \to 0$ is a left $\mathcal{V}$-resolution of M, then the sequence is exact at V_i for all $i \geq d - 1$ where $V_{-1} = M$.

Proof: Since *inj. dim $D = d$, proj. dim $V \leq d$* for all $V \in \mathcal{V}$ and so $\text{Ext}^i(V, M) = 0$ for all $i \geq d + 1$ for each $V \in \mathcal{V}$. But then (1) follows from Lemma 3.3 above.

$\text{Hom}(-, -)$ is left balanced on $Mod \times Mod$ by $\mathcal{V} \times \mathcal{V}$ where Mod denotes the category of R-modules. So (2) follows by an argument similar to that in Enochs-Jenda [2, Theorem 4.4] since *inj. dim $D = d$* implies that *right $\mathcal{V}$-dim $R = d$* by Enochs-Jenda [4]. $\square$

Theorem 3.5 *Let M be an R-module. Then*

1) If $0 \to M \to V^0 \to V^1 \to \cdots$ is an exact right $\mathcal{V}$-resolution of M and $C^i = \text{Ker}(V^i \to V^{i+1})$. then C^i is D-Gorenstein injective for each $i \geq d$.

2) If $\cdots \to V_1 \to V_0 \to M \to 0$ is a left $\mathcal{V}$-resolution and $C_i = Coker(V_{i+1} \to V_i)$, then C_i is D-Gorenstein injective for each $i \geq d - 1$.

Proof: If $i \geq d$, then the exact sequence $0 \to C^i \to V^i \to V^{i+1} \to \cdots$ is such that $0 \to \text{Hom}(V, C^i) \to \text{Hom}(V, V^i) \to \text{Hom}(V, V^i) \to \cdots$ is exact for all $V \in \mathcal{V}$ by Proposition 3.4 above. But any left $\mathcal{V}$-resolution $\cdots \to V_1 \to V_0 \to C^i \to 0$ of C^i is exact by the same proposition. Pasting these sequences together, we get an exact sequence $\cdots \to V_1 \to V_0 \to V^0 \to V^1 \to \cdots$ which remains exact when $\text{Hom}(V, -)$ is applied to it for any $V \in \mathcal{V}$. Hence C^i is D-Gorenstein injective. (2) follows similarly. $\square$

Remark 3.6 If $M \in \mathcal{G}_0(R)$, then every right $\mathcal{V}$-resolution of M is exact by Remark 3.2. So if $M \in \mathcal{G}_0(R)$, then an *ith* cosyzygy in a right $\mathcal{V}$-resolution of M is D-Gorenstein injective for each $i \geq d$ by the theorem above.

Corollary 3.7 *The following are equivalent for an R-module $M \in \mathcal{G}_0(R)$.*

(1) M is D-Gorenstein injective.

(2) $D \otimes M$ is Gorenstein injective.

(3) $\text{Ext}^i(V, M) = 0$ for all $V \in \mathcal{V}$ and all $i \geq 1$.

Proof: $1 \Leftrightarrow 2$ is Proposition 2.2.

$1 \Rightarrow 3$ is trivial since $\text{Ext}^i(V, V') = 0$ for all $V, V' \in \mathcal{V}$ and all $i \geq 1$ from Lemma 3.3.

$3 \Rightarrow 1$. (3) implies that the exact right $\mathcal{V}$-resolution of M is a left $\mathcal{V}$-resolution and so M is D-Gorenstein injective by the theorem.   □

Remark 3.8 There are similar results in [4] for D-Gorenstein projective and D-Gorenstein flat R-modules.

References

[1] E. E. Enochs, Injective and flat covers, envelopes and resolvents, Israel J. Math 39 (1981), 189-209.

[2] E. E. Enochs and O. M. G. Jenda, Balance functors applied to modules, J. Algebra, 92 (1985), 303-310.

[3] E. E. Enochs and O. M. G. Jenda, Gorenstein injective and projective modules, Math Z., 220 (1995), 611-633.

[4] E. E. Enochs and O. M. G. Jenda, D-Gorenstein covers and envelopes, submitted.

[5] E. E. Enochs, O. M. G. Jenda, and J. Xu, Foxby duality and Gorenstein injective and projective modules, Trans. Amer. Math. Soc., 348 (1996), 3223-3234.

[6] E. E. Enochs, O. M. G. Jenda, and B. Torrecillas, Gorenstein flat modules, J. Nanjing University, 10 (1993), 1-9.

[7] H.-B. Foxby, Gorenstein dimensions over Cohen-Macaulay rings, Proceedings of the international conference on commutative algebra, W. Bruns (editor), Universität Osnabrück, 1994.

Skew Polynomial Rings
with Wide Family of Prime Ideals

Miguel Ferrero
Universidade Federal do Rio Grande do Sul
91509-900 Porto Alegre, Brasil
e.mail: ferrero@mat.ufrgs.br

Jerzy Matczuk
Warsaw University
Banacha 2, 02-097 Warsaw, Poland
e.mail: jmatczuk@mimuw.edu.pl

Introduction

The recent progress in investigations of quantum groups renews the interest in skew polynomial rings $R[t; S, D] = T$ of general type, where S is an endomorphism and D an S-derivation of a ring R. One of the central problems is to describe the prime spectrum of such ring extensions. During the last few years essential progress has been made. The work [2] of Goodearl and Letzter and its list of references can serve as a good source on the subject.

One of the key properties of the prime spectrum of T in several particular cases is that chains of R-disjoint prime ideals are of length not greater than two. It is not known whether this property holds in general even in the case when S is an automorphism but it does hold under some extra assumptions imposed either on the ring R or the maps S and D (c.f. [1], [2], [3], [5], [6], [8]). In particular, it holds when R is a prime symmetrically closed ring and S is an automorphism [6] or when $T = R[t, S]$ is a skew polynomial ring of endomorphism type and either S is an automorphism [8] or R is a commutative ring [3]. Thus if both R and $R[t, S]$ are prime rings, 0 and (t) are prime ideals of $R[t, S]$ and there are no primes between them.

The aim of this paper is to present an interesting curiosity: we give examples of prime symmetrically closed rings R and skew polynomial rings $R[t, S]$ of endomorphism type, such that the family of prime ideals of $R[t, S]$ between 0 and (t) is very far from being finite. In fact we show that these families can contain chains of primes of arbitrary ordinal type, as well can contain a set of incomparable primes of arbitrarily big cardinality, when the ring R is properly chosen.

---

*This research was supported by Conselho Nacional de Desenvolvimento Científico e Tecnológico (CNPq), Brasil.

Those examples show that though the prime spectrum of $R[t, S]$ is well understood in some cases, e.g. when R is commutative or S is an automorphism, the general case is far from being solved, and at the moment looks rather hopeless.

In the first section we introduce the semigroup F_λ whose elements are "infinite words" in letters x_i, $i \geq 0$. We do not know whether this semigroup has been already introduced earlier in the literature but it seems that could be interesting to study its properties in more detail. We prove only technical results which are necessary for later use.

In Section 2 we study the skew polynomial ring $R_\lambda[t, S]$ where R_λ is the semigroup algebra KF_λ over a field K and we obtain the main results of the paper.

1   The semigroup F_λ

Let λ denote an infinite cardinal number (i.e. λ is the smallest ordinal number of cardinality λ). Notice that this choice of λ implies that $\alpha + \beta < \lambda$ for all ordinal numbers $\alpha, \beta < \lambda$. In this section we construct the semigroup F_λ and present its basic properties which will be used in the next section. Let $X = \{x_0, x_1, \ldots\}$. Elements from X are called letters. For convenience we use Ω_α for the ordered set α, where $0 \leq \alpha < \lambda$.

W_α stands for the set $\{f : \Omega_\alpha \to X \mid f(\Omega_\alpha) \text{ is a finite set }\}$. Elements of W_α are called words of length α. One can think about an element of W_α as a well-ordered set of type α whose elements range over a finite subset of X.

A word $f \in W_\alpha$ is finite if Ω_α is a finite set. In this case we write $f = x_{t_1} \cdot \ldots \cdot x_{t_n}$ where $\Omega_\alpha = \{0, \ldots, n-1\}$ and $f(j) = x_{t_{j+1}}$ for $j \in \Omega_\alpha$.

We say that a letter $x \in X$ appears in $f \in W_\alpha$ if $x \in f(\Omega_\alpha)$.

Now we are in position to construct a semigroup F_λ. Its underlying set is $\bigcup_{\alpha < \lambda} W_\alpha$ and for $f \in W_\alpha$, $g \in W_\beta$, the product fg is a word of length $\alpha + \beta < \lambda$ defined as follows. There is a unique isomorphism of well-ordered sets between $\Omega_{\alpha+\beta}$ and the disjoint union $\Omega_\alpha \dot\cup \Omega_\beta$ (where in the latter set $\gamma < \delta$ for all $\gamma \in \Omega_\alpha$ and $\delta \in \Omega_\beta$). Using this isomorphism we can identify $\Omega_{\alpha+\beta}$ with $\Omega_\alpha \dot\cup \Omega_\beta$. Then for any $\gamma \in \Omega_{\alpha+\beta}$ we put

$$fg(\gamma) = \begin{cases} f(\gamma) & \text{if } \gamma \in \Omega_\alpha \\ g(\gamma) & \text{if } \gamma \in \Omega_\beta \end{cases}$$

Since the addition of ordinal numbers is associative, the product defined above is also associative. Thus F_λ is a monoid where the unit element, denoted by 1, is the empty map.

Notice that if $\lambda = \omega_0$, where ω_0 denotes the first limit ordinal number, then F_λ is the free semigroup on the set X.

A subset Δ of Ω_α is called a segment if there are ordinal numbers β, δ such that $\Delta = \{\gamma \mid \beta \leq \gamma < \delta\}$.

We say that the segment Δ of Ω_α is of type β if Δ is a well-ordered set of type β (with the order induced from Ω_α).

Let $f \in W_\alpha$ and $g \in W_\beta$. We say that g is a subword of f if there is a segment $\Delta \subseteq \Omega_\alpha$ of type β such that, while identifying Δ with Ω_β, $f|_\Delta = g$. In other words, there exist words $g_1, g_2 \in F_\lambda$ such that $f = g_1 g g_2$.

For a word $f \in F_\lambda$, we say that $f(\gamma)$ is defined if γ belongs to the domain of f.

Let $1 \neq f \in F_\lambda$ be a finite word. For any $\beta < \lambda$ we define the element f^β as follows:

1. if $\beta = 0$ then $f^\beta = 1$.

2. if $\beta = \gamma + 1$ for some ordinal γ, then $f^\beta = f^\gamma f$.

3. if β is a limit ordinal then f^β is the word of length β such that for $\gamma \in \Omega_\beta$, $f^\beta(\gamma) = f^\rho(\gamma)$, where ρ is the smallest ordinal number such that $f^\rho(\gamma)$ is defined. Notice that such ρ do exist since $\gamma + 1 < \beta$ and, by transfinite induction, $f^{\gamma+1}(\gamma)$ is defined.

In the following lemma we collect basic properties of elements f^β.

Lemma 1.1 *Let $f \in F_\lambda$ be a finite word of length $n \geq 1$ and $\alpha, \beta, \gamma < \lambda$. Then:*

1. *Suppose $f^\beta(\gamma)$ is defined. Then for any $\alpha \geq \beta$ $f^\alpha(\gamma) = f^\beta(\gamma)$ and if γ is a limit ordinal number, then $f^\beta(\gamma) = f(0)$.*

2. *Suppose $f^\beta(\gamma + m)$ is defined and $f^\beta(\gamma) = f(i)$ for some $0 \leq m < \omega_0$ and $i \in \Omega_n$. Then $f^\beta(\gamma + m) = f((i + m)_{\mathrm{mod}\, n})$.*

3. *$f^\alpha f^\beta = f^{\alpha+\beta}$.*

4. *If $f^\alpha = g f^\beta$ for some $g \in F_\lambda$, then $g = f^\delta$ for some δ, provided either the length of g is a limit ordinal or $f(i) \neq f(j)$ for all $i, j \in \Omega_n$ with $i \neq j$.*

Proof. (1). We can suppose that β is the smallest ordinal number such that $f^\beta(\gamma)$ is defined.

We may assume that $\alpha > \beta$. If $\alpha = \delta + 1$, then $\beta \leq \delta < \alpha$. Thus, by transfinite induction, $f^\delta(\gamma) = f^\beta(\gamma)$ and $f^\alpha(\gamma) = f^{\delta+1}(\gamma) = f^\delta f(\gamma) = f^\delta(\gamma) = f^\beta(\gamma)$. If α is a limit ordinal, then $f^\alpha(\gamma) = f^\beta(\gamma)$ by definition. This completes the proof of the first statement from (1).

Now if γ is a limit ordinal then $f^\beta(\gamma) = f^{\gamma+1}(\gamma) = f^\gamma f(\gamma) = f(0)$ as γ is the first element which is not in the domain of f^γ.

(2). Clearly it is enough to prove the statement for $m = 1$. Let α be the smallest ordinal number such that $f^\alpha(\gamma)$ is defined. Then α is not a limit ordinal, say $\alpha = \delta + 1$. Let Ω and Ω' denote the domains of f^δ and f^α, respectively. Then Ω' is isomorphic to the disjoint union $\Omega \dot{\cup} \{0, \ldots, n-1\}$. Because $\gamma \notin \Omega$, γ is identified with some $i \in \Omega_n$. Now it is clear that $f^\alpha(\gamma) = f(i)$ and $f^\alpha(\gamma + 1) = f((i+1)_{\mathrm{mod}\, n})$. This yields the statement (2).

(3). We proceed by transfinite induction with respect to β. If $\beta = 0$ then $f^\beta = 1$, by definition.

If $\beta = \delta + 1$, then $f^{\alpha+\beta} = f^{\alpha+\delta+1} = f^{\alpha+\delta}f = f^\alpha f^\delta f = f^\alpha f^{\delta+1} = f^\alpha f^\beta$

Suppose that β is a limit ordinal. Thus $\alpha + \beta$ is also a limit ordinal, so $f^{\alpha+\beta}$ is of length $\alpha + \beta$. Let $\gamma \in \Omega_{\alpha+\beta}$ and choose $\delta < \alpha + \beta$ such that $f^\delta(\gamma)$ is defined. We may assume that $\delta = \alpha + \rho$ for some $\rho < \beta$. Then, making use of the statement (1) and the transfinite induction hypothesis, we get $f^{\alpha+\beta}(\gamma) = f^{\alpha+\rho}(\gamma) = f^\alpha f^\rho(\gamma) = f^\alpha f^\beta(\gamma)$ and $f^{\alpha+\beta} = f^\alpha f^\beta$ follows.

(4). Suppose that the length δ of g is a limit ordinal. Then for any $\gamma < \delta$ we have $f^\delta(\gamma) = f^\alpha(\gamma) = g(\gamma)$. This implies $g = f^\delta$ in this case, as both words are of length δ.

Now suppose that $f(i) \neq f(j)$ for all $i, j \in \Omega_n$, with $i \neq j$ and δ is arbitrary. Then there exists ordinals μ and m such that $\delta = \mu + m$, where μ is either 0 or a limit ordinal and $m \geq 0$ is finite.

Then $f(m_{\mathrm{mod}n}) = f^\alpha(\mu + m) = gf^\beta(\mu + m) = f^\beta(0) = f(0)$ and the assumption imposed on f yields that $m \equiv 0 \pmod n$. Thus $m = kn$ for some k and now it is easy to see that $g = f^{\mu+k}$ in this case. ∎

Notice that if $f = x^2$, with $x \in X$, then $f^{\omega_0} = xf^{\omega_0}$. Thus the statement (4) from the above lemma does not hold without some additional assumptions. This example shows also that the semigroup F_λ is not cancellative when $\lambda > \omega_0$. However, as the following lemma shows, we can cancel finite words.

Lemma 1.2 *Let $f_i \in W_{\alpha_i}$, $g_i \in W_{\beta_i}$, $i = 1, 2$, and $x \in X$. Then:*

1. *If either $f_1 x = f_2 x$ or $xf_1 = xf_2$, then $f_1 = f_2$.*

2. *If x appears neither in f_1 nor in g_1 and $f_1 x g_1 = f_2 x g_2$, then $f_1 = f_2$ and $g_1 = g_2$.*

3. *If $f_1 g_1 = f_2 g_2$ and f_1 and f_2 have the same lenght, then $f_1 = f_2$ and $g_1 = g_2$.*

Proof. The prove is very easy and is left as an exercise. ∎

The following two lemmas are very technical but they form the core for the results in the next section. In particular, the first one has been inspired by Lemma 13.2 from [7] and is necessary for proving that the semigroup ring KF_λ is symmetrically closed for any field K.

Lemma 1.3 *Let B be a nonempty finite subset of F_λ and $\hat{B} = xBy$, where $x, y \in X$ stand for different letters such that both x and y do not appear in any word from B. Then for any $\hat{u} = xuy, \hat{v} = xvy \in \hat{B}$ and $f, g \in F_\lambda$ we have:*

1. *If $1 \neq w \in F_\lambda$, $\hat{u} = wu'$ and $\hat{v} = v'w$ for some $u', v' \in F_\lambda$, then $\hat{u} = w = \hat{v}$.*

2. *If $\hat{u}f = \hat{v}g$ then $\hat{u} = \hat{v}$ and $f = g$.*

3. *If $f\hat{u} = g\hat{v}$ then $\hat{u} = \hat{v}$ and $f = g$.*

4. If $\hat{u}f = g\hat{v}$ and either $f \neq 1$ or $g \neq 1$, then $f = w\hat{v}$ and $g = \hat{u}w$ for some $w \in F_\lambda$.

Proof. **(1).** Suppose $1 \neq w \in F_\lambda$, $\hat{u} = wu'$ and $\hat{v} = v'w$ for some $u', v', \in F_\lambda$. Since $xvy = \hat{v} = v'w$, w has the last letter y. Now, because $xuy = \hat{u} = wu'$ and y does not appear in xu, we must have $u' = 1$ and $\hat{u} = w$. Hence $\hat{v} = v'\hat{u}$ and since x appears in $\hat{v}$ just once and as the first letter, we obtain $\hat{v} = \hat{u} = w$.

(2). The length of $\hat{u}$ is $\rho + 1$, where ρ is the minimal ordinal such that $(\hat{u}f)(\rho) = y$. Thus $\hat{u}$ and $\hat{v}$ have the same length and hence $\hat{u} = \hat{v}$ and $f = g$ by (3) of Lemma 1.2.

(3). Since x does not appear in uy and $\hat{u} = xuy$, there exists the biggest $\rho < \alpha$ such that $f\hat{u}(\rho) = x$. Now the proof is similar to the one of (2).

(4). Suppose $\hat{u}f = g\hat{v} = h$ is a word of length α.

Case 1: $f \neq 1$. Let $\gamma < \alpha$ be the smallest ordinal such that $h(\gamma) = y$. Since $\hat{u} = xuy$ and $f \neq 1$, γ is in the domain of g. Therefore, as y does not appear in xu, $\hat{u} = xuy = h|_{[0,\gamma+1)} = g|_{[0,\gamma+1)}$. This means $g = \hat{u}w$ for some $w \in F_\lambda$. Hence $\hat{u}f = \hat{u}w\hat{v}$ and, by (2), $f = w\hat{v}$ follows.

Case 2: $g \neq 1$. Let $\gamma < \alpha$ be the biggest ordinal such that $h(\gamma) = x$ (γ does exist, since $\hat{v} = xvy$ and x does not appear in vy). Since $g \neq 1$, γ is in the domain of f. Hence, similarly as in the first case, we get $f = w\hat{v}$ for some $w \in F_\lambda$ and the statement (3) yields $g = \hat{u}w$.   ∎

Let S denote the endomorphism of F_λ induced by the condition $S(x_i) = x_{i+2}$ for all $x_i \in X$, i.e. if $f \in W_\alpha$ then $S(f) \in W_\alpha$ and for any $\gamma \in \Omega_\alpha$ $S(f)(\gamma) = x_{i+2}$ when $f(\gamma) = x_i$. Then for any $\omega_0 > n > 0$, $S^n(F_\lambda)$ is the subsemigroup of F_λ consisting of all words from F_λ in letters $x_{2n}, x_{2n+1}, \ldots$. This subsemigroup is denoted by $F_{\lambda,2n}$.

Let us fix $0 < \alpha < \lambda$. For any $1 \leq N \leq k < \omega_0$ we define subsets $I_{N,k}$ of F_λ as follows:

$$I_{N,k} = F_\lambda(x_0x_1)^\alpha F_{\lambda,2N} \cup F_\lambda(x_2x_3)^\alpha F_{\lambda,2N+2} \cup \ldots \cup F_\lambda(x_{2(k-N)}x_{2(k-N)+1})^\alpha F_{\lambda,2k}$$

Thus

$$I_{N,k} = \{f \in F_\lambda \mid f = g(x_{2i}x_{2i+1})^\alpha h \text{ for some } g \in F_\lambda, \ h \in F_{\lambda,2(N+i)} \text{ and } 0 \leq i \leq k - N\}$$

In the following lemma we collect basic properties of sets $I_{N,k}$ which are necessary for later use.

Lemma 1.4 *Let $1 \leq N \leq k < \omega_0$ and $l > 1$ be an odd natural number. Then for any $f, g \in F_\lambda$ we have:*

1. $I_{N,k+1} = I_{N,k} \cup F_\lambda S(I_{N,k})$.

2. $S^{-1}(I_{N,N}) = \emptyset$ and $S^{-1}(I_{N,k+1}) = I_{N,k}$.

3. *If $f \notin I_{N,k}$ then $S^i(f) \notin I_{N,k+i}$ for any $i \geq 0$.*

4. *If $f \notin I_{N,k}$ then $x_0 x_l f \notin I_{N,k}$.*

5. If $n, m \geq N$ and $f \notin I_{N,n}$, $g \notin I_{N,m}$ then $f x_{2n} x_{2n+l} S^n(g) \notin I_{N,n+m}$.

6. If $1 \leq n \leq N - 1$ and $0 \leq i \leq n - 1$ then $f x_{2n} x_{2n+l} S^n(g) \notin I_{N,N+i}$.

7. If $0 \leq n < N \leq m$ and $g \notin I_{N,m}$ then $f x_{2n} x_{2n+l} S^n(g) \notin I_{N,n+m}$.

8. If $0 \leq m < N \leq n$ and $f \notin I_{N,n}$ then $f x_{2n} x_{2n+l} S^n(g) \notin I_{N,n+m}$.

Proof. The statement (1) is clear.

(2). Since x_0 appears in every word from $I_{N,N}$, $S^{-1}(I_{N,N}) = \emptyset$. Now $S^{-1}(I_{N,k+1}) = I_{N,k}$ follows easily from (1) and the fact that S is an injective map.

(3) is a direct consequence of (2).

(4). Suppose $h = x_0 x_l f \in I_{N,k}$ for some $f \in F_\lambda$. Then $h = u(x_{2i} x_{2i+1})^\alpha w$ for some $u \in F_\lambda$, $w \in F_{\lambda, 2(N+1)}$ and $0 \leq i \leq k - N$. Since l is an odd integer, x_l is not the first letter of $(x_{2i} x_{2i+1})^\alpha$ and because $l > 1$, $x_0 x_l$ is not a subword of $(x_{2i} x_{2i+1})^\alpha$. Therefore $u = x_0 x_l u'$ for some $u' \in F_\lambda$ and $x_0 x_l f = h = x_0 x_l u'(x_{2i} x_{2i+1})^\alpha w$ This together with Lemma 1.2(1) yields $f \in I_{N,k}$ and completes the proof of (4).

(5). Suppose that $n, m \geq N$ and $f \notin I_{N,n}$, $g \notin I_{N,m}$. By (4), $x_0 x_l g \notin I_{N,m}$. Now (3) implies that $S^n(x_0 x_l g) \notin I_{N,n+m}$. Assume that $h = f S^n(x_0 x_l g) = f x_{2n} x_{2n+l} S^n(g) \in I_{N,n+m}$. It means that $h = u(x_{2i} x_{2i+1})^\alpha w$ for some $0 \leq i \leq n + m - N$, $u \in F_\lambda$ and $w \in F_{\lambda, 2(N+1)}$. Since $x_{2n} x_{2n+l} S^n(g) \notin I_{N,n+m}$ and $l > 1$, $(x_{2i} x_{2i+1})^\alpha$ has to be a subword of f. This means that x_{2n} appears in $w \in F_{\lambda, 2(N+1)}$. Therefore $N + i \leq n$ and $f \in I_{N,n}$ follows. This contadiction shows that $f x_{2n} x_{2n+l} S^n(g) \notin I_{N,n+m}$.

(6). Suppose that $1 \leq n \leq N - 1$ and $0 \leq i \leq n - 1$. Since $I_{N,k} \subseteq I_{N,k+1}$ for all $k \geq N$, it is enough to show that $h = f x_{2n} x_{2n+l} S^n(g) \notin I_{N,N+n-1}$ for any $f, g \in F_\lambda$.

Assume that $h \in I_{N,N+n-1}$ for some $f, g \in F_\lambda$. Then $h = u(x_{2i} x_{2i+1})^\alpha w$ for some $0 \leq i \leq n - 1$, $u \in F_\lambda$ and $w \in F_{\lambda, 2(N+1)}$. Since $i < n < N$, x_{2n} appears neither in $(x_{2i} x_{2i+1})^\alpha$ nor in w. Therefore x_{2n} appears in u. This yields that $(x_{2i} x_{2i+1})^\alpha$ is a subword of $x_{2n+l} S^n(g) \in F_{\lambda, 2n}$, which is impossible, because $i < n$. This contradicts our assumption and (6) follows.

(7). Suppose $0 \leq n < N \leq m$ and $g \notin I_{N,m}$. Then, by conditions (4) and (3), $x_{2n} x_{2n+l} S^n(g) \notin I_{N,n+m}$.

Assume $h = f x_{2n} x_{2n+l} S^n(g) \in I_{N,n+m}$. Then $h = u(x_{2i} x_{2i+1})^\alpha w$ for some $0 \leq i \leq n + m - N$, $u \in F_\lambda$ and $w \in F_{\lambda, 2(N+1)}$. Because $x_{2n} x_{2n+l} S^n(g) \notin I_{N,n+m}$ and $l > 1$, $(x_{2i} x_{2i+1})^\alpha$ is a subword of f. Hence x_{2n} appears in $w \in F_{\lambda, 2(N+1)}$ and $N + i \leq n$ follows. This is impossible since, by assumption, $n < N$. This gives $h \notin I_{N,n+m}$ and completes the proof of (7).

(8). Suppose $0 \leq m < N \leq n$ and $f \notin I_{N,n}$. Since $I_{N,k} \subseteq I_{N,k+1}$ for any $k \geq N$, it is enough to show that $h = f x_{2n} x_{2n+l} S^n(g) \notin I_{N,N+n-1}$.

Assume $h \in I_{N,N+n-1}$. Then $h = u(x_{2i} x_{2i+1})^\alpha w$ for some $0 \leq i \leq n - 1$, $u \in F_\lambda$ and $w \in F_{\lambda, 2(N+1)}$. Because $i < n$, both x_{2i} and x_{2i+1} do not appear in $x_{2n} x_{2n+l} S^n(g)$. Hence $(x_{2i} x_{2i+1})^\alpha$ has to be a subword of f. It means that x_{2n} appears in $w \in F_{\lambda, 2(N+1)}$ and $N + i \leq n$ follows. This shows that $f \in I_{N,n}$, which contradicts the assumption. Therefore $h \notin I_{N,N+n-1}$. ∎

2   The example

Let R denote an associative ring with unity and S stand for an endomorphism of R.

The skew polynomial ring $R[t, S]$ is a ring whose elements are polynomials in t with coefficients written on the left and the multiplication in $R[t, S]$ is given by multiplication in R and the condition $ta = S(a)t$ for $a \in R$.

For subsets $A_i \subseteq R$, $i = 1, 2, \ldots$, $\sum_{i=0}^{\infty} A_i t^i$ denotes $\{a_n t^n + \ldots + a_0 \in R[t, S] \mid a_i \in A_i, \ 0 \leq i \leq n, \ n \geq 0\}$.

Lemma 2.1 *Let $N \geq 1$ and $a \in R$. Then the ideal J of $R[t, S]$ generated by at^N is equal to $\sum_{k=N}^{\infty} J_{N,k} t^k$ where $J_{N,k} = RaS^N(R) + RS(a)S^{N+1}(R) + \ldots + RS^{k-N}(a)S^k(R)$ for all $k \geq N$.*

Proof. Let $\hat{J} = \sum_{k=N}^{\infty} J_{N,k} t^k$ where $J_{N,k}$'s are defined as above. Then clearly $J_{N,k+1} = J_{N,k} + RS(J_{N,k})$. Notice also that for any $k \geq N$, $J_{N,k}$ has a natural structure of an $(R - S^k(R))$–bimodule. Using this, it is easy to check that $\hat{J}$ is an ideal of $R[t, S]$ containing at^N. Thus $J \subseteq \hat{J}$.

The inclusion $Rat^N R \subseteq J$ shows that $J_{N,N} t^N \subseteq J$. Assume that $J_{N,k} t^k \subseteq J$, where $k \geq N$. Let $b \in J_{N,k+1}$. Then there exist $n \geq 1$ and $r_i \in R$, $c, b_i \in J_{N,k}$ with $1 \leq i \leq n$ such that $b = c + \sum_{i=1}^{n} r_i S(b_i)$. By the inductive assumption, $ct^k, b_i t^k \in J$ for $1 \leq i \leq n$. Therefore $bt^{k+1} = ct^k t + \sum_{i=1}^{n} r_i t b_i t^k \in J$. This shows that $J_{N,k+1} t^{k+1} \subseteq J$ and, consequently, $J = \hat{J}$ follows. ∎

Let $R_\lambda = KF_\lambda$ be the semigroup ring of the semigroup F_λ over a field K. The support of an element $a \in R_\lambda$ is denoted by $Supp(a)$, i.e. for $a = \sum_{f \in F_\lambda} k_f f$ where all but finitely many $k_f \in K$ are equal to 0, $Supp(a) = \{f \in F_\lambda \mid k_f \neq 0\}$.

Lemma 2.2 *Let a, b be non-zero elements from R_λ and $x \in X$ a letter such that x appears neither in any word from $Supp(a)$ nor in any word from $Supp(b)$. Then:*

1. *$Supp(axb) = Supp(a) \, x \, Supp(b)$.*

2. *$axb \neq 0$*

In particular R_λ is a prime ring.

Proof. The statement (1) is an easy application of Lemma 1.2(2), while the statement (2) is a consequence of (1). ∎

The endomorphism S of F_λ, defined in Section 1, can be extended in a unique way to a K-endomorphism of R_λ. The extension is also denoted by S.

Now we are in position to prove the following main result.

Theorem 2.3 *Let $1 \leq N, M < \omega_0$, $1 \leq \alpha, \beta < \lambda$ and $J(\alpha, N)$ denote the ideal of $R_\lambda[t, S]$ generated by $(x_0 x_1)^\alpha t^N$. Then*

1. $J(\alpha, N)$ *is an R_λ-disjoint prime ideal of $R_\lambda[t, S]$.*

2. *The following conditions are equivalent:*

   (i) $J(\beta, M) \subseteq J(\alpha, N)$.

   (ii) $N \leq M$ *and there exists an ordinal number γ such that $\beta = \gamma + \alpha$.*

Proof. (1) Let $J = J(\alpha, N)$. Clearly J is R_λ-disjoint. By Lemma 2.1, $J = \sum_{k=N}^{\infty} J_{N,k} t^k$, where $J_{N,k}$ denotes the K-linear subspace of R_λ spanned by $I_{N,k}$ with $I_{N,k}$'s defined as in the paragraphs prior to Lemma 1.4. For the convenience of the proof we set $J_{N,k} = 0$ for all $0 \leq k \leq N - 1$. Then $J = \sum_{k=0}^{\infty} J_{N,k} t^k$.

Notice that in order to show primeness of J it is enough to prove that for any $n, m \geq 0$ and $at^n, bt^m \notin J$, where $a, b \in R_\lambda$, there exists $c \in R_\lambda$ such that $at^n cbt^m = aS^n(c)S^n(b)t^{n+m} \notin J$. This in turn, is equivalent to the following statement:

For any $n, m \geq 0$ and $a \in R_\lambda \backslash J_{N,n}$, $b \in R_\lambda \backslash J_{N,m}$ there exists $c \in R_\lambda$ such that $aS^n(c)S^n(b) \notin J_{N,n+m}$.

Let $n, m \geq 0$ and $a, b \in R_\lambda$ be such that $a \notin J_{N,n}$, $b \notin J_{N,m}$. Choose an odd number $l > 1$ such that x_{2n+l} does not appear in every word from $Supp(a)$ and x_l does not appear in every word from $Supp(b)$. We claim that $aS^n(x_0 x_l)S^n(b) = ax_{2n}x_{2n+l}S^n(b) \notin J_{N,n+m}$. The proof is divided into five cases.

Case 1: $n + m < N$. In this case $J_{N,n+m} = 0$ and Lemma 2.2(2) together with the choice of l yield $ax_{2n}x_{2n+l}S^n(b) \neq 0$.

Case 2: $n, m \geq N$. We can choose $f \in Supp(a)$ and $g \in Supp(b)$ such that $f \notin I_{N,n}$, $g \notin I_{N,m}$. By Lemma 1.4(5), $fx_{2n}x_{2n+l}S^n(g) \notin I_{N,n+m}$. Now Lemma 2.2(1) yields the thesis in this case.

Case 3: $n + m \geq N$ and $1 \leq n, m \leq N - 1$. In this case $J_{N,n} = J_{N,m} = 0$, so a, b are arbitrary non-zero elements from R_λ. Let $f \in Supp(a)$ and $g \in Supp(b)$. Applying Lemma 1.4(6) with $i = n + m - N$ we obtain $fx_{2n}x_{2n+l}S^n(g) \notin I_{N,n+m}$ and Lemma 2.2(1) gives the thesis in this case.

Case 4: $0 \leq n < N \leq m$. Since $b \notin J_{N,m}$, we can choose $g \in Supp(b)$ such that $g \notin I_{N,m}$. Let $f \in Supp(a)$. By Lemma 1.4(7), $fx_{2n}x_{2n+l}S^n(g) \notin I_{N,n+m}$ and Lemma 2.2(1) yields the thesis.

Case 5: $0 \leq m < N \leq n$. Since $a \notin J_{N,n}$, we can choose $f \in Supp(a)$ such that $f \notin I_{N,n}$. By Lemma 1.4(8), $fx_{2n}x_{2n+l}S^n(g) \notin I_{N,n+m}$ and, as above, the thesis follows.

(2) Suppose that (ii) holds. Then, using Lemma 1.1(3), we have $(x_0 x_1)^\beta t^M = (x_0 x_1)^\gamma (x_0 x_1)^\alpha t^N t^{M-N} \in J(\alpha, N)$ and (i) follows.

Suppose now that $J(\beta, M) \subseteq J(\alpha, N)$. Then $(x_0 x_1)^\beta t^M \in J(\alpha, N)$. Hence clearly $N \leq M$ and by Lemma 2.1 $(x_0 x_1)^\beta \in I_{N,M}$, where the notation is as in the first part of the proof. This means that $(x_0 x_1)^\beta = g(x_{2i} x_{2i+1})^\alpha h$ for some $0 \leq i \leq M - N$, $g \in F_\lambda$ and $h \in F_{2(N+i)}$. Therefore we get $i = 0$, $h = 1$ and $(x_0 x_1)^\beta = g(x_0 x_1)^\alpha$ for some $g \in F_\lambda$. Now, Lemma 1.1(4) implies that $g = (x_0 x_1)^\gamma$ for some ordinal γ and $\beta = \gamma + \alpha$.

This completes the proof of the theorem. ∎

Remark 2.4 The argumentation given in Case 1 in the proof of the above theorem yields also that $R_\lambda[t, S]$ is a prime ring.

One can repeat the construction of the semigroup F_λ replacing the well-ordered sets Ω_α by their anti-isomorphic copies Ω_α^{op}. As a result we get a new semigroup F_λ^-. Notice that in the case of F_λ^-, we identify the disjoint union $\Omega_\alpha^{op} \dot\cup \Omega_\beta^{op}$ (with its natural order) with $\Omega_{\beta+\alpha}^{op}$. Thus if $f \in W_\alpha^-$ and $g \in W_\beta^-$ then the product $fg \in W_{\beta+\alpha}^-$.

For a finite word $f \in F_\lambda^-$ and $\beta < \lambda$, one can define f^β in the same way as it was done in F_λ but replacing the definition for non limit ordinal numbers by: if $\beta = \gamma + 1$ then $f^\beta = f f^\gamma$. Then, making minor modifications in proofs, one can check that $f^\alpha f^\beta = f^{\beta+\alpha}$ for all ordinals $\alpha, \beta < \lambda$ and that Lemmas 1.2 and 1.4 hold while replacing F_λ by F_λ^-.

Let R_λ^- denote the semigroup ring $K R_\lambda^-$, where K is a field. Then we have the following:

Theorem 2.5 *Let* $1 \leq N, M < \omega_0$, $1 \leq \alpha, \beta < \lambda$ *and* $J(\alpha, N)^-$ *denote the ideal of* $R_\lambda^-[t, S]$ *generated by* $(x_0 x_1)^\alpha t^N$. *Then*

1. $J(\alpha, N)^-$ *is an* R_λ^-*-disjoint prime ideal of* $R_\lambda^-[t, S]$.

2. The following conditions are equivalent:

(i) $J(\beta, M)^- \subseteq J(\alpha, N)^-$.

(ii) $N \leq M$ *and* $\alpha \leq \beta$.

Proof. The proof of the statement (1) is exactly the same as the one of Theorem 2.3(1).

(2). Suppose (ii) holds. Then $\beta = \alpha + \gamma$ for some ordinal number γ and $(x_0 x_1)^\beta t^M = (x_0 x_1)^{\alpha+\gamma} t^M = (x_0 x_1)^\gamma (x_0 x_1)^\alpha t^N t^{M-N} \in J(\alpha, N)^-$. This gives $J(\beta, M)^- \subseteq J(\alpha, N)^-$.

Suppose (i) holds. Then clearly $N \leq M$ and, as in the proof of Theorem 2.3, $(x_0 x_1)^\beta = g(x_0 x_1)^\alpha$ for some $g \in F_\lambda^-$. Therefore the length of $(x_0 x_1)^\beta$ is not greater than the length of $(x_0 x_1)^\alpha$. This implies $\beta \leq \alpha$. ∎

Let $L(N)$, where $1 \leq N < \omega_0$, denote the family of ideals $\{J(\alpha, N) \mid 1 \leq \alpha < \lambda\}$ of $R_\lambda[t, S]$. It is known (c.f. Ex.4., pp.252 [4]) that for any ordinal α there exists only a finite number of β such that $\alpha = \gamma + \beta$ for some ordinal γ. Thus, by Theorem 2.3(2), for any $J(\alpha, N) \in L(N)$ there exists only a finite number of ideals $J(\beta, N) \in L(N)$ such that $J(\alpha, N) \subset J(\beta, N)$. Therefore infinite chains in $L(N)$ are of type ω_0 and, if λ is uncountable, the cardinality of the set of such incomparable chains is equal to λ. Let us notice also that if γ is a limit ordinal such that $\alpha \leq \gamma < \lambda$, then $J(\gamma + \alpha, N) \subset J(\alpha, N)$ and for different γ's the ideals $J(\gamma + \alpha, N)$ are different from each other.

Let $L(N)^-$, where $1 \leq N < \omega_0$, denote the family of ideals $\{J(\alpha, N)^- \mid 1 \leq \alpha < \lambda\}$ of $R_\lambda^-[t, S]$. By Theorem 2.3(2), $L(N)^-$ is a chain of type λ.

When $\lambda = \omega_0$, then R_λ is the free K-algebra on the set X and R_λ is symmetrically closed by Theorem 13.4 of [7]. One can extend the arguments from the proof of this theorem to obtain the following:

Proposition 2.6 *For any λ, the ring R_λ is symmetrically closed.*

Proof. Let q denote an element from the symmetric Martindale quotient ring of R_λ and I be a non-zero ideal of R_λ such that $Iq, qI \subseteq R_\lambda$. For a nonzero element $a \in I$ we can choose two letters $x, y \in X$ which do not appear in all words from $Supp(a)$. Then, by Lemma 1.3 (2) and (3), $xay \in I$ is regular in R_λ. Now one can follow the proof of Theorem 13.4 of [7], using Lemma 1.3 when necessary, to obtain $q \in R_\lambda$. ∎

As we have said in the introduction, if R is a prime symmetrically closed ring, then every chain of R-disjoint prime ideals of $R[t; \sigma, D]$ is of length not greater than two, provided σ is an automorphism of R (Theorem 2.10 of [6]). This observation together with the above proposition give the following

Corollary 2.7 *Suppose that σ is an automorphism and D a σ-derivation of R_λ. Then every chain of R_λ-disjoint prime ideals of $R_\lambda[t; \sigma, D]$ is of length at most 2.*

Let us remark that both in the Proposition 2.6 and Corollary 2.7 one can replace the ring R_λ by R_λ^- similarily as it was done in the case of Theorem 2.5.

References

[1] M. Ferrero, K. Kishimoto, *On differential rings and skew polynomials*, Comm. Algebra 13(2) (1985), 285-304.

[2] K. R. Goodearl, E. S. Letzter, *Prime ideals in skew and q-skew polynomial rings*, Memoirs of the Amer. Math. Soc. 109 (521), (1994).

[3] R. S. Irving, *Prime ideals of Ore extensions over commutative rings*, J. Algebra 56 (1979), 315-342.

[4] K. Kuratowski, A. Mostowski, *Set theory*, North-Holland Publ. Company, Amsterdam and Polish Scientific Publ., Warsaw, (1968).

[5] T. Y. Lam, K. H. Leung, A. Leroy, J. Matczuk, *Invariant and semi-invariant polynomials in skew polynomial rings*, Israel Math. Conference Proc., 1 (1989), 247-261.

[6] A. Leroy, J. Matczuk, *Prime ideals of Ore extensions*, Comm. Algebra 19(7) (1991), 1893-1907.

[7] D. S. Passman, *Infinite crossed products*, Academic Press, Inc., (1989).

[8] K. R. Pearson, W. Stephenson, J. F. Watters, *Skew polynomial rings and Jacobson rings*, Proc. London Math. Soc. 42 (1981), 559-576.

On Auslander–Reiten Systems via Tensor Product

Odile Garotta

Institut Fourier, Laboratoire de Mathématiques,
UMR 5582 (UJF-CNRS), BP 74, F-38402 St
Martin d'Hères Cedex, France, *e-mail:* Odile.Garotta@ujf-grenoble fr

Introduction

Let G be a finite group, k an algebraically closed field whose characteristic p divides the order of G. In [G1] we introduced the notion of Auslander–Reiten system of G over a given symmetric interior G-algebra (that is a k-algebra A together with a symmetrizing form τ and a homomorphism $\phi : G \to A^{\times}$), as a generalization in terms of idempotents of the notion of almost split sequences of kG-modules (see (1.5), this particular case arises when A is the algebra of k-endomorphisms of a kG-module). A special interesting case is (up to embedding) the algebra kG itself.

It is a well-known result of Benson, and Auslander–Carlson that the tensor product of any indecomposable kG-module M with the Auslander–Reiten sequence terminating in the trivial kG-module k is either split or almost split up to an injective factor, this last case happening if and only if $p \nmid \dim M$ ([Au-Ca], [Be-Ca]). In [G2], we considered the more general tensor product of a G-fixed *idempotent* i of A with the Auslander–Reiten system $\mathcal{L}_k$ associated to the trivial kG-module (1.5) : using "relative trace" methods and [G1], we obtained sufficient conditions for pullbacks of the resulting system to be split or a sum of Auslander–Reiten and split systems (SARS for short), conditions which proved to be equivalences when G is a p-group and A has a G-stable basis; this applied to the source algebra of a block of kG, yielding an explicit expression in that case (on the other hand Auslander and Carlson's theorem for kG-modules was recovered in very specific cases only). Note that as can be read between the lines of [G2], *contrary to the case of modules, the system $i \otimes \mathcal{L}_k$ need not in general be SARS* (see (2.4)).

In this note we clarify and complete the approach presented in [G2]: as our starting point, we use [G1] to characterize those pullbacks from the system $i \otimes \mathcal{L}_k$ which are split or SARS; Auslander and Carlson's theorem for kG-modules ([Au-Ca] 3.6.) immediately follows (see also [Be-Ca], [Br-Ro]); a quite different example is the case of the principal block. Suppose now i is primitive with defect group $P \neq 1$: whenever P is not a Sylow p-subgroup of G, the system $i \otimes \mathcal{L}_k$ is split [G1], nevertheless a "relative trace" version of our characterization holds (2.2), so one may look for cases when a generator of the socle of our extension bimodule (which yields the Auslander–Reiten system ending with i, see (1.7)) is the image by $\overline{\mathrm{Tr}_P^G}$ of an appropriate tensor product. In section 3, we show this holds in particular if either A has a P-stable basis, or else in the case of modules whenever p does not divide the dimension of the source : indeed the general properties of bilinear forms on G-algebras developed in [Br-Ro] enable us to characterize in both cases the elements yielding such a type of generator. Finally, in section 4 we turn to the question whether the socle of our extension bimodule is the image by $\overline{\mathrm{Tr}_P^G}$ of the corresponding socle for a source l of i : such conditions have been investigated in whole generality by Thévenaz; we obtain from [T1] that the answer is positive *for all* sources l if and only if the associated multiplicity module is simple; our remark is that however in all cases there *exists* a (good) source for which the statement is true. This may be combined with our preceding results, giving then "explicit" generators as soon as a good source is known: in particular we obtain a complete answer for the block idempotents (using [G1]) and the Scott modules.

1. Notation and preliminaries

Our notation mainly follows that in [Br-Ro] and [G1]. For a k-algebra A (always associative with unity), we write $A^\times$ for its group of units and $J(A)$ for its Jacobson radical. If A is symmetric we denote by $\mathring{S}(A)$ its (left or right) socle. An algebra is called *local* provided its unity element is a primitive idempotent. All tensor products are taken over k, and we let SG denote the element $\sum_{g \in G} g$ in kG.

(1.1) If A is a G-algebra, and H, K are subgroups of G with $K \leq H$, recall A^H denotes the $N_G(H)$-algebra of all elements of A fixed by H, and the relative trace map $\mathrm{Tr}_K^H : A^K \to A^H$ is defined by $\mathrm{Tr}_K^H(a) = \sum a^x$, where x runs over a right transversal of H modulo K, its image is the two-sided ideal A_K^H of A^H.

Furthermore we set $\overline{A^H} = A^H/A_1^H$, and for a in A^H we let $\bar{a}$ denote its image in $\overline{A^H}$. Idempotents in A_1^H are called *projective*.

If P is a p-subgroup of G, we set $\overline{N}(P) = N_G(P)$, we denote by $A(P)$ the quotient of A^P by its ideal $\sum_{Q < P} A_Q^P$ and by $\mathrm{Br}_P : A^P \to A(P)$ the canonical epimorphism : this is a morphism of $\overline{N}(P)$-algebras, called the *Brauer morphism* (see [Br-Ro]). Recall that (see [Br]):

$$\mathrm{Br}_P \circ \mathrm{Tr}_P^G = \mathrm{Tr}_1^{\overline{N}(P)} \circ \mathrm{Br}_P .$$

Thus any G-stable k-linear form τ on A satisfies $\ker \mathrm{Br}_P \subset \ker \tau$, therefore it induces a linear form $\tau_P : A(P) \to k$ which is $\overline{N}(P)$-stable.

(1.2) Assume now $A^G = A_P^G$, so that $\mathrm{Br}_P(A^G) = A(P)^{\overline{N}(P)} = A(P)_1^{\overline{N}(P)}$; if $\tau : A \to k$ is a G-stable linear form such that $\tau(aa') = \tau(a'a)$ for all a, a' in A, then applying [Br-Ro] (1.), we have a linear form $\tau^{[G,P]}$ on A^G defined by $a \mapsto \tau(a')$, where a' is any element such that $\mathrm{Tr}_P^G(a') = a$. This satisfies

$\tau^{[G,P]} = \tau_P^{[\overline{N}(P),1]} \circ \mathrm{Br}_{P|A^G}$; in case (A,τ) is a symmetric algebra and $P = 1$ it is a symmetrizing form on A^G.

(1.3) Following Puig, a *point* α of A is an $A^\times$-conjugacy class of primitive idempotents of A. Points are in natural bijection with maximal two-sided ideals (there exists a unique such $\mathcal{M}$ not containing α), and with the isomorphism classes of simple A-modules (we have $A/\mathcal{M} = \mathrm{End}_k(V)$, for V the simple module $Ai/J(A)i$, some $i \in \alpha$; this is the *simple quotient associated to α*).

Assume now A^G is local. A *defect group* of 1_A in G is a minimal subgroup such that $1 \in A_P^G$, or equivalently a maximal subgroup such that $A(P) \neq 0$, see [P1] 1.2, or [Br-Ro] 1; as shown there, there exist then primitive idempotents l of A^P, called *sources* of 1_A, such that 1 belongs to the two-sided ideal $\mathrm{Tr}_P^G(A^P l A^P)$. The corresponding points of A^P are the *source points* of 1, they form a $\overline{N}(P)$-conjugacy class.

From now on we **let A be a finite dimensional symmetric interior G-algebra**, endowed with the map $\phi : G \to A^\times$ and **with $\tau : A \to k$ a symmetrizing form**. For a in A and x in G we put $a^x = \phi(x^{-1})a\phi(x)$, and this induces a structure of G-algebra on A.

We recall the necessary basic definitions from [G1]: given three mutually orthogonal idempotents i, i° and i' of A^G, and elements $d \in iA^G i^\circ$ and $d' \in i^\circ A^G i'$, we say that $\mathcal{S} = (i, i^\circ, i', d, d')$ is a *system* of G over A if $dd' = 0$ and if there exist elements $s \in i^\circ Ai$ and $s' \in i'Ai^\circ$ satisfying the conditions: $i = ds$, $i^\circ = sd + d's'$ and $i' = s'd'$. In analogy to short exact sequences we write :

$$\mathcal{S}: \quad i' \xrightarrow{\ d'\ } i^\circ \xrightarrow{\ d\ } i \ ;$$

accordingly, we say i is the end term, i° the middle term and i' the left hand term of $\mathcal{S}$. We call $\mathcal{S}$ *trivial* if $i = 0$, and *split* if $i \in dA^G$.

(1.4) In case $i^\circ \in A_1^G$, $\mathcal{S}$ is a *Heller system* and i' a *Heller translate* of i. For $H \leq G$ we define the linear form $\overline{\tau_{\mathcal{S},H}} : \overline{i'A^H i} \to k$ by $\bar{a} \mapsto \tau^{[H,1]}(dad')$ (see (1.2)): as shown in [G1], it satisfies $\overline{\tau_{\mathcal{S},H}} = \overline{\tau_{\mathcal{S},G}} \circ \overline{\mathrm{Tr}_H^G}$ and induces a duality between the $\overline{i'A^H i'} - \overline{iA^H i}$ bimodules $\overline{i'A^H i}$ and $\overline{iA^H i} \simeq \overline{i'A^H i'}$ (this last isomorphism is obtained via the commuting algebra of $\mathcal{S}$). The right (or left) socle of $\overline{i'A^H i}$, which is isomorphic to $\overline{iA^H i}/J(\overline{iA^H i})$, will be denoted simply by $\mathring{\mathcal{S}}_{i,H}$.

(1.5) Recall an *embedding* of interior G-algebras $f : A \to C$ is a homomorphism of interior G-algebras that is one-to-one and satisfies $\mathrm{Im}\, f = f(1)Cf(1)$. We call $\mathcal{S}$ an *Auslander–Reiten system* (*AR system* for short) if it is a non split system, if i and i' are primitive in A^G, and if for every embedding of interior G-algebras $f : A \to C$, where C is symmetric, we have $f(i)J(C^G) \subset f(d)C^G$ (equivalently, the exact sequence of right

$A[G] = A \otimes kG$-modules $0 \rightarrow i'A \xrightarrow{d'} i^\circ A \xrightarrow{d.} iA \rightarrow 0$ induced by S should be almost split, where we put $(ia) \cdot a' \otimes g = \phi(g^{-1})iaa'$. We use the notation $\mathcal{L}_i$ for an Auslander–Reiten system ending with i (recall that whenever $i \notin A_1^G$ this is uniquely defined up to embedding of A into a symmetric interior G-algebra and to conjugacy by invertible G-fixed elements of that algebra). As shown in [G1], the left hand term of $\mathcal{L}_i$ is a double Heller translate of i, furthermore almost split sequences of kG-modules correspond exactly to AR systems over the interior G-algebra of k-endomorphisms of the direct sum of their three terms (all endomorphism algebras of kG-modules will be endowed with the symmetrizing form $\tau = \mathrm{tr}$, the usual trace map). Therefore tensoring a primitive non projective idempotent i of A^G with the AR system $\mathcal{L}_k$ gives the same "double Heller translate" relation between the end terms (up to a projective part), as in $\mathcal{L}_i$, and it is natural to ask when $i \otimes \mathcal{L}_k$ is split, or is $\mathcal{L}_i$ up to a trivial system (and embedding).

(1.6) Now it is convenient to fix the following notation : we let E be the symmetric interior G-algebra $\mathrm{End}_k[k \oplus \Pi k \oplus \Omega k \oplus \Pi(\Omega k) \oplus \Omega^2 k \oplus R_k]$, (where ΠM, resp. ΩM, denote a projective cover, resp. a Heller translate of the kG-module M, and R_k denotes the middle term of the almost split sequence $\mathcal{R}_k$ ending with k), and write e, e°, e', e'°, e'' and r for the orthogonal idempotents of E^G equal to the projections on the summands k, Πk, Ωk, $\Pi(\Omega k)$, $\Omega^2 k$ and R_k respectively. We get Heller systems $\mathcal{H}_k = (e, e^\circ, e', h, h')$ and $\mathcal{H}_{\Omega k} = (e', e'^\circ, e'', m, m')$ of G over E, aswell as the AR system $\mathcal{L}_k$ corresponding to $\mathcal{R}_k$; for $H \leq G$ we choose u_H an element of $e'E^H e$ such that $\overline{u_H}$ generates the one-dimensional k-space $e'E^H e$.

We denote by C the symmetric interior G-algebra $A \otimes E$, so tensor product by e defines an embedding of interior G-algebras $A \rightarrow C$. We write $\mathcal{H}$, resp. $\mathcal{K}$ for the Heller systems $1 \otimes \mathcal{H}_k$, resp. $1 \otimes \mathcal{H}_{\Omega k}$ of G over C.

(1.7) Finally, we will need to consider *pullback systems of $\mathcal{K}$* over specific elements c in $(1 \otimes e')C^G(1 \otimes e)$ (we use the obvious notation $PB_c(\mathcal{K})$), this is a system of G over some symmetric interior G-algebra C' where C embeds; it is uniquely defined up to embedding and conjugacy, see [G1]); as shown in [G1], the system $PB_c(\mathcal{K})$ is split if and only if $\bar{c} = 0$, and is SARS if and only if $\bar{c} \in \mathring{S}_{1 \otimes e, G}$ (1.4). For $H \leq G$, we let $\mathcal{L}_k^{[H]}$ denote the system $PB_{u_H}(\mathcal{H}_{\Omega k})$ of H, so $\mathcal{L}_k^{[G]} = \mathcal{L}_k$.

2. The basic results

(2.1) Proposition *Let a be in A^G.*

   (i) *The system $PB_{a \otimes e}(1 \otimes \mathcal{L}_k)$ is split over G if and only if $\tau(aA^G) = 0$.*

   (ii) *The system $PB_{a \otimes e}(1 \otimes \mathcal{L}_k)$ is SARS if and only if $\tau(aJ(A^G)) = 0$.*

(2.2) Proposition *Let a be in A^H for some non trivial subgroup H of G, put $c = \mathrm{Tr}_H^G(a \otimes u_H)$.*

   (i) *The system $PB_c(\mathcal{K})$ is split over G if and only if $\tau(aA^G) = 0$.*

(ii) *The system* $PB_c(\mathcal{K})$ *is SARS if and only if* $\tau(aJ(A^G)) = 0$.

Proof of (2.1). Clearly this is a special case of (2.2) for $H = G$.

Proof of (2.2). Set $i = 1 \otimes e$. Using (1.7) and (1.4), we know the system $PB_c(\mathcal{K})$ is split (resp. SARS) if and only if $\overline{\tau_{\mathcal{H},G}}(\bar{c} \cdot \overline{iC^G i}) = 0$ (resp. $\overline{\tau_{\mathcal{H},G}}(\bar{c} \cdot J(\overline{iC^G i})) = 0$. Note that $J(iC^G i) = J((A \otimes eEe)^G) = J(A^G) \otimes e$ (similarly $iC^G i = A^G \otimes e$), so, using (1.4) : $\overline{\tau_{\mathcal{H},G}}(\bar{c} \cdot J(\overline{iC^G i})) = \overline{\tau_{\mathcal{H},H}}(\overline{a \otimes u_H} \cdot J(\overline{iC^G i})) = \overline{\tau_{\mathcal{H},H}}(\overline{aJ(A^G) \otimes u_H}) = \tau(aJ(A^G)) \cdot \overline{\mathrm{tr}_{\mathcal{H}_k,H}}(\overline{u_H})$ (similarly $\overline{\tau_{\mathcal{H},G}}(\bar{c} \cdot \overline{iC^G i}) = \tau(aA^G) \cdot \overline{\mathrm{tr}_{\mathcal{H}_k,H}}(\overline{u_H})$). But $\overline{\mathrm{tr}_{\mathcal{H}_k,H}}(\overline{u_H}) \neq 0$ since it generates the k-space $\overline{\mathrm{tr}_{\mathcal{H}_k,H}}(\overline{u_H \cdot eE^H e})$ and $\overline{u_H} \neq 0$, so we can conclude.

(2.3) Application to modules Take $A = \mathrm{End}_k(M)$ where M is a kG-module, set $a = \mathrm{id}_M$. Nilpotency implies that $\mathrm{tr}(J(A^G)) = 0$, so after reformulation (2.1) states that $M \otimes \mathcal{R}_k$ is always a sum of split and almost split sequences, and that if M is indecomposable, it fails to split if and only if $\mathrm{tr}(A^G) \neq 0$, which is equivalent to $p \nmid \dim M$. Thus we recover the well-known result of Benson and Auslander–Carlson ([Au-Ca] 3.6).

(2.4) Application to the principal block : (see also (4.3)) Let $A = kGB_0$ be the principal block of kG and set $a = SG$.

(i) *The system* $PB_{a \otimes e}(B_0 \otimes \mathcal{L}_k)$ *is the sum of an AR system terminating in* $B_0 \otimes e$ *and a trivial system.*

(ii) *The system* $B_0 \otimes \mathcal{L}_k$ *is NOT SARS.*

Proof. As usual we set $\tau : \sum_{g \in G} \lambda_g g \mapsto \lambda_1$. By definition the primitive central idempotent B_0 of kG acts trivially on the trivial kG-module, that is we have $\epsilon(B_0) = 1$ where ϵ is the augmentation map of kG. The relation $y \cdot SG = SG \cdot y = \epsilon(y) SG$ for any $y \in kG$ shows that SG lies in $A^G \cap \ker \epsilon$. But since $\epsilon(A^G) = k$, this is a maximal ideal of the local algebra A^G. Therefore we deduce (i) from (2.1), having both $SG \cdot J(A^G) = SG \cdot (A^G \cap \ker \epsilon) = 0$ and $\tau(SG \cdot A^G) \ni \tau(SG) \neq 0$. Also (ii) follows since $SG \in J(A^G)$.

3. Two specific cases

In this section we start by considering the following hypothesis, for P a non trivial p-subgroup of G :

(SYM$_P$): *the form* τ_P *makes* $A(P)$ *into a symmetric algebra.*

We first observe that

(SYM$_P$) holds whenever A has a P-stable basis, therefore in particular when $A = kGb$ is a block algebra (recall that $(kG)(P) = kC_G(P)$), or when $A = \mathrm{End}_k(M)$ for M an endo-permutation module (then $A(P)$ is simple, see [P2] 5.8). The claim follows from statement (3.1), which can be read between the lines of [Br] (2.4) (note that it also follows from lemma 2 in [G2] combined with (2.1)); notation is standard ([Br]) :

(3.1) [Br] *Let ρ be a non-degenerate P-stable bilinear form on M a p-permutation kP-module. Then the induced bilinear form $\rho_P \colon M(P) \times M(P) \to k$ is non-degenerate.*

Proof. Let $\mathcal{B}$ be a P-stable basis of M. Its dual basis $\mathcal{B}'$ with respect to ρ is P-stable aswell and the subsets $\mathcal{B}^P$ and $\mathcal{B}'^P$ correspond dually; in the quotient $M(P)$ those map bijectively to bases which are mutually dual with respect to ρ_P.

Using τ_P to reformulate (2.1) we now get an improved version of Prop. 2 in [G2] (here we assume $G = P$ so $\mathcal{L}_k^{[P]}$ is a system over "our" E) :

(3.2) Proposition *Let a be in A^P and assume (SYM$_P$) holds.*

The system $PB_{a \otimes e}(1 \otimes \mathcal{L}_k^{[P]})$ is split if and only if $\mathrm{Br}_P(a) = 0$; it is SARS if and only if $\mathrm{Br}_P(a) \in \mathring{S}(A(P))$.

(3.3) From now on we assume A^G **is local and $P \neq 1$ is a defect group** of 1_A in G (see (1.3); note that if Q is any p-subgroup of G such that $1 \otimes \mathcal{L}_k^{[Q]} = PB_{1 \otimes u_Q}(\mathcal{K})$ is *not* split, (this condition is necessary in order for $PB_c(\mathcal{K})$ not to split in (2.2) for $H = Q$) then we have $A(Q) \neq 0$, therefore $Q \leq P^x$, some $x \in G$).

Having $A^G = A_P^G$, (1.2) applies and also $\mathrm{Br}_P(J(A^G)) = J(A(P)^{\overline{N}(P)})$.

(3.4) Therefore we get for a in A^P :

$$\tau(aA^G) = \tau^{[G,P]}(\mathrm{Tr}_P^G(a)A^G) = \tau_P^{[\overline{N}(P),1]}[\mathrm{Br}_P(\mathrm{Tr}_P^G(a))A(P)^{\overline{N}(P)}],$$
similarly
$$\tau(aJ(A^G)) = \tau_P^{[\overline{N}(P),1]}[\mathrm{Br}_P(\mathrm{Tr}_P^G(a))J(A(P)^{\overline{N}(P)})] \quad .$$

Consider now the condition

(SYM') : *the form $\tau_P^{[\overline{N}(P),1]}$ makes $A(P)^{\overline{N}(P)}$ into a symmetric algebra.*

It clearly holds whenever (SYM$_P$) does. Moreover since $A(P)^{\overline{N}(P)}$ is local (1.2), (SYM') implies that the socle $\mathring{S}[A(P)^{\overline{N}(P)}]$ is 1-dimensional.

(3.5) Proposition *Assume (SYM') holds (this is the case in particular if A has a P-stable basis). Let $c = \mathrm{Tr}_P^G(a \otimes u_P)$, where $a \in A^P$.*

The system $PB_c(\mathcal{K})$ of G is the direct sum of an AR system $\mathcal{L}_{1 \otimes e}$ and a trivial system if and only if the element $\mathrm{Br}_P(\mathrm{Tr}_P^G(a))$ of $A(P)^{\overline{N}(P)}$ generates the socle $\mathring{S}[A(P)^{\overline{N}(P)}]$.

In particular there always exist such elements a yielding the AR system $\mathcal{L}_{1 \otimes e}$.

Proof. Since 1 is primitive non projective in A^G, the system $PB_c(\mathcal{K})$ of G is the direct sum of an AR system $\mathcal{L}_{1 \otimes e}$ and a trivial system if and only if it is both SARS and non split. So the result follows from (2.2) for $H = P$, rewritten using (3.4) and (SYM').

Remark. Assume P is a Sylow subgroup of G. Then we may take $u_P = u_G$, moreover since $\overline{N}(P)$ is a p'-group, we have $J[A(P)^{\overline{N}(P)}] = J(A(P))^{\overline{N}(P)}$ (see [P3] 2.9.4). Therefore we obtain :

Assume (SYM'), let $a \in A^G$; the system $PB_{a \otimes e}(1 \otimes \mathcal{L}_k)$ is the direct sum of $\mathcal{L}_{1 \otimes e}$ and a trivial system if and only if the element $\mathrm{Br}_P(a)$ of $A(P)^{\overline{N}(P)}$ generates the socle $\mathring{S}[A(P)^{\overline{N}(P)}]$. Moreover if (SYM$_P$) holds, then this socle is equal to $\mathring{S}(A(P))^{\overline{N}(P)}$.

(3.6) We here fix some more notation (see (1.3)): let $\gamma \subset A^P$ be a source point of 1 in G, $N = N_G(P, \gamma)$ its stabilizer in $N_G(P)$, and set $\overline{N} = N/P$; we denote by $s_\gamma : A^P \to S_\gamma$ the canonical surjection onto the simple quotient associated to γ (this is a $\overline{N}$-morphism), and by tr_γ the trace map on S_γ. Then (see [P1] 1.3) $s_\gamma(A_P^G) = (S_\gamma)_1^{\overline{N}}$ contains the unity, therefore we get a symmetric algebra $S_\gamma^{\overline{N}} = (S_\gamma)_1^{\overline{N}} = s_\gamma(A^G)$ (1.2) with 1-dimensional socle.

We consider also the *multiplicity module* of γ introduced by Puig (see e.g. [T1] 9.); this is a module V over a twisted algebra $\widetilde{k\overline{N}}$, such that $S_\gamma = \mathrm{End}_k(V)$, and $S_\gamma^{\overline{N}} = \mathrm{End}_{\widetilde{k\overline{N}}}(V)$ (so V is projective indecomposable).

Note that if $l \in \gamma$ it is easy to show using (2.1), τ_P and the fact that all sources are conjugate, that

(3.7) *the system $1 \otimes \mathcal{L}_k^{[P]}$ splits if and only if $l \otimes \mathcal{L}_k^{[P]}$ does.*

(3.8) Case of modules Suppose now that $A = \mathrm{End}_k(M)$, where M is a kG-module. For $l \in \gamma$ (3.6), the kP-module $L = l \cdot M$ is a source of M, and N is its inertia group. It follows from (3.7) and (2.3) that the system $1 \otimes \mathcal{L}_k^{[P]}$ splits if and only if $p | \dim L$. We here deduce from the "local" calculation [Br-Ro] (1.8) that in all other cases we can find a generator of the socle $\mathring{S}_{1_M, G}$ of the desired form (note that the quoted local formula also shows that condition (SYM$_Q$) (where $Q \le P$) does not hold here generally, but precisely if and only if $A(Q)$ is semi-simple and $p \nmid \dim N$ for all summands N of $\mathrm{Res}_Q(M)$ having vertex Q; also (SYM') is here equivalent to having both $p \nmid \dim L$ and $J(A(P))^{\overline{N}(P)} = 0$).

Proposition *(notations above) Assume $p \nmid \dim L$ and let $c = \mathrm{Tr}_P^G(a \otimes u_P)$, where $a \in \mathrm{End}_{kP}(M)$. Denoting by $\mathcal{P}$ the short exact sequence giving a projective cover of Ωk over kG,*

the exact sequence obtained by pullback of c from $M \otimes \mathcal{P}$ is the almost split sequence $\mathcal{R}_M$ up to an injective factor if and only if the element $s_\gamma(\mathrm{Tr}_P^G(a))$ of $S_\gamma^{\overline{N}}$ generates its socle $\mathring{S}(S_\gamma^{\overline{N}})$, (or equivalently if and only if the image of the endomorphism $s_\gamma(\mathrm{Tr}_P^G(a))$ of V is $\mathrm{soc}V$).

In particular there exist such elements a yielding the AR system $\mathcal{L}_{1 \otimes e}$.

Proof. The formula (1.8) in [Br-Ro], which is in force in the module case, may be rewritten here as : $\tau^{[G,P]} = \mathrm{tr}(l) \cdot \mathrm{tr}_\gamma^{[\overline{N},1]} \circ s_{\gamma|A^G}$, and $\mathrm{tr}(l) = \dim L \ne 0$ in k. Moreover $s_\gamma(J(A^G)) = J(S_\gamma^{\overline{N}})$ (see (3.6)), so we get a rewriting of (3.4), from which our statement follows via (2.2) (the second equivalence has been shown by Thévenaz in [T2] (3.4)).

4. Good sources

We keep throughout the notations of (3.3), (3.6). For $l \in \gamma$ we set $i = 1 \otimes e$, $j = l \otimes e$; these are the end terms of the Heller systems $\mathcal{H}$ and $l \otimes \mathcal{H}_k$ of G (resp P) over C (1.6). Using the notation in (1.4), we say the source l of 1_A is *good* if

$$\mathring{S}_{i,G} = \overline{\mathrm{Tr}}_P^G(\mathring{S}_{j,P}) \ .$$

Parts of the following proposition are already contained in [T1] (4.7, 9.3):

(4.1) Proposition (i) *The source l is good if and only if we have $\overline{lJ(A^G)l} \subset J(\overline{lA^Pl})$, that is $s_\gamma(l)J(S_\gamma^{\overline{N}})s_\gamma(l) = 0$.*

(ii) *There always exists a good source in the point γ. (More precisely, any $l \in \gamma$ such that $\mathrm{Im}\, s_\gamma(l) \subset \mathrm{soc}V$, is good and satisfies $\mathrm{Im}\, s_\gamma(\mathrm{Tr}_P^G(l)) = \mathrm{soc}V$.)*

(iii) *[T1] The condition that every source in γ is good holds if and only if $S_\gamma^{\overline{N}} \simeq k$, that is the multiplicity module of γ is simple.*

Proof. (i) Let $\overline{I_P^G} \colon \overline{C^G} \to \overline{C^P}$ be the map induced by inclusion. First note that by (1.4) both socles above have dimension 1. Now we show $\overline{\mathrm{Tr}}_P^G(\mathring{S}_{j,P})$ is never 0 : indeed using $\overline{\tau_{\mathcal{H},G}}$ (1.4) this would mean $\overline{\tau_{\mathcal{H},P}}(\mathring{S}_{j,P} \cdot \overline{I_P^G}(iC^Gi)) = 0$, but this is false since $\mathring{S}_{j,P} \cdot \overline{i} = \mathring{S}_{j,P} \cdot \overline{j}$. Therefore l is good if and only if we have $\overline{\mathrm{Tr}}_P^G(\mathring{S}_{j,P}) \subset \mathring{S}_{i,G}$; using (1.4) again this condition is equivalent to $\overline{\tau_{\mathcal{H},P}}(\mathring{S}_{j,P} \cdot \overline{I_P^G(\overline{J(iC^Gi)})}\overline{j}) = 0$, which in turn holds if and only if we have $\overline{jJ(iC^Gi)j} \subset J(\overline{jC^Pj})$, that is $\overline{lJ(A^G)l} \subset J(\overline{lA^Pl})$. But $\overline{lA^Pl}$ is local, so the condition reduces to $s_\gamma(lJ(A^G)l) = 0$, that is (see (3.6)) : l is good if and only if $s_\gamma(l)J(S_\gamma^{\overline{N}})s_\gamma(l) = 0$.

(ii) To obtain a good source it is enough to find a primitive idempotent $s \ (= s_\gamma(l))$ in $S_\gamma = \mathrm{End}_k(V)$ (3.6) such that $J(S_\gamma^{\overline{N}}) \cdot s = 0$. Indeed we may take s to be any projection onto a line included in $\mathrm{soc}V$, so that the left composition with any non injective endomorphism is 0. The last part of the statement follows from the facts that $\mathrm{soc}V$ is $\overline{N}$-stable, that the maps $s_\gamma \circ \mathrm{Tr}_P^G$ and $\mathrm{Tr}_1^{\overline{N}} \circ s_\gamma$ coincide on the ideal generated by γ (see [P1] (1.3)), and that $\mathrm{Tr}_1^{\overline{N}}(s) \neq 0$ since $\mathrm{Tr}_1^{\overline{N}}$ is selfadjoint with respect to tr (see (1.2), (3.6)).

(iii) Using (i) this is now [T1] (9.3), (a,b,i,i').

(4.2) Remark. Statement (ii) may be combined with our preceding results: using (2.3) and (3.2) for $G = P$ and $A = lAl$, we obtain more explicit elements $a \in A^P$ such that $\mathrm{Tr}_P^G(a \otimes u_P)$ generates $\mathring{S}_{i,G}$: in the case of modules with $p \nmid \dim L$, any good source works, whereas if (SYM_P) holds, we may choose a in $lA^P l$ where l is a good source and $\mathrm{Br}_P(a)$ generates $\mathring{S}((lAl)(P))$.

(4.3) Case of blocks Assume $A = kGb$ is a block. Then it has been shown that the module V is simple ([Ba] 1.2), therefore using (4.1)(iii) and our formula in [G2], we see that *the element* $a = SZ(P) \cdot l$, *where* l *is any source of* b, *is an element of the type looked for in (4.2)*.

(4.4) Scott modules Let $M = Sc_G(P)$ be the Scott module with vertex P (see [Br] or [Bu]; this is a p-permutation module, we have $\mathrm{Br}_P = s_\gamma$, and $V = M(P)$ is the $k\overline{N}(P)$-module Πk). As shown in [Bu] theorem 1, the (unique) trivial kG-submodule of M is a summand of $\mathrm{Res}_P(M)$; we claim *the corresponding idempotent* l *of* A^P *is a good source* : indeed (see [Br] §3) we have $0 \neq \mathrm{Im}(\mathrm{Br}_P(l)) = \mathrm{Br}_P^M(M^G) \subset (\Pi k)^{\overline{N}(P)} = \mathrm{soc}\Pi k \simeq k$.

References

[Au-Ca] M. Auslander and J.F. Carlson, Almost split sequences and group rings, *J. Algebra* **103** (1986), 122–140.

[Ba] L. Barker, Modules with simple multiplicity modules, *J. Algebra* **172** (1995), 152–158.

[Be-Ca] D.J. Benson and J.F. Carlson, Nilpotent elements in the Green Ring, *J. Algebra* **104** (1986), 329–350

[Br] M. Broué, On Scott modules and p-permutation modules. an approach through the Brauer morphism, *Proc. Amer. Math. Soc.* **93** (1985), 401–408.

[Br-Ro] M. Broué and G R. Robinson, Bilinear Forms on G-Algebras, *J. Algebra* **104** (1986), 377–396.

[Bu] D.W. Burry, Scott modules and lower defect groups, *Comm. alg.* **10** (1982), 1855–1872

[G1] O. Garotta, Suites presque scindées d'algèbres intérieures et algèbres intérieures des suites presque scindées, *Publ. Math. de l'Université Paris VII* **34** (1994), 137–237

[G2] O Garotta, On Auslander–Reiten systems, *S.M.F. Astérisque* **181–182** (1990), 191–194.

[P1] L. Puig, Pointed groups and construction of characters, *Math. Z.* **176** (1981), 265–292

[P2] L. Puig, Nilpotent blocks and their source algebras, *Invent. math.* **93** (1988), 77–116.

[P3] L. Puig, Pointed groups and construction of modules, *J. Algebra* **116** (1988), 7–129.

[T1] J. Thévenaz, Duality in G-algebras, *Math. Z.* **200** (1988), 47–85

[T2] J Thévenaz, G-algebras, Jacobson radical and almost split sequences, *Invent. math.* **93** (1988), 131–159.

On a Projective Version of Maranda's Theorem

Yuval Ginosar
Department of Mathematics
Brandeis University
Waltham, Massachusetts 00254-9110

Uri Onn
Department of Mathematics
Technion, Haifa 32000
Israel

ABSTRACT. In this note we compare a lifting theorem for ordinary representations due to Maranda to a possible analogue for projective representations. We give an example in which a generalization of the theorem to twisted group rings does not hold.

Recall that a projective representation of a group G over a commutative ring R is a map of sets

$$\eta : G \to GL_n(R)$$

such that its composition with the natural map $\pi : GL_n(R) \to PGL_n(R)$ is a group homomorphism. Any projective representation determines a 2-cocycle $f : G \times G \to R^*$ defined by

$$f(\sigma, \tau) = \eta(\sigma)\eta(\tau)\eta(\sigma\tau)^{-1} \quad \forall \sigma, \tau \in G$$

Similar to the ordinary case, in which a representation determines an RG- lattice, every projective representation corresponds to a lattice of rank n over the twisted group ring $R^f G$. The 2-cocycle f which is associated with this projective representation twists the multiplication in $R^f G$ by

$$u_\sigma u_\tau = f(\sigma, \tau)u_{\sigma\tau} \quad \forall \sigma, \tau \in G$$

189

Let $\phi : S \to R$ be a map of rings. A *lifting* of a projective representation η from R to S is a projective representation

$$\psi : G \to GL_n(S)$$

such that $\eta = \hat{\phi} \circ \psi$ where $\hat{\phi} : GL_n(S) \to GL_n(R)$. In this case the associated 2-cocycle of η is in the image of the 2-cocycle associated with ψ under the induced map $\tilde{\phi} : Z^2(G, S^*) \to Z^2(G, R^*)$, where $Z^2(G, A)$ denotes the 2-cocycles group.

Consider the inverse system defined by the projections $\phi_s : \mathbb{Z}_{p^{s+1}} \to \mathbb{Z}_{p^s}$ (p-prime). It might be possible to construct representations over $\mathbb{O}_p$ (the p-adic integers) by steps of consecutive liftings of a given representation over $\mathbb{Z}_{p^s}$. A representation may lift to $\mathbb{O}_p$ (e.g. the trivial representation) or may fail to lift at some stage. For example, take the 1-dimensional ordinary representation of $C_2 = \langle \sigma \rangle$ over $\mathbb{Z}_8$ given by $\sigma \longmapsto (5)$. This representation has no lifting to an ordinary representation over $\mathbb{Z}_{16}$, however, it can be lifted to ordinary representations over $\mathbb{Z}_{16}$ after going one step down to $\mathbb{Z}_4$.

This is an example of the following general criterion due to J.M. Maranda. The theorem deals with general p-modular systems. Here we formulate it for the rings $\mathbb{Z}_{p^s}$.

Theorem 1 (Maranda) ([M], [B] 3.7.8).

Suppose G is a finite group of order $p^r q$ where q is coprime to p. If M is a $\mathbb{Z}_{p^a} G$ lattice with $a \geq 2r + 1$, then there is an $\mathbb{O}_p G$ lattice $\hat{M}$ ($\mathbb{O}_p$ the p-adic integers) with $\hat{M}/p^{a-r}\hat{M} \simeq M/p^{a-r}M$.

Note that the representation above, which fails to lift as an ordinary representation, does have a lifting to $\mathbb{Z}_{16}$ as a projective representation with the associated

2-cocycle f given by: $f(\sigma,\sigma) = 9, f(1,1) = f(1,\sigma) = f(\sigma,1) = 1$.

It is not difficult to give an example of an ordinary representation which does

not lift even projectively. As an example take again an ordinary representation of

$C_2 = \langle\sigma\rangle$ over $\mathbb{Z}_8$ where σ is sent to $\begin{pmatrix} 1 & 0 \\ 0 & 5 \end{pmatrix}$.

This representation has no lifting to any projective representation over $\mathbb{Z}_{16}$.

Nevertheless, as in the ordinary case, there exists an analogue of Maranda's

Theorem to projective representations. The formulation is again over the rings $\mathbb{Z}_{p^s}$.

Theorem 2 (Aljadeff, McCarthy)([A,M]).

Suppose G is a finite group of order $p^r q$ where q is coprime to p. Then if

$$\psi : G \to GL_n(\mathbb{Z}_{p^a})$$

is a projective representation where $a \geq 2r + 1$, then there is a projective represen-

tation

$$\tilde{\psi} : G \to GL_n(\mathbb{O}_p)$$

that coincide with ψ at level p^{a-r}.

Note that in terms of twisted group rings, theorem 2 provides a lifting condition

of the corresponding lattice over the twisted group ring $\mathbb{Z}_{p^a}^g G$ to a lattice over *some*

twisted group ring $\mathbb{Z}_{p^{a+1}}^f G$ such that $g \equiv f \mod p^{a-r}$. One may pose a more

delicate generalization of Maranda's theorem for given twisted group rings.

Let $\mathbb{Z}_{p^{a+1}}^f G$ and $\mathbb{Z}_{p^a}^g G$ be twisted group rings such that $g \equiv f \mod p^a$ and let

M be a $\mathbb{Z}_{p^a}^g G$ lattice. Does there exist a $\mathbb{Z}_{p^{a+1}}^f G$ lattice $\hat{M}$ such that $\hat{M}/p^{a-r}\hat{M} \simeq$

$M/p^{a-r}M$ *for some* $r = r(G) \geq 0$?

The following example shows that there is no hope of such a generalization.

Example: Let $G = C_2 \times C_2 = \langle \sigma \rangle \times \langle \tau \rangle$ the Klein 4-group. For every $s \geq 2$ define the following projective representation:

$$\eta : C_2 \times C_2 \to GL_4(\mathbb{Z}_{2^s})$$

$$\sigma \longmapsto \begin{pmatrix} 0 & -1 & 0 & 0 \\ 1 & 0 & 0 & 0 \\ 0 & 0 & 0 & -1 \\ 0 & 0 & 1 & 0 \end{pmatrix} \quad \tau \longmapsto \begin{pmatrix} 0 & -1 & -1 & -1 \\ -1 & 0 & -1 & 1 \\ 1 & 1 & 1 & 0 \\ 1 & -1 & 0 & -1 \end{pmatrix} \quad \sigma\tau \longmapsto \begin{pmatrix} 1 & 0 & 1 & -1 \\ 0 & -1 & -1 & -1 \\ -1 & 1 & 0 & 1 \\ 1 & 1 & 1 & 0 \end{pmatrix}$$

The projective representation η defines a nonprojective lattice M_η over $\mathbb{Z}_{2^s}^g.C_2 \times C_2$. The 2-cocycle g which is associated to η represents an element in $H^2(C_2 \times C_2, Z_{2^s}^*)$ which is determined by the relations: $u_\sigma u_\tau = -u_\tau u_\sigma$, $u_\sigma^2 = -1$, $u_\tau^2 = -1$. Now, let f be a 2-cocycle representing the class in $H^2(C_2 \times C_2, Z_{2^{s+1}}^*)$ determined by the relations: $u_\sigma u_\tau = (2^s - 1)u_\tau u_\sigma$, $u_\sigma^2 = -1$, $u_\tau^2 = -1$. Clearly, $g \equiv f \mod 2^s$. By ([AGO], Proposition 3.5), $\mathbb{Z}_{2^{s+1}}^f C_2 \times C_2$ admits only projective lattices (in these rings all projective lattices are free). We claim that a $\mathbb{Z}_{2^s}^\alpha.C_2 \times C_2$ lattice M is projective (free) if and only if for every $m \leq s$ $M/2^m M$ is projective (free) over $\mathbb{Z}_{2^m}^{\bar{\alpha}} C_2 \times C_2$ where $\bar{\alpha} \equiv \alpha \mod 2^m$. This proves that neither M_η nor any of its quotients $M_\eta/2^m M_\eta$, which are nonprojective, can be lifted to a $\mathbb{Z}_{2^{s+1}}^f C_2 \times C_2$ lattice.

Let us prove the claim: Assume M is an indecomposable $\mathbb{Z}_{2^s}^\alpha.C_2 \times C_2$ lattice which is not free. In ([AGO],1.1) it is shown that M does not contain a free sublattice.

It follows that every element in M is annihilated by the unique minimal ideal $\{2^{s-1}(1+u_\sigma+u_\tau+u_{\sigma\tau}),0\}$ in $\mathbb{Z}_{2^s}^\alpha C_2 \times C_2$. Consequently every element in $M/2^m M$ is annihilated by the corresponding minimal ideal in $\mathbb{Z}_{2^m}^{\overline{\alpha}} C_2 \times C_2$ and hence not free.

The other direction of the claim follows by change of scalars.

REFERENCES

[AGO] E. Aljadeff, Y. Ginosar and U. Onn, *Projective Representations and Relative Semisimplicity*, J. Algebra, to appear.

[AM] E. Aljadeff and R. McCarthy, *Liftings of Projective Representations: Homological Classifications and Obstructions*, Preprint.

[B] D. J. Benson, *Representation and Cohomology I*, Cambridge studies in advanced Mathematics **30**, Cambridge University Press.

[M] J.M. Maranda, *On p-adic Integral Representations of Finite Groups*, Canad. J. Math. 5 **5** (1953), 344–355.

Chain Conditions on Direct Summands and Pure Quotient Modules

José L. Gómez Pardo

Departamento de Alxebra, Universidade de Santiago,

15771 Santiago de Compostela, Spain

Pedro A. Guil Asensio

Departamento de Matemáticas, Universidad de Murcia,

30100 Espinardo, Murcia, Spain

1   Introduction.

It was shown in [5] that a finitely presented pure-injective module M has an indecomposable decomposition if and only if M is completely pure-injective, that is, every pure quotient of M is pure-injective. In particular, a right pure-injective ring R is semiperfect if and only if R_R is completely pure-injective. A semiperfect ring R need not be right pure-injective but R_R still has an indecomposable decomposition. Thus one would expect that this fact is related to some weak injectivity property of the pure quotients of R_R which are still well-behaved in this case —indeed, they are direct summmmands of R_R ([15, 36.4]).

In this paper we deal with this more general situation by considering a condition weaker than pure-injectivity which, when satisfied by the pure quotients of a finitely presented module M, is still strong enough to guarantee the existence of an indecomposable decomposition for M. Recall that a module M is called quasi-continuous whenever i) every submodule of M is essential in a direct summand of M, and ii) if U, V are direct summands of M and $U \cap V = 0$, then $U \oplus V$ is also a direct summand of M. We will consider a sort of quasi-continuity with respect to pure-essential sequences and so we will say that M is a pqc-module when i) every pure submodule of M is purely essential (see definition below) in a direct summand of M, and ii) if U, V are direct summands of M such that $U \cap V = 0$ and $U \oplus V$ is pure in M, then $U \oplus V$ is a direct summand of M. Similarly, M will be called a completely pqc-module when every pure quotient of M is a pqc-module. This concept will be used merely as a technical device to state our main result which shows that if M is a finitely presented completely pqc-module, then M has ACC (and DCC) on direct summands and so it is a (finite) direct sum of indecomposable

*Work partially supported by the DGES (PB96-0961, Spain). The second author was also partially supported by the Fundación Séneca (PB16FS97).

modules. Since the modules such that each pure submodule is a direct summand are completely pqc-modules it is clear that completely pqc-modules need not be pure-injective, so that this extends the result of [5] mentioned above.

This result applies, in particular, to any semiperfect ring, which is a right (and a left) completely pqc-module over itself. We obtain that R is semiperfect if and only if it is semiregular (F-semiperfect) and a (right) completely pqc-module.

Throughout this paper, all rings R will be associative and with identity, and Mod-R will denote the category of right R-modules. By a module we will usually mean a right R-module and, whenever we want to emphasize the fact that M is a right R-module, we will write M_R. We refer to [1], [8], and [15], for all undefined notions used in the text.

2   Results.

Let $u : K \to M$ be a pure monomorphism. We will say that u is a *purely essential* monomorphism if for each homomorphism $g : M \to N$ such that $g \circ u$ is a pure monomorphism, g is also a pure monomorphism. This definition was used in [7] and is stronger than the usual definition of pure-essential (see, e.g. [2]) which only requires that g be a monomorphism. If the inclusion $u : K \to M$ is purely essential, then we say that K is a purely essential submodule of M (or that M is a purely essential extension of K) and we denote this fact by $K \subseteq_{pe} M$. One of the main advantages of purely essential monomorphisms with respect to pure-essential monomorphisms is their good behaviour under composition, which is given by the following lemma whose proof is straightforward.

Lemma 2.1 *Let $u : K \to L$, $v : L \to M$ be pure monomorphisms. Then $v \circ u$ is purely essential if and only if u and v are purely essential.*

In fact, the transitivity property of the preceding lemma characterizes purely essential extensions for we have:

Lemma 2.2 *Let R be a ring. The following properties of the category of right R-modules are equivalent:*

(i) *Every pure-essential extension is purely essential.*

(ii) *The property of being pure-essential is transitive.*

Proof. The implication (i)$\Rightarrow$(ii) follows from Lemma 2.1. For the converse, suppose that pure-essentiality is transitive and let $u : M \to N$ be a pure-essential monomorphism. Let $v : M \to E$ be the pure-injective hull of N. Then $v \circ u : M \to E$ is pure-essential and so E is also the pure-injective hull of M. Then it is well known that $v \circ u$ is actually a purely essential monomorphism (see, e.g., [9, Theorem 3]). It is easily checked that this, together with the fact that v is pure, implies that u is a purely essential monomorphism. $\square$

The next example, communicated to the authors by L. Fuchs, shows that a pure-essential extension need not be purely essential.

Example 2.3 *([3])* Let R be a Prüfer domain which is not a Dedekind domain and D a divisible R-module. Then the pure-injective hull $PE(D)$ coincides with the injective hull E of D. If D is not injective and B/D is a nonzero cyclic submodule of E/D, then D is pure and essential (hence pure-essential) in B. But B is not divisible, so it is not pure in E. Therefore D is not purely essential in B.

Remarks. The fact that all the pure-essential extensions are not necessarily purely essential also shows that one of the usual proofs of existence of pure-injective hulls (the one given, for example, in [13]) does not work. This proof purports to show that (absolutely) maximal pure-essential extensions of a module K exist by embedding K as a pure submodule of a pure-injective module N and considering a maximal pure-essential extension of K within N. But such an extension need not be an absolutely maximal pure-essential extension as it can be seen by taking K a pure-essential submodule of M which is not pure-essential in the pure-injective hull $N = PE(M)$ of M (cf. Example 2.3) and E a maximal pure-essential extension of K in N containing M. If E were an absolutely maximal pure-essential extension of K, then E would have to be pure-injective ([13, Corollary to Proposition 4.4]) and hence a pure-injective hull of K. Then it would follow that E is a direct summand of N ([13, Proposition 4.4]) and the inclusion of M in E is pure-essential, so that E would also be a pure-injective hull of M and hence $E = N$. But this contradicts the fact that K is not pure-essential in N.

A different proof of the existence of absolutely maximal pure-essential extensions, using a cardinality argument, is given, for example, in [2] and [9].

If $K \subseteq_{pe} M$, then K is said to be *purely essentially closed* in M when it has no proper purely essential extensions in M, i.e., when $K \subseteq_{pe} L \subseteq M$ implies $K = L$. A purely essentially closed purely essential extension L of K in M is called a *purely essential closure* of K in M. Observe that by Lemma 2.1 a purely essential closure of K in M is the same as a maximal purely essential extension of K in M and this gives the existence of purely essential closures:

Proposition 2.4 *Let K be a pure submodule of M. Then K has a (not necessarily unique) purely essential closure in M.*

Proof. Using Zorn's Lemma it is enough to show that if $\{L_i\}_I$ is a chain of purely essential extensions of K in M, then the union $L = \cup_I L_i$ is also a purely essential extension of K. For each $i \in I$, let $v_i : K \to L_i$ and $q_i : L_i \to L$ be the inclusions which are pure monomorphisms and $q : K \to L$ the inclusion. Suppose that $g : L \to N$ is a homomorphism such that $g \circ q$ is a pure monomorphism. Then, for each $i \in I$, $g \circ q = g \circ q_i \circ v_i$ and, since each v_i is purely essential, we see that $g \circ q_i$ is a pure monomorphism. It is clear that $g = g \circ 1_L = g \circ \varinjlim(q_i) = \varinjlim(g \circ q_i)$ and, since a direct limit of pure monomorphisms is pure, we are done. $\square$

The next lemma will play an important role in the proof of our main result.

Lemma 2.5 *Let M be a right R-module and Q, T submodules of M such that $Q \cap T = 0$ and $Q \oplus T$ is pure in M. If P_Q, P_T are purely essential extensions of Q, T, respectively, in M, then $P_Q \cap P_T = 0$ and $P_Q \oplus P_T$ is pure in M.*

Proof. We do the proof in several steps.

1) We first show that $P_Q \cap T = 0$. Since $(P_Q \cap T) \cap Q = Q \cap T = 0$ and $Q \subseteq_{pe} P_Q$, it is enough to prove that $Q \cong (Q + (P_Q \cap T))/(P_Q \cap T) \subseteq_p P_Q/(P_Q \cap T)$. Since $Q \oplus T \subseteq_p M$ by hypothesis, we have that the canonical morphism $M \to M/(Q+T)$ is a pure epimorphism that factors in the form $M \to M/T \to M/(Q+T)$ and so $M/T \to M/(Q+T)$ is also a pure epimorphism. The kernel of this morphism is

$$(Q+T)/T \cong (Q + (P_q \cap T))/(P_Q \cap T) \to P_Q/(P_Q \cap T) \to M/T$$

and since it is pure we obtain that $(Q + (P_Q \cap T))/(P_Q \cap T) \subseteq_p P_Q/(P_Q \cap T)$.

2) Next we show that $(P_Q + T)/T \subseteq_p M/T$. We have just seen that $Q \cong (Q+T)/T \subseteq_p M/T$. Thus we have a pure monomorphism

$$Q \cong (Q+T)/T \to P_Q \cong (P_Q + T)/T \to M/T$$

where the first morphism is purely essential because $Q \subseteq_{pe} P_Q$. Therefore $(P_Q + T)/T \subseteq_p M/T$.

3) We now show that $T \cong (P_Q + T)/P_Q \subseteq_p M/P_Q$. We have that $P_Q \cong (P_Q + T)/T \to M/T$ is pure and its cokernel is the canonical morphism $M/T \to M/(P_Q + T)$ which is then a pure epimorphism. Further, $M \to M/T$ is pure by hypothesis and so $M \to M/(P_Q + T)$ is pure. This morphism factors as

$$M \to M/P_Q \to M/(P_Q + T)$$

and so the epimorphism $M/P_Q \to M/(P_Q + T)$ is pure. Thus its kernel $T \cong (P_Q + T)/P_Q \to M/P_Q$ is pure.

4) We show that $P_Q \cap P_T = 0$. Since $T \subseteq_{pe} P_T$ and, as we have seen in 1) $(P_Q \cap P_T) \cap T = P_Q \cap T = 0$, it is enough to show that $T \cong ((P_Q \cap P_T) + T)/(P_Q \cap P_T) \subseteq_p P_T/(P_Q \cap P_T)$. We have seen that $((P_Q \cap P_T) + T)/(P_Q \cap P_T) \cong T \cong (P_Q + T)/P_Q$ is pure in M/P_Q. This morphism factors as

$$((P_Q \cap P_T) + T)/(P_Q \cap P_T) \to P_T/(P_Q \cap P_T) \to M/P_Q$$

and so the first one is pure.

5) Finally, we show that $P_Q \oplus P_T$ is a pure submodule of M. We have seen in Step 3) that $T \subseteq_p M/P_Q$ and so the following monomorphism is pure:

$$T \cong (P_Q + T)/P_Q \to P_T \cong (P_Q + P_T)/P_Q \to M/P_Q$$

Since the first one is purely essential we deduce that the second is pure, that is, $(P_Q + P_T)/P_Q \subseteq_p M/P_Q$. Its cokernel $M/P_Q \to M/(P_q + P_T)$ is then a pure epimorphism and, since $P_Q \subseteq_p M$, so is $M \to M/P_Q$. Their composition is the canonical morphism $M \to M/(P_Q + P_T)$ which is, therefore, pure, and hence $P_Q + P_T \subseteq_p M$. $\square$

We next introduce pqc-modules.

Definition 2.6 *A module M is called a pqc-module whenever every purely essentially closed submodule of M is a direct summand and, if U, V are direct summands of M such that $U \cap V = 0$ and $U \oplus V$ is pure in M, then $U \oplus V$ is a direct summand of M.*

Observe that if M is a module such that every submodule is pure, then M is a pqc-module if and only if it is quasi-continuous. It is also clear that both pure-injective modules and the modules such that every pure submodule is a direct summand are pqc-modules but the latter need not be pure-injective.

We will make use of the following basic property of pqc-modules.

Lemma 2.7 *If L is a direct summand of M and K a pure submodule of L, then a purely essential closure of K in L is also a purely essential closure of K in M. In particular, every direct summand of a pqc-module is a pqc-module.*

Proof. Let X be a purely essential closure of K in L and suppose that X is not purely essentially closed in M. Let Z be a purely essential closure of X in M. Call $j : X \to L$, $v : X \to Z$, $q : Z \to M$ and $u : L \to M$ to the inclusions and observe that, since L is a direct summand, u has a left inverse t such that $t \circ u = 1_L$. Then $j = t \circ u \circ j = t \circ q \circ v$. Since j is a pure monomorphism and v is purely essential, we deduce that $t \circ q$ is a pure monomorphism and so X has a proper purely essential extension in L which is a contradiction and completes the proof. $\square$

Recall that a pure-injective hull of a module K is a pure-essential extension E of K such that E is pure-injective. It is well known that pure-injective envelopes always exist [2]. Since E is a direct summand of every pure extension of K, it follows that the pure-injective hull can also be described as a purely essential extension of K which is pure-injective. The next lemma gives additional information about purely essential closures in a pqc-module.

Lemma 2.8 *Let M be a pqc-module and Q, T submodules of M such that $Q \cap T = 0$ and $Q \oplus T$ is pure in M. If P_Q, P_T are purely essential closures of Q, T, respectively, in M, then $P_Q \oplus P_T$ is a purely essential closure of $Q \oplus T$ in M.*

Proof. By Lemma 2.5, $P_Q \oplus P_T$ is a pure submodule of M and so it is a direct summand of M because M is a pqc-module. Thus it is enough to prove that $P_Q \oplus P_T$ is a purely essential extension of $Q \oplus T$. This in turn will be a consequence of the fact that the direct sum of two purely essential monomorphisms is purely essential. Indeed, if $PE(Q)$ and $PE(T)$ are pure-injective envelopes of Q and T respectively, then it is clear that they are also pure-injective envelopes of P_Q and P_T. Let U be the direct sum of a set of representatives of the isomorphism classes of finitely presented left R-modules and T the corresponding functor ring. Then by [15, 52.3] we have a fully faithful functor $- \otimes_R U :$ Mod-$R :\to$ Mod-T which takes pure-injective envelopes into injective envelopes and so it is clear that $(PE(Q) \oplus PE(T)) \otimes U = (PE(Q) \otimes U) \oplus (PE(T) \otimes U) = E(Q \otimes U) \oplus E(T \otimes U) = E((Q \otimes U) \oplus (T \otimes U)) = E((Q \oplus T) \otimes U)$, which by naturality shows that $PE(Q) \oplus PE(T)$ is a pure-injective envelope of $Q \oplus T$ in Mod-R. Thus the monomorphism $Q \oplus T \to P_Q \oplus P_T \to PE(Q) \oplus PE(T)$ is purely essential and, by Lemma 2.1 we see that $Q \oplus T \subseteq_p P_Q \oplus P_T$. $\square$

A module M will be called a completely pqc-module when every pure quotient of M is a pqc-module. Our main result below gives indecomposable decompositions of these modules and extends [5, Theorem 2.2] where a similar result was proved for completely pure-injective modules.

Theorem 2.9 *Let M be a finitely presented completely pqc-module. Then M has ACC on direct summands and, in particular, it is a direct sum of indecomposable modules.*

Proof. It is well known that M has ACC on direct summands if and only if the endomorphism ring $S = End(_R M)$ has no infinite sets of orthogonal idempotents. Suppose, on the contrary, that there exists an infinite set $\{e_i\}_I$ of orthogonal idempotents in S. Then each $e_i M$ is a direct summand of M and $L = \oplus_I e_i M$ is a local direct summand and hence a pure submodule of M. Consider now a nonempty subset A of I and let $A' = I - A$. Since the idempotents $\{e_i\}_{i \in A}$ are orthogonal, it is clear that $\oplus_A(e_i M)$ and $\oplus_{A'}(e_i M)$ are submodules of M that satisfy the hypotheses of Lemma 2.8. Thus there exist purely essential closures P_A of $\oplus_A(e_i M)$ and $P_{A'}$ of $\oplus_{A'}(e_i M)$ and a submodule T of M such that

$$M = P_A \oplus P_{A'} \oplus T$$

Then there exists an idempotent element $e_A \in S$ such that $P_A = e_A M$ and $P_{A'} \oplus T = (1 - e_A)M$. Furthermore, we have that for each $i \in A$, and each $x \in M$, $e_i x \in P_A = e_A M$ and so $e_A e_i x = e_i x$, which gives $e_A e_i = e_i$ for each $i \in A$. Similarly, for each $i \in A'$ we have that $e_i x \in (1 - e_A)M$ for each $x \in M$ and hence $(1 - e_A)e_i x = e_i x$, from wich it follows that $(1 - e_A)e_i = e_i$ and $e_A e_i = 0$.

Let us now write $I = \cup_{A \in \mathcal{A}} A$ as an infinite union of infinite pairwise disjoint subsets. As in [11] we have that, by Zorn's lemma, there exists a maximal subset $\mathcal{K} \subseteq 2^I$ with respect to the properties: 1) $\mathcal{A} \subseteq \mathcal{K}$; 2) $|A| \geq \aleph_0$ for each $A \in \mathcal{K}$; 3) $|A \cap B| < \aleph_0$ for each $A, B \in \mathcal{K}, A \neq B$.

Let N be the submodule of M defined by $N = \Sigma_{A \in \mathcal{K}} e_A M$. We claim that N is a pure submodule of M. To see this consider $A, B \in \mathcal{K}$, $A \neq B$. Then $A \cap B$ is finite, say $A \cap B = \{i_1, \ldots, i_r\}$. Observe that if $C = B - \{i_1, \ldots, i_r\}$, then $C \cap A = \emptyset$ and $\oplus_B e_i M = e_{i_1} M \oplus \ldots \oplus e_{i_r} M \oplus (\oplus_C e_i M) \subseteq e_B M$. Since $e_B M$ is a direct summand of M we have by Lemma 2.7 that purely essential closures in $e_B M$ of pure submodules are the same as purely essential closures in M, and by Lemma 2.8 a purely essential closure of $\oplus_B e_i M$ can be obtained as the direct sum $e_{i_1} M \oplus \ldots \oplus e_{i_r} M \oplus X_B$, where $e_{i_1} M \oplus \ldots \oplus e_{i_r} M$ is a purely essential closure of itself (because it is a direct summand) and X_B is a purely essential closure of $\oplus_C e_i M$ in $e_B M$ and hence in M. Since $e_B M$ is also a purely essential closure of $\oplus_B e_i M$, we see that $e_B M = e_{i_1} M \oplus \ldots \oplus e_{i_r} M \oplus X_B$.

Since $e_A M$ is a pure-injective hull of $\oplus_A(e_i M)$, it follows from Lemma 2.5 that $e_A M \cap X_B = 0$ and that $e_A M + e_B M = e_A M + X_B$ is a direct summand of M and, actually, by Lemma 2.8, a purely essential closure of $\oplus_{A \cup B} e_i M$ in M. By induction we obtain that if $A_1, \ldots, A_n \in \mathcal{K}$, then $e_{A_1} M + \ldots + e_{A_n} M$ is a direct summand of M. Thus N is a directed union of these direct summands and hence a pure submodule of M. This implies, in particular, that N/L is a pure submodule of M/L.

Next, consider the quotient module $N/L = \Sigma_{A \in \mathcal{K}}((e_A M + L)/L)$. We claim that this sum is direct. Suppose, then, that $A, B_1, \ldots, B_s \in \mathcal{K}$ are different. As we have just seen $\Sigma_{j=1}^s e_{B_j} M$ is a direct summand of M, so that we may write $\Sigma_{j=1}^s e_{B_j} M = gM$ with $g \in S$ an idempotent. Furthermore, gM is a pure essential closure of $\oplus_B(e_i M)$ with $B = \cup_{j=1}^s B_j$. To show that $e_A M \cap gM \subseteq L$ observe that

$A \cap B = A \cap (\cup_{j=1}^{s} B_j) = \cup_{j=1}^{s}(A \cap B_j)$ is a finite set, say $A \cap B = \{k_1, \ldots, k_r\}$. Then we have as before that $e_B M = e_{k_1} R \oplus \ldots \oplus e_{k_r} M \oplus X_B$, where $e_A M \cap X_B = 0$. Since $e_{k_1} M \oplus \ldots \oplus e_{k_r} M \subseteq e_A M$ we obtain by modularity that

$$e_A M \cap gM = e_A M \cap (e_{k_1} R \oplus \ldots \oplus e_{k_r} M \oplus X_B) = e_{k_1} M \oplus \ldots \oplus e_{k_r} M \subseteq L$$

Thus the sum is indeed direct.

Now, we finish our argument as in the proof of [12, Theorem A]. Let $\mathcal{B} = \mathcal{K} - \mathcal{A}$, $N_{\mathcal{A}} = \Sigma_{A \in \mathcal{A}} e_A M$, $N_{\mathcal{B}} = \Sigma_{A \in \mathcal{B}} e_A M$ and observe that $N/L = ((N_{\mathcal{A}} + L)/L) \oplus ((N_{\mathcal{B}} + L)/L)$. Since this is a pure submodule of M/L we may apply Lemma 2.8 to the pqc-module M/L to obtain purely essential closures U, V of $(N_{\mathcal{A}} + L)/L$ and $(N_{\mathcal{B}} + L)/L$, respectively, such that $U \oplus V$ is a direct summand of M/L. We may write $M/L = U \oplus V \oplus X$ and, replacing $V \oplus X$ by V if necessary, assume that $M/L = U \oplus V$ with $(N_{\mathcal{A}} + L)/L \subseteq_{pe} U$ and $(N_{\mathcal{B}} + L)/L \subseteq V$.

Let now $\pi : M/L \to M/L$ be the morphism induced by the projection over the direct summand U and $v : M \to M/L$ the canonical projection. Since L is pure in M and M is pure-projective (because it is finitely presented), there exists a homomorphism $\lambda : M \to M$ such that $\pi \circ v = v \circ \lambda$. For each $A \in \mathcal{A}$, we have that $(e_A M + L)/L \subseteq U$, so that $\pi((e_A M + L)/L) = (e_A M + L)/L$. Therefore $(\pi \circ v)|_{e_A M} = v|_{e_A M}$ and so $e_A M \subseteq Ker(v \circ \lambda - v)$ or, equivalently, $v((\lambda e_A)(x) - e_A(x)) = 0$ for every $x \in M$. This means that $Im(\lambda e_A - e_A) \subseteq L$ for each $A \in \mathcal{A}$ and, since M is finitely generated by hypothesis, this in turn implies that, for almost all $i \in I$, $e_i \lambda e_A = e_i e_A$. Now, since $e_A e_i = e_i$ for each $i \in A$, we have that for almost all $i \in A \in \mathcal{A}$, $e_i \lambda e_i = e_i \lambda e_A e_i = e_i^2 = e_i$.

Similarly, for each $B \in \mathcal{B}$. $(\pi \circ v)|_{e_B M} = 0$. Thus $(v \circ \lambda)(e_B M) = 0$ and so $Im(\lambda e_B) \subseteq L$ and hence $e_i \lambda e_B = 0$ for almost all $i \in I$. Since $e_B e_i = e_i$ for $i \in B$, we see that for almost all $i \in B$, $e_i \lambda e_i = e_i \lambda e_B e_i = 0$.

Now, choose for each $A \in \mathcal{A}$ an index $i_A \in A$ such that $e_{i_A} \lambda e_{i_A} = e_{i_A}$ and let $C = \{i_A | A \in \mathcal{A}\}$. By the maximality of $\mathcal{K}$, C must have infinite intersection with some $D \in \mathcal{K}$ and this D cannot belong to $\mathcal{A}$, so that it must belong to $\mathcal{B}$ and this contradicts the fact that $e_i \lambda e_i = 0$ for almost all $i \in D \in \mathcal{B}$, completing the proof. $\square$

Remarks. The hypotheses of the theorem are not optimal and can be substantially weakened; we chose not to do so in order to avoid excessive technicalities. For example, the fact that M is finitely presented is used only to obtain a lifting to M of a homomorphism from M to M/L (and a special one at that). Recall from [6] that a submodule L of M is called directly split whenever L is a directed union of direct summands of M. Directly split submodules are in particular pure submodules and the module L constructed in the proof of Theorem 2.9 is a directly split submodule of M. Thus in Theorem 2.9, instead of requiring that M be finitely presented, it suffices to assume that M is finitely generated and has the projective property with respect to directly split epimorphisms with domain M. Of course, this property is automatically satisfied if every directly split submodule of M is a direct summand which, for M finitely generated, is equivalent to M having ACC on direct summands.

On the other hand, it is clear that instead of assuming that M is a completely pqc-module, it suffices to require that every directly split quotient of M is a pqc-module.

In a slightly different direction, the theorem can be proved in a more general way if, instead of working with pure submodules we deal with S-pure submodules in the sense of [14, 3], where S is a class of modules contained in the class of finitely presented modules. This is a proper class, so that by [13, 9] S-pure injective hulls exist and we can reproduce all the previous results leading to the theorem, including Lemma 2.8, which also holds by the results in [3]. Thus we obtain, with the obvious definitions an analogous result to theorem 2.9 for completely S-pqc-modules. This applies, for instance to RD-purity ([14]) and also, taking $S = \{R\}$ to the case in which the S-pure submodules are all the submodules and so we obtain that a finitely presented quasi-continuous module M such that all its directly split quotients are quasi-continuous has ACC on direct summands and hence is finite-dimensional.

Recall that a ring R is called semiregular (or F-semiperfect) when R/J is Von Neumann regular and idempotents lift modulo J.

Corollary 2.10 *Let R be a ring. R is semiperfect if and only if it is semiregular and R_R is a completely pqc-module.*

Proof. For the sufficiency just observe that if R is semiperfect then every pure right ideal is a direct summand by [15, 36.4 (1)] and so R_R is a pqc-module. For the necessity, applying Theorem 2.9 we see that R has no infinite sets of orthogonal idempotents, and it is well known that this condition implies that a semiregular ring is semiperfect. $\square$

References

[1] F.W. Anderson and K.R. Fuller, *Rings and categories of modules*, Second Edition, Springer-Verlag, New York, 1992.

[2] J. Dauns, *Modules and rings*, Cambridge Univ. Press, Cambridge, 1994.

[3] A. Facchini, *Relative injectivity and pure-injective modules over Prüfer rings*, J. Algebra 110 (1987), 380-406.

[4] L. Fuchs, L. Salce, and P. Zanardo, *Note on the transitivity of pure-essential extensions*, Colloq. Math. 78 (1998), 283-291.

[5] J.L. Gómez Pardo and P.A. Guil Asensio, *Indecomposable decompositions of finitely presented pure-injective modules*, J. Algebra 192 (1997), 200-208.

[6] J.L. Gómez Pardo and P.A. Guil Asensio, *On the Goldie dimension of injective modules*, Proc. Edinburgh Math. Soc. 41 (1998), 265-275..

[7] F. Héaulme, *Modules purs-semi-simples*, Comm. Algebra 17 (1989), 33-58.

[8] C.U. Jensen and H. Lenzing. *Model theoretic algebra*, Gordon and Breach, New York, 1989.

[9] R. Kielpinski, *On Γ-pure injective modules*, Bull. Acad. Pol. Sci. XV (1967), 127-131.

[10] S.H. Mohamed and B.J. Müller, *Continuous and discrete modules*, Cambridge Univ. Press, Cambridge, 1990.

[11] B.L. Osofsky, *Noninjective cyclic modules*, Proc. Amer. Math. Soc. 19 (1968), 1383-1384.

[12] B.L. Osofsky, *Non-quasi-continuous quotients of finitely generated quasi-continuous modules*, in Ring Theory (S.K. Jain and S.T. Rizvi, Eds.), 259-275, World Scientific, Singapore, 1993.

[13] B. Stenström, *Pure submodules*, Arkiv for Mat. 7 (1967), 159-171.

[14] R.B. Warfield, Jr., *Purity and algebraic compactness for modules*, J. Algebra 28 (1969), 699-719.

[15] R. Wisbauer, *Foundations of Module and Ring Theory*, Gordon and Breach, Reading, 1991.

Prime Spectra of Quantized Coordinate Rings

K. R. Goodearl

Department of Mathematics, University of California, Santa Barbara, CA 93106, USA

E-mail address: goodearl@math.ucsb.edu

This paper is partly a report on current knowledge concerning the structure of (generic) quantized coordinate rings and their prime spectra, and partly propaganda in support of the conjecture that since these algebras share many common properties, there must be a common basis on which to treat them. The first part of the paper is expository. We survey a number of classes of quantized coordinate rings, as well as some related algebras that share common properties, and we record some of the basic properties known to occur for many of these algebras, culminating in stratifications of the prime spectra by the actions of tori of automorphisms. As our main interest is in the generic case, we assume various parameters are not roots of unity whenever convenient. In the second part of the paper, which is based on [**20**], we offer some support for the conjecture above, in the form of an axiomatic basis for the observed stratifications and their properties. At present, the existence of a suitable supply of normal elements is taken as one of the axioms; the search for better axioms that yield such normal elements is left as an open problem.

I. Quantized Coordinate Rings and Related Algebras

This part of the paper is an expository account of the prime ideal structure of algebras on the "quantized coordinate ring" side of the theory of quantum groups – quantizations of the coordinate rings of affine spaces, matrices, semisimple groups, symplectic and Euclidean spaces, as well as a few related algebras – quantized enveloping algebras of Borel and nilpotent subalgebras of semisimple Lie algebras, and quantized Weyl algebras. These algebras occur widely throughout the quantum groups literature, and different papers often investigate different versions. Thus, we begin by giving definitions of the most general versions of which we are aware; in quoting results from the literature, we will specify which version is under consideration. The

This research was partially supported by NATO Collaborative Research Grant 960250 and National Science Foundation research grant DMS-9622876.

reader should bear in mind that many authors, when studying one version of a quantized coordinate algebra, use results proved for slightly different versions, on the understanding that the proofs carry over. This is especially prevalent with regard to quantized coordinate rings of semisimple groups; a detailed development covering the most general case would be a welcome addition to the literature.

Fix a base field k throughout. It need not be algebraically closed, and may have arbitrary characteristic.

1. Descriptions

This section is designed to be a reference source for descriptions of the main classes of the standard quantized coordinate rings currently studied, as well as some related algebras with similar properties. (The reader lacking a strong stomach for generators and relations should just skim this section and refer back to it as necessary.) For the sake of uniformity, and to emphasize that these algebras are deformations of classical coordinate rings, we label all quantized coordinate rings using notations of the form $\mathcal{O}_\bullet(\mathcal{C})$, where $\bullet$ records one or more parameters and $\mathcal{C}$ records the name of the classical object. Thus, $\mathcal{O}_q(k^n)$ refers to a one-parameter quantization of the coordinate ring of affine n-space over a field k, while $\mathcal{O}_{\lambda,p}(GL_n(k))$ refers to a multiparameter quantization of the coordinate ring of $GL_n(k)$, and so on. Many different labels are used throughout the literature for these algebras, and we do not attempt to list the alternates here.

1.1. Quantum affine spaces. These are meant to be viewed as deformations of the coordinate rings $\mathcal{O}(k^n)$. Recall that $\mathcal{O}(k^n)$ is a (commutative) polynomial ring $k[x_1,\ldots,x_n]$, where x_i is the i-th coordinate function on k^n. In the present case, one deforms $\mathcal{O}(k^n)$ in a rather straightforward manner, by altering the commutativity relations $x_i x_j = x_j x_i$ to "commutativity up to scalars", that is, $x_i x_j = q_{ij} x_j x_i$. In order to prevent degeneracy, one needs some assumptions on the q_{ij}. In particular, they should be nonzero, and q_{ji} should equal q_{ij}^{-1} in order to prevent the pair of relations $x_i x_j = q_{ij} x_j x_i$ and $x_j x_i = q_{ji} x_i x_j$ from implying $x_i x_j = 0$. Thus, one assumes that $q = (q_{ij})$ is a *multiplicatively antisymmetric* $n \times n$ matrix over k, that is, $q_{ii} = 1$ and $q_{ji} = q_{ij}^{-1}$ for all i,j. Given q, the corresponding *multiparameter coordinate ring of quantum affine n-space over k* is the k-algebra $\mathcal{O}_q(k^n)$ generated by elements $x_1,\ldots,x_n$ subject only to the relations $x_i x_j = q_{ij} x_j x_i$ for $i,j = 1,\ldots,n$. The standard single parameter version occurs when the q_{ij} for $i < j$ are all equal to a fixed nonzero scalar q; in this case, we denote the algebra $\mathcal{O}_q(k^n)$.

Quantum affine spaces occur already in the work of Manin [**37**, Section 1, §2 and Section 4, §5]; a superalgebra version is discussed in [**38**, §1.4].

1.2. Quantum matrices. The set $M_n(k)$ of $n \times n$ matrices over k, viewed as an algebraic variety, is just affine n^2-space, and its coordinate ring $\mathcal{O}(M_n(k))$ is a polynomial ring in n^2 indeterminates. Hence, one might expect to deform this algebra exactly as in (1.1). However, there is more structure in this case, and one seeks to preserve it as far as possible. Namely, $\mathcal{O}(M_n(k))$ is a bialgebra with a comultiplication $\Delta : \mathcal{O}(M_n(k)) \to \mathcal{O}(M_n(k)) \otimes \mathcal{O}(M_n(k))$ which is effectively the transpose of the map $M_n(k) \times M_n(k) \to M_n(k)$ given by matrix multiplication: If X_{ij} denotes the i,j-th coordinate function on $M_n(k)$, then $\Delta(X_{ij}) = \sum_{l=1}^{n} X_{il} \otimes X_{lj}$. In particular, one would like deformations of $\mathcal{O}(M_n(k))$, with appropriate sets of generators X_{ij}, which are also bialgebras in which the comultiplication of the X_{ij} is given by the equation above.

In addition, multiplication of matrices with row or column vectors from k^n induces on $\mathcal{O}(k^n)$ structures of left and right comodule over $\mathcal{O}(M_n(k))$, and one would like some quantum affine spaces $\mathcal{O}_q(k^n)$ to be left and right co-modules over the deformation of $\mathcal{O}(M_n(k))$, with comodule structure maps behaving as in the classical case. In the single parameter case, there is a deformation $\mathcal{O}_q(M_n(k))$ over which $\mathcal{O}_q(k^n)$ becomes both a left and a right comodule in the desired manner (see [**38**] and [**52**]). However, if one tries to make a multiparameter quantum affine space $\mathcal{O}_q(k^n)$ into a left <u>and</u> right comodule over the desired type of deformation of $\mathcal{O}(M_n(k))$, the latter defor-mation will usually be degenerate. To obtain nondegenerate deformations, one must use different multiparameter quantum affine spaces as left and right comodules, as discovered in [**2**] and [**57**], to which we refer the reader for a more complete discussion. The resulting algebras can be described as follows.

Let $\boldsymbol{p} = (p_{ij})$ be a multiplicatively antisymmetric $n \times n$ matrix over k, and let λ be a nonzero element of k not equal to -1. The corresponding *multiparameter coordinate ring of quantum $n \times n$ matrices over k* is the k-algebra $\mathcal{O}_{\lambda,\boldsymbol{p}}(M_n(k))$ generated by elements X_{ij} (for $i, j = 1, \ldots, n$) subject only to the following relations:

$$X_{\ell m} X_{ij} = \begin{cases} p_{\ell i} p_{jm} X_{ij} X_{\ell m} + (\lambda - 1) p_{\ell i} X_{im} X_{\ell j} & (\ell > i, \ m > j) \\ \lambda p_{\ell i} p_{jm} X_{ij} X_{\ell m} & (\ell > i, \ m \leq j) \\ p_{jm} X_{ij} X_{\ell m} & (\ell = i, \ m > j). \end{cases}$$

(We assume $\lambda \neq 0$ to avoid obvious degeneracies in the second relation above. The assumption $\lambda \neq -1$ is needed to ensure that $\mathcal{O}_{\lambda,\boldsymbol{p}}(M_n(k))$ has the same Hilbert series as a commutative polynomial algebra in n^2 indeterminates [**2**,

Theorem 1].) The standard single parameter version, denoted $\mathcal{O}_q(M_n(k))$, occurs when the p_{ij} for $i > j$ are all equal to a fixed nonzero scalar q, and $\lambda = q^{-2}$. In some references, such as [49] and [56], the roles of q and q^{-1} are interchanged, and/or q is squared.

Quantized coordinate rings for rectangular matrices are defined as subalgebras of those for square matrices. For instance, if $m < n$ then $\mathcal{O}_{\lambda,p}(M_{m,n}(k))$ is defined to be the k-subalgebra of $\mathcal{O}_{\lambda,p}(M_n(k))$ generated by those X_{ij} with $i \leq m$. There is a k-algebra retraction of $\mathcal{O}_{\lambda,p}(M_n(k))$ onto $\mathcal{O}_{\lambda,p}(M_{m,n}(k))$ whose kernel is generated by the X_{ij} with $i > m$; thus $\mathcal{O}_{\lambda,p}(M_{m,n}(k)) \cong \mathcal{O}_{\lambda,p}(M_n(k))/\langle X_{ij} \mid i > m \rangle$. Hence, results for $\mathcal{O}_{\lambda,p}(M_n(k))$ easily extend to the rectangular case; we shall so extend results from the literature without explicit mention.

1.3. Quantum general linear groups. Within the algebraic variety $M_n(k)$, the set $GL_n(k)$ of invertible matrices forms an open subvariety, the complement of the variety defined by the vanishing of the determinant function D. The coordinate ring $\mathcal{O}(GL_n(k))$ thus has the form $\mathcal{O}(M_n(k))[D^{-1}]$. To construct a quantum analog, one inverts a "quantum determinant", call it $D_{\lambda,p}$, in $\mathcal{O}_{\lambda,p}(M_n(k))$. In order for the inversion process to work smoothly and avoid degeneracies, the powers of $D_{\lambda,p}$ should form an Ore set. This occurs because the chosen $D_{\lambda,p}$ turns out to be a normal element. (Recall that a *normal element* in a ring R is an element c such that $cR = Rc$.)

As in the classical case, the quantum determinant $D_{\lambda,p}$ can be defined as a linear combination of products $X_{1,\pi(1)}X_{2,\pi(2)} \cdots X_{n,\pi(n)}$ as π runs through the symmetric group S_n. However, the coefficients are taken to be certain products of elements $-p_{ij}$ rather than powers of -1. Specifically,

$$D_{\lambda,p} = \sum_{\pi \in S_n} \left(\prod_{\substack{1 \leq i < j \leq n \\ \pi(i) > \pi(j)}} (-p_{\pi(i),\pi(j)}) \right) X_{1,\pi(1)}X_{2,\pi(2)} \cdots X_{n,\pi(n)}.$$

The motivation for this choice of $D_{\lambda,p}$ comes from the construction of a quantized version of the exterior algebra on k^n; the n-th "quantum exterior power" of k^n is 1-dimensional, spanned by $D_{\lambda,p}$ (see [2] for details). In the single parameter case, the quantum determinant D_q in $\mathcal{O}_q(M_n(k))$ can be expressed as

$$D_q = \sum_{\pi \in S_n} (-q)^{\ell(\pi)} X_{1,\pi(1)}X_{2,\pi(2)} \cdots X_{n,\pi(n)},$$

where $\ell(\pi)$ denotes the *length* of the permutation π, that is, the minimum length of an expression for π as a product of adjacent transpositions $(i, i+1)$. (Cf. [49, Lemma 4.1.1] but interchange q and q^{-1}.)

It has been computed that $D_{\lambda,\boldsymbol{p}}$ is a normal element of $\mathcal{O}_{\lambda,\boldsymbol{p}}(M_n(k))$ [2, Theorem 3]; in fact,

$$D_{\lambda,\boldsymbol{p}}X_{ij} = \lambda^{j-i}\left(\prod_{l=1}^{n} p_{jl}p_{li}\right)X_{ij}D_{\lambda,\boldsymbol{p}}$$

for all i,j. In the single parameter case, D_q is central (e.g., [49, Theorem 4.6.1]). It follows that the set of nonnegative powers of $D_{\lambda,\boldsymbol{p}}$ is a right and left Ore set in $\mathcal{O}_{\lambda,\boldsymbol{p}}(M_n(k))$. The *multiparameter coordinate ring of quantum* $GL_n(k)$ is now defined to be the localization $\mathcal{O}_{\lambda,\boldsymbol{p}}(GL_n(k)) = \mathcal{O}_{\lambda,\boldsymbol{p}}(M_n(k))[D_{\lambda,\boldsymbol{p}}^{-1}]$; in the single parameter case, $\mathcal{O}_q(M_n(k))[D_q^{-1}]$ is denoted $\mathcal{O}_q(GL_n(k))$.

1.4. Quantum special linear groups. Note that the set $SL_n(k)$ of $n \times n$ matrices with determinant 1 forms a closed subvariety of $M_n(k)$, defined by the single equation $D = 1$. The coordinate ring $\mathcal{O}(SL_n(k))$ thus has the form $\mathcal{O}(M_n(k))/(D - 1)$. To construct a quantum analog, we would thus factor out from $\mathcal{O}_{\lambda,\boldsymbol{p}}(M_n(k))$ the ideal generated by $D_{\lambda,\boldsymbol{p}} - 1$. However, unless $D_{\lambda,\boldsymbol{p}}$ is central, such a factor would be degenerate. Hence, quantum special linear groups are only defined for special choices of the parameters.

From the normality relations for $D_{\lambda,\boldsymbol{p}}$, we see that $D_{\lambda,\boldsymbol{p}}$ is central if and only if

$$\lambda^i \prod_{l=1}^{n} p_{il} = \lambda^j \prod_{l=1}^{n} p_{jl}$$

for all i,j. In this case, the *multiparameter coordinate ring of quantum* $SL_n(k)$ is defined to be the factor algebra

$$\mathcal{O}_{\lambda,\boldsymbol{p}}(SL_n(k)) = \mathcal{O}_{\lambda,\boldsymbol{p}}(M_n(k))/\langle D_{\lambda,\boldsymbol{p}} - 1\rangle.$$

In the single parameter case, where D_q is automatically central, the factor $\mathcal{O}_q(M_n(k))/\langle D_q - 1\rangle$ is denoted $\mathcal{O}_q(SL_n(k))$.

1.5. Quantum semisimple groups. A systematic development of quantized coordinate rings for simple algebraic groups corresponding to the Dynkin diagrams A_n, B_n, C_n, D_n was given by Reshetikhin, Takhtadzhyan, and Fadeev in [53]. These algebras were expressed in terms of the entries of certain "R-matrices" attached to quantizations of the corresponding simple Lie algebras. Calculations of generators and relations for these quantized coordinate rings can be found in, e.g., [58]. Quantized coordinate rings for the exceptional groups, however, were not computed, due to the lack of explicit R-matrices in those cases (cf. [53, Remark 14, p. 212]; the G_2 case is considered in [**55**]). The now standard approach to these algebras involves a quantization of the classical duality between the coordinate ring of a semisimple algebraic group and the enveloping algebra of its Lie algebra. Thus, one constructs quantized enveloping algebras of semisimple Lie algebras and *defines* the quantized coordinate rings of the corresponding semisimple groups as certain Hopf algebra duals. We outline this procedure.

(a) Let $\mathfrak{g}$ be a complex semisimple Lie algebra, and let G be a connected complex semisimple algebraic group with Lie algebra $\mathfrak{g}$. Both $\mathfrak{g}$ and G play symbolic roles; they serve mainly as suggestive labels. The only data needed from these objects are: a (symmetrized) Cartan matrix $(d_i a_{ij})$ for $\mathfrak{g}$; root and weight lattices $Q \subseteq P$ corresponding to choices of Cartan subalgebra $\mathfrak{h}$ and root system for $\mathfrak{g}$; a lattice L lying between Q and P corresponding to the character group of a maximal torus of G with Lie algebra $\mathfrak{h}$; choices of simple roots $\alpha_1, \ldots, \alpha_n$ and fundamental weights $\omega_1, \ldots, \omega_n$; and the (unique) bilinear pairing $(-,-) : P \times Q \to \mathbb{Z}$ such that $(\omega_i, \alpha_j) = \delta_{ij} d_i$ for all i, j. There is a unique extension of $(-,-)$ to a symmetric bilinear form $P \times P \to \mathbb{Q}$, which we also denote by $(-,-)$.

Different authors base quantized coordinate algebras of G on different quantizations of the enveloping algebra $U(\mathfrak{g})$. Thus we first describe a general one-parameter quantization of $U(\mathfrak{g})$, as in [**11**, §0.3]. Let M be a lattice lying between Q and P (independent of L). Then let $q \in k$ be a nonzero scalar, set $q_i = q^{d_i}$ for $i = 1, \ldots, n$, and assume that these $q_i \neq \pm 1$. (In much of the literature on quantized enveloping algebras, k is taken to be $\mathbb{C}$, or a rational function field $\mathbb{C}(q)$ or $\mathbb{Q}(q)$, or the algebraic closure of one of these rational function fields.) Let U^0 be a copy of the group algebra kM, written as the k-algebra with basis $\{k_\lambda \mid \lambda \in M\}$ where $k_0 = 1$ and $k_\lambda k_\mu = k_{\lambda+\mu}$ for $\lambda, \mu \in M$. The *single parameter quantized enveloping algebra of* $\mathfrak{g}$ associated with the above choices is the k-algebra $U_q(\mathfrak{g}, M)$ generated by U^0 and elements $e_1, \ldots, e_n$, $f_1, \ldots, f_n$ satisfying the following relations for $\lambda \in M$ and $i, j = 1, \ldots, n$:

$$k_\lambda e_i k_\lambda^{-1} = q^{(\lambda,\alpha_i)} e_i \qquad \text{and} \qquad k_\lambda f_i k_\lambda^{-1} = q^{-(\lambda,\alpha_i)} f_i$$

$$e_i f_j - f_j e_i = \delta_{ij} \frac{k_{\alpha_i} - k_{\alpha_i}^{-1}}{q_i - q_i^{-1}}$$

$$\sum_{l=1}^{1-a_{ij}} (-1)^l \binom{1-a_{ij}}{l}_{q_i} e_i^{1-a_{ij}-l} e_j e_i^l = 0 \qquad (i \neq j)$$

$$\sum_{l=1}^{1-a_{ij}} (-1)^l \binom{1-a_{ij}}{l}_{q_i} f_i^{1-a_{ij}-l} f_j f_i^l = 0 \qquad (i \neq j),$$

where $\binom{1-a_{ij}}{l}_{q_i}$ is a q_i-binomial coefficient. The most commonly studied case is $U_q(\mathfrak{g}, Q)$, denoted just $U_q(\mathfrak{g})$. This algebra is generated by e_i, f_i, k_i for $i = 1, \ldots, n$, where $k_i = k_{\alpha_i}$. At the other extreme, $U_q(\mathfrak{g}, P)$ is often denoted $\check{U}_q(\mathfrak{g})$.

The algebra $U_q(\mathfrak{g}, M)$ is in fact a Hopf algebra, with comultiplication Δ, counit ϵ, and antipode S such that

$$\Delta(k_\lambda) = k_\lambda \otimes k_\lambda \qquad \epsilon(k_\lambda) = 1 \qquad S(k_\lambda) = k_\lambda^{-1}$$
$$\Delta(e_i) = e_i \otimes 1 + k_{\alpha_i} \otimes e_i \qquad \epsilon(e_i) = 0 \qquad S(e_i) = -k_{\alpha_i}^{-1} e_i$$
$$\Delta(f_i) = f_i \otimes k_{\alpha_i}^{-1} + 1 \otimes f_i \qquad \epsilon(f_i) = 0 \qquad S(f_i) = -f_i k_{\alpha_i}.$$

Hence, one can define the *finite dual* $U_q(\mathfrak{g}, M)^\circ$ as in, e.g., [41, Definition 1.2.3]; this is a k-linear subspace of $U_q(\mathfrak{g}, M)^*$ which becomes a Hopf algebra using the transposes of the multiplication, comultiplication, and antipode of $U_q(\mathfrak{g}, M)$ (e.g., [41, Theorem 9.1.3]).

(b) Single-parameter quantizations of $\mathcal{O}(G)$ are defined as subalgebras of $U_q(\mathfrak{g}, M)^\circ$ generated by the "coordinate functions" of certain "highest weight" $U_q(\mathfrak{g}, M)$-modules. It turns out that the resulting algebras are independent of the choice of M. Hence, we first describe quantizations of $\mathcal{O}(G)$ as subalgebras of $U_q(\mathfrak{g})^\circ$; this has the advantage that no roots of q are required. For descriptions in terms of $U_q(\mathfrak{g}, M)^\circ$, see part (c). To avoid problems with certain calculations, one assumes that $\mathrm{char}(k) \neq 2$, and also $\mathrm{char}(k) \neq 3$ in case $\mathfrak{g}$ has a component of type G_2.

Here we only give a quantization of $\mathcal{O}(G)$ for the case that q is not a root of unity. The root of unity case requires defining a suitable algebra over a Laurent polynomial ring $k_0[t^{\pm 1}]$ and then specializing t to q (see, e.g., [35, Sections 7,8] or [12, Section 4]).

For each $\lambda \in P^+$, there is a finite dimensional simple $U_q(\mathfrak{g})$-module $V(\lambda)$ with *highest weight* λ, that is, $V(\lambda)$ is generated by an element u_λ satisfying $k_\mu u_\lambda = q^{(\mu,\lambda)} u_\lambda$ for all $\mu \in Q$ and $e_i u_\lambda = 0$ for $i = 1, \ldots, n$ (see, e.g., [26, Theorem 5.10]). For $v \in V(\lambda)$ and $f \in V(\lambda)^*$, let $c_{f,v}^{V(\lambda)} \in U_q(\mathfrak{g})^\circ$ denote the *coordinate function* defined by the rule $c_{f,v}^{V(\lambda)}(u) = f(uv)$. The *single parameter quantized coordinate ring of G* is the k-subalgebra $\mathcal{O}_q(G, L)$ of $U_q(\mathfrak{g})^\circ$ generated by the $c_{f,v}^{V(\lambda)}$ for $\lambda \in L^+$, $f \in V(\lambda)^*$, and $v \in V(\lambda)$. There is some redundancy in the notation $\mathcal{O}_q(G, L)$, since L is determined by G, but we prefer to emphasize L in this way. (Thus, the pair (G, L) is used as a label for that connected semisimple group with Lie algebra $\mathfrak{g}$ and weight lattice L.) The most commonly studied case corresponds to a *simply* *connected* group. This is the case $L = P$, and in this case we write $\mathcal{O}_q(G)$ for $\mathcal{O}_q(G, P)$.

When $G = SL_n(\mathbb{C})$, one obtains the algebra $\mathcal{O}_q(M_n(\mathbb{C}))/\langle D_q - 1 \rangle$ as in (1.4) (e.g., [**22**, Theorem 1.4.1]; replace q^2 by q). Thus, there is no ambiguity in the notation $\mathcal{O}_q(SL_n(\mathbb{C}))$.

(c) In order to exhibit $\mathcal{O}_q(G, L)$ as a subalgebra of $U_q(\mathfrak{g}, M)^\circ$, one needs sufficient roots of q in k so that $q^{\langle M, L \rangle} \subset k$. For $\lambda \in L^+$, there is then a simple finite dimensional $U_q(\mathfrak{g}, M)$-module $V(M, \lambda)$ with highest weight λ, where now the highest weight vector u_λ must satisfy $k_\mu u_\lambda = q^{\langle \mu, \lambda \rangle} u_\lambda$ for all $\mu \in M$. Let us write $\mathcal{O}_q(G, L, M)$ for the k-subalgebra of $U_q(\mathfrak{g}, M)^\circ$ generated by the coordinate functions of the $V(M, \lambda)$ for $\lambda \in L^+$. Each $V(M, \lambda)$ is also simple as a $U_q(\mathfrak{g})$-module; thus $V(M, \lambda)$ becomes $V(\lambda)$ by restriction, and coordinate functions of $V(M, \lambda)$ map to coordinate functions of $V(\lambda)$ by restriction. Therefore the restriction map $U_q(\mathfrak{g}, M)^\circ \to U_q(\mathfrak{g})^\circ$ induces a k-algebra homomorphism $\mathcal{O}_q(G, L, M) \to \mathcal{O}_q(G, L)$. This map is an isomorphism, because

$$\mathcal{O}_q(G, L, M) = \bigoplus_{\lambda \in L^+} C^{V(M, \lambda)} \qquad \text{and} \qquad \mathcal{O}_q(G, L) = \bigoplus_{\lambda \in L^+} C^{V(\lambda)}$$

where $C^{V(M, \lambda)}$ and $C^{V(\lambda)}$ denote the k-linear spans of the coordinate functions on $V(M, \lambda)$ and $V(\lambda)$, respectively (cf. [**29**, §2.2], [**30**, §1.4.13], [**24**, §3.3]). One also needs the fact that $C^{V(M, \lambda)}$ and $C^{V(\lambda)}$ have the same k-dimension, since both are isomorphic to $V(\lambda) \otimes V(\lambda)^*$.

For the reasons sketched above, we can – and do – choose to work with $\mathcal{O}_q(G, L)$. In [**24**, §3.3], for instance, the corresponding algebra – there denoted $\mathbb{C}_q[G]$ – is defined as $\mathcal{O}_q(G, L, L)$.

(d) Multiparameter versions of $\mathcal{O}_q(G, L)$ are obtained by twisting the multiplication. Let $p : L \times L \to k^\times$ be an *alternating bicharacter*, that is,

$$p(\lambda, \lambda) = 1$$
$$p(\mu, \lambda) = p(\lambda, \mu)^{-1}$$
$$p(\lambda, \mu + \mu') = p(\lambda, \mu)p(\lambda, \mu')$$

for $\lambda, \mu, \mu' \in L$. There is a natural $(L \times L)$-bigrading on $\mathcal{O}_q(G, L)$ (cf. [**24**, §3.3]), and the multiplication on $\mathcal{O}_q(G, L)$ can be twisted via p as in [**24**, §2.1]; the new multiplication, call it $\cdot$, is determined by the rule

$$a \cdot b = p(\lambda, \lambda')p(\mu, \mu')^{-1} ab \qquad \text{for } a \in \mathcal{O}_q(G, L)_{\lambda, \mu} \text{ and } b \in \mathcal{O}_q(G, L)_{\lambda', \mu'}.$$

The vector space $\mathcal{O}_q(G, L)$, equipped with this new multiplication, is called a *multiparameter quantized coordinate ring of* G, denoted $\mathcal{O}_{q,p}(G, L)$. In the case $L = P$, we write simply $\mathcal{O}_{q,p}(G)$.

1.6. Quantized enveloping algebras of Borel and nilpotent subalgebras. Each of the quantized enveloping algebras $U_q(\mathfrak{g}, M)$ contains subalgebras which are viewed as quantizations of the enveloping algebras of Borel or nilpotent subalgebras of $\mathfrak{g}$. Unlike $U_q(\mathfrak{g}, M)$, these algebras behave much like $\mathcal{O}_q(G)$, and they play prominent roles in the study of $\mathcal{O}_q(G)$. Hence, we include them here under the rubric of "algebras similar to quantized coordinate rings", along with quantized Weyl algebras (see the following subsection).

Carry over the objects and notation from (1.5a), in particular the Lie algebra $\mathfrak{g}$, lattices $Q \subseteq M \subseteq P$, the powers $q_\iota = q^{d_\iota} \neq \pm 1$, and the group algebra $U^0 \subseteq U_q(\mathfrak{g}, M)$. Let $\mathfrak{b}^+$ and $\mathfrak{n}^+$ denote the positive Borel and nilpotent subalgebras of $\mathfrak{g}$ corresponding to the choices above. The *single parameter quantized enveloping algebra of* $\mathfrak{b}^+$ associated with the above choices is the k-algebra $U_q(\mathfrak{b}^+, M)$ generated by U^0 and $e_1, \ldots, e_n$. The corresponding *single parameter quantized enveloping algebra of* $\mathfrak{n}^+$ is the k-subalgebra $U_q(\mathfrak{n}^+)$ of $U_q(\mathfrak{b}^+, M)$ generated by $e_1, \ldots, e_n$; this algebra is often denoted $U_q(\mathfrak{g})^+$ (it is, of course, independent of the choice of M).

There is an $(M \times M)$-bigrading on the algebra $U_q(\mathfrak{b}^+, M)$ such that $k_\lambda \in U_q(\mathfrak{b}^+, M)_{-\lambda, \lambda}$ for $\lambda \in M$ and $e_\iota \in U_q(\mathfrak{b}^+, M)_{-\alpha_\iota, 0}$ for $i = 1, \ldots, n$ (see [24, Corollary 3.3]). Obviously, $U_q(\mathfrak{n}^+)$ is a bigraded subalgebra. Given an alternating bicharacter $p : M \times M \to k^\times$, we can twist the multiplications in $U_q(\mathfrak{b}^+, M)$ and $U_q(\mathfrak{n}^+)$ via p just as for $\mathcal{O}_q(G, L)$ above. The resulting algebras are the *multiparameter quantized enveloping algebras of* $\mathfrak{b}^+$ *and* $\mathfrak{n}^+$, denoted $U_{q,p}(\mathfrak{b}^+, M)$ and $U_{q,p}(\mathfrak{n}^+)$. As in (1.5a), we omit M from the notation in the case $M = Q$.

There are two cases in which a quantized Borel is known to be a homomorphic image of a quantum semisimple group. Namely, when q is transcendental over $\mathbb{Q}$, the algebra $U_q(\mathfrak{b}^+, P)$ is isomorphic to a factor algebra of $\mathcal{O}_q(G)$ [30, Corollary 9.2.12], and for $q \in \mathbb{C}$ not a root of unity, $U_{q,p}(\mathfrak{b}^+, M)$ is a homomorphic image of $\mathcal{O}_{q,p^{-1}}(G, M)^{\mathrm{op}}$ [24, Proposition 4.6].

1.7. Quantized Weyl algebras. Recall that the Weyl algebra $A_n(k)$ can (at least in characteristic zero) be viewed as the algebra of polynomial differential operators on affine n-space, i.e., the algebra of all linear partial differential operators with polynomial coefficients on the coordinate ring $\mathcal{O}(k^n)$. Quantized versions of $A_n(k)$ arose in Maltsiniotis' development of "quantum differential calculus" [36], and can be viewed as algebras of "partial q-difference operators" on quantum affine spaces.

Let $Q = (q_1, \ldots, q_n)$ be a vector in $(k^\times)^n$, and let $\Gamma = (\gamma_{\iota j})$ be a multiplicatively antisymmetric $n \times n$ matrix over k. The *multiparameter quantized Weyl algebra of degree n over k* is the k-algebra $A_n^{Q, \Gamma}(k)$ generated by elements $x_1, y_1, \ldots, x_n, y_n$ subject only to the following relations:

$$y_i y_j = \gamma_{ij} y_j y_i \qquad \qquad \text{(all } i,j)$$

$$x_i x_j = q_i \gamma_{ij} x_j x_i \qquad \qquad (i < j)$$

$$x_i y_j = \gamma_{ji} y_j x_i \qquad \qquad (i < j)$$

$$x_i y_j = q_j \gamma_{ji} y_j x_i \qquad \qquad (i > j)$$

$$x_j y_j = 1 + q_j y_j x_j + \sum_{l<j} (q_l - 1) y_l x_l \qquad \text{(all } j).$$

See [**28**, §2.9] for a discussion of how to view elements of $A_n^{Q,\Gamma}(k)$ as q-difference operators on $\mathcal{O}_\Gamma(k^n)$.

1.8. Quantum symplectic spaces. The classical linear, symplectic, and orthogonal groups all act on the same affine space, k^n. We have already mentioned above that the relations in quantum matrices are partly chosen so that quantum affine spaces become comodules over quantum matrices, in a "standard" manner. It follows that quantum affine spaces are also comodules over quantum special linear groups. In order to develop an analogous situation for quantum symplectic and orthogonal groups, one needs, as it turns out, different deformations of $\mathcal{O}(k^n)$ than quantum affine spaces. These new algebras are called quantum symplectic and Euclidean spaces; they have (to our knowledge) only been defined in single parameter versions to date.

Let q be a nonzero scalar in k and n a positive integer. Set

$$(1, 2, \ldots, n, n', (n-1)', \ldots, 1') = (1, 2, \ldots, n, n+1, \ldots, 2n)$$

$$(\rho(1), \rho(2), \ldots, \rho(2n)) = (n, n-1, \ldots, 1, -1, -2, \ldots, -n)$$

$$(\epsilon_1, \ldots, \epsilon_n, \epsilon_{n+1}, \ldots, \epsilon_{2n}) = (1, \ldots, 1, -1, \ldots, -1).$$

The one-parameter *coordinate ring of quantum symplectic 2n-space over k* (cf. [**53**, Definition 14] or [**43**, §1.1]) is the k-algebra $\mathcal{O}_q(\mathfrak{sp}\, k^{2n})$ generated by elements $x_1, \ldots, x_{2n}$ satisfying the following relations:

$$x_i x_j = q x_j x_i \qquad \qquad (i < j;\ j \neq i')$$

$$x_i x_{i'} = x_{i'} x_i + (1-q^2) \sum_{l=1}^{i'-1} q^{\rho(i')-\rho(l)} \epsilon_{i'} \epsilon_l x_l x_{l'} \qquad (i \leq n)$$

(cf. [**53**, p. 210]). A simpler set of relations was found by Musson [**43**, §1.1] (cf. [**45**, §1.1]):

$$x_i x_j = q x_j x_i \qquad \qquad (i < j;\ j \neq i')$$

$$x_i x_{i'} = q^2 x_{i'} x_i + (q^2 - 1) \sum_{l=1}^{i-1} q^{l-i} x_l x_{l'} \qquad (i \leq n).$$

1.9. **Quantum Euclidean spaces.** Let $n \geq 2$ be an integer and q a nonzero scalar in k, such that q has a square root in k in case n is odd. Set $m = \lfloor n/2 \rfloor$, the integer part of $n/2$, and set $i' = n + 1 - i$ for $1 \leq i \leq n$. Further, set

$$
\begin{aligned}
&(\rho(1), \ldots, \rho(n)) \\
&= \begin{cases} (m - \frac{1}{2}, m - \frac{3}{2}, \ldots, \frac{1}{2}, 0, -\frac{1}{2}, -\frac{3}{2}, \ldots, -m + \frac{1}{2}) & (n \text{ odd}) \\ (m - 1, m - 2, \ldots, 1, 0, 0, -1, -2, \ldots, -m + 1) & (n \text{ even}). \end{cases}
\end{aligned}
$$

The one-parameter *coordinate ring of quantum Euclidean n-space over k* (cf. [**53**, Definition 12] or [**43**, §2.1]) is the k-algebra $\mathcal{O}_q(\mathfrak{o}\, k^n)$ generated by elements $x_1, \ldots, x_n$ satisfying the following relations:

$$ x_i x_j = q x_j x_i \qquad\qquad (i < j; \; j \neq i') $$

$$ x_i x_{i'} = x_{i'} x_i + \frac{1 - q^2}{1 + q^{n-2}} \left(q^{n-2} \sum_{l=1}^{i'-1} q^{\rho(i')-\rho(l)} x_l x_{l'} - \sum_{l=i'}^{n} q^{\rho(i')-\rho(l)} x_l x_{l'} \right) $$

$$ (i \leq m). $$

A simpler set of relations was given in [**43**, §§2.1, 2.2]:

$$ x_i x_j = q x_j x_i \qquad\qquad (i < j; \; j \neq i') $$

$$ x_i x_{i'} = x_{i'} x_i + (1 - q^2) \sum_{l=i+1}^{m} q^{l-i-2} x_l x_{l'} $$

$$ (i \leq m; \; n \text{ even}) $$

$$ x_i x_{i'} = x_{i'} x_i + (1 - q) q^{m-i-(1/2)} x_{m+1}^2 + (1 - q^2) \sum_{l=i+1}^{m} q^{l-i-2} x_l x_{l'} $$

$$ (i \leq m; \; n \text{ odd}). $$

For an alternative presentation of $\mathcal{O}_q(\mathfrak{o}\, k^n)$, see [**46**, Example 5; **47**, Definition 3.1].

2. SOME COMMON PROPERTIES

We record some fundamental properties common to the algebras discussed in Section 1. The cases we list are those for which we have located references in which a result appears, or from which it follows readily. Many cases which we conjecture should be included must be omitted at present, because they have not (to our knowledge) been addressed in the literature. We leave it to the reader to identify appropriate "missing" cases, and to formulate the corresponding conjectures. Perhaps more important than filling in specific cases is the problem of developing general theorems which could verify properties for all these algebras simultaneously.

2.1. Noetherian domains.

Theorem. *All of the following algebras are noetherian domains:* $\mathcal{O}_q(k^n)$; $\mathcal{O}_{\lambda,p}(M_{m,n}(k))$; $\mathcal{O}_{\lambda,p}(GL_n(k))$; $\mathcal{O}_{\lambda,p}(SL_n(k))$; $\mathcal{O}_{q,p}(G,L)$ *for q transcendental over $\mathbb{Q}$ or $q \in \mathbb{C}$ not a root of unity;* $U_{q,p}(\mathfrak{b}^+)$; $U_{q,p}(\mathfrak{n}^+)$; $A_n^{Q,\Gamma}(k)$; $\mathcal{O}_q(\mathfrak{sp}\, k^{2n})$; $\mathcal{O}_q(\mathfrak{o}\, k^n)$.

Proof. In the cases $\mathcal{O}_q(k^n)$; $\mathcal{O}_{\lambda,p}(M_{m,n}(k))$; $A_n^{Q,\Gamma}(k)$; $\mathcal{O}_q(\mathfrak{sp}\, k^{2n})$; $\mathcal{O}_q(\mathfrak{o}\, k^n)$, one has only to observe that the algebra is an iterated skew polynomial ring over k. This is clear for $\mathcal{O}_q(k^n)$; for the other cases, see [**2**, pp. 890-891], [**28**, §§2.1, 2.8], [**43**, §§1.2, 2.3]. It follows that the localization $\mathcal{O}_{\lambda,p}(GL_n(k))$ is a noetherian domain, and that the factor algebra $\mathcal{O}_{\lambda,p}(SL_n(k))$ is noetherian. That the latter is a domain is proved in [**34**, Corollary].

For $\mathcal{O}_q(G)$ in the stated cases, see [**29**, Lemma 3.1, Proposition 4.1] and [**30**, Lemma 9.1.9, Proposition 9.2.2]. (An alternate proof of noetherianity is given in [**4**, Corollary 5.6].) The desired properties of $\mathcal{O}_q(G,L)$ are proved by the same arguments, and they carry over to $\mathcal{O}_{q,p}(G,L)$ by graded ring methods [**24**, Remark, p. 80] or by twisting results [**60**, Propositions 5.1, 5.2].

By [**10**, Corollary 1.8], $U_q(\mathfrak{g})$ is a domain, and so $U_q(\mathfrak{b}^+)$ and $U_q(\mathfrak{n}^+)$ are domains. There are several ways to see that $U_q(\mathfrak{b}^+)$ and $U_q(\mathfrak{n}^+)$ are noetherian. For instance, it follows from the existence of a PBW basis for $U_q(\mathfrak{g})$ (e.g., [**26**, Theorem 4.21]) that $U_q(\mathfrak{g})$ is free as a right $U_q(\mathfrak{b}^+)$-module and as a right $U_q(\mathfrak{n}^+)$-module. Hence, the poset of left ideals of either subalgebra embeds in the poset of left ideals of $U_q(\mathfrak{g})$, via $I \mapsto U_q(\mathfrak{g})I$. This shows that $U_q(\mathfrak{b}^+)$ and $U_q(\mathfrak{n}^+)$ are left noetherian. There exists an antiautomorphism τ on $U_q(\mathfrak{g})$ such that $\tau(e_i) = e_i$ for $i = 1, \ldots, n$ and $\tau(k_\lambda) = k_\lambda^{-1}$ for $\lambda \in Q$ (e.g., [**26**, Lemma 4.6]). Since τ obviously stabilizes $U_q(\mathfrak{b}^+)$ and $U_q(\mathfrak{n}^+)$, these subalgebras must be right as well as left noetherian.

On the other hand, in view of [**10**, Proposition 1.7], given either $A_0 = U_q(\mathfrak{b}^+)$ or $A_0 = U_q(\mathfrak{n}^+)$, there is a sequence of algebras $A_0, A_1, \ldots, A_N$ such that each A_{i+1} is the associated graded ring of A_i with respect to some $\mathbb{Z}^+$-filtration, and such that A_N is an iterated skew polynomial ring over either U^0 or k. Since A_N is noetherian, it follows from standard filtered/graded techniques (e.g., [**39**, Theorem 1.6.9]) that A_0 is noetherian.

Finally, the twisting results mentioned above ([**60**, Propositions 5.1, 5.2]) imply that $U_{q,p}(\mathfrak{b}^+)$ and $U_{q,p}(\mathfrak{n}^+)$ are noetherian domains. $\square$

2.2. Complete primeness of prime factors.
For any set $\{\alpha_i\}$ of nonzero scalars in k, let $\langle \alpha_i \rangle$ denote the multiplicative subgroup of $k^\times$ generated by the α_i.

Theorem. *Let A be one of the following algebras: $\mathcal{O}_q(k^n)$ with $\langle q_{ij} \rangle$ torsionfree; $\mathcal{O}_{\lambda,p}(M_{m,n}(k))$ or $\mathcal{O}_{\lambda,p}(GL_n(k))$ or $\mathcal{O}_{\lambda,p}(SL_n(k))$ with $\langle \lambda, p_{ij} \rangle$ torsionfree; $\mathcal{O}_q(G)$ or $U_q(\mathfrak{b}^+)$ with $q \in \mathbb{C}$ not a root of unity; $U_q(\mathfrak{n}^+)$ with q transcendental over $\mathbb{Q}$; $A_n^{Q,\Gamma}(k)$ with $\langle q_i, \gamma_{ij} \rangle$ torsionfree; $\mathcal{O}_q(\mathfrak{sp}\, k^{2n})$ or $\mathcal{O}_q(\mathfrak{o}\, k^n)$ with q not a root of unity. Then all prime ideals of A are completely prime, i.e., all prime factor rings of A are domains.*

Proof. The cases $\mathcal{O}_q(k^n)$; $\mathcal{O}_{\lambda,p}(M_{m,n}(k))$; $A_n^{Q,\Gamma}(k)$; $\mathcal{O}_q(\mathfrak{sp}\, k^{2n})$; $\mathcal{O}_q(\mathfrak{o}\, k^n)$ follow from a general result about prime ideals of certain iterated skew polynomial rings: [**18**, Theorem 2.3]. See [**18**, Theorem 2.1], [**16**, Theorem 5.1], [**8**, Proposition II.1.2], [**43**, Corollary 1.2, Theorem 2.3] for further details. The cases $\mathcal{O}_{\lambda,p}(GL_n(k))$ and $\mathcal{O}_{\lambda,p}(SL_n(k))$ follow immediately. In addition, $U_q(\mathfrak{n}^+)$ is an iterated q-skew polynomial ring over k (see [**54**, Section 5] for the case $k = \mathbb{Q}(q)$ and extend scalars), and so in this case also the result follows from [**18**, Theorem 2.3] – see [**54**, Corollary to Theorem 3]. Finally, the cases $\mathcal{O}_q(G)$ and $U_q(\mathfrak{b}^+)$ are proved in [**29**, Theorems 11.4, 11.5]. $\square$

2.3. Division rings of fractions.

Theorem. *Let A be one of the algebras $\mathcal{O}_{\lambda,p}(M_{m,n}(k))$; $U_q(\mathfrak{sl}_n(k))^+$ with q not a root of unity; $\mathcal{O}_q(G)$ or $U_q(\mathfrak{b}^+, P)$ or $U_q(\mathfrak{n}^+)$ with $q \in \mathbb{C}$ not a root of unity; $A_n^{Q,\Gamma}(k)$; $\mathcal{O}_q(\mathfrak{sp}\, k^{2n})$; $\mathcal{O}_q(\mathfrak{o}\, k^n)$. Then $\operatorname{Fract} A \cong \operatorname{Fract} \mathcal{O}_q(K^t)$ for some q, some field $K \supseteq k$, and some t.*

Now let A be either $\mathcal{O}_q(M_n(k))$ with q not a root of unity, or $A_n^{Q,\Gamma}(k)$ with $\langle q_i, \gamma_{ij} \rangle$ torsionfree. If P is any prime ideal of A, then $\operatorname{Fract}(A/P) \cong \operatorname{Fract} \mathcal{O}_q(K^t)$ for some q, some field $K \supseteq k$, and some t.

Proof. For the first set of cases, see [**42**, Theorem 1.24], [**1**, Théorème 2.15], [**25**, Theorem 3.5], [**7**, Theorems 3.3, 3.1], [**28**, §3.1], [**43**, Theorems 1.3, 2.4]. The case of $\operatorname{Fract} \mathcal{O}_q(M_n(k)) = \operatorname{Fract} \mathcal{O}_q(GL_n(k))$ was treated earlier in [**9**, Proposition 5] and [**48**, Theorem 3.8]. The case of $U_q(\mathfrak{n}^+)$ for q transcendental over $\mathbb{C}$ was also obtained in [**31**]. For the cases of prime factors of $\mathcal{O}_q(M_n(k))$ and $A_n^{Q,\Gamma}(k)$, see [**8**, Théorèmes II.2.1, III.3.2.1]. $\square$

2.4. Catenarity.

Recall that (the prime spectrum of) a ring A is *catenary* if for any comparable prime ideals $P \supset Q$ in A, all saturated chains of prime ideals from P to Q have the same length. Our discussion of this property will involve the following concept. The ring A, or its prime spectrum spec A, is said to have *normal separation* provided the following condition holds: For any proper inclusion $P \supset Q$ of prime ideals of A, the factor P/Q contains a nonzero element which is normal in A/Q. In other words, there must exist an element $c \in P \setminus Q$ such that $c + Q$ is normal in A/Q; we refer to the latter condition by saying that c is *normal modulo Q*.

Theorem. *The following algebras are catenary:* $\mathcal{O}_q(k^n)$; $\mathcal{O}_q(GL_n(\mathbb{C}))$ *and* $\mathcal{O}_q(SL_n(\mathbb{C}))$ *with* $q \in \mathbb{C}$ *not a root of unity;* $U_q(\mathfrak{n}^+)$ *with* q *transcendental over* $\mathbb{Q}$; $A_n^{Q,\Gamma}(k)$ *with no* q_ι *a root of unity;* $\mathcal{O}_q(\mathfrak{sp}\, k^{2n})$ *and* $\mathcal{O}_q(\mathfrak{o}\, k^n)$ *with* q *not a root of unity.*

Proof. These cases all follow from a general theorem based on Gabber's methods: If A is an affine, noetherian, Auslander-Gorenstein, Cohen-Macaulay algebra with finite Gelfand-Kirillov dimension, and if spec A is normally separated, then A is catenary [**17**, Theorem 1.6]. See [**17**, Theorems 2.6, 3.13, 4.5, 4.8] and [**46**, Corollaries 12, 13] for the individual cases listed. □

2.5. Normal separation. As indicated in (2.4), normal separation plays a key role in proving catenarity. Normal separation in a ring A also implies Jategaonkar's strong second layer condition (cf. [**27**, Theorem 3.3.16, Proposition 8.1.7, discussion p. 225]), from which it follows, in particular, that every module has a finite filtration such that associated primes in adjacent layers are linked in spec A (cf. [**27**, Theorem 9.1.2]). Other influences of the second layer condition on the representation theory of A are discussed in [**5**] and [**27**]. Finally, we mention that normal separation, in conjunction with several other properties typical of quantum coordinate rings, leads to a description of links and cliques in terms of lattices of automorphisms [**3**, Section 3].

Theorem. *The following algebras have normal separation:* $\mathcal{O}_q(k^n)$; $\mathcal{O}_q(G)$ *with* q *transcendental over* $\mathbb{Q}$; $\mathcal{O}_{q,p}(G,L)$ *with* $q \in \mathbb{C}$ *not a root of unity;* $U_q(\mathfrak{b}^+, P)$ *and* $U_q(\mathfrak{n}^+)$ *over a rational function field* $\mathbb{C}(q)$; $U_q(\mathfrak{b}^+, P)$ *with* q *transcendental over* $\mathbb{Q}$; $U_{q,p}(\mathfrak{b}^+, M)$ *with* $q \in \mathbb{C}$ *not a root of unity;* $A_n^{Q,\Gamma}(k)$ *with no* q_ι *a root of unity;* $\mathcal{O}_q(\mathfrak{sp}\, k^{2n})$ *and* $\mathcal{O}_q(\mathfrak{o}\, k^n)$ *with* q *not a root of unity.*

Proof. For $\mathcal{O}_q(k^n)$, $A_n^{Q,\Gamma}(k)$, $\mathcal{O}_q(\mathfrak{sp}\, k^{2n})$ and $\mathcal{O}_q(\mathfrak{o}\, k^n)$, see [**17**, Corollary 2.4, Theorem 3.12] and [**46**, Theorem 10]. The algebras $U_q(\mathfrak{n}^+)$ and $U_q(\mathfrak{b}^+, P)$ over $\mathbb{C}(q)$ are actually *polynormal*, i.e., every ideal has a normalizing sequence of generators [**6**, Corollaires 3.2, 3.3].

Normal separation for $A = \mathcal{O}_q(G)$ or $A = \mathcal{O}_{q,p}(G,L)$ is a consequence of results in [**30**] and [**24**], as follows (cf. [**3**, Theorem 5.8]). In these cases, there is a partition

$$\text{spec}\, A = \bigsqcup_{w \in W \times W} \text{spec}_w A,$$

where W is the Weyl group of G, described in [**30**, Corollary 9.3.9] and [**24**, Corollary 4.5]. For each w, there are an ideal I_w of A and an Ore set $\mathcal{E}_w$ of nonzero normal elements in A/I_w such that

$$\text{spec}_w A = \{P \in \text{spec}\, A \mid P \supseteq I_w \text{ and } (P/I_w) \cap \mathcal{E}_w = \varnothing\}$$

[30, Proposition 10.3.2], [24, Theorem 4.4]. In particular, normal separation follows for primes $P \supset Q$ such that $Q \in \mathrm{spec}_w A$ but $P \notin \mathrm{spec}_w A$. Further, the localization $(A/I_w)[\mathcal{E}_w^{-1}]$ has *central separation*: for any primes $P' \supset Q'$, there exists a central element in $P' \setminus Q'$. In one of our cases, this follows from a sequence of results in [30], as shown in [3, §5.7]; in the other case, it is known that all ideals of $(A/I_w)[\mathcal{E}_w^{-1}]$ are centrally generated [24, Theorem 4.15]. Central separation in $(A/I_w)[\mathcal{E}_w^{-1}]$ together with normality of the elements of $\mathcal{E}_w$ yields normal separation for primes $P \supset Q$ in the same $\mathrm{spec}_w A$ [3, Proposition 1.7]. An alternate proof of normal separation for $\mathcal{O}_q(G)$, based on Hopf algebra technology, is given in [33, Proposition 2.4].

Finally, the second $U_q(\mathfrak{b}^+, P)$ case and the $U_{q,p}(\mathfrak{b}^+, M)$ case follow from the cases of $\mathcal{O}_q(G)$ and $\mathcal{O}_{q,p}(G, L)$ (see the last paragraph of (1.6)). $\square$

2.6. The Dixmier-Moeglin equivalence. Recall that a prime ideal P of a noetherian k-algebra A is *rational* provided the center of $\mathrm{Fract}(A/P)$ is algebraic over k. Recall also that P is *locally closed* in $\mathrm{spec}\, A$ provided the singleton $\{P\}$ is closed in some Zariski-neighborhood, i.e., the intersection of the primes properly containing P is larger than P. One says that the algebra A satisfies the *Dixmier-Moeglin equivalence* if the sets of rational prime ideals, locally closed prime ideals, and primitive ideals all coincide (cf. [51]). The advantage of this equivalence is that when it holds, the primitive ideals of A can be identified without constructing any irreducible representations.

Theorem. *The Dixmier-Moeglin equivalence holds in the following algebras:* $\mathcal{O}_q(k^n)$; $\mathcal{O}_{\lambda, \mathbf{p}}(M_{m,n}(k))$; $\mathcal{O}_{\lambda, \mathbf{p}}(GL_n(k))$; $\mathcal{O}_{\lambda, \mathbf{p}}(SL_n(k))$; $\mathcal{O}_q(G)$ *with* q *transcendental over* $\mathbb{Q}$ *and* k *algebraically closed;* $\mathcal{O}_{q,p}(G, L)$ *with* $q \in \mathbb{C}$ *not a root of unity;* $U_q(\mathfrak{b}^+, P)$ *with* q *transcendental over* $\mathbb{Q}$; $U_{q,p}(\mathfrak{b}^+, M)$ *with* $q \in \mathbb{C}$ *not a root of unity;* $A_n^{Q, \Gamma}(k)$ *with the* q_ι *not roots of unity;* $\mathcal{O}_q(\mathfrak{sp}\, k^{2n})$ *and* $\mathcal{O}_q(\mathfrak{o}\, k^n)$ *with* q *not a root of unity.*

Proof. For $\mathcal{O}_q(k^n)$, $\mathcal{O}_{\lambda, \mathbf{p}}(M_{m,n}(k))$, $A_n^{Q, \Gamma}(k)$, $\mathcal{O}_q(\mathfrak{sp}\, k^{2n})$ and $\mathcal{O}_q(\mathfrak{o}\, k^n)$, see [19, Corollary 2.5, Theorem 3.2] and [20, Theorems 5.3, 5.5, 5.8, 5.11]. The cases $\mathcal{O}_{\lambda, \mathbf{p}}(GL_n(k))$ and $\mathcal{O}_{\lambda, \mathbf{p}}(SL_n(k))$ follow directly. The case $\mathcal{O}_{q,p}(G, L)$ follows from the results of [24] exactly as in the case $\mathcal{O}_q(SL_n(\mathbb{C}))$, which was given explicitly in [23, Theorem 4.2]. The case $\mathcal{O}_q(G)$ follows from results in [30], as noted in [20, §2.4]. Finally, the cases $U_q(\mathfrak{b}^+, P)$ and $U_{q,p}(\mathfrak{b}^+, M)$ follow from the previous cases. $\square$

3. STRATIFIED SPECTRA

In this section, we give a more extensive discussion of another feature common to many quantized coordinate algebras – a stratification of the prime spectrum in which each stratum is a classical scheme, in fact the prime spectrum of a (commutative) Laurent polynomial algebra. Such a stratification was first discovered in $\operatorname{spec} \mathcal{O}_q(SL_3(\mathbb{C}))$ by Hodges and Levasseur [22]; they soon extended it to $\mathcal{O}_q(SL_n(\mathbb{C}))$ [23]. This was generalized by Joseph to $\mathcal{O}_q(G)$ [29, 30], and finally extended to $\mathcal{O}_{q,p}(G, L)$ by Hodges, Levasseur, and Toro [24]. In all these cases, the result followed from a long sequence of involved calculations specific to the algebra at hand. Later, it was noticed that some features of these stratifications could be tied to the action of a torus of automorphisms (cf. [3, Section 5 and Proposition 1.9]), and similar stratifications were observed in other quantized algebras (e.g., [3, Section 4], [19, §2.1, Theorem 2.3]). A general development of this type of stratification was begun in [20].

We begin with an outline of the main features of the stratification of $\operatorname{spec} \mathcal{O}_{q,p}(G, L)$, followed by a discussion of the axiomatic framework into which this stratification nicely fits. Then we exhibit appropriate tori of automorphisms for the algebras from Section 1, and indicate the cases in which the details of the stratification have been worked out.

3.1. Let A be either the algebra $\mathcal{O}_q(G)$ with q transcendental over $\mathbb{Q}$, or $\mathcal{O}_{q,p}(G, L)$ with $q \in \mathbb{C}$ not a root of unity. As noted in (2.5), there is a partition

$$\operatorname{spec} A = \bigsqcup_{w \in W \times W} \operatorname{spec}_w A,$$

where W is the Weyl group of G, described in [30, Corollary 9.3.9] and [24, Corollary 4.5]. For each w, there are an ideal I_w of A and an Ore set $\mathcal{E}_w$ of nonzero normal elements in A/I_w such that

$$\operatorname{spec}_w A = \{P \in \operatorname{spec} A \mid P \supseteq I_w \text{ and } (P/I_w) \cap \mathcal{E}_w = \varnothing\}.$$

In fact, $\mathcal{E}_w$ is finitely generated, and so if e_w is the product (in some order) of a finite set of generators for $\mathcal{E}_w$, we can write

$$\operatorname{spec}_w A = \{P \in \operatorname{spec} A \mid P \supseteq I_w\} \cap \{P \in \operatorname{spec} A \mid e_w \notin P\},$$

an intersection of a closed and an open set. Thus $\operatorname{spec}_w A$ is a locally closed subset of $\operatorname{spec} A$. Let $A_w = (A/I_w)[\mathcal{E}_w^{-1}]$. In the case of $\mathcal{O}_{q,p}(G, L)$ over $\mathbb{C}$, the center $Z(A_w)$ is a Laurent polynomial ring over k [24, Theorem 4.14], the ideals of A_w are centrally generated [24, Theorem 4.15], and it follows that $\operatorname{spec}_w A$ is homeomorphic to $\operatorname{spec} Z(A_w)$. That such a homeomorphism also exists in the case of $\mathcal{O}_q(G)$ over $k \supseteq \mathbb{Q}(q)$ is not proved in [30] (a Laurent polynomial subalgebra $Z_w \subseteq Z(A_w)$ is studied instead [30, §10.3.3]), but this can be deduced from the results there together with what we shall prove in Part II.

If prim A denotes the *primitive spectrum* of A (the set of primitive ideals, equipped with the Zariski topology), we obtain a corresponding partition

$$\text{prim } A = \bigsqcup_{w \in W \times W} \text{prim}_w A,$$

where each $\text{prim}_w A = \text{prim } A \cap \text{spec}_w A$ is a locally closed subset of $\text{prim } A$. Further, $\text{prim}_w A$ is precisely the set of maximal elements of $\text{spec}_w A$ (this is given explicitly in [30, Theorem 10.3.7] for the first case, and is implicit in [24, Section 4] for the second; see also [3, Corollary 1.5]). Over an algebraically closed base field, a maximal torus H of G acts naturally as automorphisms of A [30, §10.3.8], [24, §3.3], and the sets $\text{prim}_w A$ are precisely the H-orbits in $\text{prim } A$ [30, Theorem 10.3.8], [24, Theorem 4.16].

3.2. The picture of $\text{spec } A$ outlined in (3.1) can be conveniently organized in terms of the action of the group H. There is a unique minimal element J_w in $\text{spec}_w A$ for each w ([30, Proposition 10.3.5]; implicit in [24, Section 4]), and it is easily checked that J_w is H-stable (cf. [3, §5.4]). Over an algebraically closed base field, it then follows that the J_w are precisely the H-stable prime ideals of A, and that each $\text{spec}_w A$ consists precisely of those prime ideals P such that the intersection of the H-orbit of P equals J_w [3, Proposition 1.9].

These observations provide a framework that can be set up with respect to any group of automorphisms of any ring, as follows. As we shall see later, the special properties of this framework in the cases discussed in (3.1) appear in many other algebras as well.

3.3. H-stratifications. Let H be a group acting as automorphisms on a ring A. Recall that an *H-prime ideal* of A is any proper H-stable ideal J of A such that a product of H-stable ideals is contained in J only when one of the factors is contained in J. We shall write H-$\text{spec } A$ for the set of H-prime ideals of A. For any ideal I of A, let $(I : H)$ denote the intersection of the H-orbit of I, that is,

$$(I : H) = \bigcap_{h \in H} h(I).$$

Alternatively, $(I : H)$ is the largest H-stable ideal of A contained in I.

Observe that if P is a prime ideal of A, then $(P : H)$ is an H-prime ideal. On the other hand, if A is noetherian and I is an H-prime ideal, it is easily checked that I is semiprime and that the prime ideals minimal over I form a single H-orbit (cf. [15, Remarks 4*,5*, p. 338]). In this case, $I = (P : H)$ for any prime P minimal over I.

The *H-stratum* of any *H*-prime ideal J is the set

$$\text{spec}_J A = \{P \in \text{spec}\, A \mid (P : H) = J\}.$$

One defines *H*-strata in any subset of spec A by intersection with these strata. In particular, the *H*-stratum of J in prim A is the set

$$\text{prim}_J A = \text{prim}\, A \cap \text{spec}_J A.$$

Since $(P : H) \in H\text{-spec}\, A$ for each $P \in \text{spec}\, A$, the *H*-strata partition spec A, that is,

$$\text{spec}\, A = \bigsqcup_{J \in H\text{-spec}\, A} \text{spec}_J A.$$

We refer to this partition as *the H-stratification of* spec A. Similarly, the partition of prim A into *H*-strata $\text{prim}_J A$ is called the *H*-stratification of prim A.

3.4. Finite *H*-stratifications enjoy the minimal topological properties required of stratifications in algebraic and differential geometry, as follows. A *stratification* of an algebraic variety X is defined in [**32**, p. 56] to be a partition of X into a disjoint union of finitely many locally closed subsets (called *strata*) such that the closure of each stratum is a union of strata.

Lemma. *Let H be a group acting as automorphisms on a ring A, and assume that H-spec A is finite.*

(a) *Each H-stratum in* spec A *is locally closed.*

(b) *The closure of each H-stratum in* spec A *is a union of H-strata.*

(c) *For each nonnegative integer d, the union of the H-strata corresponding to the elements of height at most d in H-spec A is open in* spec A.

Proof. (a) Let $J \in H\text{-spec}\, A$, and let J' be the intersection of the *H*-primes properly containing J (take $J' = A$ if J is maximal in $H\text{-spec}\, A$). Since J is *H*-prime and $H\text{-spec}\, A$ is finite, $J' \supsetneq J$. Thus

$$\text{spec}_J A = \{P \in \text{spec}\, A \mid P \supseteq J\} \cap \{P \in \text{spec}\, A \mid P \not\supseteq J'\}.$$

(b) If $J \in H\text{-spec}\, A$, the closure of $\text{spec}_J A$ is just the set $V(J)$ of those prime ideals of A containing J, and $V(J)$ equals the union of the *H*-strata $\text{spec}_K A$ as K runs over those *H*-primes containing J.

(c) Let Y be the set of elements of $H\text{-spec}\, A$ of height greater than d, that is, those *H*-primes K from which there descends a chain $K = K_0 \supsetneq K_1 \supsetneq \cdots \supsetneq K_{d+1}$ in $H\text{-spec}\, A$. The union of the *H*-strata $\text{spec}_J A$ for J of height at most d in $H\text{-spec}\, A$ is the set

$$\{P \in \text{spec}\, A \mid P \not\supseteq \cap Y\},$$

which is open in spec A. $\square$

3.5. Tori of automorphisms. Whether or not our base field k is algebraically closed, we shall refer to any finite direct product of copies of the multiplicative group $k^\times$ as an *(algebraic) torus*. For each of the algebras A discussed in Section 1, there is a naturally occurring torus $\mathcal{H}$ acting as k-algebra automorphisms on A, as follows. The action of an element $h \in \mathcal{H}$ on an element $a \in A$ will be denoted $h.a$; we leave it to the reader to verify in each case that there exist k-algebra automorphisms $h.(-)$ as described, and that the rule $h \mapsto h.(-)$ is a group homomorphism from $\mathcal{H}$ to $\operatorname{Aut} A$.

Many of these tori arise as groups of 'winding automorphisms' of a bialgebra or Hopf algebra, as follows. Suppose that A is a bialgebra over k, and let $A^\wedge$ be the set of linear characters on A, that is, k-algebra homomorphisms $A \to k$. This set forms a monoid under convolution, the counit of A being the identity element of $A^\wedge$. For $\chi \in A^\wedge$, there are k-algebra endomorphisms θ^l_χ and θ^r_χ on A given by the rules

$$\theta^l_\chi(a) = \sum_{(a)} a_1 \chi(a_2) \qquad \text{and} \qquad \theta^r_\chi(a) = \sum_{(a)} \chi(a_1) a_2.$$

When χ is invertible in $A^\wedge$, say with inverse χ', the endomorphisms θ^l_χ and θ^r_χ are automorphisms of A, with inverses $\theta^l_{\chi'}$ and $\theta^r_{\chi'}$ respectively, and we refer to them as the left and right *winding automorphisms* of A associated with χ. The maps $\chi \mapsto \theta^l_\chi$ and $\chi \mapsto \theta^r_\chi$ are a homomorphism and an antihomomorphism, respectively, from the group of units of $A^\wedge$ to $\operatorname{Aut} A$. The facts outlined above are well known when A is a Hopf algebra, in which case $A^\wedge$ is a group under convolution (see, e.g., [**30**, §1.3.4]).

If C is a right comodule algebra over the bialgebra A (e.g., see [**41**, Definition 4.1.2]), with structure map $C \to C \otimes A$ written as $c \mapsto \sum_{(c)} c_0 \otimes c_1$, then each (invertible) character $\chi \in A^\wedge$ induces a k-algebra endomorphism (automorphism) of C according to the rule $c \mapsto \sum_{(c)} c_0 \chi(c_1)$. The analogous statements hold for a left comodule algebra over A.

    1. Quantum affine spaces. For $A = \mathcal{O}_q(k^n)$, take $\mathcal{H} = (k^\times)^n$ and

$$(\alpha_1, \ldots, \alpha_n).x_i = \alpha_i x_i$$

for $i = 1, \ldots, n$. These automorphisms can be viewed as winding automorphisms arising when A is made into a right comodule algebra over the bialgebra $\mathcal{O}_{\lambda,q}(M_n(k))$, or into a left comodule algebra over $\mathcal{O}_{\lambda,\lambda^{-1}q}(M_n(k))$.

    2. Quantum matrices. For the algebra $A = \mathcal{O}_{\lambda,p}(M_{m,n}(k))$, take $\mathcal{H} = (k^\times)^m \times (k^\times)^n$ and

$$(\alpha_1, \ldots, \alpha_m, \beta_1, \ldots, \beta_n).X_{ij} = \alpha_i \beta_j X_{ij}$$

for $i = 1, \ldots, m$ and $j = 1, \ldots, n$. In the case $m = n$, the automorphisms above are compositions of left and right winding automorphisms. Namely, for $\gamma = (\gamma_1, \ldots, \gamma_n) \in (k^\times)^n$ there is a character $\chi_\gamma \in A^\wedge$ such that $\chi_\gamma(X_{ij}) = \delta_{ij}\gamma_i X_{ij}$ for all i, j, and the automorphism displayed above can be expressed as $\theta^l_{\chi_\beta}\theta^r_{\chi_\alpha}$ where $\beta = (\beta_1, \ldots, \beta_n)$ and $\alpha = (\alpha_1, \ldots, \alpha_n)$. The map $\gamma \mapsto \chi_\gamma$ embeds the torus $(k^\times)^n$ in the group of units of $A^\wedge$; it is an isomorphism when λ and p are sufficiently generic.

3. Quantum general linear groups. For $A = \mathcal{O}_{\lambda,p}(GL_n(k))$, we can take the torus $\mathcal{H} = (k^\times)^n \times (k^\times)^n$ acting on $\mathcal{O}_{\lambda,p}(M_n(k))$ as above, and extend the given automorphisms from $\mathcal{O}_{\lambda,p}(M_n(k))$ to A, because $D_{\lambda,p}$ is an $\mathcal{H}$-eigenvector.

4. Quantum special linear groups. For $A = \mathcal{O}_{\lambda,p}(SL_n(k))$ (assuming a suitable choice of λ, p so that $D_{\lambda,p}$ is central), we must take a subgroup of the torus used for $\mathcal{O}_{\lambda,p}(M_n(k))$, namely the stabilizer of $D_{\lambda,p}$:

$$\mathcal{H} = \{(\alpha_1, \ldots, \alpha_n, \beta_1, \ldots, \beta_n) \in (k^\times)^n \times (k^\times)^n \mid \alpha_1\alpha_2 \cdots \alpha_n\beta_1\beta_2 \cdots \beta_n = 1\}.$$

The automorphisms of $\mathcal{O}_{\lambda,p}(M_n(k))$ corresponding to elements of $\mathcal{H}$ all fix $D_{\lambda,p} - 1$ and therefore induce automorphisms of A.

5. Quantum semisimple groups. Let $A = \mathcal{O}_{q,p}(G, L)$ as described in (1.5). In the case $k = \mathbb{C}$, one works with a maximal torus of G with Lie algebra $\mathfrak{h}$ and character group L [24, §3.1]. In the general case, it is simplest to take the set $\mathcal{H}$ of group homomorphisms $L \to k^\times$; this is a group under pointwise multiplication, and it is a torus because L is a lattice. An action of $\mathcal{H}$ on A derives from an embedding of $\mathcal{H}$ into the character group $A^\wedge$. Namely, there is a monic group homomorphism $\iota : \mathcal{H} \to A^\wedge$ such that

$$\iota(h)(c_{f,v}^{V(\lambda)}) = h(\mu)f(v)$$

for $\lambda \in P^+$, $\mu \in P$, $v \in V(\lambda)_\mu$, $f \in V(\lambda)^*$ (see [24, §3.3]).

Therefore there is a group homomorphism $\mathcal{H} \to \operatorname{Aut} A$ such that $h \mapsto \theta^l_{\iota(h)}$, which provides an action of $\mathcal{H}$ on A. This is the action used in [30] and [24]. For some purposes, however, an action of $\mathcal{H} \times \mathcal{H}$ is needed, given by $(h_1, h_2) \mapsto \theta^l_{\iota(h_1)}\theta^r_{\iota(h_2)}$ (see [3, §5.2]).

6. Quantized enveloping algebras of Borel and nilpotent subalgebras. For $A = U_{q,p}(\mathfrak{n}^+)$, take $\mathcal{H} = (k^\times)^n$ and

$$(\alpha_1, \ldots, \alpha_n).e_i = \alpha_i e_i$$

for $i = 1, \ldots, n$. For $A = U_{q,p}(\mathfrak{b}^+, M)$, let $\mathcal{H}_M$ be the group of group homomorphisms $M \to k^\times$ under pointwise multiplication and take $\mathcal{H} =$

$\mathcal{H}_M \times (k^\times)^n$, with

$$(h, \alpha_1, \ldots, \alpha_n).k_\mu = h(\mu)k_\mu$$
$$(h, \alpha_1, \ldots, \alpha_n).e_i = \alpha_i e_i$$

for $\mu \in M$ and $i = 1, \ldots, n$.

7. Quantized Weyl algebras. For $A = A_n^{Q,\Gamma}(k)$, take $\mathcal{H} = (k^\times)^n$ and

$$(\alpha_1, \ldots, \alpha_n).x_i = \alpha_i x_i$$
$$(\alpha_1, \ldots, \alpha_n).y_i = \alpha_i^{-1} y_i$$

for $i = 1, \ldots, n$.

8. Quantum symplectic spaces. For $A = \mathcal{O}_q(\mathfrak{sp}\, k^{2n})$, take $\mathcal{H} = (k^\times)^n$ and

$$(\alpha_1, \ldots, \alpha_n).x_i = \alpha_i x_i$$
$$(\alpha_1, \ldots, \alpha_n).x_{i'} = \alpha_i^{-1} x_{i'}$$

for $i = 1, \ldots, n$.

9. Quantum Euclidean spaces. For $A = \mathcal{O}_q(\mathfrak{o}\, k^n)$, set $m = \lfloor n/2 \rfloor$ and take $\mathcal{H} = (k^\times)^m$, with

$$(\alpha_1, \ldots, \alpha_m).x_i = \alpha_i x_i \qquad (i = 1, \ldots, m)$$
$$(\alpha_1, \ldots, \alpha_m).x_{i'} = \alpha_i^{-1} x_{i'} \qquad (i = 1, \ldots, m)$$
$$(\alpha_1, \ldots, \alpha_m).x_{m+1} = x_{m+1} \qquad (\text{if } n \text{ is odd})$$

3.6. We close this part of the paper by recording cases in the literature where the prime spectrum of an algebra A from Section 1, stratified by a torus $\mathcal{H}$ from (3.5), fits into the pattern discussed in (3.1) and (3.2). Although the results quoted from [20] do not provide Ore sets of $\underline{normal}$ elements, they can be modified to do so in all but one of the cases of interest here, as we shall show in Part II.

Theorem. *Let A be one of the following algebras: $\mathcal{O}_q(k^n)$; $\mathcal{O}_{\lambda,\mathbf{p}}(M_{m,n}(k))$ or $\mathcal{O}_{\lambda,\mathbf{p}}(GL_n(k))$ or $\mathcal{O}_{\lambda,\mathbf{p}}(SL_n(k))$ with λ not a root of unity; $\mathcal{O}_q(G)$ with q transcendental over $\mathbb{Q}$ and k algebraically closed; $\mathcal{O}_{q,\mathbf{p}}(G, L)$ with $q \in \mathbb{C}$ not a root of unity; $A_n^{Q,\Gamma}(k)$ with the q_i not roots of unity; $\mathcal{O}_q(\mathfrak{sp}\, k^{2n})$ or $\mathcal{O}_q(\mathfrak{o}\, k^n)$ with q not a root of unity. Let $\mathcal{H}$ be the corresponding torus acting on A as in (3.5). Then $\mathcal{H}$-spec A is finite, and for each $J \in \mathcal{H}$-spec A the following hold:*

(a) *There exists an Ore set $\mathcal{E}_J$ of regular elements in A/J such that $\mathrm{spec}_J\, A$ is homeomorphic to $\mathrm{spec}\, Z((A/J)[\mathcal{E}_J^{-1}])$ via localization and contraction.*

(b) *$Z((A/J)[\mathcal{E}_J^{-1}])$ is a commutative Laurent polynomial algebra over an extension field of k.*

(c) *$\mathrm{prim}_J\, A$ equals the set of maximal elements of $\mathrm{spec}_J\, A$.*

(d) *If k is algebraically closed, then $\mathrm{prim}_J\, A$ consists of a single $\mathcal{H}$-orbit.*

Proof. For $\mathcal{O}_q(k^n)$, see [**19**, Theorem 2.3, Proposition 2.11, Corollary 1.5]. For $\mathcal{O}_{\lambda,\boldsymbol{p}}(M_{m,n}(k))$, $A_n^{Q,\Gamma}(k)$, $\mathcal{O}_q(\mathfrak{sp}\,k^{2n})$ and $\mathcal{O}_q(\mathfrak{o}\,k^n)$, see [**20**, Theorems 5.3, 5.5, 5.8, 5.11, 6.6, 6.8]. The cases $\mathcal{O}_{\lambda,\boldsymbol{p}}(GL_n(k))$ and $\mathcal{O}_{\lambda,\boldsymbol{p}}(SL_n(k))$ follow directly. For $\mathcal{O}_q(G)$ and $\mathcal{O}_{q,p}(G,L)$, see (3.1) and (3.2). $\square$

II. An Axiomatic Development of Stratification Properties

This part of the paper is a variation on a theme developed in joint work with E. S. Letzter [**20**]. The goal is to show that, in the case of a torus $\mathcal{H}$ acting *rationally* on a noetherian k-algebra A, the $\mathcal{H}$-stratification of $\mathrm{spec}\, A$ enjoys analogs of all the properties of the stratification presented in Theorem 3.6, under relatively mild hypotheses on A and $\mathcal{H}$. Both developments (here and in [**20**]) include – at present – one hypothesis which requires substantial work to verify in the examples of interest. In [**20**, Section 6], that hypothesis is the complete primeness of the $\mathcal{H}$-prime ideals of A. Here, we rely on a type of normal separation in $\mathcal{H}$-spec A. One disadvantage is the "loss" of one example, namely $\mathcal{O}_{\lambda,\boldsymbol{p}}(M_{m,n}(k))$, for which normal separation has not been verified. (It is conjectured to hold, and is easily checked for the case $m = n = 2$.) However, the development here provides a better fit with the work of Hodges-Levasseur, Joseph, and Hodges-Levasseur-Toro on $\mathcal{O}_q(G)$ and $\mathcal{O}_{q,p}(G,L)$ in which the stratified picture of $\mathrm{spec}\, A$ first arose. For example, the localizations $(A/J)[\mathcal{E}_J^{-1}]$ in the present development coincide with the ones obtained in [**22**, **23**, **29**, **30**, **24**] in the appropriate cases, whereas the localizations used in [**20**, Section 6] are larger. Here we also obtain a few additional properties, namely that $(A/J)[\mathcal{E}_J^{-1}]$ is an affine k-algebra in appropriate circumstances, and that normal separation in $\mathcal{H}$-spec A by $\mathcal{H}$-eigenvectors implies normal separation in $\mathrm{spec}\, A$.

The development of our theme is based on the equivalence between rational actions of $(k^\times)^r$ and $\mathbb{Z}^r$-gradings (see (5.1) for details). We begin by working out the basic properties of a stratification relative to graded-prime ideals for $\mathbb{Z}^r$-graded rings satisfying an appropriate normal separation condition. This may have some independent interest in the context of group-graded rings. Since it is at this stage where our variation differs in certain aspects from that in [**20**, Section 6], we provide full details. This is the content of Section 4. In Section 5, we translate these results into the context of $(k^\times)^r$-actions. That process is essentially the same as in [20, Section 6].

4. GRADED-NORMAL SEPARATION

Here we build a stratification in the context of rings graded by a free abelian group, with a normal separation hypothesis. This hypothesis will allow us to construct certain localizations which are graded-simple (i.e., have no nontrivial homogeneous ideals). Hence, we begin by analyzing the prime spectra of some graded-simple rings.

Lemma 4.1. *Let R be a graded-simple ring graded by an abelian group G.*

(a) *The center $Z(R)$ is a homogeneous subring of R, strongly graded by the subgroup $G_Z = \{x \in G \mid Z(R)_x \neq 0\}$ of G.*

(b) *Every nonzero homogeneous element of $Z(R)$ is invertible.*

(c) *As $Z(R)$-modules, $Z(R)$ is a direct summand of R.*

(d) *Suppose that G_Z is a free abelian group of finite rank. Choose a basis $\{g_1, \ldots, g_n\}$ for G_Z, and choose a nonzero element $z_j \in Z(R)_{g_j}$ for each j. Then $Z(R) = Z(R)_1[z_1^{\pm 1}, \ldots, z_n^{\pm 1}]$, a Laurent polynomial ring over the field $Z(R)_1$.*

Proof. It is obvious that $Z(R)$ is a homogeneous subring of R, graded by the set G_Z. Part (b) follows from the graded-simplicity of R, and then it is clear that G_Z is a subgroup of G and that $Z(R)$ is strongly G_Z-graded. This proves part (a). Part (d) is routine (see the proof of [20, Lemma 6.3(c)]).

Note that $S = \bigoplus_{x \in G_Z} R_x$ is a homogeneous subring of R containing $Z(R)$. Further, S is a left S-module direct summand of R, with complement $M = \bigoplus_{x \in G \setminus G_Z} R_x$. On the other hand, as in [20, Lemma 6.3(c)], S is a free $Z(R)$-module with a basis including 1, whence $Z(R)$ is a $Z(R)$-module direct summand of S. Part (c) follows. $\square$

Proposition 4.2. *Let R be a graded-simple ring graded by an abelian group G. Then there exist bijections between the sets of ideals of R and $Z(R)$, given by contraction and extension, that is,*

$$I \mapsto I \cap Z(R) \qquad \text{and} \qquad J \mapsto JR.$$

Proof. It is clear from Lemma 4.1(c) that $JR \cap Z(R) = J$ for every ideal J of $Z(R)$. It remains to show that $(I \cap Z(R))R = I$ for any ideal I of R. Set $J = I \cap Z(R)$, and suppose that $I \neq JR$. Pick an element $x \in I \setminus JR$ of minimal length, say length n. Then $x = x_1 + \cdots + x_n$ for some nonzero elements $x_i \in R_{g_i}$, where $g_1, \ldots, g_n$ are distinct elements of G.

Now Rx_1R is a nonzero homogeneous ideal of R, so $Rx_1R = R$ by graded-simplicity and $\sum_j a_j x_1 b_j = 1$ for some $a_j, b_j \in R$. Express each a_j, b_j as a sum of homogeneous elements, and substitute these expressions in the terms $a_j x_1 b_j$. This expands the sum $\sum_j a_j x_1 b_j$ in the form $\sum_t c_t x_1 d_t$ with all c_t, d_t homogeneous. Hence, after relabelling, we may assume that the a_j and b_j are all homogeneous, say of degrees e_j and f_j, respectively.

Comparing identity components in the equation $\sum_j a_j x_1 b_j = 1$, we obtain

$$\sum_{e_j g_1 f_j = 1} a_j x_1 b_j = 1.$$

Thus, after deleting all other $a_j x_1 b_j$ terms, we may assume that $e_j g_1 f_j = 1$ for all j. Since G is abelian, it follows that $e_j f_j = g_1^{-1}$ for all j.

Set $x' = \sum_j a_j x b_j$, and note that $x' \in I$. Further,

$$x' = \sum_{i=1}^{n} \sum_{j} a_j x_i b_j$$

where $\sum_j a_j x_i b_j \in R_{g_1^{-1} g_i}$ for each i. As a result, x' is an element whose support is contained in $\{1, g_1^{-1} g_2, \ldots, g_1^{-1} g_n\}$ and whose identity component is 1. Hence, $x_1 x'$ is an element of I whose support is contained in $\{g_1, \ldots, g_n\}$ and whose g_1-component is x_1. Comparing this element with x, we see that $x - x_1 x'$ is an element of I whose support is contained in $\{g_2, \ldots, g_n\}$. By the minimality of n, we must have $x - x_1 x' \in JR$, and so $x' \notin JR$. Therefore, after replacing x by x', there is no loss of generality in assuming that $g_1 = 1$ and $x_1 = 1$.

<u>Claim</u>: There do not exist $g \in G$ and a nonzero element $y \in I$ whose support is properly contained in the set $\{g g_1, \ldots, g g_n\}$.

Suppose there do exist such g and y. Say the $g g_s$-component of y is nonzero. As above, there exists an element of the form $y' = \sum_j a_j y b_j$ whose support is properly contained in $\{g_s^{-1} g_1, \ldots, g_s^{-1} g_n\}$ and whose identity component is 1. Then $y' \in I$, and $x - x_s y'$ is an element of I whose support is properly contained in $\{g_1, \ldots, g_n\}$. Since y' and $x - x_s y'$ are elements of I of length less than n, they must lie in JR, by the minimality of n. But then $x \in JR$, contradicting our assumptions. Therefore the claim is proved.

Finally, consider any homogeneous element $r \in R$, say $r \in R_g$. Then $rx - xr$ is an element of I with support contained in $\{g g_2, \ldots, g g_n\}$, and so $rx - xr = 0$ by the claim. It follows that $x \in Z(R)$ and so $x \in J$, contradicting our assumption that $x \notin JR$.

Therefore $I = JR$. $\square$

Corollary 4.3. *If R is a graded-simple ring graded by an abelian group, then contraction and extension provide mutually inverse homeomorphisms between* spec R *and* spec $Z(R)$. □

4.4. Graded-normal separation. Let R be a group-graded ring. Recall that a *graded-prime* ideal of R is any proper homogeneous ideal P such that whenever I, J are homogeneous ideals of R with $IJ \subseteq P$, then either $I \subseteq P$ or $J \subseteq P$. We say that R has *graded-normal separation* provided that for any proper inclusion $P \supset Q$ of graded-prime ideals of R, there exists a homogeneous element $c \in P \setminus Q$ which is normal modulo Q.

Theorem 4.5. *Let R be a right noetherian ring graded by an abelian group G, and assume that R has graded-normal separation.*

Let J be a graded-prime ideal of R, set

$$\mathcal{S}_J = \{P \in \text{spec } R \mid J \text{ is the largest homogeneous ideal contained in } P\},$$

and let $\mathcal{E}_J$ be the multiplicative set of all the nonzero homogeneous normal elements in the ring R/J.

(a) *$\mathcal{E}_J$ is a right and left denominator set of regular elements in R/J.*

(b) *The localization map $R \to R/J \to R_J = (R/J)[\mathcal{E}_J^{-1}]$ induces a homeomorphism of $\mathcal{S}_J$ onto* spec R_J.

(c) *Contraction and extension induce mutually inverse homeomorphisms between* spec R_J *and* spec $Z(R_J)$.

(d) *If G is free abelian of rank $r < \infty$, then $Z(R_J)$ is a commutative Laurent polynomial ring over the field $Z(R_J)_1$ (the identity component of $Z(R_J)$ in the induced G-grading), in r or fewer indeterminates.*

Proof. Without loss of generality, $J = 0$. Consequently, R is now a graded-prime ring, i.e., 0 is a graded-prime ideal of R.

(a) For $x \in \mathcal{E}_J$, observe that $\text{r.ann}_R(x)$ is a homogeneous ideal such that $(xR)\,\text{r.ann}_R(x) = 0$. Since R is graded-prime and $x \neq 0$, we must have $\text{r.ann}_R(x) = 0$. Similarly, $\text{l.ann}_R(x)=0$; thus all elements of $\mathcal{E}_J$ are regular. The Ore condition follows directly from normality.

(b) If $P \in \text{spec } R$ and P_0 is the largest homogeneous ideal contained in P, then P_0 is a graded-prime ideal. Because of graded-normal separation, P_0 is nonzero precisely when $P_0 \cap \mathcal{E}_J$ is nonempty. Hence, $\mathcal{S}_J$ consists precisely of those prime ideals of R disjoint from $\mathcal{E}_J$. Standard localization theory (e.g., [**21**, Theorem 9.22]) now yields part (b).

(c) Since the elements of $\mathcal{E}_J$ are homogeneous, the G-grading on R extends (uniquely) to a G-grading on R_J. Because of the noetherian hypothesis on R, two-sided ideals I in R induce two-sided ideals IR_J in R_J (e.g., [**21**, Theorem

9.20(a)]). The standard arguments concerning contractions of prime ideals (e.g., [**21**, Theorem 9.20(c)]) now show that any graded-prime ideal of R_J must contract to a graded-prime ideal of R. Since every nonzero graded-prime ideal of R meets $\mathcal{E}_J$, we thus see that R_J has no nonzero graded-prime ideals. On the other hand, any maximal proper homogeneous ideal of R_J is graded-prime. Therefore R_J must be graded-simple. Part (c) now follows from Corollary 4.3.

(d) This follows from Lemma 4.1(d). $\square$

Corollary 4.6. *Let R be a right noetherian ring graded by an abelian group G. If R has graded-normal separation, then R also has normal separation.*

Proof. Let $P \supset Q$ be a proper inclusion of prime ideals of R. Let P_0 and Q_0 be the largest homogeneous ideals contained in P and Q, respectively, and note that P_0 and Q_0 are graded-prime ideals of R. Obviously $P_0 \supseteq Q_0$.

If $P_0 \neq Q_0$, then by assumption there exists a homogeneous element $c \in P_0 \setminus Q_0$ which is normal modulo Q_0. In particular, $c \in P$ and c is normal modulo Q. Since RcR is a homogeneous ideal of R, not contained in Q_0, we see that $RcR \not\subseteq Q$. Thus in this case we are done.

Now assume that $P_0 = Q_0$. It is harmless to pass to R/Q_0, and hence there is no loss of generality in assuming that $P_0 = Q_0 = 0$. Consequently, 0 is now a graded-prime ideal of R.

Let S denote the localization $R_0 = R[\mathcal{E}_0^{-1}]$, in the notation of Theorem 4.5. By part (b) of the theorem, $PS \supset QS$ is a proper inclusion of prime ideals of S. Part (c) then implies that there exists a central element $z \in PS \setminus QS$. Write $z = cx^{-1}$ for some $c \in P$ and $x \in \mathcal{E}_0$, and note that $c \notin Q$. Since $xR = Rx$, we have $zxR = zRx$ in S. But z commutes with R, and so $zxR = Rzx$, that is, $cR = Rc$. Therefore c is a normal element of R, and normal separation is proved. $\square$

5. STRATIFICATION UNDER RATIONAL TORUS ACTIONS

We now turn to rational actions by tori as automorphisms of noetherian algebras. Under suitable normal separation hypotheses, the results of the previous section apply, yielding a picture of the corresponding stratifications that incorporates the features of the examples discussed in Section 3.

Throughout this section, we assume that our base field k is infinite. In the examples of interest, this is automatic due to the presence of non-roots of unity.

5.1. Rational torus actions. Let A be a k-algebra, and let $\mathcal{H}$ be a torus over k, say $\mathcal{H} = (k^\times)^r$. An action of $\mathcal{H}$ on A by k-algebra automorphisms is said to be *rational* provided A is a directed union of finite dimensional

$\mathcal{H}$-stable subspaces V_i such that the induced group homomorphisms $\mathcal{H} \to GL(V_i)$ are morphisms of algebraic varieties. Let $\widehat{\mathcal{H}}$ denote the set of *rational characters* of $\mathcal{H}$, that is, algebraic group morphisms $\mathcal{H} \to k^\times$. Then $\widehat{\mathcal{H}}$ is an abelian group under pointwise multiplication. Since k is infinite, $\widehat{\mathcal{H}}$ is a lattice of rank r, with a basis given by the component projections $(k^\times)^r \to k^\times$.

The rationality of an $\mathcal{H}$-action implies that A is spanned by $\mathcal{H}$-eigenvectors [**44**, Chapter 5, Corollary to Theorem 36], and that the eigenvalues of these eigenvectors are rational. This yields a k-vector space decomposition $A = \bigoplus_{x \in \widehat{\mathcal{H}}} A_x$, where A_x denotes the x-eigenspace of A. (Up to this point, everything holds for a rational action by k-linear transformations.) Since $\mathcal{H}$ acts by automorphisms, $A_x A_y \subseteq A_{xy}$ for all $x, y \in \widehat{\mathcal{H}}$. Thus we have a grading of A by the free abelian group $\widehat{\mathcal{H}}$, such that the homogeneous elements are the $\mathcal{H}$-eigenvectors in A.

Conversely, any grading of A by $\mathbb{Z}^r$ arises in this fashion, as noted in [**50**, p. 784]. To see this, identify $\mathbb{Z}^r$ with $\widehat{\mathcal{H}}$, and let $A = \bigoplus_{x \in \widehat{\mathcal{H}}} A_x$ be an $\widehat{\mathcal{H}}$-grading. Let each $h \in \mathcal{H}$ act on A as the k-linear transformation with eigenvalue $x(h)$ on A_x for all x, that is, $h.a = x(h)a$ for $a \in A_x$. This yields a rational action of $\mathcal{H}$ on A by k-algebra automorphisms, such that each A_x is the x-eigenspace. The reader may take this as the definition of a rational $\mathcal{H}$-action if so desired.

In case k is algebraically closed and A is noetherian, a remarkable theorem of Moeglin-Rentschler and Vonessen [**40**, Théorème 2.12(ii)], [**59**, Theorem 2.2] states that $\mathcal{H}$ acts transitively on the $\mathcal{H}$-strata of rational ideals in A, that is, each $\mathcal{H}$-stratum of rational ideals is a single $\mathcal{H}$-orbit. This holds, in fact, for any algebraic group $\mathcal{H}$, not just tori. If, further, the Dixmier-Moeglin equivalence holds, then each $\mathcal{H}$-stratum in prim A is a single $\mathcal{H}$-orbit. For the very special case where $\mathcal{H}$ is a torus and all $\mathcal{H}$-prime ideals of A are completely prime, a direct proof of this transitivity result is given in [**20**, Theorem 6.8]. Here we shall derive such a result using normal separation in place of complete primeness – see Theorem 5.5(c).

5.2. Normal separation in $\mathcal{H}$-spec. Assume that we have a torus $\mathcal{H}$ acting rationally on a k-algebra A by k-algebra automorphisms. In this situation, we shall assume that A has been equipped with the natural $\widehat{\mathcal{H}}$-grading as in (5.1). With respect to this grading, the homogeneous ideals of A are precisely the $\mathcal{H}$-stable ideals, and so the graded-prime ideals coincide with the $\mathcal{H}$-prime ideals. We say that $\mathcal{H}$-spec A has *normal $\mathcal{H}$-separation* provided that for any proper inclusion $P \supset Q$ of $\mathcal{H}$-prime ideals, there exists an $\mathcal{H}$-eigenvector $c \in P \setminus Q$ which is normal modulo Q. This condition is just graded-normal separation with respect to the $\widehat{\mathcal{H}}$-grading as defined in

(4.4).

The term '$\mathcal{H}$-normal separation' we would reserve for a slightly stronger condition involving '$\mathcal{H}$-normal' elements. If Q is an $\mathcal{H}$-stable ideal of A, an element $c \in A$ is said to be $\mathcal{H}$-*normal modulo* Q in case there exists $h \in \mathcal{H}$ such that $ca - h(a)c \in Q$ for all $a \in A$. It is easily checked that each of the homogeneous components of such an element c is also $\mathcal{H}$-normal modulo Q. Now $\mathcal{H}$-spec A has $\mathcal{H}$-*normal separation* provided that for any proper inclusion $P \supset Q$ of $\mathcal{H}$-prime ideals of A, there exists an element $c \in P \setminus Q$ which is $\mathcal{H}$-normal modulo Q. This strengthening of normal $\mathcal{H}$-separation is not needed in our proofs, but it does hold in almost all the examples discussed in Sections 2 and 3, and is conjectured to hold in the remaining one, namely $\mathcal{O}_{\lambda,p}(M_n(k))$.

We note also that since the homogeneous ideals of A coincide with the $\mathcal{H}$-stable ideals, the largest homogeneous ideal of A contained in a given ideal I is just $(I : \mathcal{H})$. Hence, the strata of spec A with respect to graded-prime ideals, as in Theorem 4.5, coincide with the $\mathcal{H}$-strata. Thus Theorem 4.5 and Corollary 4.6 yield the following results.

Theorem 5.3. *Let A be a right noetherian k-algebra, and let $\mathcal{H}$ be a torus of rank r, acting rationally on A by k-algebra automorphisms. Assume that $\mathcal{H}$-spec A has normal $\mathcal{H}$-separation. Then spec A has normal separation.*

Now let J be an $\mathcal{H}$-prime ideal of A, let $\mathcal{E}_J$ be the multiplicative set of all nonzero normal $\mathcal{H}$-eigenvectors in A/J, and set $A_J = (A/J)[\mathcal{E}_J^{-1}]$.

(a) The localization map $A \to A/J \to A_J$ induces a homeomorphism of $\mathrm{spec}_J A$ onto spec A_J.

(b) Contraction and extension induce mutually inverse homeomorphisms between spec A_J and spec $Z(A_J)$.

(c) The ring $Z(A_J)$ is a commutative Laurent polynomial ring over the field $Z(\mathrm{Fract}\, A/J)^{\mathcal{H}}$ (the fixed subfield of $Z(\mathrm{Fract}\, A/J)$ under the induced action of $\mathcal{H}$), in r or fewer indeterminates.

Proof. These statements follow immediately from Theorem 4.5 and Corollary 4.6 except for the description of the coefficient field in part (c). According to Theorem 4.5(d), this field equals $Z(A_J)_1$. By definition of the $\widehat{\mathcal{H}}$-grading, $Z(A_J)_1 = Z(A_J)^{\mathcal{H}}$, which is clearly contained in $Z(\mathrm{Fract}\, A/J)^{\mathcal{H}}$. Thus, it only remains to prove the reverse inequality.

As in the proof of Theorem 4.5(c), A_J is graded-simple with respect to the $\widehat{\mathcal{H}}$-grading, and so it is $\mathcal{H}$-simple. Given any $u \in Z(\mathrm{Fract}\, A/J)^{\mathcal{H}}$, observe that the set $I = \{a \in A_J \mid au \in A_J\}$ is a nonzero $\mathcal{H}$-stable ideal of A_J. By $\mathcal{H}$-simplicity, $I = A_J$, whence $u \in A_J$. Therefore $u \in Z(A_J)^{\mathcal{H}}$, as desired. $\square$

5.4. To compare Theorem 5.3 with [20, Theorem 6.6], note that the latter theorem only applies to completely prime $\mathcal{H}$-prime ideals of A. That restriction is due to the lack of any known graded Goldie theorem for graded-prime rings (cf. [20, §6.1]), whereas one can easily prove a graded Ore theorem [20, Lemma 6.2]. We have finessed the graded Goldie problem here by assuming a suitable supply of normal elements. Other advantages of this assumption include the following result.

Proposition. *Let A, $\mathcal{H}$, J, $\mathcal{E}_J$, A_J be as in Theorem 5.3. If $\mathcal{H}$-spec A is finite and A is an affine k-algebra, then A_J is an affine k-algebra.*

Proof. After passing to A/J, we may assume that $J = 0$. If $\mathcal{E}_J$ consists of units, then $A_J = A$ and we are done. Otherwise, A contains some proper nonzero $\mathcal{H}$-stable ideals (generated by nonzero normal $\mathcal{H}$-eigenvectors which are not units). Hence, the maximal proper $\mathcal{H}$-stable ideals of A are nonzero, and these are, in particular, $\mathcal{H}$-prime. In other words, 0 is not the only $\mathcal{H}$-prime ideal of A.

Let $J_1, \ldots, J_n$ be the nonzero $\mathcal{H}$-prime ideals of A. By our normal $\mathcal{H}$-separation assumption, each J_i contains a nonzero normal $\mathcal{H}$-eigenvector c_i. Set $c = c_1 c_2 \cdots c_n$, a normal $\mathcal{H}$-eigenvector contained in all nonzero $\mathcal{H}$-primes of A. Since $J = 0$ is $\mathcal{H}$-prime, $c \neq 0$.

The action of $\mathcal{H}$ on A extends naturally to an action by k-algebra automorphisms on the localization $B = A[c^{-1}]$. Since any $\mathcal{H}$-prime of $A[c^{-1}]$ contracts to an $\mathcal{H}$-prime of A, we see that 0 is the only $\mathcal{H}$-prime of B. Thus B is $\mathcal{H}$-simple.

If $e \in \mathcal{E}_J$, then since eA is an ideal of A and A is right noetherian, eB is an ideal of B. But eB is nonzero and $\mathcal{H}$-stable, whence $eB = B$. Thus e is right invertible in B, and hence invertible. Therefore $A_J = B$, which is clearly affine. $\square$

5.5. Recall that a noetherian k-algebra A satisfies the *Nullstellensatz (over k)* [39, Chapter 9] provided that A is a Jacobson ring and the endomorphism rings of all simple A-modules are algebraic over k. In that case, it follows from the Jacobson condition that every locally closed prime of A is primitive, and from [14, Lemma 4.1.6] that every primitive ideal of A is rational. As in [20, Section 6], the presence of the Nullstellensatz yields the following refinements to Theorem 5.3.

Theorem. *Let A be a noetherian k-algebra satisfying the Nullstellensatz, and let $\mathcal{H}$ be a torus acting rationally on A by k-algebra automorphisms. Assume that $\mathcal{H}$-spec A is finite and satisfies normal $\mathcal{H}$-separation.*

(a) A satisfies the Dixmier-Moeglin equivalence, and the primitive ideals of A are precisely the maximal elements of the $\mathcal{H}$-strata in spec A.

(b) *The fields $Z(\text{Fract } A/J)^{\mathcal{H}}$ occurring in Theorem 5.3(c) are all algebraic over k.*

(c) *If k is algebraically closed, each $\mathcal{H}$-stratum in $\text{prim } A$ consists of a single $\mathcal{H}$-orbit.*

Proof. Except for minor modifications, the proofs are the same as in [20, Theorem 6.8, Corollary 6.9].

Since $\mathcal{H}$-$\text{spec } A$ is finite, maximal elements of $\mathcal{H}$-strata are locally closed in $\text{spec } A$ [20, §2.2(ii)]. Hence, to prove part (a) it remains to show that rational ideals are maximal within their $\mathcal{H}$-strata in $\text{spec } A$.

Let $J \in \mathcal{H}$-$\text{spec } A$, and define $\mathcal{E}_J$ and A_J as in Theorem 5.3. Then $Z(A_J)$ is a Laurent polynomial ring of the form $k_J[z_1^{\pm 1}, \ldots, z_n^{\pm 1}]$, where $k_J = Z(\text{Fract } A/J)^{\mathcal{H}}$. By the previous paragraph, any maximal element of $\text{spec}_J A$ is locally closed, and such primes are rational because of the Nullstellensatz. Hence, there exist rational ideals in $\text{spec}_J A$.

Now let P be any rational ideal in $\text{spec}_J A$, and set $Q = PA_J \cap Z(A_J)$. Then $Z(A_J)/Q$ embeds in $Z(\text{Fract } A/P)$, which is algebraic over k by assumption. Hence, k_J is algebraic over k (thus establishing part (b)), and Q is a maximal ideal of $Z(A_J)$. It follows that P is maximal in $\text{spec}_J A$, which completes part (a).

If k is algebraically closed, one checks that the induced action of $\mathcal{H}$ on $Z(A_J)$ incorporates all automorphisms such that $z_1 \mapsto \beta_1 z_1, \ldots, z_n \mapsto \beta_n z_n$ for arbitrary $\beta_i \in k^{\times}$. Hence, $\mathcal{H}$ acts transitively on $\max Z(A_J)$, and part (c) follows. $\square$

5.6. Theorems 5.3 and 5.5 can be used to derive all cases of Theorem 3.6 except that of $\mathcal{O}_{\lambda,p}(M_{m,n}(k))$. All the necessary hypotheses except for normal separation can be verified fairly readily from the basic descriptions of the algebras. Further, these hypotheses are also known to hold in $\mathcal{O}_{\lambda,p}(M_{m,n}(k))$, where normal separation is conjectured. Thus, we conclude by focusing on normal separation as a key problem:

Problem. *Find general hypotheses on a noetherian k-algebra A equipped with a rational action of a torus $\mathcal{H}$ by k-algebra automorphisms which*

(a) *imply that $\mathcal{H}$-$\text{spec } A$ is finite and satisfies normal $\mathcal{H}$-separation;*

(b) *are readily proved for the algebras described in Section 1.*

ACKNOWLEDGEMENT

We thank Ken Brown, Iain Gordon, Tim Hodges, David Jordan and Thierry Levasseur for helpful correspondence and references.

REFERENCES

1. J. Alev and F. Dumas, *Sur les corps de fractions de certaines algèbres de Weyl quantiques*, J. Algebra **170** (1994), 229-265.
2. M. Artin, W. Schelter, and J. Tate, *Quantum deformations of GL_n*, Comm. Pure Appl. Math. **44** (1991), 879-895.
3. K. A. Brown and K. R. Goodearl, *Prime spectra of quantum semisimple groups*, Trans. Amer. Math. Soc. **348** (1996), 2465-2502.
4. _____, *A Hilbert basis theorem for quantum groups*, Bull. London Math. Soc. **29** (2) (1997), 150-158.
5. K. A. Brown and R. B. Warfield, Jr., *The influence of ideal structure on representation theory*, J. Algebra **116** (1988), 294-315.
6. P. Caldero, *Étude des q-commutations dans l'algèbre $U_q(\mathbf{n}^+)$*, J. Algebra **178** (1995), 444-457.
7. _____, *On the Gelfand-Kirillov conjecture for quantum algebras* (to appear).
8. G. Cauchon, *Quotients premiers de $O_q(m_n(k))$*, J. Algebra **180** (1996), 530-545.
9. G. Cliff, *The division ring of quotients of the coordinate ring of the quantum general linear group*, J. London Math. Soc. (2) **51** (1995), 503-513.
10. C. De Concini and V. G. Kac, *Representations of quantum groups at roots of 1*, in Operator Algebras, Unitary Representations, Enveloping Algebras, and Invariant Theory (Paris 1989), Birkhäuser, Boston, 1990, pp. 471-506.
11. C. De Concini, V. Kac, and C. Procesi, *Quantum coadjoint action*, J. Amer. Math. Soc. **5** (1992), 151-189.
12. C. De Concini and V. Lyubashenko, *Quantum function algebras at roots of 1*, Advances in Math. **108** (1994), 205-262.
13. C. De Concini and C. Procesi, *Quantum groups*, in D-Modules, Representation Theory, and Quantum Groups (Venezia, June 1992) (G. Zampieri, and A. D'Agnolo, eds.), Lecture Notes in Math. 1565, Springer-Verlag, Berlin, 1993, pp. 31-140.
14. J. Dixmier, *Enveloping Algebras*, The 1996 printing of the 1977 English translation, Amer. Math. Soc., Providence, 1996.
15. A. W. Goldie and G. O. Michler, *Ore extensions and polycyclic group rings*, J. London Math. Soc. (2) **9** (1974), 337-345.
16. K. R. Goodearl, *Uniform ranks of prime factors of skew polynomial rings*, in Ring Theory, Proc. Biennial Ohio State – Denison Conf. 1992 (S. K. Jain and S. T. Rizvi, eds.), World Scientific, Singapore, 1993, pp. 182-199.
17. K. R. Goodearl and T. H. Lenagan, *Catenarity in quantum algebras*, J. Pure Appl. Alg. **111** (1996), 123-142.
18. K. R. Goodearl and E. S. Letzter, *Prime factor algebras of the coordinate ring of quantum matrices*, Proc. Amer. Math. Soc. **121** (1994), 1017-1025.
19. _____, *Prime and primitive spectra of multiparameter quantum affine spaces*, in Trends in Ring Theory. Proc. Miskolc Conf. 1996 (V. Dlab and L. Márki, Eds.), Canad. Math. Soc. Conf. Proc. Series **22** (1998), 39-58.
20. _____, *The Dixmier-Moeglin equivalence in quantum coordinate rings and quantized Weyl algebras*, Trans. Amer. Math. Soc. (to appear).
21. K. R. Goodearl and R. B. Warfield, Jr., *An Introduction to Noncommutative Rings*, Cambridge Univ. Press, Cambridge, 1989.
22. T. J. Hodges and T. Levasseur, *Primitive ideals of $C_q[SL(3)]$*, Comm. Math. Phys. **156** (1993), 581-605.
23. _____, *Primitive ideals of $C_q[SL(n)]$*, J. Algebra **168** (1994), 455-468.

24. T. J. Hodges, T. Levasseur, and M. Toro, *Algebraic structure of multi-parameter quantum groups*, Advances in Math. **126** (1997), 52-92.

25. K. Iohara and F. Malikov, *Rings of skew polynomials and Gelfand-Kirillov conjecture for quantum groups*, Comm. Math. Phys. **164** (1994), 217-237.

26. J. C. Jantzen, *Lectures on Quantum Groups*, Grad. Studies in Math. 6, Amer. Math. Soc., Providence, 1996.

27. A. V. Jategaonkar, *Localization in Noncommutative Noetherian Rings*, London Math. Soc. Lecture Note Series 98, Cambridge Univ. Press, Cambridge, 1986.

28. D. A. Jordan, *A simple localization of the quantized Weyl algebra*, J. Algebra **174** (1995), 267-281.

29. A. Joseph, *On the prime and primitive spectra of the algebra of functions on a quantum group*, J. Algebra **169** (1994), 441–511.

30. _____, *Quantum Groups and Their Primitive Ideals*, Ergebnisse der Math. (3) 29, Springer-Verlag, Berlin, 1995.

31. _____, *Sur une conjecture de Feigin*, C. R. Acad. Sci. Paris, Sér. I **320** (1995), 1441-1444.

32. G. Kempf, F. Knudsen, D. Mumford, and B. Saint-Donet, *Toroidal Embeddings I*, Lecture Notes in Math. 339, Springer-Verlag, Berlin, 1973.

33. E. S. Letzter, *Remarks on the twisted adjoint representation of $R_q[G]$*, Comm. Algebra (to appear).

34. T. Levasseur and J. T. Stafford, *The quantum coordinate ring of the special linear group*, J. Pure Appl. Algebra **86** (1993), 181-186.

35. G. Lusztig, *Quantum groups at roots of 1*, Geom. Dedicata **35** (1990), 89-114.

36. G. Maltsiniotis, *Calcul différentiel quantique*, Groupe de travail, Université Paris VII (1992).

37. Yu. I. Manin, *Some remarks on Koszul algebras and quantum groups*, Ann. Inst. Fourier (Grenoble) **37** (1987), 191-205.

38. _____, *Multiparametric quantum deformation of the general linear supergroup*, Comm. Math. Phys. **123** (1989), 163–175.

39. J. C. McConnell and J. C. Robson, *Noncommutative Noetherian Rings*, Wiley-Interscience, Chichester-New York, 1987.

40. C. Moeglin and R. Rentschler, *Orbites d'un groupe algébrique dans l'espace des idéaux rationnels d'une algèbre enveloppante*, Bull. Soc. Math. France **109** (1981), 403–426.

41. S. Montgomery, *Hopf Algebras and Their Actions on Rings*, CBMS Regional Conf. Series in Math. 82, Amer. Math. Soc., Providence, 1993.

42. V. G. Mosin and A. N. Panov, *Division rings of quotients and central elements of multiparameter quantizations*, Sbornik: Mathematics **187**:6 (1996), 835-855.

43. I. M. Musson, *Ring theoretic properties of the coordinate rings of quantum symplectic and Euclidean space*, in Ring Theory, Proc. Biennial Ohio State–Denison Conf., 1992 (S. K. Jain and S. T. Rizvi, eds.), World Scientific, Singapore, 1993, pp. 248-258.

44. D. G. Northcott, *Affine Sets and Affine Groups*, London Math. Soc. Lecture Note Series 39, Cambridge Univ. Press, Cambridge, 1980.

45. S.-Q. Oh, *Primitive ideals of the coordinate ring of quantum symplectic space*, J Algebra **174** (1995), 531-552.

46. _____, *Catenarity in a class of iterated skew polynomial rings*, Comm. Alg. **25** (1997), 37-49.

47. S.-Q. Oh and C.-G. Park, *Primitive ideals in the coordinate ring of quantum Euclidean space*, Bull. Austral. Math. Soc. **58** (1998), 57-73.

48. A. N. Panov, *Skew fields of twisted rational functions and the skew field of rational functions on $GL_q(n, K)$*, St. Petersburg Math. J. **7** (1996), 129-143.

49. B. Parshall and J.-p. Wang, *Quantum linear groups*, Mem. Amer. Math. Soc. **439** (1991).

50. Z. Reichstein and N. Vonessen, *Torus actions on rings*, J. Algebra **170** (1994), 781-804.

51. R. Rentschler, *Primitive ideals in enveloping algebras (general case)*, in Noetherian Rings and their Applications (Oberwolfach, 1983) (L. W. Small, ed.), Math. Surveys and Monographs 24, Amer. Math. Soc., Providence, 1987, pp. 37-57.

52. N. Yu. Reshetikhin, *Multiparameter quantum groups and twisted quasitriangular Hopf algebras*, Lett. Math. Phys. **20** (1990), 331-335.

53. N. Yu. Reshetikhin, L. A. Takhtadzhyan, and L. D. Faddeev, *Quantization of Lie groups and Lie algebras*, Leningrad Math. J. **1** (1990), 193-225.

54. C. M. Ringel, *PBW bases of quantum groups*, J. reine angew. Math. **470** (1996), 51-88.

55. N. Sasaki, *Quantization of Lie group and algebra of G_2 type in the Faddeev-Reshetikhin-Takhtajan approach*, J. Math. Phys. **36** (1995), 4476-4488.

56. S. P. Smith, *Quantum groups: an introduction and survey for ring theorists*, in Noncommutative Rings (S. Montgomery and L. Small, eds.), MSRI Publ. 24, Springer-Verlag, New York, 1992, pp. 131-178.

57. A. Sudbury, *Consistent multiparameter quantisation of GL(n)*, J. Phys. A **23** (1990), L697-L704.

58. M. Takeuchi, *Quantum orthogonal and symplectic groups and their embedding into quantum GL*, Proc. Japan Acad., Ser. A **65** (1989), 55-58.

59. N. Vonessen, *Actions of algebraic groups on the spectrum of rational ideals, II*, J. Algebra **208** (1998), 216-261.

60. J. J. Zhang, *Twisted graded algebras and equivalences of graded categories*, Proc. London Math. Soc. (3) **72** (1996), 281-311.

Prime Ideals in Certain Quantum Determinantal Rings

K. R. GOODEARL AND T. H. LENAGAN

DEPARTMENT OF MATHEMATICS, UNIVERSITY OF CALIFORNIA, SANTA BARBARA, CA 93106, USA
E-mail address: goodearl@math.ucsb.edu

DEPARTMENT OF MATHEMATICS, J.C.M.B., KINGS BUILDINGS, MAYFIELD ROAD, EDINBURGH EH9 3JZ, SCOTLAND
E-mail address: tom@maths.ed.ac.uk

ABSTRACT. The ideal $\mathcal{I}_1$ generated by the 2×2 quantum minors in the co-ordinate algebra of quantum matrices, $\mathcal{O}_q(M_{m,n}(k))$, is investigated. Analogues of the First and Second Fundamental Theorems of Invariant Theory are proved. In particular, it is shown that $\mathcal{I}_1$ is a completely prime ideal, that is, $\mathcal{O}_q(M_{m,n}(k))/\mathcal{I}_1$ is an integral domain, and that $\mathcal{O}_q(M_{m,n}(k))/\mathcal{I}_1$ is the ring of coinvariants of a coaction of $k[x, x^{-1}]$ on $\mathcal{O}_q(k^m) \otimes \mathcal{O}_q(k^n)$, a tensor product of two quantum affine spaces. There is a natural torus action on $\mathcal{O}_q(M_{m,n}(k))/\mathcal{I}_1$ induced by an $(m + n)$-torus action on $\mathcal{O}_q(M_{m,n}(k))$. We identify the invariant prime ideals for this action and deduce consequences for the prime spectrum of $\mathcal{O}_q(M_{m,n}(k))/\mathcal{I}_1$.

INTRODUCTION

Let k be a field and let $q \in k^{\times}$. The *coordinate ring of quantum $m \times n$ matrices*, $\mathcal{A} := \mathcal{O}_q(M_{m,n}(k))$, is a deformation of the classical coordinate ring of $m \times n$ matrices, $\mathcal{O}(M_{m,n}(k))$. As such it is a k-algebra generated by mn indeterminates X_{ij}, for $1 \leq i \leq m$ and $1 \leq j \leq n$, subject to the relations

$$
\begin{aligned}
X_{ij}X_{lj} &= qX_{lj}X_{ij} && \text{when } i < l; \\
X_{ij}X_{is} &= qX_{is}X_{ij} && \text{when } j < s; \\
X_{is}X_{lj} &= X_{lj}X_{is} && \text{when } i < l \text{ and } j < s; \\
X_{ij}X_{ls} - X_{ls}X_{ij} &= (q - q^{-1})X_{is}X_{lj} && \text{when } i < l \text{ and } j < s.
\end{aligned}
$$

In some references (e.g., [6, §3.5]), q is replaced by q^{-1}. When $q = 1$ we recover $\mathcal{O}(M_{m,n}(k))$, which is the commutative polynomial algebra $k[X_{ij}]$.

This research was partially supported by National Science Foundation research grant DMS-9622876 and NATO Collaborative Research Grant 960250.

When $m = n$, the algebra $\mathcal{A}$ possesses a special element, the *quantum determinant*, D_q, defined by

$$D_q := \sum_{\sigma \in S_n} (-q)^{l(\sigma)} X_{1,\sigma(1)} X_{2,\sigma(2)} \cdots X_{n,\sigma(n)},$$

where $l(\sigma)$ denotes the number of inversions in the permutation σ. The quantum determinant D_q is a central element of $\mathcal{A}$ (see, for example, [**6**, Theorem 4.6.1]), and the localization $\mathcal{A}[D_q^{-1}]$ is the *coordinate ring of the quantum general linear group*, denoted $\mathcal{O}_q(GL_n(k))$.

If $I \subseteq \{1, \ldots, m\}$ and $J \subseteq \{1, \ldots, n\}$ with $|I| = |J| = t$, let $D(I,J)$ denote the $t \times t$ quantum minor obtained as the quantum determinant of the subalgebra of $\mathcal{A}$ obtained by deleting generators X_{ij} from the rows outside I and from the columns outside J. We write $\mathcal{I}_t$ for the ideal generated by the $(t+1) \times (t+1)$ quantum minors of $\mathcal{A}$. In [**3**] it is proved that $\mathcal{A}/\mathcal{I}_t$ is an integral domain, for each $1 \leq t \leq \min\{m, n\}$. Independently, Rigal [**7**] has shown that $\mathcal{A}/\mathcal{I}_1$ is a domain; he also shows that $\mathcal{A}/\mathcal{I}_1$ is a maximal order in its division ring of fractions.

There is an action of the torus $\mathcal{H} := (k^\times)^m \times (k^\times)^n$ by k-algebra automorphisms on $\mathcal{A}$ such that

$$(\alpha_1, \ldots, \alpha_m, \beta_1, \ldots, \beta_n) \cdot X_{ij} := \alpha_i \beta_j X_{ij}$$

for all i, j. The ideals $\mathcal{I}_t$ are easily seen to be invariant under $\mathcal{H}$; so there is an induced action of $\mathcal{H}$ on the factor algebras $\mathcal{A}/\mathcal{I}_t$. In this paper, we study the prime ideal structure in the algebra $\mathcal{A}/\mathcal{I}_1$, paying particular attention to the $\mathcal{H}$-invariant prime ideals.

1. COMPLETE PRIMENESS OF $\mathcal{I}_1$

We give a direct derivation of the fact that $\mathcal{A}/\mathcal{I}_1$ is a domain. Although this is already established in both [**3**] and [**7**], the proof we give here is so much simpler and more transparent than either of the previous proofs that we think it will be useful to have it in a published form.

The *coordinate ring of quantum affine n-space*, denoted $\mathcal{O}_q(k^n)$, is defined to be the k-algebra generated by elements $y_1, \ldots, y_n$ subject to the relations $y_i y_j = q y_j y_i$ for each $1 \leq i < j \leq n$. It is well known that $\mathcal{O}_q(k^n)$ is an iterated Ore extension, and thus, in particular, $\mathcal{O}_q(k^n)$ is a domain. Our strategy is to produce a homomorphism of $\mathcal{A}$ into $\mathcal{O}_q(k^m) \otimes \mathcal{O}_q(k^n)$. This latter algebra can also be presented as an iterated Ore extension and thus is a domain. We show that $\mathcal{I}_1$ is the kernel of this map and so $\mathcal{A}/\mathcal{I}_1$ is a domain.

1.1. Theorem. *The algebra $\mathcal{O}_q(M_{m,n}(k))/\mathcal{I}_1$ is isomorphic to a subalgebra of the tensor product $\mathcal{O}_q(k^m) \otimes \mathcal{O}_q(k^n)$. In particular, $\mathcal{I}_1$ is a completely prime ideal of $\mathcal{O}_q(M_{m,n}(k))$.*

Proof. Let $\mathcal{O}_q(k^m) = k[y_1, \ldots, y_m]$ and $\mathcal{O}_q(k^n) = k[z_1, \ldots, z_n]$ be the coordinate rings of quantum affine m-space and n-space, respectively. We define an algebra homomorphism $\theta : \mathcal{A} \to \mathcal{O}_q(k^m) \otimes \mathcal{O}_q(k^n)$ such that $\theta(X_{ij}) = y_i \otimes z_j$ for all i, j. In order that this does extend to a well-defined algebra homomorphism, we must check that the elements $y_i \otimes z_j$ satisfy at least the relations defining $\mathcal{A}$. These are routine verifications; for example, if $i < l$ and $j < s$ then

$$(y_i \otimes z_j)(y_l \otimes z_s) = y_i y_l \otimes z_j z_s = y_i y_l \otimes q z_s z_j = q(y_i \otimes z_s)(y_l \otimes z_j),$$

while

$$(y_l \otimes z_s)(y_i \otimes z_j) = y_l y_i \otimes z_s z_j = q^{-1} y_i y_l \otimes z_s z_j = q^{-1}(y_l \otimes z_s)(y_i \otimes z_j).$$

Thus,

$$(y_i \otimes z_j)(y_l \otimes z_s) - (y_l \otimes z_s)(y_i \otimes z_j) = (q - q^{-1})(y_i \otimes z_s)(y_l \otimes z_j),$$

so that the fourth relation of the introduction holds. One can also obtain θ as the composition of the comultiplication $\mathcal{A} \to \mathcal{A} \otimes \mathcal{A}$ with the tensor product of the quotient maps from $\mathcal{A}$ to $\mathcal{A}/\langle X_{ij} \mid i > 1 \rangle$ and $\mathcal{A}/\langle X_{ij} \mid j > 1 \rangle$. We shall pursue the latter point of view in the next section.

Thus, there exists a unique k-algebra homomorphism

$$\theta : \mathcal{A} \to \mathcal{O}_q(k^m) \otimes \mathcal{O}_q(k^n)$$

such that $\theta(X_{ij}) = y_i \otimes z_j$ for all i, j. If $i < l$ and $j < s$ then the above calculations also show that $\theta(X_{ij}X_{ls} - qX_{is}X_{lj}) = 0$; thus $\mathcal{I}_1 \subseteq \ker(\theta)$.

Now, $\mathcal{O}_q(k^m) \otimes \mathcal{O}_q(k^n)$ is a domain, since it can be viewed as a (multiparameter) quantum affine $(m + n)$-space with respect to the generators $y_1 \otimes 1, \ldots, y_m \otimes 1, 1 \otimes z_1, \ldots, 1 \otimes z_n$. Hence, $\ker(\theta)$ is a completely prime ideal of $\mathcal{A}$. We show that $\mathcal{I}_1 = \ker(\theta)$, so that $\mathcal{I}_1$ is completely prime. It remains to show that the induced map $\bar{\theta} : \mathcal{A}/\mathcal{I}_1 \to \mathcal{O}_q(k^m) \otimes \mathcal{O}_q(k^n)$ is injective. Let $\mathcal{S}$ denote the set of monomials $X_{i_1 j_1} X_{i_2 j_2} \ldots X_{i_l j_l}$ in $\mathcal{A}$ such that $i_1 \geq i_2 \geq \cdots \geq i_l$ and $j_1 \leq j_2 \leq \cdots \leq j_l$. (We allow the monomial to be equal to 1 when $l = 0$.) We claim that the set $\bar{\mathcal{S}}$ of images forms a spanning set of $\mathcal{A}/I_1$.

It suffices to show that an arbitrary monomial C in $\mathcal{A}$ is congruent modulo $\mathcal{I}_1$ to a linear combination of monomials from $\mathcal{S}$. We proceed by induction on the index sets, where row index sequences $(i_1, i_2, \ldots, i_l)$ are ordered lexicographically with respect to $\geq$, column index sequences $(j_1, j_2, \ldots, j_l)$ are ordered lexicographically with respect to $\leq$, and pairs of sequences are ordered lexicographically.

If the claim fails, then it fails for a monomial $C = X_{i_1 j_1} X_{i_2 j_2} \cdots X_{i_l j_l}$ whose index set is minimal with respect to the ordering given in the previous paragraph. In particular, $C \notin \mathcal{S}$. Let r be the first subindex such that either $i_r < i_{r+1}$ or $j_r > j_{r+1}$.

If $i_r < i_{r+1}$ and $j_r \geq j_{r+1}$ then $C = \lambda C'$, where λ is either 1 or q and C' is obtained from C by switching $X_{i_r j_r}$ and $X_{i_{r+1} j_{r+1}}$. However,

$$(i_1, , \ldots, i_{r-1}, i_{r+1}, i_r, i_{r+2}, , \ldots, i_l) < (i_1, i_2, \ldots, i_l)$$

in our ordering, so C' is congruent modulo $\mathcal{I}_1$ to a linear combination of elements of $\mathcal{S}$. Then C is congruent to such a linear combination, contradicting our assumptions. A similar contradiction occurs if $i_r \leq i_{r+1}$ and $j_r > j_{r+1}$: this time, the row indices might not change, but

$$(j_1, \ldots, j_{r-1}, j_{r+1}, j_r, j_{r+2}, \ldots, j_l) < (j_1, \ldots, j_l),$$

so again we have a contradiction. Therefore, we must either have $i_r < i_{r+1}$ and $j_r < j_{r+1}$ or $i_r > i_{r+1}$ and $j_r > j_{r+1}$.

Suppose that $i_r < i_{r+1}$ and $j_r < j_{r+1}$. In this case, we have

$$X_{i_r j_r} X_{i_{r+1} j_{r+1}} - q X_{i_{r+1} j_r} X_{i_r j_{r+1}} \in \mathcal{I}_1,$$

so that $C - qC' \in \mathcal{I}_1$, where

$$C' = X_{i_1 j_1} \cdots X_{i_{r-1} j_{r-1}} X_{i_{r+1} j_r} X_{i_r j_{r+1}} X_{i_{r+2} j_{r+2}} \cdots X_{i_l j_l}.$$

We obtain a contradiction as above.

The final case is $i_r > i_{r+1}$ and $j_r > j_{r+1}$, where we have

$$X_{i_r j_r} X_{i_{r+1} j_{r+1}} - q^{-1} X_{i_r j_{r+1}} X_{i_{r+1} j_r} \in \mathcal{I}_1.$$

Thus, $C - q^{-1} C' \in \mathcal{I}_1$, where

$$C' = X_{i_1 j_1} \cdots X_{i_{r-1} j_{r-1}} X_{i_r j_{r+1}} X_{i_{r+1} j_r} X_{i_{r+2} j_{r+2}} \cdots X_{i_l j_l},$$

and once again we reach a contradiction. This finishes the proof of the claim and establishes that $\overline{\mathcal{S}}$ spans $\mathcal{A}/\mathcal{I}_1$.

Now, observe that in $\mathcal{O}_q(k^m) \otimes \mathcal{O}_q(k^n)$ we have

$$\theta(X_{i_1j_1}X_{i_2j_2}\dots X_{i_lj_l}) = y_{i_1}y_{i_2}\dots y_{i_l} \otimes z_{j_1}z_{j_2}\dots z_{j_l}.$$

The monomials $y_{i_1}y_{i_2}\dots y_{i_l}$ with $i_1 \geq i_2 \geq \dots \geq i_l$ are linearly independent over k, and, likewise, the monomials $z_{j_1}z_{j_2}\dots z_{j_l}$ with $j_1 \leq j_2 \leq \dots \leq j_l$ are linearly independent over k. Hence, θ maps S bijectively to a linearly independent set in $\mathcal{O}_q(k^m) \otimes \mathcal{O}_q(k^n)$, so that $\overline{S}$ is a linearly independent set in $\mathcal{A}/\mathcal{I}_1$. Therefore, the map $\overline{\theta} : \mathcal{A}/\mathcal{I}_1 \to \mathcal{O}_q(k^m) \otimes \mathcal{O}_q(k^n)$ maps the k-basis $\overline{S}$ bijectively onto a linearly independent set, so that $\overline{\theta}$ is injective. $\square$

2. Coinvariants

Theorem 1.1 has an invariant theoretic interpretation, which we discuss in this section. First, we outline what happens in the classical ($q = 1$) case.

2.1. Let $M_{u,v}(k)$ denote the algebraic variety of $u \times v$ matrices over k. Fix positive integers m, n and $t \leq \min\{m, n\}$. The general linear group $GL_t(k)$ acts on $M_{m,t}(k) \times M_{t,n}(k)$ via

$$g \cdot (A, B) := (Ag^{-1}, gB).$$

Matrix multiplication yields a map

$$\mu : M_{m,t}(k) \times M_{t,n}(k) \to M_{m,n}(k),$$

the image of which is the variety of $m \times n$ matrices with rank at most t, and there is an induced map

$$\mu_* : \mathcal{O}(M_{m,n}(k)) \to \mathcal{O}\big(M_{m,t}(k) \times M_{t,n}(k)\big) = \mathcal{O}(M_{m,t}(k)) \otimes \mathcal{O}(M_{t,n}(k)).$$

The First Fundamental Theorem of invariant theory identifies the fixed ring of the coordinate ring $\mathcal{O}\big(M_{m,t}(k) \times M_{t,n}(k)\big)$ under the induced action of $GL_t(k)$ as precisely the image of μ_*. The Second Fundamental Theorem states that the kernel of μ_* is $\mathcal{I}_t$, the ideal generated by the $(t+1) \times (t+1)$ minors of $\mathcal{O}(M_{m,n}(k))$, so that the coordinate ring of the variety of $m \times n$ matrices of rank at most t is $\mathcal{O}(M_{m,n}(k))/\mathcal{I}_t$. As a consequence, since this variety is irreducible, the ideal $\mathcal{I}_t$ is a prime ideal of $\mathcal{O}(M_{m,n}(k))$.

2.2. We now proceed to explain the connection between Theorem 1.1 and the above invariant theoretic point of view.

The analog of μ_* is the k-algebra homomorphism

$$\theta_t : \mathcal{O}_q(M_{m,n}(k)) \to \mathcal{O}_q(M_{m,t}(k)) \otimes \mathcal{O}_q(M_{t,n}(k))$$

induced from the comultiplication on $\mathcal{O}_q(M_{m,n}(k))$, that is,

$$\theta_t(X_{ij}) = \sum_{l=1}^{t} X_{il} \otimes X_{lj}$$

for $1 \leq i \leq m$ and $1 \leq j \leq n$. The comultiplications on $\mathcal{O}_q(M_{m,t}(k))$ and $\mathcal{O}_q(M_{t,n}(k))$ yield k-algebra homomorphisms

$$\rho_t : \mathcal{O}_q(M_{m,t}(k)) \rightarrow \mathcal{O}_q(M_{m,t}(k)) \otimes \mathcal{O}_q(M_t(k))$$
$$\rightarrow \mathcal{O}_q(M_{m,t}(k)) \otimes \mathcal{O}_q(GL_t(k))$$
$$\lambda_t : \mathcal{O}_q(M_{t,n}(k)) \rightarrow \mathcal{O}_q(M_t(k)) \otimes \mathcal{O}_q(M_{t,n}(k))$$
$$\rightarrow \mathcal{O}_q(GL_t(k)) \otimes \mathcal{O}_q(M_{t,n}(k))$$

which make $\mathcal{O}_q(M_{m,t}(k))$ into a right $\mathcal{O}_q(GL_t(k))$-comodule and $\mathcal{O}_q(M_{t,n}(k))$ into a left $\mathcal{O}_q(GL_t(k))$-comodule. Since $\mathcal{O}_q(GL_t(k))$ is a Hopf algebra, the right comodule $\mathcal{O}_q(M_{m,t}(k))$ becomes a <u>left</u> $\mathcal{O}_q(GL_t(k))$-comodule on composing ρ_t with $1 \otimes S$ followed by the flip (where S denotes the antipode). Finally, the tensor product of the two left $\mathcal{O}_q(GL_t(k))$-comodules $\mathcal{O}_q(M_{m,t}(k))$ and $\mathcal{O}_q(M_{t,n}(k))$ becomes a left $\mathcal{O}_q(GL_t(k))$-comodule via the multiplication map on $\mathcal{O}_q(GL_t(k))$. This comodule structure map,

$$\gamma_t : \mathcal{O}_q(M_{m,t}(k)) \otimes \mathcal{O}_q(M_{t,n}(k)) \rightarrow \mathcal{O}_q(GL_t(k)) \otimes \mathcal{O}_q(M_{m,t}(k)) \otimes \mathcal{O}_q(M_{t,n}(k)),$$

can be described (using the Sweedler summation notation) as follows:

$$\gamma_t(a \otimes b) = \sum_{(a)} \sum_{(b)} S(a_1) b_{-1} \otimes a_0 \otimes b_0$$

where $\rho_t(a) = \sum_{(a)} a_0 \otimes a_1$ and $\lambda_t(b) = \sum_{(b)} b_{-1} \otimes b_0$ for $a \in \mathcal{O}_q(M_{m,t}(k))$ and $b \in \mathcal{O}_q(M_{t,n}(k))$. Note that for $t > 1$, the map γ_t is not an algebra homomorphism, since neither the antipode nor the multiplication map on $\mathcal{O}_q(GL_t(k))$ is an algebra homomorphism. On the other hand, γ_1 is a k-algebra homomorphism.

Recall that the *coinvariants* of the coaction γ_t are the elements x in the tensor product $\mathcal{O}_q(M_{m,t}(k)) \otimes \mathcal{O}_q(M_{t,n}(k))$ such that $\gamma_t(x) = 1 \otimes x$. Quantum analogs of the First and Second Fundamental Theorems would be the following:

Conjecture 1. The set of coinvariants of γ_t equals the image of θ_t.

Conjecture 2. The kernel of θ_t is the ideal $\mathcal{I}_t$.

We have proved Conjecture 2 in [**3**, Proposition 2.4] (essentially; the cited result covers the case $m = n$, and the general case follows easily by the method of [**3**, Corollary 2.6]). However, Conjecture 1 is open at present. Here we shall establish it in the case $t = 1$.

2.3. Note that $\mathcal{O}_q(M_{m,1}(k))$ and $\mathcal{O}_q(M_{1,n}(k))$ are quantum affine spaces on generators $X_{11}, X_{21}, \ldots, X_{m1}$ and $X_{11}, X_{12}, \ldots, X_{1n}$, respectively. In studying the case $t = 1$, it is convenient to replace $\mathcal{O}_q(M_{m,1}(k))$ and $\mathcal{O}_q(M_{1,n}(k))$ by $\mathcal{O}_q(k^m) = k[y_1, \ldots, y_m]$ and $\mathcal{O}_q(k^n) = k[z_1, \ldots, z_n]$, respectively. Then θ_1 becomes the k-algebra homomorphism

$$\theta : \mathcal{O}_q(M_{m,n}(k)) \to \mathcal{O}_q(k^m) \otimes \mathcal{O}_q(k^n), \qquad X_{ij} \mapsto y_i \otimes z_j$$

used in the proof of Theorem 1.1. Next, the (quantum) coordinate ring of 1×1 matrices is just a polynomial ring $k[x]$, and the (quantum) coordinate ring of the 1×1 general linear group is the localization $k[x, x^{-1}]$. Thus, in the present case the coaction γ_1 becomes the k-algebra homomorphism

$$\gamma : \mathcal{O}_q(k^m) \otimes \mathcal{O}_q(k^n) \to k[x^{\pm 1}] \otimes \mathcal{O}_q(k^m) \otimes \mathcal{O}_q(k^n),$$
$$y_i \otimes 1 \mapsto x^{-1} \otimes y_i \otimes 1, \qquad 1 \otimes z_j \mapsto x \otimes 1 \otimes z_j.$$

2.4. Theorem. *The set of coinvariants of γ is exactly the image of the algebra $\mathcal{O}_q(M_{m,n}(k))$ in $\mathcal{O}_q(k^m) \otimes \mathcal{O}_q(k^n)$ under θ.*

Proof. Clearly $\gamma\theta(X_{ij}) = 1 \otimes y_i \otimes z_j = 1 \otimes \theta(X_{ij})$ for all i, j. Since θ and γ are k-algebra homomorphisms, it follows that the image of θ is contained in the coinvariants of γ.

The algebra $\mathcal{O}_q(k^m) \otimes \mathcal{O}_q(k^n)$ has a basis consisting of pure tensors $Y \otimes Z$ where Y is an ordered monomial in the y_i and Z is an ordered monomial in the z_j. Note that $\gamma(Y \otimes Z) = x^{s-r} \otimes Y \otimes Z$ where r and s are the total degrees of Y and Z, respectively. Hence, the images $\gamma(Y \otimes Z)$ are k-linearly independent, and a linear combination $\sum_{l=1}^{d} \alpha_l Y_l \otimes Z_l$ of distinct monomial tensors is a coinvariant for γ if and only if each $Y_l \otimes Z_l$ is a coinvariant.

Thus, we need only consider a single term

$$Y \otimes Z = y_{i_1} y_{i_2} \cdots y_{i_r} \otimes z_{j_1} z_{j_2} \cdots z_{j_s}.$$

If $Y \otimes Z$ is a coinvariant, then because $\gamma(Y \otimes Z) = x^{s-r} \otimes Y \otimes Z$ we must have $r = s$. Therefore

$$Y \otimes Z = \theta(X_{i_1 j_1} X_{i_2 j_2} \cdots X_{i_r j_r}),$$

which shows that $Y \otimes Z$ is in the image of θ, as desired. $\quad\square$

3. $\mathcal{H}$-INVARIANT PRIME IDEALS OF $\mathcal{O}_q(M_{m,n}(k))/\mathcal{I}_1$

Under the mild assumption that our ground field k is infinite, we identify the $\mathcal{H}$-invariant prime ideals of the domain $\mathcal{A}/\mathcal{I}_1 = \mathcal{O}_q(M_{m,n}(k))/\mathcal{I}_1$. (Recall that $\mathcal{H}$ denotes the torus $(k^\times)^m \times (k^\times)^n$, acting on $\mathcal{A}$ as described in the introduction.) This identifies the minimal elements in a stratification of $\operatorname{spec}\mathcal{A}/\mathcal{I}_1$, and yields a description of this prime spectrum as a finite disjoint union of commutative schemes corresponding to Laurent polynomial rings.

3.1. Let H be a group acting as automorphisms on a ring A. We refer the reader to [1] for the definition of the H-*stratification of* $\operatorname{spec} A$, and here recall only that the H-stratum of $\operatorname{spec} A$ corresponding to an H-prime ideal J is the set
$$\operatorname{spec}_J A := \{P \in \operatorname{spec} A \mid (P : H) = J\}.$$

In the case of the algebra $\mathcal{A}/\mathcal{I}_1$, we shall (assuming k infinite) identify the $\mathcal{H}$-prime ideals – they turn out to be the same as the $\mathcal{H}$-invariant primes – and thus pin down the minimum elements of the $\mathcal{H}$-strata. Further, we shall see that each $\mathcal{H}$-stratum of $\operatorname{spec}\mathcal{A}/\mathcal{I}_1$ is homeomorphic to the spectrum of a Laurent polynomial ring over an algebraic extension of k. This pattern is also known to hold for $\operatorname{spec}\mathcal{A}$ itself (at least when q is not a root of unity), but there the $\mathcal{H}$-prime ideals have not yet been completely identified.

3.2. It turns out that if a generator X_{ij} lies in an $\mathcal{H}$-prime ideal P of $\mathcal{A}$ containing $\mathcal{I}_1$, then either all the generators from the same row, or all the generators from the same column must also lie in P. This leads us to make the following definition.

For subsets $I \subseteq \{1, \dots, m\}$ and $J \subseteq \{1, \dots, n\}$, set

$$P(I, J) := \mathcal{I}_1 + \langle X_{ij} \mid i \in I \rangle + \langle X_{ij} \mid j \in J \rangle.$$

Obviously, $P(I, J)$ is an $\mathcal{H}$-invariant ideal of $\mathcal{A}$. We shall show that $P(I, J)$ is (completely) prime, and hence $\mathcal{H}$-prime.

Lemma. *The factor algebra $\mathcal{A}/P(I, J)$ is isomorphic to $\mathcal{O}_q(M_{m',n'}(k))/\mathcal{I}_1'$, where $m' = m - |I|$ and $n' = n - |J|$, and $\mathcal{I}_1'$ is the ideal generated by the 2×2 quantum minors of $\mathcal{O}_q(M_{m',n'}(k))$. Hence, $P(I, J)$ is a completely prime ideal of $\mathcal{A}$.*

Proof. The second statement follows immediately from the first statement and Theorem 1.1.

Set $I' := \{1, \dots, m\} \backslash I$, and $J' := \{1, \dots, n\} \backslash J$, and let $\mathcal{A}'$ be the k-subalgebra of $\mathcal{A}$ generated by the X_{ij} for $i \in I'$ and $j \in J'$. Note that $\mathcal{A}'$ is isomorphic to $\mathcal{O}_q(M_{m',n'}(k))$. Let $\mathcal{I}_1'$ be the ideal of $\mathcal{A}'$ generated by the 2×2

quantum minors of $\mathcal{A}'$; that is, those for which both row indices are in I' and both column indices are in J'. Obviously, $\mathcal{I}'_1 \subseteq \mathcal{A}' \cap \mathcal{I}_1$, so that the inclusion $\mathcal{A}' \hookrightarrow \mathcal{A}$ induces a k-algebra homomorphism $f : \mathcal{A}'/\mathcal{I}'_1 \to \mathcal{A}/P(I,J)$. It suffices to show that f is an isomorphism.

The factor $\mathcal{A}/P(I,J)$ is generated by the cosets of those X_{ij} with $i \in I'$ and $j \in J'$, since $X_{ij} \in P(I,J)$ whenever $i \in I$ or $j \in J$. These cosets are all in the image of f; so f is surjective.

Observe that there exists a k-algebra homomorphism $g : \mathcal{A} \to \mathcal{A}'$ such that $g(X_{ij}) = X_{ij}$ when $i \in I'$ and $j \in J'$, and $g(X_{ij}) = 0$ otherwise. To see this, note that the only problematic relations are those of the form $X_{ij}X_{ls} - X_{ls}X_{ij} = (q - q^{-1})X_{is}X_{lj}$ for $i < l$ and $j < s$. However, if $i \notin I'$ then X_{ij} and X_{is} both map to zero, and the relation maps to $0 = 0$. Likewise, this happens in all cases except when $i, l \in I'$ and $j, s \in J'$: in this case, the relation above maps to a relation in $\mathcal{A}'$.

Consider a 2×2 quantum minor in $\mathcal{A}$ of the form $D = X_{ij}X_{ls} - qX_{is}X_{lj}$ where $i < l$ and $j < s$. If $i \notin I'$ then both X_{ij} and X_{is} are in $\ker(g)$, so that $D \in \ker(g)$. Likewise, $g(D) = 0$ when $l \notin I'$, or $j \notin J'$, or $s \notin J'$. On the other hand, $g(D) = D$ when $i, l \in I'$ and $j, s \in J'$. Further, $g(X_{ij}) = 0$ when $i \in I$ or $j \in J$. Hence, $g(P(I,J)) \subseteq \mathcal{I}'_1$.

Therefore, g induces a k-algebra homomorphism $\bar{g} : \mathcal{A}/P(I,J) \to \mathcal{A}'/\mathcal{I}'_1$. Both of these algebras are generated by the cosets corresponding to those X_{ij} such that $i \in I'$ and $j \in J'$. It follows that both $f\bar{g}$ and $\bar{g}f$ are identity maps, since both f and $\bar{g}$ preserve these cosets. Hence, f is an isomorphism. $\square$

Somewhat suprisingly, the $P(I,J)$ turn out to be the only $\mathcal{H}$-prime ideals of $\mathcal{A}$ that contain $\mathcal{I}_1$. The following lemma will be helpful in establishing this fact.

3.3. Lemma. *Let $i, s \in \{1, \ldots, m\}$ and $j, t \in \{1, \ldots, n\}$. Then there exist scalars $\alpha \in \{1, q^{\pm 1}, q^{\pm 2}\}$ and $\beta \in \{1, q^{\pm 1}\}$ such that $X_{ij}X_{st} - \alpha X_{st}X_{ij}$ and $X_{ij}X_{st} - \beta X_{it}X_{sj}$ lie in $\mathcal{I}_1$. In particular, the cosets $X_{ij} + \mathcal{I}_1$ are all normal elements of $\mathcal{A}/\mathcal{I}_1$.*

Proof. If $i = s$, then in view of the relations in $\mathcal{A}$ we can take $\alpha = \beta$ to be q, 1, or q^{-1} (depending on whether $j < t$ or $j = t$ or $j > t$). Similarly, if $j = t$, we can take $\alpha \in \{1, q^{\pm 1}\}$ and $\beta = 1$.

If $i < s$ and $j > t$, or if $i > s$ and $j < t$, then X_{ij} and X_{st} commute, and we can take $\alpha = 1$. On the other hand, one of $X_{it}X_{sj} - q^{\pm 1}X_{ij}X_{st}$ is a 2×2 quantum minor, and so we can take β to be q or q^{-1}.

Now suppose that $i < s$ and $j < t$. Then $X_{ij}X_{st} - qX_{it}X_{sj}$ is a quantum minor, and we can take $\beta = q$. But $X_{st}X_{ij} - q^{-1}X_{it}X_{sj}$ is also a quantum minor, so we have $X_{ij}X_{st} \equiv qX_{it}X_{sj} \equiv q^2 X_{st}X_{ij} \pmod{\mathcal{I}_1}$, and hence we can take $\alpha = q^2$.

The remaining case follows from the previous one by exchanging (i, j) and (s, t), and then the final statement of the lemma is clear. $\quad\square$

3.4. Proposition. *Assume that k is an infinite field. Then the $\mathcal{H}$-prime ideals of $\mathcal{O}_q(M_{m,n}(k))$ that contain $\mathcal{I}_1$ are precisely the ideals $P(I, J)$.*

Proof. By Lemma 3.2, we know that the ideals $P(I, J)$ are $\mathcal{H}$-prime. Consider an arbitrary $\mathcal{H}$-prime ideal P of $\mathcal{A}$ that contains $\mathcal{I}_1$. If all of the X_{ij} are in P then P must be the maximal ideal generated by the X_{ij}. In that case, $P = P(I, J)$, where $I = \{1, \ldots, m\}$ and $J = \{1, \ldots, n\}$. Hence, we may assume that not all X_{ij} are in P. Set

$$I = \{i \in \{1, \ldots, m\} \mid X_{ij} \in P \text{ for all } j\}$$
$$J = \{j \in \{1, \ldots, n\} \mid X_{ij} \in P \text{ for all } i\}.$$

We first show that $X_{ij} \in P$ if and only if $i \in I$ or $j \in J$. Certainly, if $i \in I$ or $j \in J$ then $X_{ij} \in P$, by the definition of I and J. Suppose that there exists an $X_{ij} \in P$ such that $i \notin I$ and $j \notin J$. Then there exists an index $s \neq i$ such that $X_{sj} \notin P$ and also there exists an index $t \neq j$ such that $X_{it} \notin P$. By Lemma 3.3, there is a nonzero scalar $\beta \in k$ such that $X_{ij}X_{st} - \beta X_{it}X_{sj} \in P$. Thus, $X_{ij} \in P$ would imply that $X_{it}X_{sj} \in P$. However, X_{it} and X_{sj} are $\mathcal{H}$-eigenvectors which, by Lemma 3.3, are normal modulo P. Hence, because P is $\mathcal{H}$-prime, $X_{it}X_{sj} \in P$ would imply $X_{it} \in P$ or $X_{sj} \in P$, contradicting the choices of s and t. Thus, we have established that $X_{ij} \in P$ if and only if $i \in I$ or $j \in J$. Now $P(I, J) \subseteq P$, and we need to establish equality.

Set $B := \mathcal{A}/P(I, J)$ and $\overline{P} = P/P(I, J)$, and note that B is a domain by Lemma 3.2. Write Y_{ij} for the image of X_{ij} in B. The claim just established implies that $Y_{ij} \notin \overline{P}$ if $i \notin I$ or $j \notin J$. Recall from Lemma 3.3 that the Y_{ij} scalar-commute among themselves.

Now, $I \neq \{1, \ldots, m\}$ and $J \neq \{1, \ldots, n\}$, since not all of the X_{ij} are in P. Let $s \in \{1, \ldots, m\} \setminus I$ and $t \in \{1, \ldots, n\} \setminus J$ be minimal, and consider the localization $C := B[Y_{st}^{-1}]$. Since $Y_{st} \notin \overline{P}$ there is an embedding of B into C, and $\overline{P}C$ is an $\mathcal{H}$-prime ideal of C such that $\overline{P}C \cap B = \overline{P}$.

Note that $Y_{ij} = 0$ if $i < s$ or $j < t$. If $i > s$ and $j > t$, then we have $Y_{st}Y_{ij} - qY_{sj}Y_{it} = 0$, so that $Y_{ij} = qY_{st}^{-1}Y_{sj}Y_{it}$ in C. Hence, C is generated as an algebra by $Y_{st}^{\pm 1}$ together with Y_{sj} for $j > t$ and Y_{it} for $i > s$. Thus, C is a homomorphic image of a localized multiparameter quantum affine space $\mathcal{O}_\lambda(k^r)[z_1^{-1}]$, for $r = m - s + n - t + 1$ and for a suitable parameter matrix λ.

The standard action of the torus $\mathcal{H}_r := (k^\times)^r$ on $\mathcal{O}_\lambda(k^r)$ has 1-dimensional eigenspaces generated by individual monomials (here, we use the fact that k is infinite). Therefore, the same holds for C. Hence, any nonzero $\mathcal{H}_r$-invariant ideal of C contains a monomial, and so any nonzero $\mathcal{H}_r$-prime ideal of C must contain one of $Y_{s+1,t}, \ldots, Y_{mt}, Y_{s,t+1}, \ldots, Y_{sn}$. Since $\overline{P}C$ contains none of these elements, to show that $\overline{P}C = 0$ it suffices to establish that $\overline{P}C$ is $\mathcal{H}_r$-prime. But $\overline{P}C$ is already $\mathcal{H}$-prime, so it is enough to see that the $\mathcal{H}_r$-invariant ideals of C are the same as the $\mathcal{H}$-invariant ideals. This will follow from showing that the images of $\mathcal{H}$ and $\mathcal{H}_r$ in $\text{aut}\, C$ coincide.

Since the Y_{ij} are $\mathcal{H}$-eigenvectors, it is clear that the image of $\mathcal{H}$ is contained in the image of $\mathcal{H}_r$. The reverse inclusion amounts to the following statement:

(*) Given any $\alpha_s, \ldots, \alpha_m, \beta_{t+1}, \ldots, \beta_n \in k^\times$, there exists $h \in \mathcal{H}$ such that $h(Y_{it}) = \alpha_i Y_{it}$ for $i = s, \ldots, m$ and $h(Y_{sj}) = \beta_j Y_{sj}$ for $j = t+1, \ldots, n$.

Now, there exists $h_1 \in \mathcal{H}$ such that $h_1(X_{ij}) = X_{ij}$ for all i, j with $i < s$, and $h_1(X_{ij}) = \alpha_i X_{ij}$ for all i, j with $i \geq s$. Also, there exists $h_2 \in \mathcal{H}$ such that $h_2(X_{ij}) = X_{ij}$ for all i, j with $j \leq t$ and $h_2(X_{ij}) = \alpha_s^{-1}\beta_j X_{ij}$ for all i, j with $j > t$. Setting $h = h_1 h_2$ gives the desired element of $\mathcal{H}$, establishing (*).

Therefore, $\overline{P}C = 0$, and so $\overline{P} = 0$. This means that $P = P(I, J)$. $\square$

3.5. Corollary. *If the field k is infinite, then $\mathcal{O}_q(M_{m,n}(k))/\mathcal{I}_1$ has precisely $(2^m - 1)(2^n - 1) + 1$ distinct $\mathcal{H}$-prime ideals, all of which are completely prime. Further, each $\mathcal{H}$-stratum of $\text{spec}\, \mathcal{O}_q(M_{m,n}(k))/\mathcal{I}_1$ is homeomorphic to the prime spectrum of a Laurent polynomial ring over an algebraic field extension of k.*

Proof. The first statement is clear from Proposition 3.4. The second statement is not actually a corollary of the Proposition, but is included to fill in the picture. It may be obtained from [1, Theorems 5.3, 5.5] (all but the algebraicity of the coefficient fields also follows from [4, Theorem 6.6]). $\square$

3.6. In particular, the corollary above explains why in the algebra $\mathcal{O}_q(M_2(k))$ there are precisely $10 = (2^2 - 1)^2 + 1$ distinct $\mathcal{H}$-primes which contain the quantum determinant. This fact was known previously by direct enumeration of these primes. The remaining $\mathcal{H}$-primes correspond to $\mathcal{H}$-primes of $\mathcal{O}_q(GL_2(k))$; there are 4 of these, as has long been known. We can display the lattice of $\mathcal{H}$-prime ideals of $\mathcal{O}_q(M_2(k))$ as in the diagram below, where the symbols $\bullet$ and $\circ$ stand for generators X_{ij} which are or are not included in a given prime, and $\square$ stands for the 2×2 quantum determinant. For example, $\left(\begin{smallmatrix}\circ&\bullet\\\bullet&\circ\end{smallmatrix}\right)$ stands for the ideal $\langle X_{12}, X_{21}\rangle$, and $(\square)$ stands for the ideal $\langle X_{11}X_{22} - qX_{12}X_{21}\rangle$.

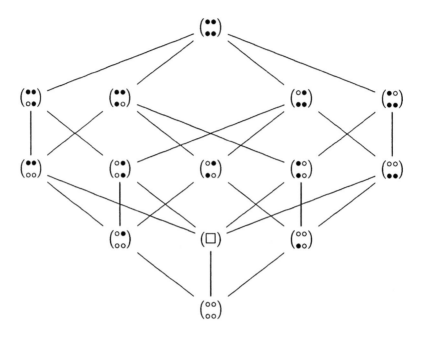

$\mathcal{H}$-spec $\mathcal{O}_q(M_2(k))$

The corresponding $\mathcal{H}$-strata in spec $\mathcal{O}_q(M_2(k))$ can be easily calculated. For instance, if q is not a root of unity, the strata corresponding to $\left(\begin{smallmatrix} \circ\circ \\ \circ\circ \end{smallmatrix}\right)$ and $\left(\begin{smallmatrix} \circ\bullet \\ \bullet\circ \end{smallmatrix}\right)$ are 2-dimensional, the strata corresponding to $\left(\begin{smallmatrix} \bullet\bullet \\ \circ\bullet \end{smallmatrix}\right)$, $\left(\begin{smallmatrix} \bullet\bullet \\ \bullet\circ \end{smallmatrix}\right)$, $\left(\begin{smallmatrix} \circ\bullet \\ \bullet\bullet \end{smallmatrix}\right)$, and $\left(\begin{smallmatrix} \bullet\circ \\ \bullet\bullet \end{smallmatrix}\right)$ are all 1-dimensional, and the remaining 8 strata are singletons.

3.7. We close with some remarks concerning catenarity. (Recall that the prime spectrum of a ring A is *catenary* provided that for any comparable primes $P \subset Q$ in spec A, all saturated chains of primes from P to Q have the same length.) It is conjectured that spec $\mathcal{O}_q(M_{m,n}(k))$ is catenary. In [**2**, Theorem 1.6], we showed that catenarity holds for any affine, noetherian, Auslander-Gorenstein, Cohen-Macaulay algebra A with finite Gelfand-Kirillov dimension, provided spec A has normal separation. All hypotheses but the last are known to hold for the algebra $\mathcal{A} = \mathcal{O}_q(M_{m,n}(k))$. We can, at least, say that the portion of spec $\mathcal{A}$ above $\mathcal{I}_1$ – that is, spec $\mathcal{A}/\mathcal{I}_1$ – is catenary: In view of Lemma 3.3, $\mathcal{A}/\mathcal{I}_1$ is a homomorphic image of a multi-parameter quantum affine space $\mathcal{O}_\lambda(k^{n^2})$, and spec $\mathcal{O}_\lambda(k^{n^2})$ is catenary by [**2**, Theorem 2.6].

REFERENCES

1. K. R. Goodearl, *Prime spectra of quantized coordinate rings*, This volume.
2. K. R. Goodearl and T. H. Lenagan, *Catenarity in quantum algebras*, J. Pure and Applied Algebra **111** (1996), 123-142.
3. K. R. Goodearl and T. H. Lenagan, *Quantum determinantal ideals*, Preprint, 1997.
4. K. R. Goodearl and E. S. Letzter, *The Dixmier-Moeglin equivalence in quantum coordinate rings and quantized Weyl algebras*, Trans. Amer. Math. Soc. (to appear).
5. J. C. McConnell and J. C. Robson, *Noncommutative Noetherian Rings*, Wiley-Interscience, Chichester-New York, 1987.
6. B. Parshall and J.-p. Wang, *Quantum linear groups*, Memoirs Amer. Math. Soc. **89** (1991).
7. L. Rigal, *Normalité de certains anneaux déterminantiels quantiques*, Proc. Edinburgh Math. Soc. (to appear).

Noetherian Semigroup Algebras: A Survey

Eric Jespers

Department of Mathematics
Vrije Universiteit Brussel
Pleinlaan 2
1050 Brussels
Belgium

Jan Okniński *

Institute of Mathematics
Warsaw University
Banacha 2, 02-097 Warsaw, Poland

Natural classes of Noetherian rings include enveloping algebras of finite dimensional Lie algebras, crossed product algebras of polycyclic-by-finite groups and finitely generated commutative algebras. The algebraic structure of these algebras is a topic of ongoing research. Although many aspects are well understood, some problems remain unsolved, such as for example a characterization of when a group algebra is Noetherian.

In the search for other classes of Noetherian rings it is natural to look at semigroup algebras. If S is an Abelian monoid and K is a field, then it is well known that the semigroup algebra $K[S]$ is Noetherian if and only if S is finitely generated ([17]). Furthermore the algebraic structure of these algebras has been well investigated. The attention is now focussed on the noncommutative semigroup algebras. In general, at present, the best one could aim for is to find a characterization of when a semigroup algebra $K[S]$ is right Noetherian in terms of Noetherian group algebras $K[G]$, where G runs through a set of groups "determined" by the semigroup S. A more realistic aim is to describe when some more concrete classes of algebras are right Noetherian. A first class is that of semigroup algebras $K[S]$ of submonoids S of a polycyclic-by-finite group. A second class is that of finitely presented algebras which are defined via homogeneous relations. Instead of focusing on special classes of semigroups, one can also investigate special classes of Noetherian semigroup algebras, such as for example those that are principal right ideal rings or those that are Noetherian maximal orders. Finally the algebraic structure of Noetherian semigroup algebras should be investigated, in particular the prime and primitive spectra.

In this survey paper we report on some recent results. The reader is referred to [36] for other results on semigroup algebras, and to [8] for background on semigroups in general. Throughout K will denote a field.

*This work is supported in part by research grant OGP0036631, NSERC Canada, Onderzoeksraad Vrije Universiteit Brussel, and a KBN research grant 2P03A 003 12, Poland. The second author would like to thank Memorial University of Newfoundland for the warm hospitality while being a visiting scientist during the academic year 1997-1998.
1991 Mathematics Subject Classification: Primary 16S36, 16P40, 20C07, 20M25, Secondary 16P60, 20F22 and 16S15.

1   Submonoids of polycyclic-by-finite groups

In this section we describe when a semigroup algebra $K[S]$ of a submonoid of a polycyclic-by-finite group G is Noetherian. Since $K[S]$ has a natural G-gradation, the following result reduces the problem at once to the semigroup S.

Theorem 1.1 (Quinn [43]) *Let $R = \oplus_{g \in G} R_g$ be a ring (with identity) which is graded by a polycyclic-by-finite group G. Then R is right Noetherian if and only if R is graded right Noetherian, that is. R satisfies the ascending chain condition on right ideals L with $L = \sum_{g \in G} L \cap R_g$.*

Corollary 1.2 ([36]) *Let S be a submonoid of a polycyclic-by-finite group. Then $K[S]$ is right Noetherian if and only if S satisfies the ascending chain condition on right ideals. Moreover, in this case, S is finitely generated.*

In general, the following remains an unsolved problem.

Problem 1: Let S be an arbitrary semigroup. Is S necessarily finitely generated if the semigroup algebra $K[S]$ is right Noetherian.

Okniński [36] showed that the answer is affirmative if either $K[S]$ is left and right Noetherian, or if $K[S]$ is right Noetherian and satisfies a polynomial identity. Note, that for semiprime rings R satisfying a polynomial identity being left Noetherian is equivalent to being right Noetherian, and it is also equivalent to having the ascending chain condition on two-sided ideals.

We note that it is easily seen that if $K[S]$ is right Noetherian, then S satisfies the ascending chain condition on right congruences. Special classes of semigroups with the latter property have been investigated by Hotzel in [20].

In order to obtain a characterization of submonoids of polycyclic-by-finite groups satisfying the ascending chain condition on right ideals the following elementary lemmas are crucial. For a monoid S we denote by $\mathcal{U}(S)$ the subgroup of invertible elements of S. For a subset A of S we denote by $\langle A \rangle$ the submonoid generated by A. If, furthermore, S is a group, then $gr(A)$ denotes the subgroup generated by A.

Lemma 1.3 *Let S be a submonoid of a group G. If S satisfies the ascending chain condition on right ideals, then*

1. *for any $a, b \in S$, there exists a positive integer n so that $a^{-1}b^n a \in S$. and furthermore, $a^{-1}b^n a \in \mathcal{U}(S)$ if $b \in \mathcal{U}(S)$.*

2. *S satisfies the right Ore condition and S has a group SS^{-1} of right quotients.*

Lemma 1.4 *Let H be subgroup of finite index in a group G and let S be a submonoid of G so that $G = gr(S)$. Then, $K[S]$ is right Noetherian if and only if $K[S \cap H]$ is right Noetherian, and in this case $K[S]$ is a Noetherian right $K[S \cap H]$-module.*

In particular, if G is a polycyclic-by-finite group, then S satisfies the ascending chain condition on right ideals if and only if $S \cap H$ does. Moreover, in this case, H is the the group of right quotients of $S \cap H$.

It is well know that a polycyclic-by-finite group G has a torsion-free normal subgroup H of finite index which is a linear group and has nilpotent commutator subgroup H'. Hence, to deal with submonoids of polycyclic-by-finite groups, we will first deal with cancellative nilpotent semigroups. We recall some definitions.

Let x, y be elements of a semigroup S and let $w_1, w_2, \ldots$ be elements of S^1. Consider the sequence of elements of S defined inductively as follows:

$$x_0 = x, \qquad y_0 = y,$$

and for $n \geq 0$

$$x_{n+1} = x_n w_{n+1} y_n, \qquad y_{n+1} = y_n w_{n+1} x_n.$$

We say the identity $X_n = Y_n$ is satisfied in S if $x_n = y_n$ for all $x, y \in S$ and $w_1, w_2, \ldots$ in S^1. A semigroup S is said to be *(Malcev) nilpotent of class n* if S satisfies the identity $X_n = Y_n$ and n is the least positive integer with this property. Note that, as in [31], the condition $X_n = Y_n$ is a bit stronger than the one required by Malcev in [34] where it is required that the elements w_i belong to S only. However the definitions agree on the class of cancellative semigroups.

It is well known [34] that a group G is nilpotent of class n (in the above sense) if and only if G is nilpotent of class n in the classical sense. Moreover, a cancellative semigroup S is nilpotent of class n if and only if S has a two-sided group of quotients which is nilpotent of class n (see also [36]). Another class of nilpotent semigroups are the power nilpotent semigroups, that is, semigroups S with zero θ such that $S^m = \{\theta\}$ for some $m \geq 1$. A third natural class of nilpotent semigroups is that of the inverse semigroups S of matrix type over a nilpotent group G (in other words, inverse completely 0-simple semigroups over nilpotent groups). So $S = \mathcal{M}^0(G, M, M, \Delta_M)$, the set of all $M \times M$ matrices (M an indexing set) over G^0 (the group G adjoined with a zero element θ) with at most one non-zero entry, and with the usual matrix multiplication; here Δ_M denotes the $M \times M$ identity matrix. The semigroup S satisfies the identity $X_{n+2} = Y_{n+2}$, where n is the nilpotency class of G. For more examples of nilpotent semigroups we refer to [23]. Some of the latter yield examples of Noetherian semigroup algebras.

Gromov [19] showed the following result for groups and later Grigorchuk [18] extended it to semigroups. For an algebra R we denote by $GK(R)$ the Gelfand-Kirillov dimension of R.

Theorem 1.5 (Gromov [19] and Grigorchuk [18]) *Let S be a finitely generated cancellative semigroup, and let K be a field. The following conditions are equivalent:*

1. *$GK(K[S])$ is finite.*

2. *S is almost nilpotent, that is, the quotient group of S is nilpotent-by-finite.*

We now describe when a semigroup algebra of a submonoid of a nilpotent-by-finite group is right Noetherian.

Theorem 1.6 ([26]) *Let S be a submonoid of a finitely generated nilpotent-by-finite group. Then S satisfies the ascending chain condition on right ideals if and only if the quotient group G of S has a normal subgroup H of finite index and a normal subgroup $F \subseteq H$ such that $S \cap H$ is a finitely generated semigroup, $F \subseteq S$ and H/F is Abelian.*

Corollary 1.7 *Let S be a submonoid of a finitely generated nilpotent-by-finite group. Then the following conditions are equivalent:*

1. *S satisfies the ascending chain condition on right ideals;*

2. *S satisfies the ascending chain condition on left ideals;*

3. *$K[S]$ is right Noetherian;*

4. *$K[S]$ is left Noetherian.*

Furthermore, if any of these equivalent conditions holds, then S is finitely generated.

The following example shows that in Theorem 1.6 one cannot weaken the condition that $S \cap H$ is finitely generated to S being finitely generated. Let G be the group generated by a, b and c subject to the following relations: $ca = bc$, $cb = ac$, $ab = ba$ and $c^2 = 1$. Clearly the subgroup A generated by a and b is Abelian, normal in G and of finite index. Let S be the submonoid of G generated by a and ac. It is easily verified that $S \cap A = \{a^i \mid i \geq 0\} \cup \{a^k b^l \mid l \geq 1, \, k \geq 1\}$. Since $a^n b \in S \cap A$, for any $n \geq 1$, it follows that $S \cap A$ is not finitely generated. So $K[S \cap A]$ is not right Noetherian. Because of Lemma 1.4 it follows that $K[S]$ is not right Noetherian.

In general one cannot expect the full unit group $\mathcal{U}(S)$ to be normal in G if S satisfies the ascending chain condition on right ideals. Indeed, let $G = Z \times F$, where $Z = gr(z)$ is an infinite cyclic group and F is a finite group with a non-normal subgroup H. Let $S = \langle zF, H \rangle$. Clearly, $S \cap Z = \langle z \rangle$ is finitely generated and commutative. Hence, S satisfies the ascending chain condition on left and right ideals. On the other hand, $\mathcal{U}(S) = H$ is not a normal subgroup of the group $G = SS^{-1}$.

We note that even if S is a finitely generated submonoid of a nilpotent group one does not obtain, in general, that the semigroup algebra $K[S]$ is left or right Noetherian. Indeed, let G be the group generated by a, b and c such that c is central and $ab = bac$; that is, G is the Heisenberg group. Let S be the submonoid generated by a and b. Using the principal right ideals $a^n bS$ one can construct easily a strict increasing chain of right ideals of S. So $K[S]$ is not right Noetherian. Similarly one shows $K[S]$ is not left Noetherian.

A characterization of Noetherian semigroup algebras of arbitrary Malcev nilpotent semigroups is yet unknown. Hence the following problem.

Problem 2: Characterize right Noetherian semigroup algebras $K[S]$ of arbitrary Malcev nilpotent semigroups.

That such semigroup algebras are necessarily finitely generated has been established. We state this result in a more general context. For an algebra R we denote by $\mathcal{B}(R)$ the prime radical of R. Recall that if the latter is nilpotent (for example when R is Noetherian) then $GK(R) < \infty$ if and only if $GK(R/\mathcal{B}(R)) < \infty$.

Theorem 1.8 ([26, 38]) *Assume that $K[S]$ is right Noetherian. Then S is finitely generated if $GK(K[S])$ is finite. In particular this is the case if S is a Malcev nilpotent semigroup.*

The second part of the statement follows from Theorem 1.5 and [38]. In the latter it was shown that $GK(K[S]/\mathcal{B}(K[S])) = GK(K[T])$, for some cancellative subsemigroup T of S. Though, with little evidence, one might even ask the following question.

Problem 3: Is $GK([S]) = GK(K[S]/\mathcal{B}(K[S]))$ for all right Noetherian semigroup algebras $K[S]$?

Using the description in Theorem 1.6 of Noetherian semigroup algebras of nilpotent cancellative semigroups, one can deal with left and right Noetherian semigroup algebras of submonoids of polycyclic-by-finite groups.

Theorem 1.9 ([27]) *Let S be a submonoid of a polycyclic-by-finite group. The following conditions are equivalent.*

1. *$K[S]$ is left and right Noetherian.*

2. *S satisfies the ascending chain condition on left and right ideals.*

3. *S has a group of quotients $G = SS^{-1}$ which contains a normal subgroup H of finite index and a normal subgroup $F \subseteq H$ such that $S \cap H$ is a finitely generated semigroup, $F \subseteq S$ and H/F is Abelian.*

There exists a finitely generated submonoid T of a solvable group such that T satisfies the ascending chain condition on left ideals, but its group of left quotients is not polycyclic-by-finite. However T is not a right Ore semigroup ([26]); in particular, T does not satisfy the ascending chain condition on right ideals. This shows that the ascending chain condition on left ideals does not imply that $K[T]$ is left Noetherian (for otherwise T has a group of quotients, say G, with $K[G]$ also Noetherian, and thus G satisfies the ascending chain condition on subgroups and therefore is polycyclic-by-finite, a contradiction).

Nevertheless one might pose the following question.

Problem 4: Is the left-right symmetric hypothesis in Theorem 1.9 required?

As a consequence of the structural characterization obtained in Theorem 1.9, we get

Corollary 1.10 *Let S be a submonoid of a polycyclic-by-finite group. If S satisfies the ascending chain condition on left and right ideals, then S is a finitely presented monoid. In particular. the semigroup algebra $K[S]$ is finitely presented.*

Corollary 1.11 *Let S be a submonoid of a polycyclic-by-finite group. If S satisfies the ascending chain condition on left and right ideals, then $K[S]$ is a Jacobson ring.*

2   Prime ideals of semigroup algebras

Let S be a submonoid of a polycyclic-by-finite group such that S satisfies the ascending chain condition on right ideals. Let $G = SS^{-1}$ be the group of right quotients of S. In this section we closely relate each prime ideal P of $K[S]$ to a prime ideal of a group algebra $K[H]$, where H is a group that is naturally associated to S and P.

For this let $\overline{S}$ be the natural image of the semigroup S in the classical ring of right quotients $Q_{cl} = Q_{cl}(K[S]/P)$ of the prime right Noetherian algebra $K[S]/P$. As $Q_{cl} \cong M_n(D)$, for some division algebra D, we therefore consider $\overline{S}$ as a (skew) linear semigroup. It has been shown by Okniński [26] and Salwa [44] (see also [39]) that such linear semigroups have an algebraic structure that, in some sense, resembles the Wedderburn-Artin decomposition for semisimple rings. More specifically, these semigroups are built of blocks that are uniform (that is, "large") subsemigroups of completely 0-simple semigroups (i.e., semigroups of the type $J = \mathcal{M}(G, X, Y, P)$ with G a group) and of some power nilpotent semigroups.

Recall that a subsemigroup U of the completely 0-simple semigroup J is said to be *uniform* if it intersects all the non-zero $\mathcal{H}$-classes of a completely 0-simple subsemigroup I of J. It is known [37] that there exists a smallest completely 0-simple subsemigroup $\hat{U}$ of J with $U \subseteq \hat{U} \subseteq I$. The semigroup $\hat{U}$ is called the *closure* of U in J.

It then follows from the structure theorem of skew linear semigroups that $\overline{S}$ has an ideal U contained in a completely 0-simple subsemigroup J of Q_{cl} in such a way that U is uniform in J and $J = \hat{U}$. Furthermore, the non-zero elements of U are the elements of minimal non-zero rank of $\overline{S}$ in $M_n(D)$. For a subset A of $M_n(D)$ we denote by $K\{A\}$ the K-subalgebra of $M_n(D)$ generated by A.

Since $K[S]$ is right Noetherian and $K[G]$ is a localization of $K[S]$, it is well known that for any twosided ideal I of $K[S]$, the extension $I^e = IK[G]$ is a twosided ideal of $K[G]$. Hence, the map $P \mapsto P^e$ determines a one-to-one correspondence between the set of prime ideals P of $K[S]$ that do not intersect S and the prime ideals of $K[G]$. Furthermore, for such a prime P, the ring $K[G]/P^e$ is a right ring of quotients of $K[S]/P$ and thus we get the equality of the classical rings of quotients $Q_{cl}(K[S]/P) = Q_{cl}(K[G]/P^e)$. We now extend these results to arbitrary primes of $K[S]$. For this, recall that a ring R is said to be a *generalised matrix ring* if

$$R = \oplus_{1 \leq i,j \leq n} R_{ij},$$

as additive subgroups, and $R_{ij}R_{kl} \subseteq \delta_{jk}R_{il}$.

If the identity of a ring R is the sum of orthogonal idempotents, say $e_1, \ldots, e_n$, then clearly $R = \oplus e_i R e_j$ is a generalised matrix ring. If S is a multiplicative subsemigroup of R such that every element $s \in S$, represented as a generalised matrix $s = (e_i s e_j)_{ij}$ in R, has at most one non zero entry in each row and column then we say S is in *block monomial form* with respect to the given idempotents. The Hirsch number of a polycyclic-by-finite group G is denoted by $h(G)$.

Proposition 2.1 ([27]) *Let S be a cancellative monoid satisfying the ascending chain condition on right ideals such that the group of right quotients $G = SS^{-1}$ is polycyclic-by-finite. Let P be a prime ideal in $K[S]$. If $P \cap S \neq \emptyset$, then*

1. *there exists an ideal I of S and a cancellative right Ore subsemigroup T of I such that the image $\overline{I}$ of I in $K[S]/P$ is a uniform subsemigroup in a completely 0-simple inverse subsemigroup $\hat{I}$ of the ring $Q_{cl}(K[S]/P)$ such that $1 = e_1 + \cdots + e_m$, for the non-zero idempotents e_i of $\hat{I}$.*

2. $h(TT^{-1}) < h(G)$.

3. *the image $R = K\{\overline{I}\}$ of $K[I]$ in $K[S]/P$ is a generalised matrix ring $R = \oplus_{i,j=1}^m R_{ij}$ with $R_{ij} = e_i Q_{cl}(K[S]/P)e_j \cap R$ and $R_{11} = K\{\overline{T}\}$.*

4. *the natural image $\overline{S}$ of S in $K[S]/P$ is in block monomial form with respect to $e_1, \ldots, e_m$.*

5. $R \subseteq R' \subseteq Q_{cl}(K[S]/P)$, where $R' \cong M_m(K[TT^{-1}]/Q)$ is a localization of R and Q is a prime ideal in $K[TT^{-1}]$. Therefore $Q_{cl}(K[S]/P)$ is also the right quotient ring of $M_m(K[TT^{-1}]/Q)$.

If in the previous proposition G is also torsion-free, then $K[S]$ is a domain (see for example [42, Theorem 37.5]). For these algebras, the following application of Proposition 2.1 and Theorem 1.9 shows that the height one prime ideals intersecting S are homogeneous for the natural G-gradation. By $X^1(K[S])$ we denote the set of height one prime ideals of $K[S]$ and by $X_h^1(K[S])$ the set of height one prime ideals of $K[S]$ intersecting S. The set of all prime ideals of S is denoted $spec(S)$, the set of all minimal prime ideals of S is denoted $X^1(S)$ and $dim(S)$ denotes the supremum of the lengths of chains of prime ideals of S.

Proposition 2.2 ([27]) *Let S be a cancellative monoid satisfying the ascending chain condition on right ideals. If the group of right quotients $G = SS^{-1}$ is a torsion-free polycyclic-by-finite group, then*

1. *$K[S \cap P]$ is a prime ideal in $K[S]$ for any prime ideal P in $K[S]$ with $P \cap S \neq \emptyset$.*

2. *$K[Q]$ is a prime ideal in $K[S]$ for any prime ideal Q in S.*

3. *the height one prime ideals of $K[S]$ intersecting S are of the form $K[Q]$ where Q is a minimal prime ideals of S. If, moreover, S satisfies the ascending chain condition on left ideals, then any ideal of the from $K[Q]$. with Q a minimal prime in S, is a height one prime ideal of $K[S]$.*

It is well known that a minimal prime ideal of a ring R graded by a torsion-free Abelian group is homogeneous. (In [22] it is shown that this result even holds for rings graded by a unique product group; we refer to [41] for the definition.) is homogeneous. Hence, if, moreover, R is a domain, then the height one prime ideals that contain a non-zero homogeneous element are also homogeneous. The previous proposition is an extension to a non-Abelian context.

For any semigroup S we define $rk(S)$ as the supremum of the ranks of the free Abelian subsemigroups contained in S. If S is a submonoid of a polycyclic-by-finite group G, then $rk(S) \leq h(G) < \infty$. For a ring R we denote by $clKdim(R)$ the classical Krull dimension of R. Note that $clKdim(K[S]) = rk(S)$ for any right Noetherian semigroup algebra $K[S]$ satisfying a polynomial identity ([36]).

Corollary 2.3 *Let S be a cancellative monoid satisfying the ascending chain condition on right and left ideals such that $G = SS^{-1}$ is polycyclic-by-finite. The following properties hold:*

1. *every prime ideal P of $K[S]$ intersecting S contains a prime ideal that belongs to $X_h^1(K[S])$, and $X_h^1(K[S])$ is a finite set.*

2. $clKdim(K[S]) = clKdim(K[F]) + rk(G/F) = clKdim(K[\mathcal{U}(S)]) + rk(G/F)$, where F is a normal subgroup of G such that $F \subseteq S$, F is of finite index in $\mathcal{U}(S)$ and G/F is abelian-by-finite.

3. If $P \in X_h^1(K[S])$, then $clKdim(K[S]/P) = clKdim(K[S]) - 1$ and $h(TT^{-1}) = h(G) - 1$, where T is the Ore subsemigroup of S associated to P in Proposition 2.1.

Proposition 2.1 relates the prime images of $K[S]$ to prime images of group algebras of polycyclic-by-finite groups G. The latter have been extensively studied and are rather well known (see for example [41, 42]). In particular Roseblade proved that $clKdim(K[G]) = pl(G) \leq h(G)$, where $pl(G)$ denotes the plinth length of G. Consequently, one obtains several applications of Proposition 2.1 to prime ideals of semigroup algebras. As an example we give the following result.

Corollary 2.4 *Let S be a cancellative monoid satisfying the ascending chain condition on left and right ideals such that the group of right quotients $G = SS^{-1}$ is a polycyclic-by-finite group. If K is an absolute field, then every irreducible $K[S]$-module is finite dimensional over K.*

Prime ideals of arbitrary semigroup algebras are in general not well understood. However, we give two more cases in which a description has been obtained. The first one is that of prime Goldie semigroup algebras and the second is that of prime ideals P of semigroup algebras $K[S]$ of arbitrary nilpotent semigroups so that $K[S]/P$ is Goldie.

Concerning the former, in [49] Zelmanov described prime semigroup algebras that satisfy a polynomial identity. In [26] this has been extended to arbitrary prime Goldie semigroup algebras. The result has a similar flavour as the Faith - Utumi theorem.

Theorem 2.5 ([26]) *Let K be a field and S a semigroup. Consider the following conditions:*

1. *the contracted semigroup algebra $K_0[S]$ is prime left and right Goldie;*

2. *S does not contain non-Abelian free subsemigroups and $K_0[S]$ is prime right Goldie;*

3. *S contains a right ideal A and a left ideal B such that AB and BA are both uniform subsemigroups in a common completely 0-simple semigroup with equal closures, $AB = \mathcal{M}(T, r, r, P)$, where T is a cancellative semigroup such that $K[T]$ is prime left and right Goldie, and the right and left annihilators of the contracted semigroup ring $K_0[AB]$ in $K_0[S]$ are trivial.*

Then, (1) and (3) are equivalent, and (2) implies (3).

Let us now consider prime ideals of semigroup algebras of arbitrary (Malcev) nilpotent semigroups. The next result shows that some prime homomorphic images are Goldie.

For an ideal I of a semigroup algebra $K[S]$ one denotes by $\sim_I$ the congruence relation on S defined by $s \sim_I t$ if $s - t \in I$.

Proposition 2.6 ([23]) *Let S be a nilpotent semigroup, and let P be a prime ideal of $K[S]$ such that $S \setminus (S \cap P)$ is a subsemigroup of S. Then the following conditions hold:*

1. *$S/\sim_P$ is a 0-cancellative semigroup.*

2. *there exists a prime ideal Q of $K[G]$, G the quotient group of $(S/\sim_P)\setminus\{\theta\}$, such that $K[G]/Q$ is isomorphic to a localization of $K[S]/P$.*

3. *if, moreover, S is finitely generated, then $K[S]/P$ is a Goldie ring of finite Gelfand-Kirillov dimension.*

For semiprime algebras of finite Gelfand-Kirillov dimension verifying the Goldie conditions is often simpler. This is because of the result of Irving and Small [21] which states that a semiprime algebra A with the ascending chain condition on right annihilator ideals and which does not contain non-commutative free subalgebras is left and right Goldie.

In Proposition 2.1 it is shown that prime ideals P of semigroup algebras $K[S]$ of submonoids S of polycyclic-by-finite groups are "determined" by the ideal of $\overline{S}$ consisting of the elements of minimal rank in the classical ring of quotients of $K[S]/P$; this provided that $P \cap S \neq \emptyset$. In Theorem 2.5 it is shown that a similar statement holds for prime Goldie (contracted) semigroup algebras of arbitrary semigroups. In the next result it is stated that this "flavour" remains valid for arbitrary prime ideals P of semigroup algebras of nilpotent semigroups such that $K[S]/P$ is right Goldie.

Theorem 2.7 ([23]) *Let S be a nilpotent semigroup and P a prime ideal of $K[S]$ such that $K[S]/P$ is right Goldie with classical ring of right quotients $M_n(D)$, D a division ring. Then the semigroup $S/\sim_P$ has an ideal chain*

$$S/\sim_P = I_r \supseteq I_{r-1} \supseteq \cdots \supseteq I_1 = I \supseteq I_0,$$

where $I_0 = \{\theta\}$ if S has a zero element, otherwise $I_0 = \emptyset$. such that:

1. *each Rees factor I_j/I_{j-1}, $1 \leq j \leq r$, is either a power nilpotent semigroup or a uniform semigroup in a completely 0-simple inverse semigroup.*

2. *I is uniform in a completely 0-simple inverse subsemigroup $\hat{I}$ of $M_n(D)$ with finitely many non-zero idempotents, say q, and $\hat{S} = (S/\sim_P) \cup \hat{I}$ is a nilpotent subsemigroup of $M_n(D)$. We denote by G a maximal subgroup of $\hat{I}$. Furthermore, $r \leq 2^q + n - 2$.*

3. $K\{I\} \subseteq K[S]/P \subseteq K\{\hat{I}\}$, where $K\{\hat{I}\}$ denotes the subalgebra of $M_n(D)$ generated by $\hat{I}$. Further $M_n(D)$ is the common classical ring of quotients of these three algebras, and $K\{\hat{I}\}$ is a localization of $K\{I\}$ with respect to an Ore set.

4. There exists a prime ideal Q of $K[G]$ such that $K[G]/Q$ is a Goldie ring and

$$M_q(K[G]/Q) \cong K\{\hat{I}\}.$$

As an application we obtain that the Jacobson and prime radicals coincide. To do so one uses the description of the respective radicals for subsemigroups of nilpotent groups.

Theorem 2.8 *Let S be a subsemigroup of a nilpotent group G. Then*

1. *(Okninski [36, Corollary 11.5]) $\mathcal{J}(K[S]) = \mathcal{B}(K[S]) = \mathcal{J}(K[G]) \cap K[S]$.*

2. *(Dyment - Zalesskiı, see for example [41, Theorem 8.4.16]) If $char(K) = p > 0$, then $\mathcal{J}(K[G])$ is the K-subspace spanned by the elements $s - t$, where $s^{p^k} = t^{p^k}$ for some $k \geq 0$. If $char(K) = 0$, then $\mathcal{J}(K[G]) = \{0\}$.*

Proposition 2.9 ([23]) *Let S be a nilpotent semigroup such that for each prime ideal P of $K[S]$ either $K[S]/P$ is right Goldie or $S \setminus (S \cap P)$ is a subsemigroup of S. Then*

$$\mathcal{J}(K[S]) = \mathcal{B}(K[S]).$$

Also a description of the radical congruence $\sim_{\mathcal{B}(K[S])}$ can be obtained. This reduces the problem of describing the radical of a semigroup algebra $K[S]$ of a nilpotent semigroup S to the case where S is embedded in $K[S]/\mathcal{J}(K[S])$.

Proposition 2.10 ([23]) *Let S be a nilpotent semigroup and assume $K[S]$ is right Noetherian. Then, for $s, t \in S$, the following conditions are equivalent:*

1. $s \sim_{\mathcal{B}(K[S])} t$

2. *for every $u \in S$ there exists $n \geq 1$ and $v \in \langle su, tu \rangle$ such that*

   (a) $(su)^n \in \langle su, tu \rangle tu \langle su, tu \rangle$,

   (b) $(tu)^n \in \langle su, tu \rangle su \langle su, tu \rangle$,

   (c) *if $char(K) = p > 0$, then $(v(su)v)^{p^k} = (v(tu)v)^{p^k}$ for some $k \geq 0$.*

   (d) *if $char(K) = 0$, then $v(su)v = v(tu)v$.*

3 Finitely presented semigroup algebras

In general it is unknown whether a right Noetherian semigroup algebra $K[S]$ is finitely presented, and as said in Section 1, it is even an open problem whether S is necessarily finitely generated. However, as mentioned earlier, if S is a submonoid of a polycyclic-by-finite group and S satisfies the ascending chain condition on left and right ideals, then we know that indeed the algebra is finitely presented. In [12, 13] Gateva-Ivanova investigates finitely presented algebras defined via quadratic relations. If one restricts these to homogeneous relations then more examples of finitely presented semigroup algebras are obtained. In the more recent papers [14, 15] Gateva-Ivanova introduced monoids S defined via a set of generators $X = \{x_1, \ldots, x_n\}$ and a set of $n(n-1)/2$ quadratic relations (one for each $n \geq j > i \geq 1$)

$$x_j x_i = x_{i'} x_{j'},$$

satisfying the following conditions

(B1.) $i' < j'$ and $i' < j$;

(B2.) as we vary (i, j), every pair (i', j') occurs exactly once;

(B3.) the overlaps $(x_k x_j)x_i = x_k(x_j x_i)$ do not give rise to new relations in S.

Conditions $(B1.)$ and $(B3.)$ imply that, as sets,

$$S = \{x_1^{\alpha_1} \cdots x_n^{\alpha_n} \mid \alpha_i \in \mathbf{N}, \ 1 \leq i \leq n\}.$$

For a field K, the semigroup algebra $K[S]$ is called a *binomial skew polynomial ring*. Condition $(B2.)$ is necessary and sufficient for $K[S]$ to be Noetherian. assuming that $(B1.)$ and $(B3.)$ are satisfied, ([14, 15])

In analogy with the terminology for the semigroup algebra, in [25] such a semigroup S is called a *binomial semigroup*.

As explained in [14, 15], binomial skew polynomial rings are a restricted class of skew polynomial rings with quadratic relations considered in earlier work by Artin and Schelter [3]. It is also shown in [14, 15] that $(B2.)$ is necessary and sufficient for the existence of a finite Gröbner basis of every one sided ideal of the binomial skew polynomial ring $K[S]$ (assuming $(B1.)$ and $(B3.)$). The proofs given for all these results are of a combinatorial nature and mainly are about the semigroup S.

Gateva-Ivanova and Van den Bergh show in [16] that binomial semigroups are also intimately connected with the following three mathematical notions: (i) set theoretic solutions of the Yang-Baxter equation [11], (ii) Bieberbach groups [5] and (iii) rings of I-type studied by Tate and Van den Bergh in [47]. It is shown in [16] that binomial semigroups S are of I-type, and thus binomial skew polynomial rings $K[S]$ are algebras of I-type.

Recall that a semigroup $S = \langle x_1, \ldots, x_n \rangle$ is a *semigroup of I-type* if there exists a bijective map $v : \mathcal{V} \to S$, where $\mathcal{V}$ is the free Abelian monoid on the free generators $u_1, \ldots, u_n$, so that $v(1) = 1$ and such that for all $a \in \mathcal{V}$.

$$\{v(u_1 a), \ldots, v(u_n a)\} = \{x_1 v(a), \ldots, x_n v(a)\}.$$

It follows (Lemma 4.1 in [16]) that the map $s \mapsto sv(a)$ for a given $a \in \mathcal{V}$ induces a bijection between S and $\{v(ua) \mid u \in \mathcal{V}\}$. So in particular, S is a right cancellative semigroup. Since also the opposite semigroup is binomial we get that S is left cancellative as well.

The main result in the work of Gateva-Ivanova and Van den Bergh is the following.

Theorem 3.1 (Gateva-Ivanova and Van den Bergh [16]) *Let S be a binomial semigroup. Then the semigroup algebra $K[S]$ is a Noetherian domain satisfying a polynomial identity and*

1. *$K[S]$ has finite global dimension,*

2. *$K[S]$ is Koszul,*

3. *$K[S]$ satisfies the Auslander condition.*

4. *$K[S]$ is Cohen-Macaulay.*

Stafford and Zhang in [46] showed that if R is a positively graded, connected, Noetherian algebra satisfying a polynomial identity and of finite global dimension then R is a domain and a maximal order. Hence because of the previous theorem a semigroup algebra $K[S]$ of a binomial semigroup is also a maximal order. Hence by Propositions 13.9.8 and 13.9.11 in [35], $K[S]$ is a finite module over its (finitely generated) centre.

In [25] binomial semigroups S have been investigated from a structural point of view. The main method used to do so is of a combinatorial nature (except for proving that S is embedded in a torsion-free group; for this, one relies on the theorem stated above).

In order to state these results one first shows that there exists a positive integer $p_1 | (n - 1)!$ so that

$$\{x_1^{p_1} x_i, \ldots, x_{i-1}^{p_1} x_i\} = \{x_i x_1^{p_1}, \ldots, x_i x_{i-1}^{p_1}\}$$

and

$$\{x_{i+1}^{p_1} x_i, \ldots, x_n^{p_1} x_i\} = \{x_i x_{i+1}^{p_1}, \ldots, x_i x_n^{p_1}\},$$

for any i. It follows that for some p with $p_1 | p$ and $p | (n - 1)!$, the monoid $A = \langle x_1^p, \ldots, x_n^p \rangle$ is free Abelian of rank n and $S = \bigcup_{f \in F} Af = \bigcup_{f \in F} fA$, where F is a finite set so that $Af \cap Af' = \emptyset$ for $f \neq f'$. Hence, the semigroup algebra $K[S]$ is a finite (left and right) module over the polynomial algebra

$K[A] = K[x_1^p, \ldots, x_n^p]$. So $K[S]$ is a left and right Noetherian and satisfies a polynomial identity. Consequently, S has a (two-sided) group of quotients G. Actually, $G = \{z^{-k}s \mid s \in S, k \in \mathbf{N}\}$, where $z = x_1^p \cdots x_n^p$ is central in S and thus also in G. Note that the above also shows that each element of S acts by conjugation on the $\{x_1^p, \ldots, x_n^p\}$. Furthermore x_i acts by conjugation on the sets $\{x_{i+1}^p, \ldots, x_n^p\}$ and $\{x_1^p, \ldots, x_{i-1}^p\}$.

Proposition 3.2 ([14],[25])   *1. $S = \bigcup_{f \in F} Af$, where $F = \{x_1^{\alpha_1} \cdots x_n^{\alpha_n} \mid 0 \leq \alpha_i < p, 1 \leq i \leq n\}$ is a transversal for $gr(A)$ in $gr(S)$.*

   2. for each $n \geq j > i \geq 1$ there exist unique $1 \leq k < l \leq n$ and $1 \leq q < r \leq n$ so that $x_j x_i = x_k x_l$ and $x_i x_j = x_r x_q$. Moreover, it follows that $j > k$, $l > i$, $r > i$, $j > q$, $x_i^{-1} x_j^p x_i = x_l^p$ and $x_j^{-1} x_i^p x_j = x_q^p$.

   3. $G = gr(S)$ is a central localization of S, and hence G-conjugates are S-conjugates.

   4. $gr(S)$ is torsion-free, $gr(S) = \bigcup_{f \in F} gr(A)f$ and $gr(A)$ is a free Abelian group.

We note that the set $F' = \{x_n^{\alpha_n} \cdots x_1^{\alpha_1} \mid 0 \leq \alpha_i < p\}$ is equal to F. In fact, the reverse order on S may be used to show that we also have $S = \bigcup_{f' \in F'} Af'$. Therefore every coset $Af', f' \in F'$, is equal to some $Af, f \in F$. Hence $F' \subseteq F$ and so $F' = F$.

Since conjugation restricts to an action on the set $\{x_1^p, \ldots, x_n^p\}$, this set is the disjoint union of conjugacy classes, say $C_1, \ldots, C_r$. For each $1 \leq i \leq r$, write $z_i = \prod_{c \in C_i} c$. Clearly z_i is a central element of S. For an element $z = x_1^{p\alpha_1} \cdots x_n^{p\alpha_n} \in A$, with each $\alpha_i \geq 0$, we denote by $c(z) = \{i \mid \alpha_i > 0\}$, the *content* of the element z.

For each $1 \leq i \leq r$, let $S_i = \langle x_j \mid j \in c(z_i) \rangle$. The monoids $S_1, \ldots, S_r$ are called the *components* of S.

Corollary 3.3   *1. $S_j S_i = S_i S_j$, for each i, j; in particular each S_i is a binomial semigroup.*

   2. every element of S has a unique representation of the form $s_1 \cdots s_r$ with $s_i \in S_i$. So $S = S_1 S_2 \cdots S_r$.

   3. $gr(S) = G_1 G_2 \cdots G_r$, where $G_i = gr(S_i)$, each element $g \in gr(S)$ can be written uniquely as $g_1 g_2 \cdots g_r$ with $g_i \in G_i$, and $G_i G_j = G_j G_i$.

   4. S is cyclic if and only if S has only one component.

   5. for each $q \geq 1$, the semigroup $S(q) = \langle x_1^q, \ldots, x_n^q \rangle$ is binomial as well.

The examples in [14] show that if $S = \langle x_1, \ldots, x_n \rangle$ with $n \leq 4$ then all components of S are Abelian. In general this is not true.

With notations as before, for $Y \subseteq S$ we denote by $\overline{Y}$ the natural image of Y in the group $\overline{S} = gr(S)/gr(A)$.

Corollary 3.4 *Let S be a binomial semigroup. Write $p = p_1 \cdots p_t$, where each p_i is a power of a different prime number. Let $c_i = p/p_i$. Then*

$$gr(S)/gr(A) = \overline{S(c_1)} \cdots \overline{S(c_t)},$$

each $\overline{S(c_i)}$ is a finite nilpotent group and $S(c_i)S(c_j) = S(c_j)S(c_i)$ for any i, j. In particular, $gr(S)/gr(A)$ and $gr(S)$ are solvable groups.

Since a finitely generated torsion-free nilpotent group is an ordered group, it also follows that $gr(S)$ is nilpotent if and only if $gr(S)$ is Abelian. The following is an interesting semigroup question.

Problem 5: Is a binomial semigroup a unique product semigroup?

We now investigate prime ideals of semigroup algebras of binomial semigroups.

For an ideal I of S we denote by $\mathcal{N}(I)$ the nil radical of I, that is the largest ideal of S that is nil modulo I. Note that because S has the ascending chain condition on left and right ideals, it follows (see for example [36, Proposition 2.13]) that $\mathcal{N}(I)$ is nilpotent modulo I. Part of the following lemma is a consequence of Proposition 2.2.

Lemma 3.5 *1. $spec(S) = \{Q \cap S \mid Q \in spec(K[S]), Q \cap S \neq \emptyset\}$.*

2. Every prime ideal of S contains a minimal prime ideal P of S, and $P = Q \cap S$ for some $Q \in X^1(K[S])$.

3. If $Q \in X^1(K[S])$ and $Q \cap S \neq \emptyset$, then $Q \cap S \in X^1(S)$ and $Q \cap K[A] = K[A]z_i$ for some $1 \leq i \leq r$.

4. The minimal prime ideals of S are all the ideals $\mathcal{N}(Sz_i)$, with $1 \leq i \leq r$. Moreover $\mathcal{N}(Sz_i) \cap A = Az_i$ and in particular, S has r distinct minimal prime ideals.

Let $n_0 = 0$ and $1 \leq n_1 = |\mathcal{C}_1| < n_2 = |\mathcal{C}_1| + |\mathcal{C}_2| < \cdots < n_r = n = |\mathcal{C}_1| + \cdots + |\mathcal{C}_r|$. To simplify notation we may assume, without loss of generality, that $S_i = \langle x_{n_{i-1}+1}, x_{n_{i-1}+2}, \ldots, x_{n_i} \rangle$ for $1 \leq i \leq r$.

An element s of S is said to be *normal* if $Ss = sS$. By N we denote the subsemigroup of all normal elements of S. A nontrivial element $s \in N$ is said to be a *maximal normal* element if $Ss \subseteq St$ for $1 \neq t \in N$ implies $s = t$.

Theorem 3.6 ([25]) *The minimal primes of S are the ideals $Sf_1, \ldots, Sf_r$, where $f_i = x_{n_{i-1}+1} \cdots x_{n_i} \in N$, for each $1 \leq i \leq r$. Moreover, the monoid N is free Abelian with $\{f_1, \ldots, f_r\}$ as an independent set of generators and $f_i^p = z_i$ for every i.*

Semigroups that are Noetherian maximal orders in their group of quotients, and more general Krull semigroups, have been introduced in [7] for Abelian semigroups (cf. Theorem 4.1) and in [48] for noncommutative cancellative semigroups. Recall also that the *normalizing class group* of S is defined as the quotient group of the group of divisorial ideals of S by the group of principal ideals generated by a normal element. So the normalizing class group is trivial if and only if the minimal primes of S are generated by a normal element.

Proposition 3.7 ([25]) *A binomial semigroup S is a Noetherian maximal order in $gr(S)$ with trivial normalizing class group. Furthermore, $K[S]$ is a Noetherian maximal order and its height one prime ideals intersecting S nontrivially are all the ideals of the form $K[Sf_i]$. In particular, these ideals of $K[S]$ are S-homogeneous.*

In [4] (see Theorem 4.2) Brown showed that a prime group algebra $K[H]$ of a polycyclic-by-finite group H is a maximal order if and only if H does not contain an infinite dihedral group which is normal in a subgroup of H of finite index. Furthermore, if $K[H]$ satisfies a polynomial identity then every height one prime ideal of $K[H]$ is generated by a normal element. Proposition 3.7 is in some sense the homogeneous version of Brown's results (Theorem 4.2 and Theorem 4.3) for semigroup algebras of binomial semigroups S.

Since $K[S]$ is a Noetherian maximal order, it follows that the height one primes of $K[S]$ are localizable and their localizations are principal ideal rings. We now describe these localizations. If M is a multiplicative Ore subset of a ring R, then we denote by $R_{(M)}$ the localization of R with respect to M.

Since the elements $x_{n_{i-1}+1}, \ldots, x_{n_i}$ are in the same component of S. we know that for each $n_{i-1} + 1 \leq j \leq n_i$ there exists an element $s_j \in S$ so that

$$s_j^{-1} x_{n_{i-1}+1}^p s_j = x_j^p.$$

Notice that for $\{s_{n_{i-1}+1}, \ldots, s_{n_i}\}$ we can choose any right transversal of the subgroup $C_G(x_{n_{i-1}+1}^p)$, the centraliser of $x_{n_{i-1}+1}^p$ in G.

In Proposition 2.1 we have "described" prime images of semigroup algebras of submonoids of polycyclic-by-finite groups. As shown in the next proposition, because of the special nature of binomial semigroups, we have better and more concrete control on the homogeneous height one primes of semigroup algebras of binomial semigroups. For a prime ideal P of $K[S]$ we denote by $C(P)$ the set of elements of $K[S]$ that are regular modulo P.

Proposition 3.8 ([25]) *Let $P = K[Sf_i]$ be a height one prime ideal of $K[S]$. Let $\{s_{n_{i-1}+1}, \ldots, s_{n_i}\} \subseteq S$ be a right transversal for $C_G(x_{n_{i-1}+1}^p)$ in G, and let $m_i = n_i - n_{i-1}$. Then*

1. *S/Sf_i contains an ideal I which is uniform in the inverse semigroup $\mathcal{M}(H_i, m_i, m_i, \Delta_i)$, where Δ_i is the identity matrix and H_i is isomorphic to a finite group extension of $gr(x_1^p, \ldots, x_{n_{i-1}}^p, x_{n_{i-1}+2}^p, \ldots, x_n^p)$.*

2. $K[H_i]$ is a domain.

3. S/Sf_i contains a subsemigroup of the form $\mathcal{M}(T_i, m_i, m_i, \Delta_i) \subseteq I$ where T_i is a subsemigroup of S which has a group of quotients isomorphic to H_i.

4. $Q_{cl}(K[S]/P) = Q_{cl}(K_0[\mathcal{M}(T_i, m_i, m_i, \Delta_i)]) \cong Q_{cl}(M_m, (K[H_i]))$ and

$$Q_{cl}(K[S]/P) = (K[S]/P)_{(\overline{M_i})}.$$

where $M_i \subseteq C(P)$ is the multiplicatively closed set generated by all elements (note that these are central) of the form

$$\prod_{s \in F} s^{-1} \left(\sum_{k=n_{i-1}+1}^{n_i} s_k^{-1} \left(\prod_{n_{i-1}+1 < j \leq n_i} x_j^p \right) g s_k \right) s,$$

where $0 \neq g \in K[x_1^p, \ldots, x_{n_{i-1}}^p, x_{n_{i-1}+2}^p, \ldots, x_n^p]$.

5. if

$$\mathcal{M}_i = M_i + (K[A] \cap Z(K[S])) z_i,$$

where $Z(K[S])$ is the centre of $K[S]$, then P is localizable and

$$K[S]_{(C(P))} = K[S]_{(\mathcal{M}_i)}.$$

It is worth mentioning that all the essential properties stated in this section remain valid in the more general context of twisted semigroup algebras.

4 Special classes of Noetherian algebras

For the commutative ring theorist, in particular the number theorist, the class of commutative semigroup algebras is a very important one as it provides a wealth of examples of rings satisfying classical arithmetical ring properties, such as for example being a (Noetherian) maximal order, in particular a Dedekind domain or a principal ideal ring. For an early survey on this topic we refer to Gilmer's book [17]. There are many more recent results, for example

Theorem 4.1 (Chouinard [7]) *A semigroup algebra $K[S]$ of a commutative cancellative monoid S is a Krull domain if and only if S is a Krull semigroup and the quotient group of S is torsion-free and satisfies the ascending chain condition on cyclic subgroups.*

Furthermore, S is a Krull semigroup if and only if $S = \mathcal{U}(S) \times S_1$, where S_1 is a submonoid of a free Abelian group F such that S_1 is the intersection of the quotient group of S_1 with the positive part of F.

Chouinard's techniques and results also allow to reduce the calculation of the class group of a commutative Krull domain $K[S]$ to that of the class group of the Krull semigroup S. This way, calculations of several classical examples become quite simple. A complete characterization of commutative semigroup algebras which are principal ideal rings has been given by Decruyenaere, Jespers and Wauters in [9]. Arithmetical properties of graded orders also have received substantial interest. Here the emphasis is to determine whether the global information is determined by the graded information and also whether graded orders with some particular structure can be fully described (for example in some cases they are necessarily twisted semigroup rings). The reader is referred for example to [2] and [10]. Also in the noncommutative setting substantial work has been done, consult for example [32].

Within general ring theory, in particular for the noncommutative ring theorist, (Noetherian) orders in (central) simple algebras form an important class of rings. Also here maximal orders have received considerable interest and it has been shown that some algebraic ring constructions yield interesting classes of maximal orders. As mentioned earlier, group algebras of polycyclic-by-finite groups are one such class. The infinite dihedral group $\langle a, b \mid ba = a^{-1}b, \ b^2 = 1 \rangle$ is denoted D_∞.

Theorem 4.2 (Brown [4]) *Let G be a polycyclic-by-finite group and K a field. A group algebra $K[G]$ is a prime maximal order if and only if*

1. *the finite conjugacy subgroup $\Delta(G)$ of G is torsion-free,*

2. *G is dihedral free, that is, if D_∞ is a subgroup of G then its normalizer in G is of infinite index.*

Note that the first condition in the theorem is equivalent with the group algebra being prime ([41, Theorem 4.2.10]). Brown also determined when the height one primes are principal ideals.

Theorem 4.3 (Brown [4]) *Let G be a polycyclic-by-finite group and K a field. If $K[G]$ is a prime maximal order, then the following conditions are equivalent for a height one prime ideal P of $K[G]$:*

1. *P is right principal;*

2. *P is invertible, that is, $Q_{cl}(K[S])$ contains a $K[S]$-bimodule J with $IJ = JI = K[S]$;*

3. *P is right projective;*

4. *$P = K[G]n = nK[G]$ for some $n \in K[\Delta(G)]$;*

5. *P contains a non-zero central element;*

6. P contains a non-zero normal element;

7. P contains an invertible ideal.

In particular, if $K[G]$ is a prime maximal order satisfying a polynomial identity, then all height one prime ideals of $K[G]$ are principal.

Brown proved his results in [4] in the more general context of group rings over commutative Noetherian domains. In Theorem 4.2 one has then to add the necessary condition that R is integrally closed. Earlier results on this topic can be found in [28, 29, 45].

Only for very few noncommutative semigroups S it has been determined when the semigroup algebra is a Noetherian maximal order. Apart from the binomial semigroups, Wauters in [48] dealt with cancellative semigroups of normal elements and with the cancellative semigroup of the regular elements of a prime Goldie ring. However, semigroup algebras that are principal ideal rings have been described in [24]. In the remainder of this section we report on these results. Note that we always assume that a principal ideal ring contains an identity element. First we state Passman's result on the group algebra case. As in [30] the result is stated in the slightly more general context of matrices over group algebras.

Proposition 4.4 (Passman [40]) *Let G be a group and $R = M_n(K)$ a matrix ring over a field K. The following conditions are equivalent:*

1. *$R[G] = M_n(K[G])$ is a principal right ideal ring;*

2. *$R[G]$ is right Noetherian and the augmentation ideal $\omega(R[G])$ is a principal right ideal;*

3. *if char $K = 0$ then G is finite or finite-by-infinite cyclic;*
   if char $K = p > 0$ then G is finite p'-by-cyclic p or G is finite p'-by-infinite cyclic.

In [30] this result was extended to semigroup algebras of cancellative monoids.

Proposition 4.5 (Jespers and Wauters [30]) *Let T be a cancellative monoid and K a field of characteristic p (possibly zero). The following conditions are equivalent:*

1. *$K[T]$ is a principal right ideal ring;*

2. *T is a semigroup satisfying one of the following conditions:*

   (a) T is a group satisfying the conditions of Proposition 4.4;

(b) T contains a finite p'-subgroup H and a nonperiodic element x such that $xH = Hx$, $T = \bigcup_{i \in \mathbf{N}} Hx^i$ and the central idempotents of $K[H]$ are central in $K[T]$.

The structure theorem of linear semigroups provides a link between a linear semigroup and some of its cancellative subsemigroups. In order to exploit this link for semigroup algebras of arbitrary semigroups that are principal ideal rings one first has to reduce the problem from arbitrary semigroups to linear semigroups.

The following two results guarantee this.

Theorem 4.6 (Anan'in [1]) *A finitely generated right Noetherian algebra that satisfies a polynomial identity is embeddable in a matrix ring over a commutative algebra.*

Note that a classical result of Malcev [33] says that a finitely generated commutative algebra is embeddable in a matrix ring over a field.

Theorem 4.7 ([24]) *Let $K[S]$ be a principal right ideal ring. Then $K[S]$ satisfies a polynomial identity.*

Using the structure theorem of linear semigroups one now can prove the following results.

Proposition 4.8 ([24]) *If $K[S]$ is a principal right ideal ring, then the Gelfand-Kirillov dimension of $K[S]$ is equal to its classical Krull dimension and it is 0 or 1. In the former case S is finite. Moreover, every prime Artinian homomorphic image of $K[S]$ is finite dimensional over K.*

Theorem 4.9 ([24]) *Let S be a semigroup and K a field of characteristic p (possibly zero). The following conditions are equivalent:*

1. $K_0[S]$ is a principal (left and right) ideal ring;

2. there exists an ideal chain

$$I_1 \subset \cdots \subset I_t = S$$

such that I_1 and every factor I_j/I_{j-1} is of the form $\mathcal{M}(T, n, n, P)$ for an invertible over $K_0[T]$ sandwich matrix P, and one of the following conditions holds:

(a) T is a group of the type described in Proposition 4.4;

(b) T is a monoid with finite group of units H such that $T = \bigcup_{i \geq 0} Hx^i$ for some $x \in T$, and either this union is disjoint or $x^n = \theta$ for some $n \geq 1$. Also $Hx = xH$, the central idempotents of $K[H]$ commute with x, and $p = 0$ or $p \nmid |H|$.

In case the equivalent conditions are satisfied it follows that

$$K_0[S] \cong K_0[I_1] \oplus K_0[I_2/I_1] \oplus \cdots \oplus K_0[I_t/I_{t-1}].$$

Moreover, $K_0[S]$ is a finite module over its centre, which is finitely generated.

It in the previous theorem we obtain a characterization of left and right principal ideal rings. One might pose the following question.

Problem 6: Is the left-right symmetric hypothesis in Theorem 4.9 required?

The above applies to finite dimensional algebras $K[S]$, since a finite dimensional algebra is a principal right ideal ring if and only if it is a principal left ideal ring. One can also show that semiprime principal right ideal semigroup algebras are necessarily principal left ideal rings as well.

Theorem 4.10 ([24]) *Let S be a semigroup and K a field of characteristic p (possibly zero). Then $K_0[S]$ is a semiprime principal right ideal ring if and only if there exists an ideal chain*

$$I_1 \subset \cdots \subset I_t = S$$

such that I_1 and every factor I_j/I_{j-1} is of the form $\mathcal{M}(T, n, n, P)$ for an invertible over $K_0[T]$ sandwich matrix P and a monoid T such that

1. *either T is a group as in Proposition 4.4 so that $K[T]$ is semiprime,*

2. *or T is a monoid with finite group of units H such that $T = \bigcup_i H x^i$ is a disjoint union, for some $x \in T$. Also $Hx = xH$, the central idempotents of $K[H]$ commute with x, and $p = 0$ or $p \nmid |H|$. Furthermore, for every primitive central idempotent $e \in K[H]$, either $K[H]ex = 0$ or $K[H]ex^i \neq 0$ for all $i \geq 1$.*

Moreover, if the equivalent conditions are satisfied, then $K_0[S]$ is a principal left ideal ring.

Corollary 4.11 *$K_0[S]$ is a prime principal right ideal ring if and only if*

$$S \cong \mathcal{M}(\{1\}, n, n, Q), \quad S \cong \mathcal{M}(\langle x \rangle, n, n, Q) \quad or \quad S \cong \mathcal{M}(\langle x, x^{-1} \rangle, n, n, Q)$$

where Q is invertible in $M_n(K)$, $M_n(K[x])$ or $M_n(K[x, x^{-1}])$ respectively. Hence, $K_0[S] \cong M_n(K), M_n(K[x])$, or $M_n(K[x, x^{-1}])$.

References

[1] A.Z. Anan'in, An intriguing story about representable algebras, Ring Theory 1989, Israel Math. Conf. Proc., 31-38, Weizmann, Jerusalem, 1989.

[2] D.D. Anderson and D.F. Anderson, Divisibility properties of graded domains, Canad. J. Math. 24 (1982). 196–215.

[3] M. Artin and W. Schelter, Graded algebras of global dimension 3. Adv. Math. 66 (1987), 171–216.

[4] K.A. Brown, Height one primes of polycyclic group rings, J. London Math. Soc. (2) 32 (1985), no.3, 426–438.

[5] S.C. Charlap, Bieberbach groups and flat manifolds, Springer-Verlag. New York, 1986.

[6] W.Chin and D.Quinn, Rings graded by polycyclic-by-finite groups. Proc. Amer. Math. Soc. 102 (1988), 235-241.

[7] L.G. Chouinard II, Krull semigroups and divisor class groups, Canad. J. Math. 23 (1981), 1459–1468.

[8] A.H. Clifford and G.B. Preston, The Algebraic Theory of Semigroups, Vol. I, Amer. Math. Soc., Providence, 1961.

[9] F.Decruyenaere, E.Jespers and P.Wauters, On commutative principal ideal semigroup rings, S.Forum 43 (1991), 367-377.

[10] F. Decruyenaere and E. Jespers, Prüfer domains and graded rings, J. Algebra 150 (1992), 308–320.

[11] V.G. Drinfeld, On some unsolved problems in quantum group theory. Quantum Groups (P.P. Kulish, ed.), 1-8, Lect. Notes Math., vol. 1510, Springer -Verlag, 1992.

[12] T. Gateva-Ivanova, On the Noetherianity of some associative finitely presented algebras, J. Algebra 138 (1991). 13–35.

[13] T. Gateva-Ivanova. Noetherian properties and growth of some associative algebras, in: Effective Methods in Algebraic Geometry, 143–158. Progr. Math. 94, Birkhäuser, Boston. 1991.

[14] T. Gateva–Ivanova, Noetherian properties of skew polynomial rings with binomial relations, Trans. Amer. Math. Soc. 343 (1)(1994), 203–219.

[15] T. Gateva–Ivanova, Skew polynomial rings with binomial relations, J. Algebra, 185 (1996), 710–753.

[16] T. Gateva Ivanova and M. Van den Bergh, Semigroups of I-type, J. Algebra, to appear.

[17] R. Gilmer, Commutative Semigroup Rings, Univ. Chicago Press, Chicago, 1984.

[18] R.I.Grigorchuk, Cancellative semigroups of power growth, Mat. Zametki 43 (1988). 305–319. (In Russian)

[19] M.Gromov, Groups of polynomial growth and expanding maps, Publ. Math. IHES 53 (1) (1981), 53–73.

[20] E.Hotzel, On semigroups with maximal conditions, Semigroup Forum, Vol. 11 (1975/76), 337–362.

[21] R.S. Irving, L.W. Small, The Goldie conditions for algebras with bounded growth, Bull. London Math. Soc. 15 (1983), 596-600.

[22] E. Jespers, J. Krempa and E.R. Puczylowski, On radicals of graded rings, Comm. Algebra 10 (17)(1982), 1849-1854.

[23] E.Jespers and Okniński, Nilpotent semigroups and semigroup algebras, J. Algebra 169 (1994), 984-1011.

[24] E.Jespers and J.Okniński, Semigroup algebras that are principal ideal rings, J. Algebra 183 (1996), 837-863.

[25] E. Jespers and J. Okniński, Binomial semigroups, J. Algebra 202 (1998), 250-275.

[26] E. Jespers and J. Okniński, Noetherian semigroup algebras, J. Algebra, to appear.

[27] E. Jespers and J. Okniński, Submonoids of polycyclic-by-finite groups and their algebras, J. Algebras and Representation Theory, to appear.

[28] E. Jespers and P.F. Smith, Group rings and maximal orders, Methods in Ring Theory (Antwerp, 1983), 185-195, Nato Adv.Sci.Inst. Ser. C: Math.Phys.Sci. 129, Reidel, Dordrecht-Boston, 1984.

[29] E. Jespers and P.F. Smith, Integral group rings of torsion-free polycyclic-by-finite groups are maximal orders, Comm. Algebra 13(3) (1985), 669-680.

[30] E.Jespers and P.Wauters, Principal ideal semigroup rings, Comm. Algebra 23 (1995), 5057-5076.

[31] G. Lallement, On nilpotency in semigroups, Pacific J. Math. 42 (1972), 693-700.

[32] L. le Bruyn, M. Van den Bergh and F. Van Oystaeyen, Graded Orders, Birkäuser, Boston, 1988.

[33] A.I.Malcev, On representation of infinite algebras, Mat. Sb. 13 (1943), 263-285 (in Russian).

[34] A.I. Malcev, Nilpotent semigroups, Uc. Zap. Ivanovsk. Ped. Inst. 4 (1953), 107-111 (in Russian).

[35] J.C. McConnell and J.C. Robson, Noncommutative Noetherian Rings, Wiley Interscience, New York, 1987.

[36] J. Okniński, Semigroup Algebras, Marcel Dekker, New York, 1991.

[37] J. Okniński, Linear representations of semigroups, in: Monoids and Semigroups with Applications, 257-277, World Sci., 1991.

[38] J.Okniński, Gelfand - Kirillov dimension of noetherian semigroup algebras, J. Algebra 162 (1993), 302-316.

[39] J. Okniński, Semigroups of Matrices, World Scientific, Singapore, 1998.

[40] D.S. Passman, Observations on group rings, Comm. Algebra 5 (1977), 1119-1162.

[41] D.S. Passman, The Algebraic Structure of Group Rings, Wiley, New York, 1977.

[42] D.S. Passman, Infinite Crossed Products, Academic Press, New York, 1989.

[43] D.Quinn, Group – graded rings and duality, Trans. Amer. Math. Soc. 292 (1985). 155-167.

[44] A. Salwa, Structure of skew linear semigroups, Int. J. Algebra and Comput. 3 (1993), 101–113.

[45] P.F. Smith, Some examples of maximal orders, Math. Proc. Cambridge Philos. Soc. 98(1) (1985), 19 32.

[46] J.T. Stafford and J.J. Zhang, Homological properties of graded Noetherian PI rings, J. Algebra 168 (1994), 988–1026.

[47] J. Tate and M. Van den Bergh, Homological properties of Sklyanin algebras. Invent. Math. 124 (1996), 619–647.

[48] P. Wauters, On some subsemigroups of noncommutative Krull rings, Comm. Algebra 12 (13-14) (1984), 1751-1765.

[49] E.I. Zelmanov, Semigroup algebras with identities, Sib. Math. J. 18 (1977). 757–798 (in Russian).

On the Recognition Problem of the Irreducibility for $A_1(k)$-Modules and Their Characteristic Varieties

Huishi Li and Lu Chen
Department of Mathematics
Shaanxi Normal University
710062 Xian, P.R. China

Abstract Let L be a left ideal of the first Weyl algebra $A_1(k)$ over an algebraically closed field k of characteristic 0. We give some algorithmic approaches to the recognition problem of the irreduciblity of the $A_1(k)$-module $M = A_1(k)/L$. For the irreducibility of the characteristic variety of M, we give a complete algorithmic criterion. All computation used in this note can be realized by several computer algebra systems such as MAS, FELIX and MACAULAY.

§1.

Let k be an algebraically closed field of characteristic 0. The first Weyl algebra $A_1(k)$ is the algebra (associative with 1) over k with generators x, y subject to the relation $yx - xy = 1$; $A_1(k)$ is isomorphic to the algebra of formal differential operators (in one variable) with polynomial coefficients.

It is well known that the representation theory of $A_1(k)$ is very important in several areas of mathematics, especially Lie algebras, and that many remarkable works concerning the irreducible representations of $A_1(k)$ have been done (e.g. [Dix1,2,3], [Blo1,2]). Excepting the classification theory, the construction of simple modules has also been studied. In [Dix3] it was proved that for $\beta \in k$, the $A_1(k)$-module $A_1(k)/A_1(k)(xy + \beta)$ is simple if and only if $\beta \notin \mathbb{Z}$; It was also proved (e.g. see [Cot]) that any twisting module $k[x]_\sigma$ of $k[x]$ (as an A_1-module under the natural action) by an automorphism σ of $A_1(k)$ is a simple module. It turns out that for any $g \in k[x]$, the module $A_1(k)/A_1(k)(g - y)$ is a simple module, in particular,

---

*Supported by NSFC.

$\{A_1(k)/A_1(k)(x^r - y) \mid r > 0\}$ form an infinite family of pairwise non-isomorphic simple modules over $A_1(k)$.

In this note, we consider the recognition problem of irreducibility for $A_1(k)$-modules and their characteristic varieties, or more clearly, we study the following

Question Let L be a left ideal of $A_1(k)$ generated by $\{g_1, ..., g_s\}$. Is there an algorithmic way to check

(i) if the left module $M = A_1(k)/L$ is irreducible or not; and

(ii) if the characteristic variety $\mathbf{V}(\mathcal{I}(M))$ (see §2 for the definition) of M is irreducible or not?

Before studying this problem, the authors would like to say that *there will be no any new algorithms developed in this note* (the authors know nothing about algorithms); instead, what the authors are going to do is, motivated by the computation methods developed for commutative algebraic geometry and noncommutative algebra (in particular for the solvable polynomial algebras) in recent years (e.g. [Sti], [AL], [K-RW], [Mor]), to try to give some algorithmic approach to the above question. More precisely, after the preliminary section (§2), in §3 we give some algorithmic approaches for recognizing the irreducibility of modules of type $A_1(k)/L$ which can also be applied to some well known examples. In §4, based on the computation method due to Eisenbud and Huneke, and an algorithm given by [Sti] we give a complete algorithmic criterion for the irreducibility of the characteristic variety of modules of type $A_1(k)/L$. All computation we used in this note can berealized by several computer algebra systems (e.g., MAS, FELIX and MACAULAY, as far as the authors know) on an ordinary computer.

§2.

Considering the Bernstein filtration (see [Bj]) on $A_1(k)$, denoted $FA_1(k)$, it is well known that the associated graded algebra of $A_1(k)$, denoted $G(A_1(k))$, is isomorphic to the polynomial algebra $k[x, y]$ where $k[x, y]$ has the gradation given by total degree of polynomials. If no confusion is possible, *we will always use $k[x, y]$ instead of $G(A_1(k))$ for the associated graded algebra of $A_1(k)$.* For any left $A_1(k)$-module M with a filtration FM: $\cdots \subset F_{-1}M \subset F_0M \subset F_1M \subset \cdots$ consisting of k-subspaces F_nM of M which is compatible with $FA_1(k)$, i.e., $F_pA_1(k)F_nM \subset F_{p+n}M$, $p, n \in \mathbb{Z}$, $p \geq 0$, and $M = \cup_{n \in \mathbb{Z}}F_nM$, the associated graded $k[x, y]$-module of M with respect to FM is by definition the graded $k[x, y]$-module $G(M) = \oplus_{n \in \mathbb{Z}}G(M)_n$ with $G(M)_n = F_nM/F_{n-1}M$. FM is said to be a *good filtration* on M, if there exist $h_1, ..., h_s \in \mathbb{Z}$ and $\xi_1, ..., \xi_s \in M$ such that $F_nM = \sum_{i=1}^{s} F_{n-h_i}A_1(k)\xi_i$, $n \in \mathbb{Z}$.

Obviously, if M has a good filtration FM then M is a finitely generated module; and if M is a finitely generated module, then M has a good filtration. Let M be a finitely generated $A_1(k)$-module with a good filtration FM. Then it is well known that the graded ideal $\mathcal{I}(M) = \sqrt{\mathrm{Ann}_{k[x,y]}G(M)}$, where $\mathrm{Ann}_{k[x,y]}G(M)$ is the annihilator ideal of $G(M)$ in $k[x,y]$, is independent of the choice of good filtration on M, and hence is defined to be the *characteristic ideal* of M. The *characteristic variety* of M is then defined to be the *affine* algebraic set $V = \mathbf{V}(\mathcal{I}(M))$ in the affine plane $\mathbf{A}_k^2$.

All results of this note will be based on the following facts.

From ([AL], [K-RW]) we know that $A_1(k)$ is a solvable polynomial algebra, so every left ideal L of $A_1(k)$ has a (left) Groebner basis with respect to a fixed monomial ordering $>$ on $\mathbb{Z}_+^2$ (or equivalently, on the set of all monomials of the form $x^{\alpha}y^{\beta}$ where $(\alpha, \beta) \in \mathbb{Z}_+^2$). Furthermore, if $>$ is fixed and L is a left ideal of $A_1(k)$ generated by $F = \{f_1, ..., f_s\}$, then there is an algorithm which can be used to produce a Groebner basis G for L from F (now it can be computed by the computer algebra systems MAS or FELIX), and $L = A_1(k)$ if and only if $1 \in G$.

Let L be a left ideal of $A_1(k)$ with the filtration FL induced by the Bernstein filtration $FA_1(k)$ on $A_1(k)$, i.e., $F_pL = F_pA_1(k) \cap L$, $p \geq 0$. If L is generated by $G = \{g_1, ..., g_s\}$, then it is known that the associated graded module $G(L)$ of L, which is also a graded ideal of $k[x,y]$, is not necessarily generated by $\sigma(G) = \{\sigma(g_1), ..., \sigma(g_s)\}$, where each $\sigma(g_i)$ is the corresponding homogeneous element of g_i in $G(L)$. However, we do have

2.1. Theorem ([LW], also see [Li]) (i) Let $>_{grlex}$ be the *graded lexicographic* ordering on $\mathbb{Z}_+^2$. If L is a left ideal of $A_1(k)$ and $G = \{g_1, ..., g_s\}$ is a Groebner basis of L with respect to $>_{grlex}$, then $\sigma(G) = \{\sigma(g_1), ..., \sigma(g_s)\}$ is a Groebner basis of $G(L)$, and consequently $G(L)$ is generated by $\sigma(G)$.

(ii) Let L and G be as in (i). Then the (affine) Hilbert polynomial of the module $M = A_1(k)/L$ can be algorithmically computed via $\sigma(G)$ exactly in the commutative case, i.e., GK.dim$M$ and the multiplicity of M, denoted $e(M)$ as in the literature, can be read out from an ordinary computer.

For any two polynomials f and g in $k[x,y]$, we write $\mathbf{V}(f)$ and $\mathbf{V}(g)$ for the curves defined by f and g, respectively.

2.2. Theorem (Classical) If f and g do not have nontrivial common factors, or in other words, if $\mathbf{V}(f)$ and $\mathbf{V}(g)$ have no common irreducible components, then $\mathbf{V}(f,g) = \mathbf{V}(f) \cap \mathbf{V}(g)$ is a finite set of points, and consequently the k-algebra

$k[x,y]/\langle f,g \rangle$ is a finite dimensional k-space, where $\langle f,g \rangle$ denotes the ideal of $k[x,y]$ generated by f and g.

2.3. Theorem (Bernstein's inequality, see [Bj]) Let M be a nonzero left $A_1(k)$-module. Then GK.dim$M \geq 1$ where GK.dim$M$ denotes the Gelfand-Kirillov dimension of M.

§3.

Let the notation be as before. Recall that if L is a proper left ideal of $A_1(k)$, then the $A_1(k)$-module $M = A_1(k)/L$ has GK.dim$M = 1$, i.e., M is a *holonomic* $A_1(k)$-module (see [Bj]). Furthermore, it is well known that *if $e(M) = 1$ then M is a simple $A_1(k)$-module* (e.g. see [Cot]). Thus, Theorem 2.1 immediately yields the following algorithmic recognition of simple $A_1(k)$-modules.

3.1. Theorem Let L be a left ideal of $A_1(k)$ and put $M = A_1(k)/L$. Then the recognition problem of ($e(M) = 1$ or not) can be realized by the following algorithmic procedure.

• Under $>_{grlex}$ produce a Groebner basis for L from any generating set of L.
• Compute the (affine) Hilbert polynomial of M via $\sigma(G)$ in the commutative case and read $e(M)$ out.

3.2. Corollary Put

$$\mathcal{H} = \{ax + by + c \mid (a,b,c) \in k^3, \ a \neq 0 \text{ or } b \neq 0\}.$$

For every $f \in \mathcal{H}$, the $A_1(k)$-module $M(f) = A_1(k)/A_1(k)f$ has $e(M(f)) = 1$. Hence $M(f)$ is simple.

The above corollary shows that if L is any proper left ideal of $A_1(k)$ containing some $f \in \mathcal{H}$, then $A_1(k)/L$ is a simple module. Our next result shows how to algorithmically know that L contains an element of $\mathcal{H}$.

3.3. Proposition With notation as above, let L be a proper left ideal of $A_1(k)$ and $G = \{g_1, ..., g_s\}$ a Groebner basis of L with respect to $>_{grlex}$. Then the following are equivalent.
(i) L contains an element of $\mathcal{H}$;
(ii) Some g_i is contained in $\mathcal{H}$.

Proof (i) $\Longrightarrow$ (ii) Suppose $f \in \mathcal{H}$ and $f \in L$. Then since G is a Groebner basis of L, f has a standard presentation $f = \sum_{i=1}^s h_i g_i$ for some $h_i \in A_1(k)$ such that

$$\text{leading term of } f \geq_{grlex} \text{leading term of } h_i g_i$$

whenever $h_i g_i \neq 0$. It follows that $\deg(f) \geq \deg(g_i)$ for $h_i g_i \neq 0$ and hence there is some $g_i \in \mathcal{H}$ because we are using ">$_{grlex}$ and L is a proper left ideal.

(ii) $\Longrightarrow$ (i) This is clear. $\qquad\qquad\qquad\qquad\qquad\qquad\qquad\qquad\qquad$ □

Let us mention an application of the above result.

For $\alpha \in k$, write $(k[y], \ x - \alpha = -\frac{d}{dy})$ for the $A_1(k)$-module of polynomials in y with y acting by multiplication and x as indicated. It is well known that this module is simple and is the unique module (up to isomorphism) for which α is an eigenvalue for x. In [Blo1] the following classification theorem of simple $A_1(k)$-modules was obtained:

Theorem ([Blo1) Theorem 2) If M is a simple $A_1(k)$-module, then either $M \cong (k[y], \ x - \alpha = -\frac{d}{dy})$ for some $\alpha \in k$ or $M \cong M(b)$ for some $b \in A_1(k)$ which is irreducible in B (= the skewpolynomial ring $k(x)[y; \ \frac{d}{dx}]$) and preserving (see [Blo1] for the definition).

It is not hard to see that the module $(k[y], \ x - \alpha = -\frac{d}{dy})$ can be obtained by using the k-vector space isomorphism $k[y] \cong A_1(k)/A_1(k)(x-\alpha)$ and identifying the action of $A_1(k)$ directly on $k[y]$. But from Theorem 3.3 we have seen that such kind of simple $A_1(k)$-modules can be recognized algorithmically.

3.4. Remark There are examples showing that Theorem 3.1 covers not only simple modules of type $A_1(k)/L$ where L contains an element of $\mathcal{H}$.

(i) Let $L = A_1(k)(xy - 1) + A_1(k)y^2$. Then it is easy to see that $\{xy - 1, y^2\}$ is a Groebner basis for L under >$_{grlex}$ by fixing $x > y$. We compute out that the Hilbert polynomial of $M = A_1(k)/L$ is $t + 3$, and hence $e(M) = 1$. But neither $xy - 1$ nor y^2 is in $\mathcal{H}$.

(ii) The simple modules given by $M = A_1(k)/A_1(k)(g - y)$, $g \in k[x]$ (see §1), always have $e(M) = 1$.

For the modules considered in 3.4(ii) above, if we change to use the *lexicographic ordering* >$_{lex}$ on $\mathbb{Z}_+^2$, we also have an algorithmic recognition for such kind of simple modules.

3.5. Proposition let L be a proper left ideal of $A_1(k)$ and $G = \{g_1, ..., g_s\}$ a Groebner basis of L with respect to >$_{lex}$ by fixing $y > x$. Then the following are equivalent.

(i) L contains an element of the form $g - y$ with $g \in k[x]$;

(ii) G contains an element of the form $g - y$ with $g \in k[x]$.

Proof (i) $\Longrightarrow$ (ii) Suppose L contains some $g - y$. Note that since we are using >$_{lex}$ by fixing $y > x$, $g - y$ has the leading term y. If we consider a standard presentation of $g - y$ by G, say $g - y = \sum_{i=1}^{s} h_i g_i$ for some $h_i \in A_1(k)$, then by comparing the leading terms of both sides of the equality we conclude that some g_i must be of the desired form.

(ii) $\Longrightarrow$ (i) This is clear. $\hfill\square$

Using the computation of Hilbert polynomial for a polynomial ideal it is easy to see that not all simple $A_1(k)$-modules have multiplicity 1. For example, the simple $A_1(k)$-modules $M = A_1(k)/A_1(k)(xy - \beta)$ with $\beta \notin \mathbb{Z}$ studied in [Dix3] have hilbert polynomial $2t + 1$, and hence $e(M) = 2 \neq 1$. But we still have a similar result as before. To get this we need to recall the notion of a reduced Groebner basis for a left ideal L in a solvable polynomial algebra [K-RW]. A Groebner basis $G = \{g_1, ..., g_s\}$ is called *reduced* , if for every $g_i \in G$ there is no monomial of g_i can be divided by some leading monomial of g_j with $i \neq j$.

3.6. Proposition let L be a proper left ideal of $A_1(k)$ and $G = \{g_1, ..., g_s\}$ a reduced Groebner basis of L with respect to $>_{grlex}$ by fixing $x > y$. Then the following are equivalent.

(i) L contains an element of the form $xy - \beta$ with $\beta \in k$;

(ii) G contains an element of the form $xy - \beta$ with $\beta \in k$.

Proof By the foregoing results we may assume that G does not contain an element of $\mathcal{H}$.

(i) $\Longrightarrow$ (ii) Suppose L contains some $f = xy + \beta$. Then by our assumption and by considering the standard presentation of f by G, G must contain one member of the form

$$(\ast) \qquad\qquad\qquad xy + ay^2 + bx + cy + d.$$

If furthermore we look at the division procedure of f by the elements of the form $(\ast)$, then (ii) follows from the reduced property of G.

(ii) $\Longrightarrow$ (i) This is clear. $\hfill\square$

Of course, the results given in this section are still far from covering all cases of simple modules. For instance, the authors do not know how to algorithmically recognize if an element b of $A_1(k)$ is irreducible in $B = k(x)[y; \frac{d}{dx}]$ and is preserving in the sense of [Blo1].

§4.

We first fix the notation once for all. Let L be a nontrivial left ideal of $A_1(k)$ with the filtration FL induced by the Bernstein filtration on $A_1(k)$, and let $M =$

$A_1(k)/L$ with the filtration $\{(F_p A_1(k)+L)/L\}_{p\geq 0}$. Then $G(M) \cong G(A_1(k))/G(L) = k[x,y]/G(L)$ and by the definition of §2 the characteristic ideal of M is $\mathcal{I}(M) = \sqrt{G(L)}$, and the characteristic variety of M is $\mathbf{V}(\mathcal{I}(M)) = \mathbf{V}(\sqrt{G(L)}) = \mathbf{V}(G(L))$. In this section we give a complete algorithmic recognition of the irreducibility of the characteristic variety $\mathbf{V}(\mathcal{I}(M))$ of M.

Let $\mathcal{H} = \{ax + by + c \mid (a,b,c) \in k^3,\ a \neq 0 \text{ or } b \neq 0\}$ be as in §3. Put $\sigma(\mathcal{H}) = \{\sigma(f) = ax + by \mid f \in \mathcal{H}\}$. Then we have

4.1. Lemma $\sigma(\mathcal{H})$ exhausts the irreducible homogeneous elements of $k[x,y]$.

Proof Since k is algebraically closed, it is well known that if $F \in k[x,y]$ is a homogeneous element, then F factors into a product of linear factors. $\qquad\square$

4.2. Corollary Let L and M be as above.
(i) L contains an element of $\mathcal{H}$ if and only if $G(L)$ contains an element of $\sigma(\mathcal{H})$.
(ii) If L contains an element of $\mathcal{H}$, then $\mathbf{V}(\mathcal{I}(M))$ is an irreducible curve (a line through $(0,0)$).

4.3. Lemma Let L and M be as before.
(i) $\mathbf{V}(\mathcal{I}(M)) \neq \emptyset$, and $\mathbf{V}(\mathcal{I}(M))$ neither is $\mathbf{A}_k^2$ nor a finite set.
(ii) Any two nonzero elements of $\sqrt{G(L)}$ has a nontrivial common divisor.
(iii) Suppose $\sqrt{G(L)}$ contains an irreducible element of $k[x,y]$, say F. Then $\sqrt{G(L)} = \langle F \rangle$, and moreover $F \in \sigma(\mathcal{H})$.

Proof Note that since $\mathbf{V}(\mathcal{I}(M)) = \mathbf{V}(\sqrt{G(L)}) = \mathbf{V}(G(L))$, k is algebraically closed, L is a proper left ideal of $A_1(k)$, and $\text{GK.dim}M = \text{GK.dim}G(M)$,
(i) follows from the Bernstein's inequality;
(ii) follows from Theorem 2.2 and (i); and
(iii) follows from (ii) and Lemma 4.1 because $\sqrt{G(L)}$ is a graded ideal of $k[x,y]$. $\quad\square$

Since $M = A_1(k)/L$ is a holonomic module, $G(L)$ is an *equi-dimensional* ideal of dimension 1 in $k[x,y]$. Based on the computation of $\sqrt{I}$ for an equi-dimensional polynomial ideal I which is due to Eisenbud and Huneke, there exists an algorithm for computing $\sqrt{I}$ via any generating set of I (see [Sti]), and this algorithm has been realized by the well known computer algebra system MACAULAY. This enable us to give our main result of this section.

4.4. Theorem Let L and M be as defined in the begining, and let $G = \{g_1, ..., g_s\}$ be a Groebner basis of L with respect to $>_{grlex}$.
(i) A homogeneous Groebner basis of $\sqrt{G(L)}$ can be produced out from $\sigma(G)$.
(ii) The characteristic variety $\mathbf{V}(\mathcal{I}(M)) = \mathbf{V}(\sqrt{G(L)})$ is irreducible if and only if the Groebner basis of $\sqrt{G(L)}$ obtained from (ii) above contains an element of $\sigma(\mathcal{H})$.

Proof (i) follows from Theorem 2.1 and [Sti]. To prove (iii), suppose that $\mathbf{V}(\mathcal{I}(M))$ is irreducible. Then $\sqrt{G(L)}$ contains an element of $\sigma(\mathcal{H})$ by Lemma 4.3, say $ax + by$ with $a \neq 0$ or $b \neq 0$. Now a similar argument as in §3 shows that $ax + by$ is contained in the Groebner basis of $\sqrt{G(L)}$ obtained from (ii). Conversely, suppose that the Groebner basis of $\sqrt{G(L)}$ obtained from (ii) contains an element of $\sigma(\mathcal{H})$. Then by Lemma 4.3(iii) we see that $\sqrt{G(L)}$ is a prime ideal and hence $\mathbf{V}(\mathcal{I}(M))$ is irreducible. This finishes the proof.   □

Of course, if M is simple then $\mathbf{V}(\mathcal{I}(M))$ is not necessarily irreducible (e.g. $M = A_1(k)/A_1(k)(xy - \beta)$ with $\beta \in k - \mathbb{Z}$). But one might expect that if $\mathbf{V}(\mathcal{I}(M))$ is irreducible, then M is irreducible. Unfortunately, this may also be false. For example, let $M = A_1(k)/A_1(k)(x^2 - 1)$. Then by taking the filtration on $L = A_1(k)(x^2 - 1)$ induced by the Bernstein filtration and the filtration on M as we used before, we have $G(M) \cong k[x, y]/\langle x^2 \rangle$. Obviously, M is not irreducible but $\mathbf{V}(\mathcal{I}(M))$ is irreducible.

Despite of the unpleasantness showing by the remark above, the final result of this note is to say that if we know by Theorem 4.4 that the characteristic variety of M is irreducible , then we can algorithmically construct a simple module which has the same characteristic variety as M does.

4.5.   Proposition Let L be a proper left ideal of $A_1(k)$ with the filtration FL induced by the Bernstein filtration on $A_1(k)$. Put $M = A_1(k)/L$. If the characteristic variety $\mathbf{V}(\mathcal{I}(M))$ of M is irreducible, then there exists an irreducible $A_1(k)$-module N such that $\mathbf{V}(\mathcal{I}(M)) = \mathbf{V}(\mathcal{I}(N))$. Moreover, N can be algorithmically constructed.

Proof If $\mathbf{V}(\mathcal{I}(M))$ is irreducible, then it follows from Lemma 4.3 that $\mathcal{I}(M)$ is generated by an element of $\sigma(\mathcal{H})$, say $ax + by$ with $a \neq 0$ or $b \neq 0$. Thus $G(L) \subset \mathcal{I}(M) = \langle ax + by \rangle$. Consider the left ideal $J = A_1(k)(ax + by)$ and put $N = A_1/J$. If we endow J with the filtration induced by the Bernstein filtration on $A_1(k)$ and endow N with the filtration $\{(F_pA_1(k) + J)/J\}_{p \geq 0}$, then by §3 the module N is simple and we have $G(J) = \langle \sigma(h) \rangle$ and $G(N) \cong k[x, y]/\langle \sigma(h) \rangle$. Hence $\mathbf{V}(\mathcal{I}(M)) = \mathbf{V}(\mathcal{I}(N))$, as desired.

Algorithmically, the above procedure can go as follows.

• Produce a Groebner basis of L from any generating set of L.

• If $G = \{g_1, ..., g_s\}$ is a Groebner basis of L, then we can produce a Groebner basis of $\mathcal{I}(M) = \sqrt{G(L)}$ from $\sigma(G)$. Now the irreduciblity of $\mathbf{V}(\mathcal{I}(M))$ and the existence of the desired irreducible homogeneous element $\sigma(h)$ follow from Theorem 4.4.   □

REFERENCES

[AL] J. Apel and W. Lassner, An extension of Buchberger's algorithm and calculations in enveloping fields of Lie algebras, *J. Symbolic Computation*, 6(1988), 361–370.

[Bj] J-E. Björk, *Rings of Differential Operators*, North-Holland Mathematical Library, Vol. 21, 1979.

[Blo1] R.E Block, The irreducible representations of the Weyl algebra A_1, *in* "Séminaire d'Algèbre Paul Dubreil (Proceedings, Paris 1977-1978)" (M.P. Malliavin, Ed.), Lecture Notes in Mathematics No. 740, pp. 69–79, Springer-Verlag, Berlin/New York, 1979.

[Blo2] R.E. Block, The irreducible representations of the Lie algebra $Sl(2)$ and of the Weyl algebra, *Adv. Math.*, 39(1981), 69–110.

[CLO] D. Cox, J. Little and D. Óshea, *Ideals, Varieties, and Algorithms*, Springer-Verlag, 1991.

[Cot] S.C. Coutinho, *A primer of algebraic D-modules*, Cambridge University Press, 1995.

[Dix1] J. Dixmier, Representations irréductibles des algèbres de Lie nilpotentes, *Anais Acad. Bras. Cienc.*, 35(1963), 491–519.

[Dix2] J. Dixmier, Sur les algèbres de Weyl, *Bull. Soc. Math. France*, 96(1968), 209–242.

[Dix3] J. Dixmier, Sur les algèbres de Weyl II, *Bull. Sci. Math.*, 94(1970), 289–301.

[K-RW] A. Kandri-Rody and V. Weispfenning, Non-commutative Groebner bases in algebras of solvable type, *J. Symbolic Computation*, 9(1990), 1–26.

[Li] Li Huishi, Hilbert polynomial of modules over the homogeneous solvable polynomial algebras, *Comm. Alg.*, 5(27)(1999).

[LW] Li Huishi and Wu Yihong, Filtered-graded transfer of Groebner basis computation in solvable polynomial algebras, *To appear in Comm. Alg.*, 1998.

[Mor] T. Mora, An introduction to commutative and noncommutative Groebner bases, *Theoretical Computer Science*, 134(1994), 131–173.

[MR] J.C. McConnell and J.C. Robson, *Noncommutative Noetherian Rings*, John Wiley & Sons, 1987.

[Sti] M. Stillman, Methods for computing in algebraic geometry and commutative algebra, *Acta Applicandae Mathematicae*, 21(1990), 77–103.

The Construction of a Generator for *R*-DMod

LEANDRO MARÍN

DEPARTAMENTO DE MATEMÁTICA APLICADA, UNIVERSIDAD DE MURCIA. 30071-MURCIA-SPAIN
E-mail address: `leandro@fcu.um.es`

ABSTRACT. Let R be an associative ring (without identity). In this paper we consider the categories DMod-R of right R-modules such that $M \otimes_R R \simeq M$ and CMod-R of right R-modules such that $M \simeq \operatorname{Hom}_R(R, M)$ and the corresponding categories of left R-modules. The category CMod-R is a quotient category, therefore it is Grothendieck. The main purpose of this paper is to study the category R-DMod. A generator is constructed and as a consequence, some properties of this category arise.

1. INTRODUCTION

In this paper we are going to fix an (associative) ring R and a ring with identity A in which R is a two-sided ideal. The existence of such a ring is clear because given R, the Dorroh's extension of R, $\mathbb{Z} \times R$, has always this property. In the following, all A-modules will be considered unitary A-modules.

Definition 1. We shall denote CMod-R the full subcategory of Mod-A with the modules M, such that the canonical homomorphism

$$\lambda_M : M \to \operatorname{Hom}_A(R, M) \quad \lambda_M(m)(r) = mr$$

is an isomorphism.

The category DMod-R is the full subcategory of Mod-A with the modules M, such that the canonical homomorphism

$$\mu_M : M \otimes_A R \to M \quad \mu_M(m \otimes r) = mr$$

is an isomorphism.

We shall use the notations R-DMod and R-CMod for the corresponding categories of left modules.

The definition of these categories does not depend on the election of the ring A if R is a two sided ideal in it, as it is proved in [3].

The category CMod-R is a quotient category, therefore it is possible to build the localization functor $\mathbf{C} : \text{Mod-}A \to \text{CMod-}R$ with very nice properties. In order to study the category DMod-R it is not possible to dualize the properties of CMod-R. We shall construct a generator of this category and with the help of it we shall prove several results about this category. We shall also prove that the properties of the right adjoint of the inclusion $\mathbf{J} : \text{DMod-}R \to \text{Mod-}A$ has not so nice properties like $\mathbf{C}$. For example, the functor $\mathbf{D} : \text{Mod-}A \to \text{DMod-}R$ is not going to preserve cokernels and therefore the category DMod-R is not going to be in general a coquotient category of Mod-A.

If $f : M \to N$ and $g : N \to K$ are homomorphisms in a category of right modules we shall use the notation $g \circ f$ to the composition $M \to N \to K$. If the

Supported by D.G.E.S. of Spain, grant PB96-0961-C02-02.

category is formed with left modules, we shall use the notation $f * g$ to denote the homomorphism $M \to N \to K$. Homomorphisms of right modules will be written on the right and the homomorphisms of left modules will be written on the left. We use mainly the categories R-DMod and CMod-R.

2. Some Previous Results

An A-module M is said to be *completely annihilated by* R if $MR = 0$. The class of modules completely annihilated by R is an hereditary pre-torsion class. Using the general construction given in [4, Section VI.2] we can define the associated torsion free class $\mathcal{F}$ given by the modules M such that $\text{Hom}_A(X, M) = 0$ for all X completely annihilated by R. This class coincide with the class of modules M such that for all $m \in M$, if $mR = 0$ then $m = 0$. The class $\mathcal{T}$ of modules M such that $\text{Hom}_A(M, F) = 0$ for all F in $\mathcal{F}$ is the corresponding torsion class and $(\mathcal{T}, \mathcal{F})$ is an hereditary torsion theory. The modules M in $\mathcal{T}$ can be characterized by the property: $\forall m \in M, \forall (r_n)_{n \in \mathbb{N}} \in R^{\mathbb{N}}, \exists n_0 \in \mathbb{N}$ such that $mr_1 \cdots r_{n_0} = 0$, see [2, Lemma 3.1]. The modules that satisfy this property will be called *eventually annihilated by R*.

Associated to the class of modules completely annihilated by R we have an idempotent preradical $\mathbf{t}$ given by $\mathbf{t}(M) = \{m \in M : mR = 0\}$. Using this preradical we can build the idempotent radical associated to $\mathcal{T}$ using the general construction given in [4, Chapter VI] by transfinite induction. For that we define $\mathbf{t}_1(M) = \mathbf{t}(M)$; given an ordinal α we define $\mathbf{t}_\alpha(M) = \{m \in M : mR \subseteq \mathbf{t}_\beta(M)\}$ if $\alpha = \beta + 1$ for some ordinal β and $\mathbf{t}_\alpha(M) = \sum_{\beta < \alpha} \mathbf{t}_\beta(M)$. Fixing a module M we can define $\mathbf{T}(M) = \sum_\alpha \mathbf{t}_\alpha(M)$ because this construction should have an end, i.e., there exists an ordinal α such that $\mathbf{T}(M) = \mathbf{t}_\beta(M)$ for all $\beta \geq \alpha$. In fact, using [2, Lemma 3.1] we know that $\mathbf{T}(M)$ is the submodule of M eventually annihilated by R, i.e.

$$\mathbf{T}(M) = \{m \in M : \forall (r_n)_{n \in \mathbb{N}} \in R^{\mathbb{N}}, \exists n_0 \in \mathbb{N}, mr_1 \cdots r_{n_0} = 0\}.$$

The class of A-modules completely annihilated by R is also closed to products and subobjects, i.e., it is a pre-torsion free class. The torsion theory cogenerated by this pre-torsion free class is formed by the class of torsion modules $\mathcal{U} = \{M \in \text{Mod-}A : MR = M\}$ that we shall call *unitary modules* and the torsion free class of modules $\mathcal{V}$ with the $M \in \text{Mod-}A$ such that $\text{Hom}_A(U, M) = 0$ for all $U \in \mathcal{U}$, these modules will be called *vanishing modules*. Associated to the class of modules completely annihilated by R there exists a radical that we shall denote $\mathbf{u}$ defined by $\mathbf{u}(M) = MR$. Using again general constructions we can define the idempotent radical associated to the torsion theory $(\mathcal{U}, \mathcal{V})$ by transfinite induction. We shall denote $\mathbf{u}^1(M) = \mathbf{u}(M)$ and for an ordinal α we have, $\mathbf{u}^\alpha(M) = \mathbf{u}(\mathbf{u}^\beta(M))$ if $\alpha = \beta + 1$ and $\mathbf{u}^\alpha(M) = \cap_{\beta < \alpha} \mathbf{u}^\beta(M)$ if α is a limit ordinal. This sequence of submodules of M should stabilize for some ordinal, so we can define $\mathbf{U}(M) = \cap_\alpha \mathbf{u}^\alpha(M)$. This $\mathbf{U}$ is the idempotent radical associated to $(\mathcal{U}, \mathcal{V})$. The module $\mathbf{U}(M)$ is the biggest unitary submodule of M. All these are general constructions given for torsion theories and can be seen in [4, Chapter VI].

3. T-GENERATOR SETS

Definition 2. Let X be a subset of R. We shall say that X is a right T-generator set of R if the right A-module R/XR is eventually annihilated by R, i.e., for any sequence $(r_n)_{n \in \mathbb{N}} \in R^{\mathbb{N}}$ exists $n_0 \in \mathbb{N}$ such that $r_1 \cdots r_{n_0} \in XR$.

It is clear that any ring has T-generator sets, the whole R is a right and left T-generator set. In some circumstances it is possible to find special T-generator sets, for example, if R is a ring with identity 1_R, the set $\{1_R\}$ is a right and left T-generator set. More in general, if R is a ring with a set of local units E in the sense of [1], this set is also a right and left T-generator set. Any right generator set of R as A-module is a T-generator set of R.

Lemma 3. Let G be an abelian group, M_A and $_A N$ two A-modules with $RN = N$. Let $\beta : M \otimes_A N \to G$ be an abelian group homomorphism. Let K be a submodule of M such that for all $k \in K$, $n \in N$, $\beta(k \otimes n) = 0$. Then if $K \subseteq L \subseteq M$ and L/K is eventually annihilated by R, then for all $l \in L$ and all $n \in N$, $\beta(l \otimes n) = 0$.

Proof. First, we are going to prove that for all $m \in M$ and all $n \in N$ with $\beta(m \otimes n) \neq 0$ exists $n' \in N$ and $r' \in R$ such that $\beta(mr' \otimes n') \neq 0$. For that we only have to write $n \in RN$ as a finite sum $\sum_{i=1}^{t} r_i' n_i'$ and using the fact that $0 \neq \beta(m \otimes n) = \sum_{i=1}^{t} \beta(mr_i' \otimes n_i')$, we deduce that some summand is not 0.

Using several times this property, we can find sequences $(r_t)_{t \in \mathbb{N}} \in R^{\mathbb{N}}$ and $(n_t)_{t \in \mathbb{N}} \in N^{\mathbb{N}}$ such that $\beta(mr_1 \cdots r_t \otimes n_t) \neq 0$ for all $t \in \mathbb{N}$. From that we deduce that for all $l \in L$ and all $n \in N$, $\beta(l \otimes n) = 0$ because $l + K$ is eventually annihilated by R. $\square$

Proposition 4. Let X be a right T-generator set of R, then

1. *A left A-module $_A M$ is unitary if and only if $XM = M$.*
2. *A right A-module M_A is eventually annihilated by R if and only if for all $(x_n)_{n \in \mathbb{N}} \in X^{\mathbb{N}}$, exists $n_0 \in \mathbb{N}$ such that $m x_1 \cdots x_{n_0} = 0$.*

Proof. (1). In general we don't know if XM is an A-module, but we are sure it is an abelian group and we can calculate the quotient group M/XM and the abelian group homomorphism $\beta : R \otimes_A M \to M/XM$ given by $\beta(r \otimes m) = rm + XM$ for all $r \in R$ and $m \in M$. For any $k \in XM$ and $m \in M$ we have $\beta(k \otimes m) = 0$ and therefore if R/XR is eventually annihilated by R, $\beta \equiv 0$ and $M = XM$. Conversely, if $M = XM$, $M = XM \subseteq RM \subseteq M$ and therefore $M = RM$.

(2). If M is eventually annihilated by R, it is eventually annihilated by X because $X \subseteq R$. On the other side, suppose M is eventually annihilated by X, and let $m \in M$ and $(r_n)_{n \in \mathbb{N}} \in R^{\mathbb{N}}$ such that for all $n \in \mathbb{N}$, $m r_1 \cdots r_n \neq 0$. Using the fact that X is a right T-generator set of R, we can find $n_1 \in \mathbb{N}$ such that $r_1 \cdots r_{n_1} = \sum_{i \in I} x_i s_i$ for some finite set I. Let $I_t = \{i \in I : m x_i s_i r_{n_1+1} \cdots r_{n_1+t} \neq 0\}$, all these sets are not empty finite subsets of I and $I_1 \supseteq I_1 \supseteq \cdots$, therefore we can find $i \in \cap_{t \in I} I_t$. We have proved that for all $m \in M$ and $(r_n)_{n \in \mathbb{N}} \in R^{\mathbb{N}}$, exists $x \in X$ and a sequence $(s_n)_{n \in \mathbb{N}} \in R^{\mathbb{N}}$ such that $m x s_1 \cdots s_n \neq 0$ for all $n \in \mathbb{N}$. Using this property several times we can find a sequence $(x_n)_{n \in \mathbb{N}} \in X^{\mathbb{N}}$ such that $m x_1 \cdots x_n \neq 0$ for all $n \in \mathbb{N}$ and this is a contradiction. $\square$

4. THE CATEGORY CMod-R

We are going to denote $\mathcal{G}$ the set of right ideals I of A such that A/I is a right A-module eventually annihilated by R. This is the Gabriel topology associated to the hereditary torsion theory $(\mathcal{T}, \mathcal{F})$. We shall denote Mod-$(A, \mathcal{G})$ the quotient category of Mod-A by the topology $\mathcal{G}$ as it is defined in [4, Page 199].

Proposition 5. *Let M_A be a right A-module. The following conditions are equivalent:*

1. M *is in* CMod-R.
2. *For all $K_A \subseteq L_A$ with L/K completely annihilated by R and all $f : K \to M$, there exists one and only one $\overline{f} : L \to M$ extending f.*
3. *For all $K_A \subseteq L_A$ with L/K eventually annihilated by R and all $f : K \to M$, there exists one and only one $\overline{f} : L \to M$ extending f.*
4. M *is in* Mod-$(A, \mathcal{G})$.

Proof. $(1 \Rightarrow 2)$. Let $f : K \to M$, using the fact that $LR \subseteq K$ we can define $\tilde{f} : L \to \mathrm{Hom}_A(R, M)$ given by $\tilde{f}(l)(r) = f(lr)$ and composing with the isomorphism $\lambda_M : M \to \mathrm{Hom}_A(R, M)$ we obtain $\overline{f} : L \to M$ that extends f. Suppose we have two homomorphisms, $\overline{f}, \overline{g} : L \to M$ extending f, then for all $l \in L$, $\overline{f}(l) - \overline{g}(l) \in \mathrm{Ker}(\lambda_M) = 0$ and therefore $\overline{f} = \overline{g}$.

$(2 \Rightarrow 3)$. For any ordinal α we shall denote $K_\alpha = \{l \in L : l + K \in \mathbf{t}_\alpha (L/K)\}$. Using the definition of the $\mathbf{t}_\alpha$ we know that $K_{\alpha+1}/K_\alpha$ is completely annihilated by R and also K_1/K. Using this fact and (2), we can define $f_1 : K_1 \to M$ extending f. Let α be and ordinal and suppose we have defined $f_\beta : K_\beta \to M$ for any ordinal $\beta < \alpha$ extending the previous ones. If α is a limit ordinal we can define $f_\alpha = \varinjlim_{\beta < \alpha} f_\beta$ and if $\alpha = \delta + 1$ we can define $f_\alpha : K_\alpha \to M$ because K_α/K_δ is completely annihilated by R. To complete the proof we only have to notice that the module L/K is in $\mathcal{T}$ and therefore there exists an ordinal α such that $\mathbf{t}_\alpha (L/K) = L/K$, the we have $K_\alpha = L$ and $f_\alpha = \overline{f}$. The uniqueness is given in each step.

$(3 \Rightarrow 4)$. We have to prove that the canonical homomorphism $\mathrm{Hom}_A(A, M) \to \mathrm{Hom}_A(I, M)$ is a isomorphism for any $I \in \mathcal{G}$, but this is clear using (3) because A/I is eventually annihilated by R.

$(4 \Rightarrow 1)$. This is trivial because $R \in \mathcal{G}$. $\square$

Using the previous result, we can construct the localization functor $\mathbf{C} : \mathrm{Mod}\text{-}A \to \mathrm{CMod}\text{-}R$ that preserves kernels and is a left adjoint of the canonical inclusion $\mathbf{I} : \mathrm{CMod}\text{-}R \to \mathrm{Mod}\text{-}A$.

5. SUPPORTS AND ASSOCIATED MODULES

Let X be a set. We shall denote $\Sigma(X)$ the set of words over X or free monoid over X with the operation given by juxtaposition. The identity of this monoid is the empty word, $\varnothing$. We shall use the notation $\overline{x} \in \Sigma(X)$ as abbreviated notation of $x_1 \cdots x_{\lambda(\overline{x})}$. The map $\lambda : \Sigma(X) \to \mathbb{N} \cup \{0\}$ gives the length of the words in $\Sigma(X)$. We shall define $\overline{x} \leq \overline{y}$ if and only if exists $\overline{z} \in \Sigma(X)$ such that $\overline{x}\overline{z} = \overline{y}$. The elements in X will be identified with the words of length 1.

In some cases, we shall consider the opposite monoid $\Sigma(X)^{\mathrm{opp}}$ with the same elements of $\Sigma(X)$ but opposite operation. As we are not using any symbol to represent this operation we shall use the following rule: An element $\overline{x} \in \Sigma(X)$ will be represented by $\underline{x}$ if we consider it in $\Sigma(X)^{\mathrm{opp}}$. Therefore we have $\overline{x}\overline{y} = \underline{y}\underline{x}$ or $\overline{x}z = z\underline{x}$ for any $\overline{x}, \overline{y} \in \Sigma(X)$ and any $z \in X$.

Definition 6. A subset $\sigma \subseteq \Sigma(X)$ will be called an *unitary support* if it satisfies the following conditions:

1. For any $\overline{x} \in \sigma$ and $\overline{y} \leq \overline{x}$ we have $\overline{y} \in \sigma$.
2. For any $\overline{x} \in \sigma$, the set $\{y \in X : \overline{x}y \in \sigma\}$ is finite and nonempty.

The set of unitary supports over X will be denoted $\Xi_U(X)$.

A trivial example of unitary support is the empty set. If a unitary support is not empty, then it is infinite as a set. Given $\sigma \in \Xi_U(X)$ and $\overline{x} \in \sigma$, we shall denote $\Delta_{\underline{x}}\sigma = \{\overline{y} \in \Sigma(X) : \overline{xy} \in \sigma\}$. It is not difficult to prove that $\Delta_{\underline{x}}\sigma$ is a unitary support and that $X \cap \Delta_{\underline{x}}\sigma = \{y \in X : \overline{x}y \in \sigma\}$.

Lemma 7. *Let $\sigma \in \Xi_U(X)$, $\sigma \neq \varnothing$. Then $X^n \cap \sigma$ is a finite and non-empty set for all $n \geq 0$.*

Proof. If $n = 0$ the result is clear because $X^0 \cap \sigma = \{\varnothing\}$ because σ is not empty. Suppose we have proved the result for n, then deduce that the set

$$X^{n+1} \cap \sigma = \{\overline{x}y : \overline{x} \in X^n \cap \sigma, y \in X \cap \Delta_{\underline{x}}\sigma\}$$

if finite because it is a finite union of finite sets and it is non-empty because for any $\overline{x} \in X^n \cap \sigma$ exists at least one $y \in X \cap \Delta_{\underline{x}}\sigma$ using the definition of unitary support. $\qquad\square$

Associated to any unitary support σ we are going to define a module in R-DMod, $\langle\!\langle\sigma\rangle\!\rangle$. For that we shall fix a map $\pi : X \to R$.

Let $\sigma \in \Xi_U(X)$. If $\sigma = \varnothing$ we define $\langle\!\langle\sigma\rangle\!\rangle = 0$. If $\sigma \neq \varnothing$ we shall denote $F_n = A^{X^n \cap \sigma}$, the free A module with index set $X^n \cap \sigma$ the is finite and non-empty for any n. For any $\overline{x} \in X^n \cap \sigma$ we shall denote $(\underline{x})f$ the element in F_n with 1_A in the $\overline{x}$-th component and 0 elsewhere. With this notations we define the homomorphisms

$$F_n \to F_{n+1} \quad (\underline{x})f \mapsto \sum_{y \in X \cap \Delta_{\underline{x}}\sigma} \pi(y)(y\underline{x})f.$$

This family of homomorphisms induce a direct system and we shall define $\langle\!\langle\sigma\rangle\!\rangle = \varinjlim_{n \in \mathbb{N}} F_n$. The elements associated to $(\underline{x})f$ sin $\langle\!\langle\sigma\rangle\!\rangle$ will be denoted by $\langle\underline{x}\rangle_\sigma$. It is not hard to prove that any element in $\langle\!\langle\sigma\rangle\!\rangle$ can be written as a linear combination of elements $\{\langle\underline{x}\rangle_\sigma : \overline{x} \in \sigma\}$ with coefficients in A.

Proposition 8. *For any $\sigma \in \Xi_U(X)$, the module $\langle\!\langle\sigma\rangle\!\rangle$ is in the category R-DMod.*

Proof. If $\sigma = \varnothing$ the result is easy because the module 0 is in the category R-DMod. Suppose $\sigma \neq \varnothing$. For any element $\overline{x} \in \sigma$ and any $a \in A$,

$$a\langle\underline{x}\rangle_\sigma = \sum_{y \in X \cap \Delta_{\underline{x}}\sigma} a\pi(y)\langle y\underline{x}\rangle_\sigma \in R\langle\!\langle\sigma\rangle\!\rangle,$$

therefore $\langle\!\langle\sigma\rangle\!\rangle = R\langle\!\langle\sigma\rangle\!\rangle$. The module $\langle\!\langle\sigma\rangle\!\rangle$ is a direct limit of free modules, therefore it is flat in the category A-Mod, and the following sequence is exact

$$0 \to R \otimes_A \langle\!\langle\sigma\rangle\!\rangle \to A \otimes_A \langle\!\langle\sigma\rangle\!\rangle \to A/R \otimes_A \langle\!\langle\sigma\rangle\!\rangle \to 0.$$

Using the fact that $(A/R)R = 0$ and $R\langle\!\langle\sigma\rangle\!\rangle = \langle\!\langle\sigma\rangle\!\rangle$ we deduce that $A/R \otimes_A \langle\!\langle\sigma\rangle\!\rangle = 0$ and therefore $R \otimes_A \langle\!\langle\sigma\rangle\!\rangle \simeq A \otimes_A \langle\!\langle\sigma\rangle\!\rangle \simeq \langle\!\langle\sigma\rangle\!\rangle$. $\qquad\square$

Proposition 9. *Let M be al left A-module such that $RM = M$, X a set and $\pi : X \to R$ a map such that $\pi(X)$ is a right T-generator set of R. Then for all $m \in M$, exists $\sigma \in \Xi_U(X)$ and a homomorphism $h : \langle\!\langle\sigma\rangle\!\rangle \to M$ such that $(\langle\varnothing\rangle_\sigma)h = m$.*

Proof. If $m = 0$ we can take any non-empty unitary support and the 0 homomorphism. Suppose $m \neq 0$. Using the fact that $\pi(X)$ is a right T-generator set of R, we deduce that $\pi(X)M = M$. We are going to make the construction of the support and the homomorphism in a recursive way. Let $V_0 = \{\varnothing\}$ and $h_0 : V_0 \to M$ the map given by $h_0(\varnothing) = m$. Suppose we have already defined $V_n \subseteq X^n$ and $h_n : V_n \to M \setminus \{0\}$. For each $\overline{x} \in V_n$, $h_n(\overline{x}) \in M = \pi(X)M$, therefore we can define a finite and non-empty set $W_{\overline{x}} \subseteq X$ and $g_{\overline{x}} : W_{\overline{x}} \to M \setminus \{0\}$ such that $h_n(\overline{x}) = \sum_{y \in W_{\overline{x}}} \pi(y) g_{\overline{x}}(y)$. We shall define $V_{n+1} = \{\overline{x}y \in X^{n+1} : \overline{x} \in V_n, y \in W_{\overline{x}}\}$ and $h_{n+1} : V_{n+1} \to M \setminus \{0\}$ such that $h_{n+1}(\overline{x}y) = g_{\overline{x}}(y)$.

We shall define $\sigma = \cup_{n \geq 0} V_n$ that is a unitary support because of the construction and $X^n \cap \sigma = V_n$ for all $n \in \mathbb{N}$. Then we can define $\hat{h}_n : F_n \to M$ by $((\underline{x})f)\hat{h}_n = h_n(\overline{x})$ for all $\overline{x} \in V_n$. These homomorphisms induce $h = \varinjlim_{n \in \mathbb{N}} \hat{h}_n : \langle\!\langle \sigma \rangle\!\rangle \to M$ such that $((\underline{x})_\sigma)h = h_{\lambda(\overline{x})}(\overline{x})$, in particular, $((\varnothing)_\sigma)h = m$. $\square$

6. A GENERATOR FOR R-DMod

Proposition 10. *Let K be a left A-module from the category R-DMod and L an A-submodule of K. Then $K/L \in R$-DMod if and only if $RL = L$.*

Proof. Consider the following diagram with exact rows

$$
\begin{array}{ccccccc}
L & \longrightarrow & K & \longrightarrow & K/L & \longrightarrow & 0 \\
\uparrow{\scriptstyle \mu_L} & & \uparrow{\scriptstyle \mu_K} & & \uparrow{\scriptstyle \mu_{K/L}} & & \\
R \otimes_A L & \longrightarrow & R \otimes_A K & \longrightarrow & R \otimes_A K/L & \longrightarrow & 0
\end{array}
$$

The module K/L satisfies $R(K/L) = K/L$ because K does, therefore $\mu_{K/L}$ is an epimorphism. The result we have to prove is that $\mu_{K/L}$ is a monomorphism if and only if μ_L is an epimorphism, but using the fact that μ_K is an isomorphism, the result is a consequence of the Snake Lemma because $\mathrm{Coker}(\mu_L)$ and $\mathrm{Ker}(\mu_{K/L})$ are isomorphic. $\square$

Corollary 11. *Let $\mathbf{J} : R$-DMod $\to A$-Mod be the canonical inclusion and $f : K \to L$ be a homomorphism in R-DMod. Then f is an epimorphism in the category R-DMod if and only if $\mathbf{J}(f)$ is an epimorphism in A-Mod.*

Proof. If $\mathbf{J}(f)$ is an epimorphism in A-Mod, the map f is surjective and therefore f is an epimorphism in R-DMod. Conversely, suppose f is an epimorphism in R-DMod. Using Proposition 10 we deduce that the A-homomorphism $L \to L/\mathrm{Im}(\mathbf{J}(f))$ is in the category R-DMod, but composed with f it is 0, therefore the morphism $L \to L/\mathrm{Im}(\mathbf{J}(f))$ is 0 and $L = \mathrm{Im}(\mathbf{J}(f))$. $\square$

Proposition 12. *Let X be a set, $\pi : X \to R$ be a map such that $\pi(X)$ is a right T-generator set of R. Then $\coprod_{\sigma \in \Xi_U(X)} \langle\!\langle \sigma \rangle\!\rangle$ is a generator of the category R-DMod.*

Proof. Let G be that coproduct. The A-module G is in the category R-DMod because $R \otimes_A -$ commutes with coproducts and $R \otimes_A \langle\!\langle \sigma \rangle\!\rangle \simeq \langle\!\langle \sigma \rangle\!\rangle$ for all $\sigma \in \Xi_U(X)$, therefore $R \otimes_A G \simeq G$.

Let $M \in R$-DMod, in particular $RM = M$ because μ_M is surjective, therefore we deduce from Proposition 9 that for any $m \in M$ there exists a homomorphism $h_m : G \to M$ such that $m \in \mathrm{Im}(\mathbf{J}(h_m))$. Using this for any element in M we can find a set I and a homomorphism $e : G^{(I)} \to M$ that is surjective and then an epimorphism. This proves that G is a generator of the category R-DMod. $\square$

7. THE FUNCTOR $\mathbf{D} : A\text{-Mod} \to R\text{-DMod}$

In this section we are going to fix a generator M of the category R-DMod. Using this generator we are going to construct a right adjoint of the canonical inclusion $\mathbf{J} : R\text{-DMod} \to A\text{-Mod}$. The uniqueness of the adjoints will prove that the construction is independent of the chosen generator.

Let N be a module in A-Mod, we shall denote $H(N) = \mathrm{Hom}_A(M, N)$ and

$$\eta_N : M^{(H(N))} \to N \quad ((m_h)_{h \in H(N)})\eta_N = \sum_{h \in H(N)} (m_h)h.$$

We shall define $\mathbf{D}(N) = M^{(H(N))}/\mathbf{U}(\mathrm{Ker}(\eta_N))$. This module is in R-DMod as a consequence of Proposition 10. The unitary module $\mathbf{U}(\mathrm{Ker}(\eta_N)) \subseteq \mathrm{Ker}(\eta_N)$ induces a homomorphism $\nu_N : \mathbf{D}(N) \to N$.

Let $f : N \to L$ be a homomorphism in A-Mod. We can define the homomorphism $M^{(H(N))} \to M^{(H(L))}$ as follows: let $m \in M$ and $h \in H(N)$, the element in $M^{(H(N))}$ such that has m in the h-th component and 0 elsewhere will go to the element of $M^{(H(L))}$ that has m in the $h*f$-th component and 0 elsewhere. This homomorphism makes commutative the following diagram

$$
\begin{array}{ccc}
M^{(H(N))} & \xrightarrow{\eta_N} & N \\
\downarrow & & \downarrow f \\
M^{(H(L))} & \xrightarrow{\eta_L} & L
\end{array}
$$

The commutativity of this diagram induces $\mathrm{Ker}(\eta_N) \to \mathrm{Ker}(\eta_L)$ and using the radical $\mathbf{U}$ we obtain by restriction $\mathbf{U}(\mathrm{Ker}(\eta_N)) \to \mathbf{U}(\mathrm{Ker}(\eta_L))$. All these constructions make possible to define $\mathbf{D}(f) : \mathbf{D}(N) \to \mathbf{D}(L)$ such that the following diagram is commutative

$$
\begin{array}{ccc}
\mathbf{D}(N) & \xrightarrow{\nu_N} & N \\
\mathbf{D}(f) \downarrow & & \downarrow f \\
\mathbf{D}(L) & \xrightarrow{\nu_L} & L
\end{array}
$$

This homomorphism is unique. Suppose there are two homomorphisms $g, g' : \mathbf{D}(N) \to \mathbf{D}(L)$ making commutative the previous diagram, then the difference $g - g'$ composed with ν_L is 0 and then $g - g'$ induces a morphism $\mathbf{D}(N) \to \mathrm{Ker}(\nu_L) = \mathrm{Ker}(\eta_L)/\mathbf{U}(()\,\mathrm{Ker}(\eta_L))$ that should be 0 because $\mathbf{D}(N) \in \mathcal{U}$ and $\mathrm{Ker}(\eta_L)/\mathbf{U}(()\,\mathrm{Ker}(\eta_L)) \in \mathcal{V}$. Using this uniqueness it is not hard to prove that $\mathbf{D}$ is a functor.

Proposition 13. *With the previous notations:*

1. *For all L in A-Mod, $\mathrm{Im}(\nu_L) = \mathbf{U}(L)$.*
2. *For all N in R-DMod, ν_L is an isomorphism.*
3. *Let $N \in R\text{-DMod}$ and $K \in A\text{-Mod}$. Then for all $f : N \to K$ there exists one and only one $\overline{f} : N \to \mathbf{D}(K)$ such that $\overline{f} * \nu_K = f$.*
4. *The functor $\mathbf{D} : A\text{-Mod} \to R\text{-DMod}$ is a right adjoint of the canonical inclusion $\mathbf{J} : R\text{-DMod} \to A\text{-Mod}$.*

Proof. (1). Let $l \in L$ and X a right T-generator set of R, $\pi : X \to R$ the inclusion. Using Proposition 9 we can find a unitary support σ and a homomorphism $f : \langle\!\langle \sigma \rangle\!\rangle \to L$ such that $\text{Im}(f) \subseteq \mathbf{U}(L)$ and $((\varnothing)_\sigma)f = l$. The module M is a generator of R-DMod, therefore we can find a set I and a epimorphism $M^{(I)} \to \langle\!\langle \sigma \rangle\!\rangle$. Using this epimorphism we can take a finite subset $F \subseteq I$ and a homomorphism $g : M^{(F)} \to \langle\!\langle \sigma \rangle\!\rangle$ such that $\langle \varnothing \rangle_\sigma \in \text{Im}(g)$. We shall denote $\equiv$ the following equivalence relation in F: Two elements $v, u \in F$ are equivalent if and only if $q_u * g * f = q_v * g * f$ with $q_t : M \to M^{(F)}$ the corresponding canonical injection. Let α be an equivalence class in $F/\equiv$, we can define $h_\alpha = q_u * g * f : M \to L$ with $u \in \alpha$ (the definition does not depend on the election of u because of the definition of the equivalent relation). Given two different classes $\alpha, \beta \in F/\equiv$ we have that $h_\alpha \neq h_\beta$. The element $l \in L$ is in $\text{Im}(g * f)$ therefore we can find elements $m_\alpha \in M$ with $\alpha \in F/\equiv$ such that $l = \sum_{\alpha \in F/\equiv}(m_\alpha)h_\alpha$. This proves that $l \in \text{Im}(\eta_L) = \text{Im}(\nu_L)$.

(2). This is a consequence of Proposition 10 because $\text{Ker}(\eta_N) = \mathbf{U}(\text{Ker}(\eta_N))$.

(3). The existence is given by $\overline{f} = \nu_N^{-1} * \mathbf{D}(f)$. The uniqueness if a consequence of the fact that $N \in \mathcal{U}$, $\text{Ker}(\nu_K) \in \mathcal{V}$ and $\text{Hom}_A(U, V) = 0$ for all $U \in \mathcal{U}$, $V \in \mathcal{V}$ because $(\mathcal{U}, \mathcal{V})$ is a torsion theory.

(4). Let $N \in R$-DMod and $K \in A$-Mod. The homomorphism

$$\alpha_{NK} : \text{Hom}_A(N, \mathbf{D}(K)) \to \text{Hom}_A(N, K) \quad \alpha_{NK}(f) = f * \nu_K.$$

This homomorphism is natural in N and K. The existence and uniqueness of (3) proves that α_{NK} is an isomorphism. $\qquad\square$

8. Some Consequences in the Category R-DMod

Proposition 14. *The category R-DMod has the following properties:*

1. *It is additive.*
2. *It has a generator and a cogenerator.*
3. *It is complete and cocomplete.*
4. *It is conormal, i.e., every epimorphism is a cokernel.*
5. *A morphism $f : M \to N$ in R-DMod is a monomorphism if and only if $\text{Ker}(\mathbf{J}(f))$ is vanishing.*
6. *Any morphism in R-DMod can be decomposed in a monomorphism followed by a epimorphism.*
7. *It is balanced, i.e., if a morphism is a monomorphism and a epimorphism, then it is an isomorphism.*

Proof. (1). The additive structure of R-DMod comes from the fact that it is a full subcategory of A-Mod.

(2). The existence of the generator is already proved. Let C be a cogenerator of the category A-Mod. We are going to prove that $\mathbf{D}(C)$ is a cogenerator of the category R-DMod. Let $f : M \to N$ be a non-zero morphism in R-DMod. This is also a non-zero morphism in A-Mod and we can find $h : N \to C$ such that $f * h \neq 0$. Using Proposition 13 we can find a unique $\overline{h} : N \to \mathbf{D}(C)$ such that $\overline{h} * \nu_C = h$. Using this uniqueness we deduce that $f * \overline{h} \neq 0$ and therefore $\mathbf{D}(C)$ is a cogenerator of R-DMod.

(3). The completeness and cocompleteness of the category R-DMod is proved dualizing the proof of [4, Proposition X.1.2] with the right adjoint of the canonical inclusion R-DMod $\to A$-Mod instead of the left adjoint that is is used in that case.

(4). Let $f : M \to N$ be an epimorphism in R-DMod. Using Proposition 10 we know that $K_0 = \text{Ker}(\mathbf{J}(f))$ is unitary, and therefore, $\nu_{K_0} : K = \mathbf{D}(K_0) \to K_0$ is surjective. The composition $K \to K_0 \to M$ is a morphism in R-DMod with cokernel $M/K_0 \simeq N$ because direct limits calculated in R-DMod are the same as the ones calculated in the category A-Mod using the dual proof of [4, Proposition X.1.2].

(5). Let $K = \mathbf{U}(\text{Ker}(\mathbf{J}(f)))$. We have to prove that f is a monomorphism in R-DMod if and only if $K = 0$. Suppose $K \neq 0$, then $\mathbf{D}(K)$ is not 0 (because $\nu_K : \mathbf{D}(K) \to K$ is surjective) and the composition $\mathbf{D}(K) \to M \to N$ is 0. Therefore f cannot be a monomorphism. On the other hand, suppose $K = 0$ and let $h : L \to M$ be a morphism in R-DMod such that $L \to M \to N$ is 0. This composition is also 0 in A-Mod and therefore $\mathbf{J}(f) : \mathbf{J}(L) \to \mathbf{J}(M)$ factors through $\text{Ker}(\mathbf{J}(f))$, but L is unitary and $\text{Ker}(\mathbf{J}(f))$ is vanishing. This proves that $\mathbf{J}(f) = 0$ and $f = 0$.

(6). Let $f : M \to N$ be a morphism in R-DMod, let $K = \mathbf{U}(\text{Ker}(\mathbf{J}(f)))$. The module K is unitary and then $M/K \in R$-DMod and we can decompose f in $M \to M/K$ and $M/K \to N$. The morphism $M \to M/K$ is clearly an epimorphism. The kernel of $M/K \to N$ calculated in A-Mod is $\text{Ker}(\mathbf{J}(f))/\mathbf{U}(\text{Ker}(\mathbf{J}(f)))$ that is a vanishing module, therefore $M/K \to N$ is a monomorphism in R-DMod.

(7). Let $f : M \to N$ be an epimorphism and monomorphism in the category R-DMod. As a consequence of being an epimorphism we know that $\text{Ker}(\mathbf{J}(f))$ is unitary (Proposition 10) and as a consequence of being monomorphism, $\text{Ker}(\mathbf{J}(f))$ is vanishing, therefore $\text{Ker}(\mathbf{J}(f)) = 0$. The homomorphism $\mathbf{J}(f)$ is also an epimorphism because f is. Then f is an isomorphism. $\qquad\square$

9. THE CATEGORY R-DMod IS NOT COGIRAUD IN GENERAL

In the case of the category CMod-R, we know that it is a Grothendieck category because it is a quotient category. More precisely, because the localizing functor $\mathbf{C}$ preserves kernels, see [4, Proposition X.1.3]. A dual proof for R-DMod is not possible because $\mathbf{D}$ does not preserve cokernels in general as we are going to prove in this section.

The example is given for the ring $R = 2\mathbb{Z}$. We shall denote $\mathbb{Q}_2 = \{r/2^\alpha \in \mathbb{Q} : \alpha \in \mathbb{N}\}$. $\mathbb{Q}_2$ is an unitary $2\mathbb{Z}$-module and $2\mathbb{Z} \subseteq \mathbb{Q}_2$. We shall use the ring with identity $A = \mathbb{Z}$.

Proposition 15. *Let M be an abelian group. The following conditions are equivalent:*

1. $M \in \text{CMod-}2\mathbb{Z}$.
2. $M \in \text{DMod-}2\mathbb{Z}$.
3. $M \in \text{Mod-}\mathbb{Q}_2$.
4. $M = M \cdot 2$ *and for all $m \in M$, if $m \cdot 2 = 0$ then $m = 0$.*

Proof. $(1 \Rightarrow 4)$. Let $m \in M$, and $f_m : 2\mathbb{Z} \to M$ with $f_m(2z) = mz$. This homomorphism is in $\text{Hom}_A(R, M)$ and therefore we can take $m' \in M$ such that $f_m = \lambda_M(m')$, in particular $m'2 = m$. Now suppose $k \in M$ such that $m2 = 0$, then $m \in \text{Ker}(\lambda_M) = 0$ and $m = 0$.

$(4 \Rightarrow 3)$. Let $m \in M$, for this element there exists one and only one $m' \in M$ such that $m'2 = m$. We shall define $m(1/2) = m'$. With this multiplication M is an unitary $\mathbb{Q}_2$-module.

$(3 \Rightarrow 2)$. We have to prove that $\mu_M : M \otimes_{\mathbb{Z}} 2\mathbb{Z} \to M$ is an isomorphism. It is surjective because $m = \mu_M(m(1/2) \otimes 2)$ for all $m \in M$. Suppose $\mu_M(m \otimes 2) = 0$, then $m2 = 0$ and $m = (m2)(1/2) = 0$. This proves that μ_M is also injective.

$(2 \Rightarrow 1)$. Let $f : 2\mathbb{Z} \to M$ be a homomorphism; $f(2) \in M = M2\mathbb{Z}$, then we can find $m' \in M$ such that $f(2) = m'2$ and then $\lambda_M : M \to \mathrm{Hom}_A(R, M)$ is surjective. Let $m \in \mathrm{Ker}(\lambda_M)$, then $m2 = 0$ and $m \otimes 2 = 0$. Using the result about the 0 in the tensor product give in [5, Seite 97] we can find $a_1, \cdots, a_k \in \mathbb{Z}$ and $m_1 \cdots m_k \in M$ such that $m = \sum_i m_i a_i$ and $a_i 2 = 0$ for all i. From this result we deduce that $a_i = 0$ for all i and $m = 0$. $\qquad \square$

The previous result proves in particular that the category DMod-$2\mathbb{Z}$ is abelian (because it coincides with the category of unitary modules over a ring with identity) and $\mathbb{Q}_2$ is a generator of this category.

Let M be an abelian group and $h : \mathbb{Q}_2 \to M$ be a homomorphism. This homomorphism is completely defined if we give a sequence $(m_\alpha)_{\alpha \in \mathbb{N}} \in M^{\mathbb{N}}$ such that $h(1/2^\alpha) = m_\alpha$. In fact it is not hard to prove that in this case, the module $\mathbf{D}(M) = (\mathbb{Q}_2)^{H(M)}/\mathbf{U}(\mathrm{Ker}(\eta_M))$ as it is defined in Section 7 is isomorphic to the module given by the sequences $(m_n)_{n \in \mathbb{N}} \in M^{\mathbb{N}}$ such that $m_n 2 = m_{n-1}$ for all $n \geq 1$ and the map $\nu_M : \mathbf{D}(M) \to M$ is given by $\nu_M((m_n)_{n \in \mathbb{N}}) = m_0$.

Counterexample 16. The functor $\mathbf{D} : A\text{-Mod} \to A\text{-DMod}$ does not preserve cokernels in general.

Proof. Let $M = \mathbb{Q}_2/2\mathbb{Z}$. The morphism $p : \mathbb{Q}_2 \to M$ is an epimorphism and a cokernel in the category Mod-$\mathbb{Z}$. The kernel of this morphism is $2\mathbb{Z}$ that is a vanishing module. If we apply the functor $\mathbf{D}$ we obtain the following commutative diagram

$$
\begin{array}{ccc}
\mathbb{Q}_2 & \xrightarrow{\ p\ } & M \\
\| & & \uparrow{\scriptstyle \nu_M} \\
\mathbb{Q}_2 & \xrightarrow{\ \mathbf{D}(p)\ } & \mathbf{D}(M)
\end{array}
$$

The module $\mathbb{Q}_2$ is in the category R-DMod and therefore $\mathbf{D}(\mathbb{Q}_2) = \mathbb{Q}_2$. We are going to prove that $\mathbf{D}(p)$ is not an epimorphism in $2\mathbb{Z}$-DMod. The elements in $\mathbf{D}(M)$ can be considered as sequences like $(m_n)_{n \in \mathbb{N}} \in M^{\mathbb{N}}$ such that $m_n 2 = m_{n-1}$. Let $(e_n)_{n \in \mathbb{N}}$ be the sequence $(0, 1, 0, 1, 0, 1, 0, 1, 0, 1, 0, 1, 0, \cdots)$ and $m_n = \sum_{k=0}^{n} e_k 2^{k-n} + 2\mathbb{Z} \in M$. The element associated to the sequence $(m_n)_{n \in \mathbb{N}}$ in $\mathbf{D}(M)$ is not in the image of $\mathbf{D}(p)$ because the elements in this image are elements associated to the sequences $(r/2^{\alpha+n} + 2\mathbb{Z})_{n \in \mathbb{N}}$ with $r \in \mathbb{Z}$ and $\alpha \in \mathbb{N}$. $\qquad \square$

REFERENCES

[1] Abrams, G.D.: Morita Equivalence for Rings with Local Units, *Communications in Algebra*, 11 (8), p. 801-837, 1983.

[2] Marín, L.: Morita Equivalence Based on Contexts for Various Categories of Modules over Associative Rings. *Journal of Pure and Applied Algebra*, 133, p. 219-232 (1998).

[3] Quillen, D.: *Module Theory over Nonunital Rings*, Preprint.

[4] Stenström, B.: *Rings of Quotients*, Springer-Verlag, Berlin-Heidelberg-New York, 1975.

[5] Wisbauer, R.: *Grundlagen der Modul- und Ringtheorie*, Verlag Reinhard Fischer, München, 1988.

A Lattice Invariant for Modules

Barbara L. Osofsky
Department of Mathematics
Rutgers University
110 Frelinghuysen Road
Piscataway, NJ 08854-8019

ABSTRACT.   We study two invariants on a subclass of the class of all right modules over a ring R with identity. These invariants are very closely related, and indeed are identical in some cases of interest. We give a collection of remarks and examples where these invariants apply in a variety of situations.

In [O2] we defined a new ordinal valued lattice invariant on the nonzero submodules of a free module over a von Neumann regular ring. The purpose of the definition there was to get an upper bound on the projective dimension of submodules of free modules. We conjectured that for finite ordinals, this invariant agrees with projective dimension.

In Section 1 of this paper we repeat the definition of this invariant and that of a related invariant and discuss some of the properties of these definitions. The invariants are defined in terms of an upper bound obtained by transfinite induction. One of them need not exist for even such nice modules as finitely generated modules over Noetherian rings. The other will always exist for coherent modules but is only of interest for infinitely generated modules. We give some examples and remarks to show that these invariants can give interesting information in a variety of cases.

In Section 2 we compute the invariants in a case where they are the same, namely uniserial modules. We then compute an upper bound for the second one of them for torsion-free modules over a Prüfer domain with countably many maximal ideals, a class of rings which includes valuation domains. This upper bound may be significantly smaller than the one used to show that the invariant exists for coherent modules.

1   BASIC DEFINITIONS, REMARKS, AND ELEMENTARY EXAMPLES

Conventions and notations

1. An empty sum is equal to 0. Thus if α is an ordinal, then $\sum_{\beta<\alpha} N_\beta = 0$ if $\alpha = 0$.

2. The symbol $\mathfrak{Ord}$ will denote the class of all ordinals $\cup \{-1\}$.

3. For an R-module M, let M be generated by some set of cardinality $\aleph_\kappa$ but no set of smaller cardinality generates M. We will denote this κ by $\mathrm{gen}\,(M)$. Set $\mathrm{gen}\,(M) = -1$ if M is finitely generated.

4. We will denote by Ω_κ the smallest ordinal of cardinality $\aleph_\kappa$.

5. The words 'set of ordinals' literally means set, not class. Hence any set of ordinals is bounded above.

6. Consistent with common notation. the supremum of a set of ordinals is the smallest proper upper bound and so not in the set. Thus $\sup\{0,1\} = 2$. When it is necessary to speak of the least upper bound (possibly the maximum) of a set A of ordinals, we will refer to $\max\sup\,(A)$.

7. Module actions are on the right unless otherwise stated. Conditions are also only on the right unless otherwise stated. Thus uniserial, Noetherian, Artinian. ideal, etc. means the same as right uniserial, right Noetherian, right Artinian, right ideal. etc.

The basic definitions

Definition 1.1 *We define a function ℓ on a subclass (possibly proper) of the class of R-modules to $\mathfrak{Ord}$ by*

(a) $\ell(M) = -1 \iff M = 0$

(b) $\ell(M) \leq \kappa \iff$ there exists a set $\{x_\alpha : \alpha \in \Omega\} \subseteq M$ such that

  (i) The indexing set Ω is well ordered.

    (ii) $M = \sum_{\alpha \in \Omega} x_\alpha R.$

    (iii) For all $\beta \in \Omega.$ $\ell \left(x_\beta R \cap \left(\sum_{\alpha < \beta} x_\alpha R \right) \right) < \kappa.$

(c) $\ell(M) = \inf \{ \kappa : \ell(M) \leq \kappa \}.$

There is a modification of this definition which turns out to be very useful. The modification $\ell'(M)$ is always defined for coherent modules and is less than or equal to $\ell(M)$ whenever $\ell(M)$ is defined.

Definition 1.2 *We define a function* ℓ' *on a subclass of the class of R-modules to* $\mathfrak{Ord}$ *by*

(a) $\ell'(M) = -1 \iff M = 0$

(b) $\ell'(M) \leq \kappa \iff$ *there exists a set* $\{ N_\alpha : \alpha \in \Omega \}$ *of finitely generated submodules of* M *such that*

    (i) The indexing set Ω *is well ordered.*

    (ii) $M = \sum_{\alpha \in \Omega} N_\alpha.$

    (iii) For all $\beta \in \Omega.$ $\ell' \left(N_\beta \cap \left(\sum_{\alpha < \beta} N_\alpha \right) \right) < \kappa.$

(c) $\ell'(M) = \inf \{ \kappa : \ell'(M) \leq \kappa \}.$

Definition 1.3 *The set* $\{ x_\alpha : \alpha \in \Omega \}$ *(respectively* $\{ N_\alpha : \alpha \in \Omega \}$*) is called an* ℓ*-bounding sequence (respectively* ℓ'*-bounding sequence) for* $M.$ *If* $\ell \left(x_\beta R \cap \sum_{\alpha < \beta} x_\alpha R \right) < \ell(M)$ *(respectively* $\ell' \left(N_\beta \cap \sum_{\alpha < \beta} N_\alpha \right) < \ell'(M)$*) for all* $\beta \in \Omega$ *it is called an* ℓ*-sequence (*ℓ'*-sequence). That is, an* ℓ*-sequence (or* ℓ'*-sequence) for* M *is an* ℓ*-bounding (or* ℓ'*-bounding) sequence with*

$$\ell(M) = \sup \left\{ \ell \left(x_\beta R \cap \sum_{\alpha < \beta} x_\alpha R \right) : \beta \in \Omega \right\}$$

$$\left(\text{respectively } \ell'(M) = \sup \left\{ \ell' \left(x_\beta R \cap \sum_{\alpha < \beta} x_\alpha R \right) : \beta \in \Omega \right\} \right).$$

Remarks and examples

Remark 1 If $\ell(M)$ is defined. then M must have some ℓ-sequence. Just look at the set of ordinals Ω for which there is a one-to-one map to a set of generators of M such that for all β in Ω. $\ell\left(x_\beta R \cap \left(\sum_{\alpha<\beta} x_\alpha R\right)\right)$ is defined. This set is nonempty. and for any element in it, the supremum of $\left\{\ell\left(x_\beta R \cap \left(\sum_{\alpha<\beta} x_\alpha R\right)\right)\right\}$ is a well defined ordinal. Taking a sequence with smallest supremum will give an ℓ-sequence. Similarly. if $\ell'(M)$ is defined. then M has an ℓ'-sequence.

Remark 2 If M is coherent. that is. if finitely generated submodules of M are finitely related. then $\ell'(M)$ is defined. A simple transfinite induction shows that it is at most the subscript of the smallest $\aleph$ of a generating set for M. This is done in Proposition 1.2.

Remark 3 There are modules M for which $\ell(M)$ is not defined, even over very nice rings.

Example 4 Let R be a Dedekind domain. By the Noetherian property, every ℓ-bounding sequence for an ideal of R must have only a finite number of distinct sums $\sum_{\alpha<\beta} x_\alpha R$. Call this finite number the length of the ℓ-bounding sequence. and throw out every x_β with $x_\beta \in \sum_{\alpha<\beta} x_\alpha R$ so the number of x_β which occur is the length of the ℓ-bounding sequence. If R is not a principal ideal domain. it possesses a non-principal ideal. Assume even one such non-principal ideal I of R has well defined $\ell(I)$. Pick any ℓ-bounding sequence for this non-principal I. There is a smallest $\alpha > 1$ for which $\sum_{\beta<a} x_\beta R$ is two-generated. Then $\sum_{3=0}^{\alpha-1} x_\beta R$ is principal. say with generator y_1. Then $\{y_1, x_\alpha\}$ is an ℓ-bounding sequence for the non-principal ideal $\sum_{\beta\leq\alpha} x_\beta R$. Then by standard properties of Dedekind domains, $y_1 R \cap x_\alpha R$ is cyclic and $(y_1 R + x_\alpha R) \oplus (r_0 R \cap x_1 R) \cong y_1 R \oplus x_\alpha R$ which implies $\sum_{\beta\leq\alpha} x_\beta R$ is cyclic, a contradiction. (See any standard reference on Dedekind domains, such as [B. Chapter VII. sections 2.3 and 4.10]).

Remark 5 It is reasonably easy to see that a module M has $\ell(M) \leq 0$ if and only if M is a direct sum of cyclics, and $\ell'(M) \leq 0$ if and only if M is a direct sum of finitely generated modules.

Remark 6 *In the case of submodules of free modules over a von Neumann regular ring. uniserial modules. or any other situation in which finitely generated submodules of the module M are cyclic or direct sums of cyclics, these two definitions agree.*

Remark 7 *One possible advantage of ℓ' over ℓ is that finitely generated is a categorical property. whereas cyclic is not. Hence if φ is a Morita equivalence between $Mod - R$ and $Mod - S$. then $\ell'(N) = \ell'(\varphi(N))$. It is not clear whether or not this property also holds for ℓ. but 2×2 matrices over a Dedekind domain may well be a counterexample.*

Remark 8 *Since a countably generated module is a sum of a linearly ordered (by inclusion) chain of finitely generated submodules. for any countably generated M. $\ell'(M) \leq 1$.*

We next look at some small values of global ℓ-dimension. Our first examples are rings where every module has ℓ-invariant at most 1.

Example 9 *If R is a Noetherian ring. then from the definition of ℓ' we see that for every R-module M. $\ell'(M) \leq 1$ since each intersection which occurs is a submodule of a finitely generated module and so finitely generated.*

Example 10 *If every ideal of R is principal. then from the definition of ℓ we see that for every R-module M. $\ell(M) \leq 1$. Moreover, if R is a p.i.d. and $M \neq 0$. then $\ell(M) = 0$ if and only if M is projective. This leaves $\ell(M) = 1$ if and only if $\mathrm{pd}(M) = 1$. and ℓ is the same as projective dimension for p.i.d.'s.*

Example 11 *Let $\mathcal{R}$ be the ring $\bigoplus_{i=0}^{\infty} \mathbb{R}_i + 1 \cdot \mathbb{R} \subseteq \prod_{i=0}^{\infty} \mathbb{R}_i$. A submodule of a cyclic $\mathcal{R}$-module is a direct sum of cyclics. Hence for any $\mathcal{R}$-module M. $\ell(M) \leq 1$.*

The property that every R-module M has $\ell(M) \leq 0$ (or $\ell'(M) \leq 0$) is a strong property. These cases have been widely studied.

Remark 12 *Assume $\ell'(M) \leq 0$ for all (right) R-modules M. It is a well known result that R must be (right) Artinian (see [C] or, in a later context. [F-W]). These rings have been extensively studied (see for example.*

[F]). They are precisely the pure semisimple rings. In this sense ℓ' can be considered a kind of pure dimension. For Artin algebras at least, these are precisely the Artin algebras of finite representation type.

Example 13 *For $R = \mathbb{Z}/n\mathbb{Z}$ with $n \neq 0$, every R-module M has $\ell(M) \leq 0$. Indeed, any (one sided) Artinian ring for which the indecomposable projectives on both sides are uniserial has $\ell(M) \leq 0$ for all modules M (see for example [F-A. Theorem 32.3]).*

Example 14 *Let R be an algebra over a field with 3 isomorphism classes of simple modules and indecomposable projectives that look like*

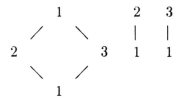

These rings are of finite representation type and have every indecomposable cyclic. They have $\ell(M) \leq 0$ for all M.

Remark 15 *Now assume $\ell(M) \leq 0$ for all right modules M. This condition is stronger than pure semisimplicity. Just as for ℓ', this implies that R is right Artinian. In addition, assume R is local. Let M be a uniform R-module with injective hull $\mathcal{E}(M)$. Then $\mathcal{E}(M)$ is indecomposable and every submodule of it is also cyclic. This plus R local forces $\mathcal{E}(M)$ to be uniserial. We then have a Morita duality between the category of finitely generated right R-modules and the category of finitely generated left modules over a ring S which is left serial. If, as has been conjectured, every ring with a Morita duality is self dual, then R itself will be left serial. But we also need serial on the right. By [D-R], if R is local with commutative residue field, then $\ell(M) \leq 0$ for all M holds precisely when R has radical square 0 and is of finite representation type with all finitely generated indecomposables cyclic.*

Example 16 *Let F be a field with an endomorphism $\sigma : F \longrightarrow F$ such that $[F : \sigma(F)] = 2$. Let R be the ring*

$$R = \left\{ \begin{bmatrix} x & y \\ 0 & x \end{bmatrix} : x, y \in F \right\}$$

with $\begin{bmatrix} x & y \\ 0 & x \end{bmatrix} \cdot \begin{bmatrix} a & b \\ 0 & a \end{bmatrix} = \begin{bmatrix} xa & xb + y\sigma\,(a) \\ 0 & xa \end{bmatrix}$. *Then* R *is local Artinian, left serial, has commutative residue field and radical square zero. Let* $\mathcal{E}\,(S)$ *denote the injective hull of the unique simple R-module S. We observe that* $\mathrm{hom}\,(R/S,\ S) \cong\ _{\sigma(F)}F_F$ *and* $\mathcal{E}\,(S)$ *is cyclic uniserial of length 2. Computations show that the only indecomposable finitely generated right R-modules are, up to isomorphism, R, R/J, and* $R \Big/ \begin{bmatrix} 0 & \sigma\,(F) \\ 0 & 0 \end{bmatrix}$. *By the results in [D-R], this ring has* $\ell\,(M) \le 0$ *for every right R-module M.*

Having seen that the ℓ'-global dimension of a ring being 0 is very close to and may possibly be the same as Artinian of finite representation type, we can look at what happens for 1. Here even ℓ-global dimension equal to 1 does not force finite representation type. Nor does it force nice properties such as local or quasi-Frobenius.

Example 17 *Let R be a semiprimary ring with radical J of square 0. Let M be an R-module. Since R is perfect, a set of generators for M/MJ will lift to a set of generators for M. So select a set of generators independent modulo MJ. The intersection of each of these with the module generated by the others is contained in MJ which is semisimple since J is. This intersection is thus a direct sum of simple (cyclic) modules and so has ℓ-invariant zero. Thus $\ell\,(M) \le 1$. But there is no reason to expect such an R to be of finite representation type.*

Example 18 *Let R be the group algebra $(\mathbb{Z}/2\mathbb{Z})\,[V]$ where V is the Klein four group (the Sylow 2 subgroup of the alternating group on 4 letters). Let M be an R-module. Let N be the sum of a maximal set of independent copies of R in M. Then N is injective, and clearly has $\ell\,(N) = 0$. If K is a complement of N in M, then K contains no cyclic submodule isomorphic to R, so K must be an R/soc\,(R)-module. As in the previous example, we have $\ell\,(K) \le 1$ for any R/soc\,(R)-module K, and $\ell\,(M) \le 1$ for any R-module M. This ring is not of finite representation type.*

Remark 19 *The definition of ℓ works for any complete lattice bounded below, with sum replaced by $\vee$, intersection replaced by $\wedge$, and submodule replaced by lattice ideal. This is significant in the initial reason for introducing*

ℓ to study the projective dimension of ideals in commutative von Neumann regular rings.

Example 20 *Let R be a commutative ring. A finite sequence of elements $\{r_1, \cdots r_n\}$ is an R-sequence if r_i is not a zero divisor on $R \big/ \sum_{j=1}^{i-1} r_j R$. This forces $\sum_{j=1}^{i} r_j R \cong r_i R \cap \sum_{j=1}^{i-1} r_j R$ under the isomorphism given by multiplication by r_i. We thus see by induction that $\ell\left(\sum_{j=1}^{n} r_j R\right) \leq n - 1$. If $n = 1$ or 2 it is easy to see that this is an equality. and I conjecture it is an equality for all $n < \infty$.*

Remark 21 *The definition of $\ell(M)$ by its very nature produces an upper bound. and invites computations by transfinite induction. Getting lower bounds on $\ell(M)$ is much harder. as there is no reason for one (well ordered) set of generators to give an upper bound which is the same as that given by another set of generators.*

In the remainder of this section, we will illustrate the comments in Remark 21 by proving two typical computations of upper bounds on $\ell(M)$. An additional upper bound computation will be found in Proposition 2.1.

Semiartinian rings and modules

Let R be any ring. The socle $\mathfrak{S}(M)$ of an R-module M is the sum of all the simple submodules of M. We define the socle series of a module M by transfinite induction by setting $\mathfrak{S}_0(M) = 0$, $\mathfrak{S}_\lambda(M) = \bigcup_{\alpha < \lambda} \mathfrak{S}_\alpha(M)$ for λ a limit ordinal. and $\mathfrak{S}_{\alpha+1}(M) = \nu^{-1}(\mathfrak{S}(M/\mathfrak{S}_\alpha(M)))$ where ν is the natural map. Because M is a set. there must be an ordinal α (or $\alpha = -1$) with $\mathfrak{S}_\alpha(M) = \mathfrak{S}_{\alpha+1}(M)$. Call the smallest such ordinal (or -1) the socle length of M, sl (M). A module M is called semiartinian if $\mathfrak{S}_{\mathrm{sl}(M)}(M) = M$. If the ring R is semiartinian as a module over itself, then every R-module is semiartinian. It is not difficult to see that M is semiartinian if and only if every quotient of M has a nonzero socle. We leave this as an exercise.

Proposition 1.1 *Let M be a semiartinian module. Then $\ell(M) \leq \mathrm{sl}(M)$.*

Proof. We use transfinite induction on sl (M). sl $(M) = -1 \iff M = 0 \iff \ell(M) = -1$.

Now assume that $\ell(N) \leq \text{sl}(N)$ whenever $\text{sl}(N) < \text{sl}(M)$. For each $\alpha < \text{sl}(M)$, $\mathfrak{S}(M/\mathfrak{S}_\alpha(M)) = \bigoplus_\beta S_\beta$ is a direct sum of simples S_β. Well order the indexing set of β's. For each β, select an $x_{\alpha,\beta} \in M$ mapping onto a non-zero element of S_β and well order the set $\{x_{\alpha,\beta}\}$ lexicographically. Now $M = \sum_{\alpha,\beta} x_{\alpha,\beta} R$ and for each $\gamma < \alpha.\beta$ we have $x_\gamma R \cap \sum_{\alpha',\beta' < \gamma} x_{\alpha',\beta'} R \subseteq \mathfrak{S}_\alpha(M)$ so by the induction hypothesis. $\ell\left(x_\gamma R \cap \sum_{\alpha',\beta' < \gamma} x_{\alpha',\beta'} R\right) \leq \text{sl}(\mathfrak{S}_\alpha(M)) = \alpha < \text{sl}(M)$. Hence by definition of $\ell(M)$, $\ell(M) \leq \text{sl}(M)$. $\square$

The upper bound here is not a lower bound unless M is semisimple. For if M contains a cyclic submodule N which is not semisimple, $\ell(N) = 0 < \text{sl}(N)$.

Cardinality of a generating set

The hypotheses of Proposition 1.2 below are precisely those in the situation being studied when the ℓ invariant was introduced.

Proposition 1.2 *Let M be a module such that the cyclic (respectively finitely generated) submodules of M form a lattice under join is sum and meet is intersection. Then $\ell(M) \leq \text{gen}(M) + 1$ (respectively $\ell'(M) \leq \text{gen}(M) + 1$). If in addition every countably generated submodule is a direct sum of cyclics (respectively finitely generated modules). then $\ell(M) \leq \text{gen}(M)$ (respectively $\ell'(M) \leq \text{gen}(M)$.*

Proof. Let $\{x_\alpha : \alpha \in \Omega_{\text{gen}(M)}\}$ be a set of generators for M. For every $\alpha \in \Omega_{\text{gen}(M)}$, $\sum_{\beta < \alpha} x_\beta R$ is generated by fewer than $\text{gen}(M)$ elements. Moreover, $x_\alpha R \cap \sum_{\beta < \alpha} x_\beta R$ is either $x_\alpha R$ or $x_\alpha R \cap \sum_{\beta < \alpha} x_\beta R$ which is generated by a set of cyclic generators for

$$\left\{ x_\alpha R \cap \sum_{i=1}^n x_{\beta_i} R : n \in \mathbb{N}, \ \beta_i < \alpha \text{ for } 1 \leq i \leq n \right\}.$$

Thus $x_\alpha R \cap \sum_{\beta < \alpha} x_\beta R$ is generated by fewer than $\text{gen}(M)$ elements.

If $\text{gen}(M) = -1$ then $\ell(M) \leq 0 = \text{gen}(M) + 1$. Else assume that the proposition is true for all modules N with $\text{gen}(N) < \text{gen}(M)$. Set $K_\alpha = x_\alpha R \cap \sum_{\beta < \alpha} x_\beta R$. Then $\ell(K_\alpha) \leq \text{gen}(K_\alpha) + 1 \leq \text{gen}(M)$, so by definition $\ell(M) \leq \text{gen}(M) + 1$.

The proof for ℓ' is the same with cyclic replaced by finitely generated. $\square$

2   UNISERIAL MODULES AND MODULES OVER PRÜFER DOMAINS

In this section we will compute precisely the invariant $\ell(M)$ for M a uniserial module over any ring. For uniserial modules, we of course have $\ell'(M) = \ell(M)$. We then compute an upper bound for torsion-free modules over a Prüfer domain. We show it is also a lower bound if the minimum number of generators of a torsion-free rank 1 module is the supremum of the minimum number of generators locally. This will happen, for example, if there are only a finite number of maximal ideals. It need not happen in general.

First we have to look at an arbitrary uniserial module over any ring.

Uniserial modules

Let M be a nonzero uniserial module, that is, the (cyclic) submodules of M are linearly ordered. We wish to compute $\ell(M)$. The computation will give $\ell(M) = \text{gen}(M) + c(M)$. where $c(M) \in \{-1, 0, 1\}$. So we must first define $c(M)$, which depends only on gen(M).

If gen$(M) = -1$, then since M is nonzero, we have $\ell(M) = 0 = \text{gen}(M) + 1$, so $c(M) = 1$.

We now define functions ϕ and v on the class of all ordinals. We set $\phi(\alpha) = \beta$ if $\alpha = \beta + 1$ and $\phi(\alpha) = \alpha$ if α is not a successor ordinal. Then set $\psi(\alpha) = \inf\{\phi^n(\alpha) : n \in \mathbb{N}\}$. Since

$$\alpha \geq \phi(\alpha) \geq \cdots \geq \phi^n(\alpha) \geq \phi^{n+1}(\alpha) \geq \cdots$$

is a descending chain of ordinals, it must be finite, and its smallest element $\psi(\alpha)$ is either 0 or a limit ordinal.

Now for any ordinal κ, define

$$c(\kappa) = \begin{cases} 0 & \text{if} \quad \psi(\kappa) \text{ is a regular limit ordinal} \\ 1 & \text{if} \quad \psi(\kappa) = 0 \\ -1 & \text{if} \quad \psi(\kappa) \text{ is a singular limit ordinal} \end{cases},$$

and for any nonzero uniserial M set $c(M) = c(\text{gen}(M))$.

We observe that, for any ordinal $\lambda = \kappa + 1$, $c(\lambda) = c(\kappa)$. Moreover, if $\psi(\kappa)$ is singular and $\alpha \in \mathfrak{Ord}$ is a regular ordinal $< \psi(\kappa) + 1$, then $\alpha + c(\alpha) < \psi(\kappa)$. If $\psi(\kappa)$ is 0 or regular and $\alpha < \psi(\kappa)$, then $\alpha + c(\alpha) <$

$\psi(\kappa) + c(\psi(\kappa))$. These observations are useful in understanding the proofs of Propositions 2.1 and 2.2.

Proposition 2.1 *Let $M \neq 0$ be uniserial. Then $\ell(M) \leq \mathrm{gen}(M) + c(M)$.*

Proof. We use induction on $\mathrm{gen}(M)$. If $\mathrm{gen}(M) = -1$ then M is cyclic so $\ell(M) = 0 = \mathrm{gen}(M) + c(M)$.

Now assume $\mathrm{gen}(M) \geq 0$ and select a set of generators $\{x_\alpha : \alpha \in \aleph_{\mathrm{gen}(M)}\}$ of M with $x_\alpha R \supseteq x_\beta R$ if $\alpha > \beta$. Then for all $\alpha < \mathrm{gen}(M)$.

$$\mathrm{gen}\left(x_\alpha R \cap \sum_{\beta < \alpha} x_\beta R\right) < \mathrm{gen}(M)$$

so by induction

$$\ell\left(x_\alpha R \cap \sum_{\beta < \alpha} x_\beta R\right) \leq \mathrm{gen}\left(\sum_{\beta < \alpha} x_\beta R\right) + c\left(\sum_{\beta < \alpha} x_\beta R\right) < \mathrm{gen}(M) + c(M).$$

Hence by definition of $\ell(M)$, $\ell(M) \leq \mathrm{gen}(M) + c(M)$. $\square$

We now show that this upper bound on $\ell(M)$ is also a lower bound.

Theorem 2.2 *Let $M \neq 0$ be uniserial. Then $\ell(M) = \mathrm{gen}(M) + c(M)$.*

Proof. Since $\ell(M)$ must be defined for any uniserial module by Proposition 2.1, M has an ℓ-sequence $\{x_\alpha : \alpha \in \Omega\}$. Take any cofinal subset Ω' of Ω. Then $\sum_{\alpha' \in \Omega'} x_{\alpha'} R = M$ and every intersection $x_{\alpha'} R \cap \sum_{\beta' < \alpha'} x_{\beta'} R$ with $\alpha', \beta' \in \Omega'$ is an intersection $x_\alpha R \cap \sum_{\beta < \alpha} x_\beta R$ with $\alpha, \beta \in \Omega$, where α is the supremum in Ω of $\{\beta' : \beta' < \alpha' \in \Omega'\}$. Then

$$\sup\left\{\ell\left(\sum_{\beta' < \alpha'} x_{\beta'} R\right) : \beta', \alpha' \in \Omega'\right\} \leq \sup\left\{\ell\left(\sum_{\beta < \alpha} x_\beta R\right) : \beta, \alpha \in \Omega'\right\} = \ell(M).$$

Now by definition,

$$\ell(M) \leq \sup\left\{\ell\left(\sum_{\beta' < \alpha'} x_{\beta'} R\right) : \beta', \alpha' \in \Omega'\right\}$$

so $\{x_{\alpha'} : \alpha' \in \Omega'\}$ is also an ℓ-sequence for M. Thus we may take $\Omega = \Omega_{\text{gen}(M)}$.

By Proposition 2.1. $\ell(M) \leq \text{gen}(M) + c(M)$. Now assume for all M' with $\text{gen}(M') < \text{gen}(M)$ we have $\ell(M') = \text{gen}(M') + c(M')$. Let $\{x_\alpha : \alpha \in \Omega_{\text{gen}(M)}\}$ be an ℓ-sequence for M.

If $\text{gen}(M) = 0$. then M is countably infinitely generated so $\ell(M) \leq 0+1$ but $\ell(M) \neq 0$ so $\ell(M) = \text{gen}(M) + c(M)$.

If $\text{gen}(M)$ is a limit ordinal then $c(M) = 0$,

$$\ell(M) = \sup\left\{\ell\left(\sum_{\beta<\alpha} x_\beta R\right) : \alpha \in \Omega_{\text{gen}(M)}\right\}$$
$$= \sup\left\{\text{gen}\left(\sum_{\beta<\alpha} x_\beta R\right) + c\left(\sum_{\beta<\alpha} x_\beta R\right) : \alpha < \Omega_{\text{gen}(M)}\right\}$$
$$= \text{gen}(M) = \text{gen}(M) + c(M).$$

Now let $\text{gen}(M) = \kappa + 1$. Then every intersection $x_\alpha \cap \sum_{\beta<\alpha} x_\beta R = \sum_{\beta<\alpha} x_\beta R$ is generated by $\aleph_\kappa$ elements.

If κ is regular, whenever $\text{gen}\left(\sum_{\beta<\alpha} x_\beta R\right) = \kappa$ we have $\ell\left(\sum_{\beta<\alpha} x_\beta R\right) = \kappa + c(\kappa)$ and whenever $\text{gen}\left(\sum_{\beta<\alpha} x_\beta R\right) < \kappa$ we have $\ell\left(\sum_{\beta<\alpha} x_\beta R\right) \leq \kappa$ so

$$\ell(M) = \sup\left\{\ell\left(\sum_{\beta<\alpha} x_\beta R\right) : \alpha \in \Omega_{\text{gen}(M)}\right\} = \kappa + c(\kappa)$$
$$= \text{gen}(M) + c(M).$$

If κ is singular, then

$$\ell(M) = \sup\left\{\ell\left(\sum_{\beta<\alpha} x_\beta R\right) : \alpha \in \Omega_{\text{gen}(M)}\right\} = \kappa$$
$$= \text{gen}(M) + c(M)$$

As this covers all cases, by transfinite induction, $\ell(M) = \text{gen}(M) + c(M)$ for all M. $\square$

Torsion-free modules over Prüfer domains

Definition 2.1 *A commutative domain R is called a Prüfer domain if every finitely generated ideal of R is projective.*

These are precisely the commutative domains such that every localization at a prime is a valuation domain.

The following lemma is introduced mainly to establish notation which will be used throughout the rest of this section. Everything in it is standard.

Lemma 2.3 (Establishing Notation) *Let M be a torsion-free right module over a Prüfer domain R with quotient field Q. Then*

(a) *M is an essential extension of a free R-module $\bigoplus_{\iota \in \mathfrak{I}} b_\iota R$.*

(b) *For each $\iota \in \mathfrak{I}$, the injective hull of $b_\iota R$ is isomorphic to the quotient field Q of R and so has every finitely generated submodule cyclic projective.*

(c) *The injective hull of M is $\bigoplus_{\iota \in \mathfrak{I}} b_\iota Q$.*

(d) *With $\bigoplus_{\iota \in \mathfrak{I}} b_\iota R$ as in statement (a), well order $\mathfrak{I}$ by the ordinal Ω. Then M is a continuous ascending union*

$$M = \bigcup_{\iota \in \Omega} M_\iota$$

where each $V_\iota = M_\iota / \bigcup_{\iota' < \iota} M_{\iota'}$ is isomorphic to a submodule of Q,

(e) *If N' is a finitely generated submodule of V_ι then there is a finitely generated submodule N of M_ι which maps isomorphically onto N'.*

Proof. Statement (a) follows from a standard Zorn's lemma argument which gives a maximal linearly independent family of cyclic submodules of M. Statements (b) and (c) are obtained by tensoring by Q, observing that any finitely generated submodule of Q is isomorphic to an ideal of R (multiply by a common denominator), applying Baer's criterion for injectivity, and using unique divisibility in a torsion-free module over a domain. For (d), let M_ι be $M \cap \sum_{\iota' \leq \iota} b_\iota R$. Then $V_\iota = M_\iota / \bigcup_{\iota' < \iota} M_{\iota'}$ is isomorphic to a submodule of $b_\iota Q$ under the restriction of the ιth projection map. Statement (e) follows from (b) since some submodule N of M_ι maps onto N', and this map must split. $\square$

Theorem 2.4 (The Upper Bound) *Let* $0 \neq M$ *be a torsion-free module over a Prüfer domain. Let*

$$\kappa = \max \sup \left\{ \ell'(V_\iota) : \iota \in \Omega \right\}.$$

Then $\ell'(M) \leq \kappa$.

Proof. Select an ℓ'-sequence $\left\{ N'_{(\iota,\alpha)} : \alpha \in \Lambda_\iota \right\}$ for V_ι for each $\iota \in \Omega$. Well order the set $\Lambda = \{ (\iota, \alpha) : \iota \in \Omega, \alpha \in \Lambda_\iota \}$ lexicographically. Let $N_{(\iota,\alpha)}$ be a finitely generated submodule of M_ι which maps isomorphically onto $N'_{\iota,\alpha}$. For $(\iota, \alpha) \in \Lambda$, look at

$$K = N_{(\iota,\alpha)} \cap \left(\sum_{\iota' < \iota} M_{\iota'} + \sum_{\beta < \alpha} N_{(\iota,\beta)} \right).$$

Under the natural map $\nu_\iota : M_\iota \to V_\iota$, K maps monomorphically into $K' = N'_{\iota,\alpha} \cap \sum_{\beta < \alpha} N'_{(\iota,\beta)}$. Let $x \in K'$. Then $x = \nu_\iota(n) = \nu_\iota(k)$ for some $n \in N_{(\iota,\alpha)}$ and $k \in \sum_{(\iota',\alpha') < (\iota,\alpha)} N_{(\iota',\alpha')}$ so $n - k \in \ker \nu_\iota = \sum_{\iota' < \iota} M_\iota$ and $n = k + (n - k) \in K$. Then $\ell'(K) < \ell'(V_\iota) \leq \kappa$. By definition, $\ell'(M) \leq \kappa$. $\square$

Definition 2.2 *For M a torsion-free R-module, call a family of modules*

$$\left\{ N_{(\iota,\alpha)} : \iota \in \Omega, \alpha \in \Lambda_\iota \right\}$$

constructed as in the proof of Theorem 2.4 a pure ℓ'-bounding sequence. If it is also an ℓ'-sequence, we will call it a pure ℓ'-sequence.

At the original presentation of this paper, I believed I had a proof that any pure ℓ'-bounding sequence is an ℓ'-sequence under the circumstances studied here. That assumption is not correct. I replace it with the conjecture that a pure ℓ'-bounding sequence for which $\sup \left\{ N_\beta \cap \sum_{\alpha < \beta} : \beta \in \omega \right\}$ is minimal must actually be an ℓ'-sequence.

References

[B]   N. Bourbaki. Commutative Algebra, (English), Springer-Verlag, New York, Berlin, Heidelberg, 1989.

[C]   S. U. Chase, *Direct products of modules*, Trans Amer Math Soc 97(1960), 457-473.

[D-R] V. Dlab and C. M. Ringel, *Decomposition of modules over right uniserial rings*. Math. Z. 129(1972), 207-230.

[F]   K. R. Fuller. *On rings whose left modules are direct sums of finitely generated modules*. Proc Amer Math Soc 54(1976), 39-44.

[F-A] F. W. Anderson and K. R. Fuller, Rings and categories of modules, Second Edition, GTM 13, Springer Verlag New York, 1992.

[F-S] L. Fuchs and L. Salce, Modules over valuation domains, Lect. Notes in Pure and Applied Math. 97, Marcel Dekker, inc., New York, 1985.

[F-W] C. Faith and E. A. Walker. *Direct sum representations of injective modules*. J. Algebra 5(1967), 203–221.

[K]   I. Kaplansky, *Projective modules*. Ann. of Math (2) 68(1958), 372–377.

[O1] B. L. Osofsky, *Projective dimension of "nice" directed unions*, J. Pure and Applied Algebra 13(1978), pp 179-219.

[O2] B. L. Osofsky, *Projective Dimension of Ideals in Von Neumann Regular Rings*, in Advances in Ring Theory, S. K. Jain and S. Tariq Rizvi editors, Birkhäuser Boston, 1997.

E-mail: osofsky@math.rutgers.edu

Unit Groups of Lie Centre-by-Metabelian Group Algebras

RICHARD ROSSMANITH

R. ROSSMANITH, MATHEMATISCHES INSTITUT, FRIEDRICH-SCHILLER-UNIVERSITAT, 07740 JENA, GERMANY
E-mail address: richard@mathematik.uni-jena.de
URL: http://www.mathematik.uni-jena.de/algebra/

ABSTRACT. We introduce Lie centre-by-metabelian group algebras, and recall their classification. Then we prove results about the solvability of their unit groups. In particular, we give examples of Lie centre-by-metabelian group algebras whose unit groups are not centre-by-metabelian, or not even solvable. This answers question 22 of A. Bovdi's survey article [3].

1. (LIE) SOLVABILITY

For subsets X, Y of a group G, we denote by (X, Y) the subgroup of G generated by all commutators of the form $(x, y) := xy(yx)^{-1}$ with $x \in X$, $y \in Y$. Then $G' := (G, G)$ is the derived subgroup of G, $G'' := (G', G')$ is the second derived subgroup of G, and, continuing in this manner, we obtain the derived series, $G \trianglerighteq G' \trianglerighteq G'' \trianglerighteq G''' \trianglerighteq \ldots \trianglerighteq G^{(n)} \trianglerighteq \ldots$, of G. If $G^{(n-1)} > G^{(n)} = 1$, then G is solvable of derived length $n =: \mathrm{dl}(G)$. In the particular case $\mathrm{dl}(G) \leq 2$, G is metabelian. We call G *centre-by-metabelian*, if $G/\mathcal{Z}(G)$ is a metabelian group, i.e., if $G'' \subseteq \mathcal{Z}(G)$.

Let L be a Lie algebra over some field $\mathbb{F}$ of characteristic $p \geq 0$. For subsets X, Y of L, we denote by $[X, Y]$ the $\mathbb{F}$-span of all Lie brackets $[x, y]$ with $x \in X$, $y \in Y$. Then $L' := [L, L]$ is the derived Lie algebra of L, and the notions of the higher derived Lie algebras, the derived series, the derived length $\mathrm{dl}(L)$, being metabelian, and being centre-by-metabelian, are defined accordingly.

Let A be an associative algebra over $\mathbb{F}$. Endowing its underlying vector space with a bracket operation, defined via $[x, y] := xy - yx$ for $x, y \in A$, results in the so-called associated Lie algebra $\mathcal{L}(A)$ of A. The algebra A is said to be Lie solvable (of Lie derived length $\mathrm{dl}(A)$, Lie metabelian, Lie centre-by-metabelian), if the associated Lie algebra $\mathcal{L}(A)$ is solvable (of derived length $\mathrm{dl}(\mathcal{L}(A))$, metabelian, centre-by-metabelian). One also speaks of the Lie derived series of A, and writes A', A'', ..., instead of $\mathcal{L}(A)'$, $\mathcal{L}(A)''$, ...

We are interested in the interconnection with regards to solvability (and, in particular, centre-by-metabelianity) between G, its group algebra $\mathbb{F}G$ over $\mathbb{F}$ (respectively the associated Lie algebra $\mathcal{L}(\mathbb{F}G)$), and the unit group $\mathcal{U}(\mathbb{F}G)$ of $\mathbb{F}G$:

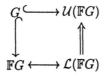

As an interpretation of the arrow between $\mathcal{L}(\mathbb{F}G)$ and $\mathcal{U}(\mathbb{F}G)$ in the diagram above, let us quote A. Bovdi [3, p. 194]: "Clearly, Lie commutators are considerably easier to calculate than group commutators, and one may think of obtaining information on the group of units by looking at the associated Lie algebra. It has turned out

313

that, in most cases, the Lie structure reflects well the characteristics of the group of units ..." E.g., M. Smirnov shows in [25] that for $p \neq 2$, Lie solvability of A implies solvability of $\mathcal{U}(A)$, such that $\mathrm{dl}(\mathcal{U}(A)) \leq 4\,\mathrm{dl}(A) + 3$. (We are going to see that this fails for $p = 2$.)

Passi-Passman-Sehgal prove in [10] that $\mathcal{L}(\mathbb{F}G)$ is solvable, if and only if either $p = 0$ and G is abelian, or $p > 2$ and G' is a finite p-group, or $p = 2$ and G contains a subgroup H of index at most 2 such that H' is a finite 2-group. (This makes the case of characteristic 0 uninteresting, so we assume $p > 0$ in the following.) This qualitative result is complemented by A. Shalev's sharp lower and upper bounds on the Lie derived length of $\mathbb{F}G$ [22, 23].

The group algebras $\mathbb{F}G$ with a solvable unit group $\mathcal{U}(FG)$ are not yet classified in general. For existing results, see Passman [11] for finite G, Bovdi-Khripta [4] for periodic G, and Bovdi [2] for nilpotent G.

Without restrictions on p, Levin-Rosenberger proved in [8] that if $\mathcal{L}(\mathbb{F}G)$ is metabelian, then so is $\mathcal{U}(\mathbb{F}G)$. (They also give conditions on G which are necessary and sufficient for $\mathcal{L}(\mathbb{F}G)$ to be metabelian; the –weaker– conditions on G which are necessary and sufficient for $\mathcal{U}(\mathbb{F}G)$ to be metabelian have been found by A. Shalev [20], Coleman-Sandling [5], and J. Kurdics [7], for the case that G is finite.) By Sharma-Srivastava [24], this holds even more generally with $\mathbb{F}G$ replaced by A; more exactly:

$$1 - \mathcal{U}(A)' \subseteq A' \cdot A, \text{ and } 1 - \mathcal{U}(A)'' \subseteq A'' \cdot A. \tag{1}$$

The aim of this paper is to show that the correspondence between $\mathcal{L}(\mathbb{F}G)$ and $\mathcal{U}(\mathbb{F}G)$ fails in general for the property of being centre-by-metabelian, and to find additional conditions under which it still holds. This will be done in section 3. Before that, we recall the classification of the Lie centre-by-metabelian group algebras in section 2.

We use the notation $\omega(\mathbb{F}G)$ for the augmentation ideal of $\mathbb{F}G$, and we denote the sum over all elements of a finite subset X of $\mathbb{F}G$ by X^+. For group elements g, h, conjugation is written as $^g h := ghg^{-1}$.

Remark.   • Most of the results quoted above, along with related ones, are also displayed (in greater detail than here) in sections 21 and 22 of A. Bovdi's survey [3].

• Notice that A', A'', ..., are not necessarily ideals of A. However, alternatively to the approach presented here, one may define A' to be the ideal generated by $[A, A]$, and A'' to be the ideal generated by $[A', A']$, etc. One then arrives at notions such as "strong" Lie solvability (which obviuosly implies Lie solvability as defined above). See chapter 5 of S. Sehgal's book [19] for details.

• Much more is known about Lie nilpotence, than about Lie solvability, of group algebras. See Passi-Passman-Sehgal [10], Levin-Sehgal [9], A. Shalev [21], Bhandari-Passi [1].

• The results proved in section 3 of this paper are taken from chapter 9 of the author's dissertation thesis [13].

2. LIE CENTRE-BY-METABELIAN GROUP ALGEBRAS

Let us begin this section by pointing out three reasons why being Lie centre-by-metabelian should be considered an interesting property of algebras:

First, after Levin-Rosenberger's results about Lie metabelian group algebras (see section 1), Lie centre-by-metabelian group algebras were the next-more complicated objects to study.

A second point might be that for matrix algebras, Lie solvability is equivalent to being Lie centre-by-metabelian: It is easily checked that $\text{Mat}(n, \mathbb{F})$ is not Lie solvable if $n \geq 2$ and $p \neq 2$. In the case $p = 2$, $\text{Mat}(n, \mathbb{F})$ is not Lie solvable for $n \geq 3$, and $\text{Mat}(2, \mathbb{F})$ is Lie centre-by-metabelian, but not Lie metabelian (cf. [10, lemma 1.2]).

Quite astounding is the following general result of Smirnov-Zaleskii [26]: The algebra A is Lie solvable, if and only if it contains a nilpotent ideal I such that A/I is Lie centre-by-metabelian. (For finite-dimensional A, this follows from the preceding paragraph; take I to be the Jacobson radical of A.) Hence Lie solvability is already very close to associative nilpotence, and differs from it only by a Lie centre-by-metabelian "head".

Let us now turn to Lie centre-by-metabelian group algebras. By Sharma-Srivastava [24], they are necessarily commutative if $p > 3$. For $p = 3$ and nonabelian G, the Lie algebra $\mathcal{L}(\mathbb{F}G)$ is centre-by-metabelian, if and only if $|G'| = 3$ (Külshammer-Sharma [6], Sahai-Srivastava [18]). Külshammer-Sharma show additionally that this implies that also $\mathcal{U}(\mathbb{F}G)$ is centre-by-metabelian. Independently, it is shown by M. Sahai in [17] that this implication is in fact an equivalence (for all $p \neq 2$).

In [3, section 22], A. Bovdi asks: "Question. Describe Lie centrally metabelian group rings of characteristic 2. Is the group of units of such a group ring a centrally metabelian group?" The answer to the first part of the question is given in [14, 15], or in [13], as follows:

Theorem 2. *Let G be a group, and let $\mathbb{F}$ be a field of characteristic 2. Then $\mathbb{F}G$ is Lie centre-by-metabelian, if and only if one of the following conditions is satisfied:*

(i) $|G'|$ *divides* 4.
(ii) G' *is central and elementary abelian of order* 8.
(iii) G *acts by element inversion on* $G' \cong Z_2 \times Z_4$, *and* $C_G(G')' \subseteq \Phi(G')$.
(iv) G *contains an abelian subgroup of index* 2.

Remark. Recently, the classification of the Lie centre-by-metabelian group algebras has been extended to group algebras over arbitrary commutative rings; see [16].

3. Unit Groups of Lie Centre-by-Metabelian Group Algebras

This section is the main part of this paper, and it is devoted to the second part of Bovdi's question. The answer is negative in general; nevertheless, let us first turn to the cases with a positive answer:

Lemma 3. *Let G be a group that satisfies either condition (i) or (ii) of theorem 2, and let $\mathbb{F}$ be a field of characteristic 2. Then $\mathcal{U}(\mathbb{F}G)$ is centre-by-metabelian.*

Proof. By [14, paragraphs 1.1–1.4], $(\mathbb{F}G)'' \cdot \mathbb{F}G \subseteq \mathcal{Z}(\mathbb{F}G)$. Equation (1) yields $1 + \mathcal{U}(\mathbb{F}G)'' \subseteq (\mathbb{F}G)'' \cdot \mathbb{F}G$. Hence $\mathcal{U}(\mathbb{F}G)'' \subseteq \mathcal{U}(\mathbb{F}G) \cap \mathcal{Z}(\mathbb{F}G) \subseteq \mathcal{Z}(\mathcal{U}(\mathbb{F}G))$.

Example 4. We now turn to the case with a negative answer; we construct a group G that satisfies both of the conditions (iii) and (iv) of theorem 2, such that $\mathcal{U}(\mathbb{F}G)$ is not centre-by-metabelian.

Let $C := \langle c, d, x \rangle \cong Z_8 \times V_4$, where $c^8 = d^2 = x^2 = 1$. Then $c \mapsto c^3$, $d \mapsto dx$, $x \mapsto x$, defines an automorphism a of C of order 2. We consider the group

$G := C \rtimes \langle a \rangle$ of order 64. There, $(a,c) = {}^a c c^{-1} = c^2 =: y$, $(a,d) = {}^a d d^{-1} = x$, $(a,x) = {}^a x x^{-1} = 1$, and $(a,y) = (a,c^2) = c^4 = y^2$. Hence $G' = \langle x, y \rangle \cong Z_2 \times Z_4$, and $C_G(G') = C$ is an abelian subgroup of index 2 in G. Moreover, ${}^a x = x = x^{-1}$ and ${}^a y = y^3 = y^{-1}$, i.e. G acts by element inversion on G'.

By theorem 2, $\mathcal{L}(\mathbb{F}G)$ is centre-by-metabelian. We show that $U := \mathcal{U}(\mathbb{F}G)$ is not. The following statements may be checked by direct expansion:

$$(1 + a + c)^{-1} = 1 + y^2 + y^3 + c + y^2 a + y^3 a + ya + y^3 ca + y^2 ca,$$

$$(1 + a + c, d) =$$
$$1 + y + y^2 + y^3 + xy + xy^2 + xy^3 + y^2 c + y^3 c + xy^2 c + xy^3 c$$
$$+ a + ya + y^2 a + xa + xya + xy^2 a + yca + xyca$$
$$= (1 + a + c, d)^{-1},$$

$$((1 + a + c, d), (a,c)) =$$
$$1 + ya + y^3 a + xya + xy^3 a + yca + y^3 ca + xyca + xy^3 ca$$
$$= ((1 + a + c, d), (a,c))^{-1},$$

$$(a, (1 + a + c, d), (a,c)) = 1 + (G')^+ ca \neq 1,$$

This shows $(U, U'') \neq 1$, as desired.

However, U still has derived length 3, as the following lemma shows.

Lemma 5. *Let G be a group that satisfies condition (iii) of theorem 2, and let $\mathbb{F}$ be a field of characteristic 2. Then $\mathcal{U}(\mathbb{F}G)'$ is nilpotent of class at most 2; in particular, $\mathcal{U}(\mathbb{F}G)$ is solvable of derived length at most 3.*

Proof. We write $\langle x, y \rangle = G' \cong Z_2 \times Z_4$ with $x^2 = 1 = y^4$, $C := C_G(G')$, $\langle aC \rangle = G/C$, and $U := \mathcal{U}(\mathbb{F}G)$. In the proof of [14, lemma 1.5], the following facts are established:

$$\omega(\mathbb{F}G')^5 = 0,$$
$$\omega(\mathbb{F}G')^4 \, \mathbb{F}G \subseteq \mathcal{Z}(\mathbb{F}G),$$
$$(\mathbb{F}G)'' \subseteq \omega(\mathbb{F}G')^4 \, \mathbb{F}G + X \, \mathbb{F}G,$$

where $X := \{\sigma h + {}^a \sigma : \sigma \in \omega(\mathbb{F}G')^2, h \in G'\}$.

We have to show that $(U', U'') = 1$, or equivalently, $[U', U''] = 0$. Equation (1) implies $1 + U' \subseteq (\mathbb{F}G)' \, \mathbb{F}G \subseteq \omega(\mathbb{F}G') \, \mathbb{F}G$ and $1 + U'' \subseteq (\mathbb{F}G)'' \, \mathbb{F}G$. Then $[U', U''] = [1 + U', 1 + U''] \subseteq [\omega(\mathbb{F}G') \, \mathbb{F}G, (\mathbb{F}G)'' \, \mathbb{F}G] \subseteq [\omega(\mathbb{F}G') \, \mathbb{F}G, \omega(\mathbb{F}G')^4 \, \mathbb{F}G] + [\omega(\mathbb{F}G') \, \mathbb{F}G, X \, \mathbb{F}G] \subseteq \omega(\mathbb{F}G)^5 \, \mathbb{F}G + [\omega(\mathbb{F}G') \, \mathbb{F}G, X \, \mathbb{F}G] = [\omega(\mathbb{F}G') \, \mathbb{F}G, X \, \mathbb{F}G]$. Therefore, it suffices to show that $[\omega(\mathbb{F}G') \, \mathbb{F}G, X \, \mathbb{F}G] = 0$.

By [12, lemma 3.1.1], $\omega(\mathbb{F}G') \, \mathbb{F}G = \{1 + x, 1 + y\} \, \mathbb{F}G$, and $\omega(\mathbb{F}G')^2 \, \mathbb{F}G = \{(1 + x)(1 + y), 1 + y^2\} \, \mathbb{F}G$. We are going to check that

$$[\tau f, (\sigma h + {}^a \sigma)g] = 0$$

for all $f, g \in G$, $h \in G'$, $\tau \in \{1 + x, 1 + y\}$, $\sigma \in \{(1 + x)(1 + y), 1 + y^2\}$.

First, since $x \in \mathcal{Z}(G)$, we have $[(1 + x)f, ((1 + x)(1 + y)h + (1 + x)(1 + y^3))g] = (1 + x)(1 + x)[f, (1 + y)hg + (1 + y^3)g] = (1 + x^2)[f, (1 + y)hg + (1 + y^3)g] = 0$.

Notice that h commutes with f modulo $\langle y^2 \rangle$, i.e. $(1 + y^2)fh = (1 + y^2)hf$. Therefore, $[(1 + x)f, ((1 + y^2)h + (1 + y^2))g] = (1 + x)(1 + y^2)[f, (h + 1)g] = (1+x)(1+y)^2(h+1)[f, g] \in \omega(\mathbb{F}G')^4 (\mathbb{F}G)' \mathbb{F}G \subseteq \omega(\mathbb{F}G')^5 \mathbb{F}G = 0.$

Observe next that for $y_1, y_2 \in \langle y \rangle$, we have $(1 + y_1)(1 + y_2) \in \{0, (1 + y^2), (1 + y^2)y\}$. So there are $i, j \in \{0, 1\}$ with $(1+y)(1+{}^f y) = (1+y^2)y^i$, and $(1+y)(1+{}^g y) = (1 + y^2)y^j$. Moreover, $(1 + y^2)y^3 = (1 + y^2)y$, and $(1 + y^2)\,{}^f h = (1 + y^2)h$. Then

$$[(1 + y)f, ((1 + x)(1 + y)h + (1 + x)(1 + y^3))g]$$
$$= (1 + x)[(1 + y)f, (1 + y)hg + (1 + y^3)g]$$
$$= (1 + x)\big((1 + y)(1 + {}^f y)\,{}^f hfg + (1 + y)(1 + {}^f y^3)fg$$
$$+ (1 + y)(1 + {}^g y)hgf + (1 + y^3)(1 + {}^g y)gf\big)$$
$$= (1 + x)\big((1 + y^2)y^i\,{}^f hfg + (1 + y)(1 + {}^f y)\,{}^f y^3 fg$$
$$+ (1 + y^2)y^j hgf + y^3(1 + y)(1 + {}^g y)gf\big)$$
$$= (1 + x)\big((1 + y^2)y^i hfg + (1 + y^2)y^i\,{}^f y^3 fg$$
$$+ (1 + y^2)y^j hgf + (1 + y^2)y^j y^3 gf\big)$$
$$= (1 + x)(1 + y^2)\,\big(y^i hfg + y^i yfg + y^j hgf + y^j ygf\big)$$
$$= (1 + x)(1 + y^2)\,\big(y^i(h + y)fg + y^j(h + y)gf\big)$$
$$= (1 + x)(1 + y^2)(h + y)(y^i fg + y^j gf)$$
$$= (1 + x)(1 + y)^2(h + y)(y^i(f, g) + y^j)gf \in \omega(\mathbb{F}G')^5 \mathbb{F}G = 0.$$

Finally, $[(1 + y)f, ((1 + y^2)h + (1 + y^2))g] = (1 + y^2)[(1 + y)f, (h + 1)g] = (1 + y^2)(1 + y)(h + 1)[f, g] \in \omega(\mathbb{F}G')^4 (\mathbb{F}G)' \mathbb{F}G \subseteq \omega(\mathbb{F}G')^5 \mathbb{F}G = 0.$

Example 6. Not every Lie centre-by-metabelian group algebra has a unit group of bounded derived length. In fact, the unit group might not even be solvable at all. This follows abstractly from the combination of theorem 2 with [4, theorem 2]. However, it is not too hard to directly compute the unit group of the concrete example $\mathbb{F}_2 D_{10}$, which turns out to be not solvable:

Write $D_{10} = \langle a, x \rangle$ with $a^2 = 1 = x^5$, ${}^a x = x^{-1}$. Then $\langle x \rangle$ is an abelian subgroup of index 2 in D_{10}, so $\mathbb{F}_2 D_{10}$ is Lie centre-by-metabelian by theorem 2. Let us examine its structure more closely:

Consider $\mathbb{F}_2 \langle x \rangle \subseteq \mathbb{F}_2 D_{10}$ first. By Maschke [28, §108], $\mathbb{F}_2 \langle x \rangle$ is semisimple, and thus, according to Wedderburn [28, §102], decomposes into a direct sum of simple $\mathbb{F}_2$-algebras, which are isomorphic to full matrix algebras over some $\mathbb{F}_2$-division algebras. But $\mathbb{F}_2 \langle x \rangle$ is commutative, so it is in fact isomorphic to a direct sum of field extensions of $\mathbb{F}_2$. Since x has multiplicative order 5, at least one of these fields contains $\mathbb{F}_{16}$. Checking dimensions, we find that $\mathbb{F}_2 \langle x \rangle \cong \mathbb{F}_2 \oplus \mathbb{F}_{16}$ as an $\mathbb{F}_2$-algebra. The adequate Wedderburn decomposition of $\mathbb{F}_2 \langle x \rangle$ is determined by the central, orthogonal idempotents $e := \langle x \rangle^+$ and $f := 1 + e$, where $e\mathbb{F}_2 \langle x \rangle = \mathbb{F}_2 e \cong \mathbb{F}_2$, and $f\mathbb{F}_2 \langle x \rangle \cong \mathbb{F}_{16}$.

Since $\langle x \rangle \trianglelefteq D_{10}$, the idempotents e, f are also central in $\mathbb{F}_2 D_{10}$, so we obtain a decomposition $\mathbb{F}_2 D_{10} = e\mathbb{F}_2 D_{10} \oplus f\mathbb{F}_2 D_{10}$ into ideals. We are going to combine this with the vector space decomposition $\mathbb{F}_2 D_{10} = \mathbb{F}_2 \langle x \rangle \oplus \mathbb{F}_2 \langle x \rangle a$.

Observe that as a vector space, $e\mathbb{F}_2 D_{10} = e\mathbb{F}_2 \langle x \rangle \oplus e\mathbb{F}_2 \langle x \rangle a = \mathbb{F}_2 e \oplus \mathbb{F}_2 ea = \mathbb{F}_2 \{e, ea\}$. Now $\{e, ea\}$ is a multiplicative group with neutral element e, hence $e\mathbb{F}_2 D_{10} \cong \mathbb{F}_2 Z_2$. It follows that $\mathcal{U}(e\mathbb{F}_2 D_{10}) \cong Z_2$. (To complete the picture, let us point out that $\mathcal{J}(e\mathbb{F}_2 D_{10}) = \mathbb{F}_2(e+ea) = \mathbb{F}_2(D_{10})^+$; we are going to see that this also is the Jacobson radical of $\mathbb{F}_2 D_{10}$.)

Let us fix an $\mathbb{F}_2$-algebra isomorphism $f\mathbb{F}_2\langle x\rangle \to \mathbb{F}_{16}$, and thus identify $f\mathbb{F}_2\langle x\rangle$ with $\mathbb{F}_{16}$ (and f with $1 = 1_{\mathbb{F}_{16}}$). Then, as a vector space, $A := f\mathbb{F}_2 D_{10} = f\mathbb{F}_2\langle x\rangle \oplus f\mathbb{F}_2\langle x\rangle a = \mathbb{F}_{16} \oplus \mathbb{F}_{16}a$, i.e. A is an $\mathbb{F}_{16}$-vector space with basis $\{1, a\}$, plus a multiplicative structure, which we are going to determine now: Conjugation by a leaves the subspace $\mathbb{F}_{16} \subseteq A$ invariant. This defines a field automorphism α: $\mathbb{F}_{16} \to \mathbb{F}_{16}$. The Galois group of $\mathbb{F}_{16}$ over $\mathbb{F}_2$ is cyclic of order 4, with generator β: $\mathbb{F}_{16} \to \mathbb{F}_{16}$, $s \mapsto s^2$. Since $\alpha(fx) = {}^a f\, {}^a x = fx^4 = (fx)^4$, we have $\alpha(s) = s^4$ for all $s \in \mathbb{F}_{16}$. Then $\alpha = \beta^2$, and $\langle\alpha\rangle = \mathrm{Gal}(\mathbb{F}_{16}|\mathbb{F}_4)$. Hence $\mathbb{F}_4$ is the centre of A. By [28, §94.3], A is (isomorphic to) the crossed product of $\mathbb{F}_{16}$ with $\mathrm{Gal}(\mathbb{F}_{16}|\mathbb{F}_4)$. In particular, A is a central simple $\mathbb{F}_4$-algebra (ibid). By Wedderburn, $A \cong \mathrm{Mat}(n, D)$ for some $n \in \mathbb{N}$ and an $\mathbb{F}_4$-division algebra D. But finite division algebras are fields, so D is in fact commutative. Since A is not commutative, we obtain $n \geq 2$. Then $2^8 = |A| = (4^{\dim_{\mathbb{F}_4} D})^{n^2}$ implies that $\dim_{\mathbb{F}_4} D = 1$ and $n = 2$, i.e. $f\mathbb{F}_2 D_{10} = A \cong \mathrm{Mat}(2, \mathbb{F}_4)$.

Together, we obtain $\mathcal{U}(\mathbb{F}_2 D_{10}) = \mathcal{U}(e\mathbb{F}_2 D_{10} \oplus f\mathbb{F}_2 D_{10}) \cong \mathcal{U}(e\mathbb{F}_2 D_{10}) \times \mathcal{U}(f\mathbb{F}_2 D_{10}) \cong Z_2 \times \mathrm{GL}(2,4)$. Since $\mathrm{PSL}(2,4) \cong A_5$ is simple, $\mathcal{U}(\mathbb{F}_2 D_{10})$ is not solvable.

Theorem 7 (summary). *We adopt the notation of theorem 2. If either (i) or (ii) is satisfied, then $\mathcal{U}(\mathbb{F}G)$ is centre-by-metabelian. If (iii) is satisfied, then $\mathcal{U}(\mathbb{F}G)$ is solvable of derived length at most 3, but not necessarily centre-by-metabelian. There are groups G that satisfy (iv) such that $\mathcal{U}(\mathbb{F}G)$ is not solvable.*

Remark. It was known before that Lie centre-by-metabelian algebras need not have a centre-by-metabelian unit group, since V. Tasić has already constructed a "counterexample" in [27]. However, his example is not a group algebra. The present paper now shows that such a behaviour cannot be expected even for the special case of group algebras.

REFERENCES

[1] A.K. Bhandari and I.B.S. Passi, Lie-nilpotency indices of group algebras, *Bull. London Math. Soc.*, **24**, (1992), 68-70.

[2] A.A. Bovdi, On group algebras with solvable unit groups, *Contemporary Math.* **131** (Part 1), (1992), 81-90.

[3] A.A. Bovdi, The group of units of a group algebra of characteristic p, *Publ. Math. Debrecen* **52/1-2**, (1998), 193-244.

[4] A.A. Bovdi and I.I. Khripta, Group algebras of periodic groups with solvable unit groups, *Math. Zametki* **22**, 3, (1977), 725-731.

[5] D. Coleman and R. Sandling, Mod 2 group algebras with metabelian unit groups, *Univ. of Manchester*, **1994/6**, (1994), (preprint, to appear in J. Pure Appl. Algebra)

[6] B. Külshammer and R.K. Sharma, Lie centrally metabelian group rings in characteristic 3, *J. Algebra* **180**, (1996), 111-120.

[7] J. Kurdics, On group algebras with metabelian unit groups, *Periodica Math. Hung.*, **32**, (1996), 183-192.

[8] F. Levin and G. Rosenberger, Lie metabelian group rings, in: "Proceedings of the international conference on group and semigroup rings at the University of Witwatersrand 1985," Mathematics studies 126, 153-161, North-Holland, Amsterdam, 1986.

[9] F. Levin and S. Sehgal, On Lie nilpotent group rings, *J. Pure Appl. Algebra*, **37**, (1985), 33-39.

[10] I.B.S. Passi, D.S. Passman, and S.K. Sehgal, Lie solvable group rings, *Canad. J. Math.* **25**, (1973), 748-757.

[11] D.S. Passman, Observations on group rings, *Comm. Algebra*, **5**, (1977), 1119-1162.

[12] D.S. Passman, "The algebraic structure of group rings", John Wiley & Sons, New York, 1977.

[13] R. Rossmanith, "Centre-by-metabelian group algebras," Dissertation, Friedrich-Schiller-Universität, Jena, 1997.

[14] R. Rossmanith, Lie centre-by-metabelian group algebras in even characteristic, I, to appear.

[15] R. Rossmanith, Lie centre-by-metabelian group algebras in even characteristic, II, to appear.

[16] R. Rossmanith, Lie centre-by-metabelian group algebras over commutative rings, to appear.

[17] M. Sahai, Group algebras with centrally metabelian unit groups, *Publ. Mat. Barcelona*, **40**, (1996), 443-456.

[18] M. Sahai and J.B. Srivastava, A note on Lie centrally metabelian group algebras, *J. Algebra* **187**, (1997), 7-15.

[19] S.K. Sehgal, "Topics in group rings," Marcel Dekker, New York, 1978.

[20] A. Shalev, Meta-abelian unit groups of group algebras are usually abelian, *J. Pure Appl. Algebra* **72**, (1991), 295-302.

[21] A. Shalev, Lie dimension subgroups, Lie nilpotency indices, and the exponent of the group of normalized units, *J. London Math. Soc.* (2) **43**, (1991), 23-36.

[22] A. Shalev, The derived length of Lie soluble group rings I, *J. Pure Appl. Algebra* **78**, (1992), 291-300.

[23] A. Shalev, The derived length of Lie soluble group rings II, *J. London Math. Soc.* **49**, (1994), 93-99.

[24] R.K. Sharma and J.B. Srivastava, Lie centrally metabelian group rings, *J. Algebra* **151** (1992), 476-486.

[25] M.B. Smirnov, The group of units of an associative ring satisfying the identity of Lie solvability, *Vestsi Akad. Nauk BSSR, Ser. Fiz.-Mat. Nauk*, **5**, (1983), 20-23.

[26] M.B. Smirnov and E. Zaleskii, Associative rings satisfying the identity of Lie solvability, *Vestsi Akad. Nauk BSSR, Ser. Fiz.-Mat. Nauk*, **2**, (1982), 15-20.

[27] V. Tasić, On unit groups of Lie centre-by-metabelian algebras, *J. Pure Appl. Algebra* **78**, (1992), 195-201.

[28] B.L. van der Waerden, Algebra II, Springer-Verlag, Berlin 1967.

[29] N, T, J, , (19), 1-2.

Covers and Envelopes in Functor Categories

Manuel Saorín and Alberto del Valle

(M. Saorín and A. del Valle) Departamento de Matemáticas. Universidad de Murcia, Aptdo. 4021. Espinardo 30.100, Murcia, Spain
E-mail address: msaorinc@fcu.um.es and alberto@fcu.um.es

ABSTRACT. Let R be a ring and let $\mathcal{C}$ be a skeletally small full subcategory of Mod_R containing the regular module R_R. In the category $(\mathcal{C}^{op}, \mathrm{Ab})$ of additive contravariant functors from $\mathcal{C}$ to the category Ab of abelian groups, we call pseudorepresentable a functor of the form $\mathrm{Hom}_R(-, M)\mid_\mathcal{C}$ with $M \in \mathrm{Mod}_R$ not necessarily in $\mathcal{C}$. We prove that every functor in $(\mathcal{C}^{op}, \mathrm{Ab})$ has a pseudorepresentable envelope that completes diagrams in a unique way, give conditions under which every functor has a pseudorepresentable (pre)cover and relate the existence of such covers to the existence of almost split sequences ending in a prescribed object of $\mathcal{C}$ with local endomorphism ring.

In particular, when $\mathcal{C} = \mathrm{mod}_R$ is the category of all finitely presented right R-modules, our results imply the fact that every functor of $\left(\mathrm{mod}_R^{op}, \mathrm{Ab}\right)$ has a pseudorepresentable (= flat, in this case) cover and the result of Auslander that every finitely presented module with local endomorphism ring is the third term in an almost split sequence in Mod_R.

1. Introduction

Throughout this paper R is a fixed arbitrary associative ring with identity. The notation M_R means that M is a right R-module, and Mod_R (resp. FMod_R, mod_R and proj_R) is the category of all (resp. all finitely generated, all finitely presented and all finitely generated projective) right R-modules. All maps between sets will act on the left of their arguments and their compositions will be written accordingly.

For a skeletally small preadditive category $\mathcal{C}$, we denote by $(\mathcal{C}^{op}, \mathrm{Ab})$ the category of all additive contravariant functors from $\mathcal{C}$ to the category Ab of abelian groups. This category, which has proved to be a very valuable tool for the study of the category $\mathcal{C}$ (see, e.g., [1] or [4]) has small Hom sets because $\mathcal{C}$ is skeletally small, and is an abelian category with direct limits, inheriting kernels, cokernels, images, direct sums and direct limits from Ab. The representable functors $\mathrm{Hom}_\mathcal{C}(-, C)$ with $C \in \mathcal{C}$ form a generating set of finitely generated projective objects in $(\mathcal{C}^{op}, \mathrm{Ab})$, and Yoneda's lemma implies that the functor $\mathcal{C} \to (\mathcal{C}^{op}, \mathrm{Ab})$ (called the Yoneda embedding) given by $C \mapsto \mathrm{Hom}_\mathcal{C}(-, C)$ is full and faithful.

In this paper, we take $\mathcal{C}$ to be a skeletally small full subcategory of Mod_R containing the regular module R_R. For any $B \in \mathrm{Mod}_R$, we denote by $(-, B)$ the

Both authors are supported by the DGES of Spain (PB96-0961-C02-02) and by the Fundación Séneca (Comunidad Autónoma de la Región de Murcia, PB/16/FS/97).

restriction to $\mathcal{C}$ of the functor $\operatorname{Hom}_R(-, B)$, and call a functor in $(\mathcal{C}^{op}, \mathrm{Ab})$ pseudorepresentable if it is naturally isomorphic to $(-, B)$ for some $B \in \operatorname{Mod}_R$. In Section 2, we prove that pseudorepresentable and flat envelopes exist in $(\mathcal{C}^{op}, \mathrm{Ab})$ and study the existence of pseudorepresentable covers in $(\mathcal{C}^{op}, \mathrm{Ab})$. We find that this study is worthy for two reasons: first, it implies the existence of flat covers in $(\operatorname{mod}_R^{op}, \mathrm{Ab})$, an interesting fact on its own right; and second, it is related to the existence of almost split sequences in Mod_R. This relation is made precise in Section 3, where we prove that, for a non-projective module $C \in \mathcal{C}$ with local endomorphism ring, the existence of a pseudorepresentable cover for the radical $r(-, C)$ of the functor $(-, C)$ is equivalent to the existence of an almost split sequence in Mod_R with third term C.

Next, we state the main definitions used in the paper.

Let $\mathcal{A}$ be any category and let $\mathcal{B}$ be any class of objects in $\mathcal{A}$. Following [5], a morphism $f : B \to A$ in $\mathcal{A}$ is a $\mathcal{B}$-**precover** for A if $B \in \mathcal{B}$ and, for every $B' \in \mathcal{B}$, the canonical map $\operatorname{Hom}_A(B', f) : \operatorname{Hom}_A(B', B) \to \operatorname{Hom}_A(B', A)$ is surjective. The morphism f is **right minimal** if the preimages of f by $\operatorname{Hom}_A(B, f)$ are all automorphisms of B. A right minimal $\mathcal{B}$-precover is called a $\mathcal{B}$-**cover**. The $\mathcal{B}$-cover f **completes the diagrams in a unique way** if $\operatorname{Hom}_A(B', f)$ is bijective for every $B' \in \mathcal{B}$. The dual concepts are $\mathcal{B}$-(pre)envelope and left minimal.

Let $A, B, C \in \operatorname{Mod}_R$; a R-homomorphism $f : B \to C$ is **right almost split** if it is not a retraction and any morphism $C' \to C$ in Mod_R which is not a retraction factors through f; dually, $g : A \to B$ is **left almost split** if it is not a section and, for any $A' \in \operatorname{Mod}_R$, each non-section $A \to A'$ factors through g. A non-split exact sequence $0 \to A \xrightarrow{g} B \xrightarrow{f} C \to 0$ in Mod_R is said to be an **almost split sequence** if g is left almost split and f is right almost split; this is equivalent to the fact that f is right almost split and right minimal (see [1, (II.4.4)]).

2. Pseudorepresentable Functors

For the rest of the paper we let $\mathcal{C}$ be a skeletally small full subcategory of Mod_R such that $R_R \in \mathcal{C}$. Given two objects $F, G \in (\mathcal{C}^{op}, \mathrm{Ab})$, we write $[F, G]$ for the set of all morphisms (i.e., natural transformations) between the functors F and G. Given any $A \in \operatorname{Mod}_R$, we write $(-, A)$ for the restriction to $\mathcal{C}$ of the contravariant functor $\operatorname{Hom}_R(-, A) : \operatorname{Mod}_R \to \mathrm{Ab}$. Also, given a morphism $f : A \to B$ in Mod_R, we shall write $(-, f) : (-, A) \to (-, B)$ for the morphism in $(\mathcal{C}^{op}, \mathrm{Ab})$ which sends $X \in \mathcal{C}$ to the map $(X, f) : (X, A) \to (X, B)$ given by $(X, f)(h) = f \circ h$.

We recall that, for any $F \in (\mathcal{C}^{op}, \mathrm{Ab})$ and any $X \in \mathcal{C}$, the abelian group FX has a natural structure of right $\operatorname{End}_R(X)$-module. Moreover, for any $\tau \in [F, G]$, the homomorphism $\tau_X : FX \to GX$ is $\operatorname{End}_R(X)$-linear [1, (I.1)]. When $X = R_R$, this gives a right R-module FR for any $F \in (\mathcal{C}^{op}, \mathrm{Ab})$ and a R-homomorphism $\tau_R : FR \to GR$ for any $\tau \in [F, G]$. We shall write $\sigma : 1_{\operatorname{Mod}_R} \to \operatorname{Hom}_R(R, -)$ for the standard natural isomorphism. Using this, it is clear that:

Lemma 2.1. *For a functor $F \in (\mathcal{C}^{op}, \mathrm{Ab})$, the following conditions are equivalent:*
(a) F is naturally isomorphic to $(-, A)$ for some $A \in \operatorname{Mod}_R$.
(b) F is naturally isomorphic to $(-, FR)$.

A functor satisfying the above conditions will be called **pseudorepresentable**, and we shall denote by $\mathcal{R}$ the class of all pseudorepresentable functors in $(\mathcal{C}^{op}, \mathrm{Ab})$.

Clearly, a pseudorepresentable functor F is representable if and only if $FR \in \mathcal{C}$. Pseudorepresentable functors are closely related to **flat** functors, i.e. functors isomorphic to direct limits of representable functors. These have been studied in [10] and [7] (see also [4]), where they are shown to have most of the properties of flat modules. By [11, (24.10) and (25.4)] we have:

Lemma 2.2. *(i) If $\mathcal{C} = \mathrm{FMod}_R$ then each pseudorepresentable functor is flat.*
(ii) If $\mathcal{C} = \mathrm{mod}_R$ then a functor is pseudorepresentable if and only if it is flat.

We start by proving the existence of pseudorepresentable envelopes in $(\mathcal{C}^{op}, \mathrm{Ab})$, and for this it is important to describe the morphisms between pseudorepresentable functors. These are easy to deal with thanks to the following result. It is easily verified and we shall use it throughout the paper without further reference.

Lemma 2.3. *Given $A, B \in \mathrm{Mod}_R$, the assignations $\tau \mapsto \sigma_B^{-1} \circ \tau_R \circ \sigma_A$ and $f \mapsto (-, f)$ define inverse isomorphisms of abelian groups between $[(-, A), (-, B)]$ and $\mathrm{Hom}_R(A, B)$.*

Thus we have a *Yoneda embedding* $\mathrm{Mod}_R \to (\mathcal{C}^{op}, \mathrm{Ab})$, i.e. a fully faithful functor given by $A \mapsto (-, A)$ which induces an equivalence of categories $\mathrm{Mod}_R \to \mathcal{R}$.

Now we prove that pseudorepresentable envelopes always exist in $(\mathcal{C}^{op}, \mathrm{Ab})$; these complete the diagrams in a unique way and admit a clear description.

Proposition 2.4. *Every functor in $(\mathcal{C}^{op}, \mathrm{Ab})$ has a pseudorepresentable envelope that completes the diagrams in a unique way.*

Proof. Let $F \in (\mathcal{C}^{op}, \mathrm{Ab})$, $X \in \mathcal{C}$ and $a \in FX$. We define a map $\phi_{X,a} : X \to FR$ as follows: for every $x \in X$, let $\lambda_x : R \to X$ be the R-homomorphism given by left multiplication by x. Then we have $F\lambda_x : FX \to FR$ in Ab, and we define $\phi_{X,a}(x) = F\lambda_x(a)$. This map $\phi_{X,a}$ is easily seen to be in $\mathrm{Hom}_R(X, FR)$, so that the assignation $a \mapsto \phi_{X,a}$ gives a map $\phi_X : FX \to (X, FR)$. Now, it is easy to check that this amounts to a natural transformation $\phi : F \to (-, FR)$, which is the desired envelope since the map $[(-, FR), (-, A)] \xrightarrow{-\circ\phi} [F, (-, A)]$ has an inverse given by $\tau \mapsto (-, \sigma_A^{-1} \circ \tau_R)$. $\square$

Remark 2.1. (1) The existence of a $\mathcal{B}$-envelope completing the diagrams in a unique way for every object of a category $\mathcal{A}$ is equivalent to the fact that $\mathcal{B}$ is a reflective subcategory of $\mathcal{A}$ (see, e.g., [6] or [9]). Thus, by [6, (36.9) and (36.13)], the class $\mathcal{R}$ of all pseudorepresentable functors of $(\mathcal{C}^{op}, \mathrm{Ab})$ is closed for direct summands and projective limits.

(2) If the category $\mathcal{C}$ has cokernels (e.g., if it is closed for cokernels in Mod_R) then one can check that every finitely presented functor in $(\mathcal{C}^{op}, \mathrm{Ab})$ has a representable envelope completing the diagrams in a unique way. Therefore, by [9, Proposition 3.4], the flat functors also form a reflective subcategory of $(\mathcal{C}^{op}, \mathrm{Ab})$ (cf. [9, Proposition 4.2]).

In the rest of this section we study pseudorepresentable (pre)covers in $(\mathcal{C}^{op}, \mathrm{Ab})$. The starting point for this investigation is a theorem of Belshoff, Enochs and Xu [2] asserting that every left module of finite flat dimension over a right coherent ring has a flat cover. We shall give a generalized version of it that fits our purposes, after introducing some notation.

If $\mathcal{B}$ is any class of objects of an abelian category $\mathcal{A}$ then we shall write $\mathcal{B}^{\perp}$ for the class of all objects K in $\mathcal{A}$ for which there are no non-split exact sequences $0 \to K \to X \to B \to 0$ in $\mathcal{A}$ with $B \in \mathcal{B}$. By Wakamatsu's lemma [12], the kernel of any $\mathcal{B}$-cover in $\mathcal{A}$ lies in $\mathcal{B}^{\perp}$. On the other hand, if $0 \to K \to B \to A \to 0$ is an exact sequence in $\mathcal{A}$ with $B \in \mathcal{B}$ and $K \in \mathcal{B}^{\perp}$ then $B \to A$ is clearly a $\mathcal{B}$-precover; following [12], an epimorphism $f : B \to A$ with $B \in \mathcal{B}$ and $\operatorname{Ker} f \in \mathcal{B}^{\perp}$ will be called a **special $\mathcal{B}$-precover** for A. Finally, we shall say that an object $A \in \mathcal{A}$ has $\mathcal{B}$-**dimension** $\leq n$ if there is an exact sequence in $\mathcal{A}$ of the form

$$0 \to B_n \to B_{n-1} \to \cdots \to B_1 \to B_0 \to A \to 0$$

with each $B_i \in \mathcal{B}$. Now we can state:

Theorem 2.5. *Let $\mathcal{A}$ be an abelian category and let $\mathcal{B}$ be a class of objects in $\mathcal{A}$ closed for extensions and satisfying the following conditions:*

  (a) *If the sequence $0 \to X \to Y \to Z \to 0$ is exact in $\mathcal{A}$ with $X, Y \in \mathcal{B}^{\perp}$ then $Z \in \mathcal{B}^{\perp}$.*

  (b) *For every $B \in \mathcal{B}$ there exists an exact sequence $0 \to B \to B_1 \to B_2 \to 0$ in $\mathcal{A}$ with $B_1 \in \mathcal{B}^{\perp}$ and $B_2 \in \mathcal{B}$.*

Then every object of finite $\mathcal{B}$-dimension of $\mathcal{A}$ has a special $\mathcal{B}$-precover.

Proof. The proof follows the lines of that of [2, Theorem 3.4]. It is clearly enough to show that, if $0 \to A' \to B \to A \to 0$ is an exact sequence in $\mathcal{A}$ such that A' has a special $\mathcal{B}$-precover and $B \in \mathcal{B}$, then A has a special $\mathcal{B}$-precover. Let then $0 \to K \to B' \to A' \to 0$ be an exact sequence in $\mathcal{A}$ with $K \in \mathcal{B}^{\perp}$ and $B' \in \mathcal{B}$, and let $0 \to B' \to B'_1 \to B'_2 \to 0$ the sequence guaranteed by (b); taking the pushout of $B' \to A'$ and $B' \to B'_1$ we get a commutative diagram with exact rows in $\mathcal{A}$

$$
\begin{array}{ccccccccc}
0 & \to & K & \longrightarrow & B' & \longrightarrow & A' & \to & 0 \\
  &   & \| &   & \downarrow &   & \downarrow & & \\
0 & \to & K & \longrightarrow & B'_1 & \longrightarrow & A'_1 & \to & 0
\end{array}
$$

and an exact sequence $0 \to A' \to A'_1 \to B'_2 \to 0$ where $A'_1 \in \mathcal{B}^{\perp}$ by (a). Taking now the pushout of $A' \to B$ and $A' \to A'_1$ we get a commutative diagram with exact rows in $\mathcal{A}$

$$
\begin{array}{ccccccccc}
0 & \to & A' & \longrightarrow & B & \longrightarrow & A & \to & 0 \\
  &   & \downarrow &   & \downarrow &   & \| & & \\
0 & \to & A'_1 & \longrightarrow & B_1 & \longrightarrow & A & \to & 0
\end{array}
$$

and an exact sequence $0 \to B \to B_1 \to B'_2 \to 0$ which shows that $B_1 \in \mathcal{B}$ and therefore the lower row of the latter diagram gives us the desired special $\mathcal{B}$-precover of A. $\square$

Remark 2.2. (1) If $\mathcal{A}$ is locally small and $\mathcal{B}$ is closed for direct limits then one can adapt the lemmas preceding Theorem 3.1 in [5] to get the existence of a $\mathcal{B}$-cover of A from any $\mathcal{B}$-precover of any object $A \in \mathcal{A}$, and therefore, in the conclusion of the theorem, we can write $\mathcal{B}$-cover instead of special $\mathcal{B}$-precover.

(2) When the category $\mathcal{A}$ has enough projective objects then condition (a) above can be replaced by: (a') For every projective object P of $\mathcal{A}$ and every epimorphism $P \to B$ with $B \in \mathcal{B}$, the kernel of $P \to B$ is in $\mathcal{B}$. Indeed, in this case the fact that

$X \in \mathcal{B}^{\perp}$ implies that $\text{Ext}^t_{\mathcal{A}}(B, X) = 0$ for every $B \in \mathcal{B}$ and every $i \geq 1$, and hence we get (a).

Next we look for conditions on $\mathcal{C}$ under which the class $\mathcal{R}$ of pseudorepresentable functors of $(\mathcal{C}^{op}, \text{Ab})$ fulfils the requirements in Theorem 2.5, and we start by relating the exactness of sequences in Mod_R to the exactness of the corresponding sequences of pseudorepresentable functors.

We shall say that a sequence $0 \to A \to B \to C \to 0$ in Mod_R is $\mathcal{C}$-**pure** (or that $A \to B$ is a $\mathcal{C}$-pure monomorphism, or that $B \to C$ is a $\mathcal{C}$-pure epimorphism) if it is exact and any $X \in \mathcal{C}$ is projective relative to it (see, e.g., [11]). When $\mathcal{C} = \text{proj}_R$ (resp. $\mathcal{C} = \text{mod}_R$) these are just the exact (resp. pure-exact) sequences. The following properties are easily checked and will be used freely in what follows.

Proposition 2.6. *Let $f : A \to B$ and $g : B \to C$ be morphisms in Mod_R. Then:*

(1) *$(-, f)$ is a monomorphism (resp. epimorphism, section or retraction) in $(\mathcal{C}^{op}, \text{Ab})$ if and only if f is a monomorphism (resp. $\mathcal{C}$-pure epimorphism, section or retraction) in Mod_R.*

(2) *The sequence $0 \to (-, A) \xrightarrow{(-, f)} (-, B) \xrightarrow{(-, g)} (-, C)$ is exact in $(\mathcal{C}^{op}, \text{Ab})$ if and only if the sequence $0 \to A \xrightarrow{f} B \xrightarrow{g} C$ is exact in Mod_R.*

(3) *The sequence $0 \to (-, A) \xrightarrow{(-, f)} (-, B) \xrightarrow{(-, g)} (-, C) \to 0$ is exact in $(\mathcal{C}^{op}, \text{Ab})$ if and only if the sequence $0 \to A \xrightarrow{f} B \xrightarrow{g} C \to 0$ is $\mathcal{C}$-pure in Mod_R.*

(4) *The sequence $0 \to (-, A) \xrightarrow{(-, f)} (-, B) \xrightarrow{(-, g)} (-, C) \to 0$ splits in $(\mathcal{C}^{op}, \text{Ab})$ if and only if the sequence $0 \to A \xrightarrow{f} B \xrightarrow{g} C \to 0$ splits in Mod_R.*

Proposition 2.7. *The class $\mathcal{R}$ of pseudorepresentable functors is closed under extensions.*

Proof. Let $0 \to (-, A) \xrightarrow{\alpha} F \xrightarrow{\beta} (-, B) \to 0$ be an exact sequence in $(\mathcal{C}^{op}, \text{Ab})$ with $A, B \in \text{Mod}_R$, and set $f = \alpha_R \circ \sigma_A$ and $g = \sigma_B^{-1} \circ \beta_R$. We then get an exact sequence $0 \to A \xrightarrow{f} FR \xrightarrow{g} B \to 0$ and, if $\phi : F \to (-, FR)$ is the $\mathcal{R}$-envelope of F, then $(-, g) \circ \phi = \beta$. Thus we get a diagram with exact rows

$$
\begin{array}{ccccccccc}
0 & \to & (-, A) & \xrightarrow{\alpha} & F & \xrightarrow{\beta} & (-, B) & \to & 0 \\
  &   & \| &   & \downarrow \phi &   & \| & & \\
0 & \to & (-, A) & \xrightarrow{(-, f)} & (-, FR) & \xrightarrow{(-, g)} & (-, B) & &
\end{array}
$$

in which the right square commutes. Next we see that so does the second. For, take $X \in \mathcal{C}$ and let us see that $\phi_X \circ \alpha_X = (X, f)$, i.e. that, for any $h : X \to A$, we have $f \circ h = \phi_{X, \alpha_X(h)}$ with the notation of Proposition 2.4. Now, for each $x \in X$, the naturality of α applied to $\lambda_x : R \to X$ gives $F\lambda_x \circ \alpha_X = \alpha_R(- \circ \lambda_x)$, so that

$$
\phi_{X, \alpha_X(h)}(x) = F\lambda_x(\alpha_X(h)) = \alpha_R(h \circ \lambda_x) = \alpha_R(\lambda_{hx}) = \alpha_R(\sigma_A(hx)) = f(hx).
$$

Thus the middle column in the diagram is an isomorphism and hence F is pseudorepresentable. $\square$

Now we can identify the pseudorepresentable functors that are in $\mathcal{R}^{\perp}$. A module $A \in \mathrm{Mod}_R$ is $\mathcal{C}$-**pure-injective** if $\mathrm{Hom}(-, A)$ carries $\mathcal{C}$-pure sequences to exact sequences, or equivalently if any $\mathcal{C}$-pure sequence starting on A splits [11, (33.7)].

Proposition 2.8. *If $A \in \mathrm{Mod}_R$, then $(-, A) \in \mathcal{R}^{\perp}$ if and only if A is $\mathcal{C}$-pure-injective.*

Proof. If A is $\mathcal{C}$-pure-injective and $0 \rightarrow (-, A) \rightarrow F \rightarrow G \rightarrow 0$ is exact with $G \in \mathcal{R}$ then also $F \in \mathcal{R}$, so that the sequence is of the form $0 \rightarrow (-, A) \rightarrow (-, B) \rightarrow (-, C) \rightarrow 0$ and therefore it comes from a $\mathcal{C}$-pure (hence split) sequence $0 \rightarrow A \rightarrow B \rightarrow C \rightarrow 0$ in Mod_R, whence the first sequence splits, thus showing that $(-, A) \in \mathcal{R}^{\perp}$. Conversely, if $(-, A) \in \mathcal{R}^{\perp}$ and the sequence $0 \rightarrow A \rightarrow B \rightarrow C \rightarrow 0$ is pure then the exact sequence $0 \rightarrow (-, A) \rightarrow (-, B) \rightarrow (-, C) \rightarrow 0$ splits and so does the first one, so that A is $\mathcal{C}$-pure-injective. $\square$

Next we see that if $\mathcal{C} \subseteq \mathrm{FMod}_R$, then $\mathcal{R}$ satisfies condition (a) in Theorem 2.5 and all objects of $(\mathcal{C}^{op}, \mathrm{Ab})$ have finite $\mathcal{R}$-dimension.

Proposition 2.9. *If each module of $\mathcal{C}$ is finitely generated (i.e., $\mathcal{C} \subseteq \mathrm{FMod}_R$) then:*

(1) *The class $\mathcal{R}$ of pseudorepresentable functors is closed under direct sums.*
(2) *Each functor of $(\mathcal{C}^{op}, \mathrm{Ab})$ is a quotient of a pseudorepresentable functor.*
(3) *Each functor of $(\mathcal{C}^{op}, \mathrm{Ab})$ has $\mathcal{R}$-dimension at most 2.*
(4) *Each projective functor of $(\mathcal{C}^{op}, \mathrm{Ab})$ is pseudorepresentable.*
(5) *The class $\mathcal{R}$ satisfies condition (a) in Theorem 2.5.*

Proof. (1) follows from [11, 24.10]. If $F \in (\mathcal{C}^{op}, \mathrm{Ab})$ and $\mathcal{C}_0$ is a skeleton of $\mathcal{C}$, then the canonical morphism $\oplus_{X \in \mathcal{C}_0} (-, X)^{([(-,X),F])} = \oplus_{X \in \mathcal{C}_0} (-, X)^{(FX)} \rightarrow F$ is an epimorphism in $(\mathcal{C}^{op}, \mathrm{Ab})$, so that (2) follows from (1). Now (3) and (4) are clear and (5) follows from Remark 2.2.(2). $\square$

Finally, we see what is needed for $\mathcal{R}$ to satisfy all conditions in Theorem 2.5.

Proposition 2.10. *Assume that $\mathcal{C} \subseteq \mathrm{FMod}_R$ and that, for every $A \in \mathrm{Mod}_R$, there exists a $\mathcal{C}$-pure-injective module $\bar{A}$ and a $\mathcal{C}$-pure monomorphism $A \rightarrow \bar{A}$. Then every functor of $(\mathcal{C}^{op}, \mathrm{Ab})$ has a special pseudorepresentable precover.*

Proof. We just need to check condition (b) to apply Theorem 2.5, but if $(-, A) \in \mathcal{R}$ then the sequence $0 \rightarrow (-, A) \rightarrow (-, \bar{A}) \rightarrow (-, \bar{A}/A) \rightarrow 0$ does the job. $\square$

Remark 2.3. (1) It is easy to see that the Yoneda embedding $\mathrm{Mod}_R \rightarrow (\mathcal{C}^{op}, \mathrm{Ab})$ takes pushout diagrams in Mod_R in which one of the source maps is a $\mathcal{C}$-pure monomorphism to pushout diagrams in $(\mathcal{C}^{op}, \mathrm{Ab})$. So, in the setting of the above proposition, the proof of Theorem 2.5 gives us a way to construct a special pseudorepresentable precover for any functor F with pseudorepresentable dimension 1. Indeed, given an exact sequence $0 \rightarrow (-, A) \xrightarrow{(-,f)} (-, B) \rightarrow F \rightarrow 0$ in $(\mathcal{C}^{op}, \mathrm{Ab})$ and a pushout diagram

$$
\begin{array}{ccc}
A & \xrightarrow{f} & B \\
\downarrow & & \downarrow \\
\bar{A} & \longrightarrow & C
\end{array}
$$

in Mod_R, where $A \to \bar{A}$ is as above, we get a commutative diagram with exact rows in $(\mathcal{C}^{op}, \text{Ab})$

$$
\begin{array}{ccccccccc}
0 & \to & (-, A) & \longrightarrow & (-, B) & \longrightarrow & F & \to & 0 \\
& & \downarrow & & \downarrow & & \| & & \\
0 & \to & (-, \bar{A}) & \longrightarrow & (-, C) & \longrightarrow & F & \to & 0
\end{array}
$$

and the morphism $(-, C) \to F$ is a special pseudorepresentable precover.

(2) From the previous results, we can derive a proof of the well known fact that, for $\mathcal{C} = \text{proj}_R$, all functors in $(\mathcal{C}^{op}, \text{Ab})$ are pseudorepresentable. Indeed, since now the Yoneda embedding preserves epimorphisms, we see that the $\mathcal{R}$-envelopes are always epic; therefore $\mathcal{R}$ is closed for subobjects [6, (37.1)] and hence the result easily follows from Propositions 2.8 and 2.10.

Corollary 2.11. *Every functor of* $(\text{mod}_R^{op}, \text{Ab})$ *has a flat cover.*

Proof. When $\mathcal{C} = \text{mod}_R$ then $\mathcal{R}$ is the class of flat functors (Lemma 2.2) and $\mathcal{C}$-purity is the usual concept of purity. Since, for every module A, the canonical map to its double character module $A \to A^{++}$ is a pure monomorphism to a pure-injective module, we can apply the above proposition and the fact that flat functors are closed under direct limits to get the result. $\square$

In view of Proposition 2.10 it is interesting, for different choices of $\mathcal{C}$, to study the existence of $\mathcal{C}$-pure monomorphisms $M \to \bar{M}$ with $\bar{M}$ $\mathcal{C}$-pure-injective for a given module M. Of course these exist for every module when $\mathcal{C} = \text{proj}_R$ or $\mathcal{C} = \text{mod}_R$. We present here a different instance of this situation.

Example 2.1. Let $S \in \text{Mod}_R$ be simple and non-projective, and let $\mathcal{C} = \text{add}\,(R \oplus S)$ be the class of all direct summands of finite direct sums of copies of $R \oplus S$. Also, for each $X \in \text{Mod}_R$, let $X\,[S]$ be the S-homogeneous component of $\text{Soc}\,(X)$. We propose the reader to check the following assertions about modules in Mod_R:

(1) An epimorphism $Y \to X$ is $\mathcal{C}$-pure if and only if the inclusion $X\,[S] \hookrightarrow X$ factors through it.

From now on assume that S is Σ-injective; this means that the inclusion $X\,[S] \hookrightarrow X$ splits for every module X.

(2) M is $\mathcal{C}$-pure-injective if and only if $\text{Ext}_R^1\,(X, M) = 0$ when $X\,[S] = 0$.

(3) If R is right pure-semisimple then direct sums of $\mathcal{C}$-pure-injective modules are again $\mathcal{C}$-pure-injective.

Now let R be the algebra of finite representation type given by the Dynkin diagram $\mathbb{A}_3$ $(\cdot \to \cdot \to \cdot)$. Its Auslander-Reiten quiver is well known to be the wing

$$
\begin{array}{ccccccc}
& & & P_1 = I_3 & & & \\
& & \nearrow & & \searrow & & \\
& P_2 & & & & I_2 & \\
& \nearrow & \searrow & & \nearrow & & \searrow \\
S_3 = P_3 & & & S_2 & & & I_1 = S_1
\end{array}
$$

where the "going up" arrows are monomorphisms, the "going down" arrows are epimorphisms and, for each $i = 1, 2, 3$, S_i is simple with projective cover P_i and injective envelope I_i. We set $\mathcal{C} = \text{add}\,(R \oplus S_1)$. By using the above claims, one checks that the indecomposable C-pure-injective modules are I_1, I_2, I_3 and S_2, and that the obvious sequences

$$0 \to P_3 \to I_3 \to I_2 \to 0 \qquad 0 \to P_2 \to I_3 \oplus S_2 \to I_2 \to 0$$

are C-pure. Therefore we have the desired map $M \to \bar{M}$ for every indecomposable M and it is then clear how to construct one such for an arbitrary M.

3. APPLICATION TO ALMOST SPLIT SEQUENCES

We still assume that C is a skeletally small subcategory of Mod_R containing R_R. For a functor $F \in (C^{op}, \text{Ab})$ we denote its **radical** (i.e., the intersection of all its maximal subfunctors) by rF.

Theorem 3.1. *Assume that $C \in C$ is not projective and has a local endomorphism ring. Then the following conditions are equivalent:*

    (a) *There exists an almost split sequence in Mod_R with third term C.*
    (b) *$r(-, C)$ has a pseudorepresentable cover in (C^{op}, Ab).*

Proof. We first note that, by [1, (II.1.9) and (II.1.5)], $r(-, C)$ is a proper subfunctor of the representable functor $(-, C)$ which contains every proper subfunctor of $(-, C)$. We write $\mu : r(-, C) \hookrightarrow (-, C)$ for be the inclusion functor.

(a)$\Rightarrow$(b). Let $f : B \to C$ be right almost split and right minimal in Mod_R. Since f is not a retraction, $(-, f)$ is not epic and therefore it factors as $(-, f) = \mu \circ \tau$ for a morphism $\tau : (-, B) \to r(-, C)$, and this is a R-cover of $r(-, C)$. Indeed, as f is right minimal, so is $(-, f)$ and hence τ, and we just have to check that, given any $D \in \text{Mod}_R$, any morphism $\sigma : (-, D) \to r(-, C)$ factors through τ. For, take $h : D \to C$ in Mod_R with $\mu \circ \sigma = (-, h)$. As h is not a retraction, it factors through f, i.e. there exists $t : D \to B$ with $h = f \circ t$, whence

$$\mu \circ \sigma = (-, h) = (-, f) \circ (-, t) = \mu \circ \tau \circ (-, t)$$

and therefore $\sigma = \tau \circ (-, t)$, as desired.

(b)$\Rightarrow$(a). Let $\tau : (-, B) \to r(-, C)$ be a pseudorepresentable cover in (C^{op}, Ab) and take $f : B \to C$ in Mod_R with $\mu \circ \tau = (-, f)$. Given a non-retraction $g : D \to C$ in Mod_R, we have $\text{Im}(-, g) \subseteq r(-, C)$ and hence $(-, g) = \mu \circ \sigma$ for a morphism $\sigma : (-, D) \to r(-, C)$. Then, by the preenveloping condition of τ, there exists $h : D \to B$ in Mod_R with $\sigma = \tau \circ (-, h)$ and therefore

$$(-, g) = \mu \circ \sigma = \mu \circ \tau \circ (-, h) = (-, f) \circ (-, h) = (-, f \circ h),$$

so that g factors through f and hence f is right almost split. Since C is not projective, f is epic. On the other hand, since τ is right minimal and μ is monic, the composition $\mu \circ \tau = (-, f)$ is right minimal, and hence so is f. $\square$

Remark 3.1. (1) If C_R is projective then it has a unique maximal submodule B and $r(-, C) = (-, B)$ is a pseudorepresentable functor, so that the above arguments also give a right almost split right minimal morphism $B \hookrightarrow C$.

(2) By the first part of the above proof, a R-precover $(-, B) \to r(-, C)$ gives a right almost split morphism $B \to C$. So, since $r(-, C)$ has R-dimension at most one, Remark 2.3 gives us a way for constructing such a morphism.

Corollary 3.2. [1, (II.5.1)] *Every nonprojective* $C \in \text{mod}_R$ *with local endomorphism ring is the third term of an almost split sequence in* Mod_R.

Proof. This follows from Theorem 3.1 and Corollary 2.11. □

Remark 3.2. Corollaries 2.11 and 3.2 admit the following partial dualization. If we consider the category $\mathbf{D}(R) = (_R \text{mod}, \text{Ab})$ of additive covariant functors $_R \text{mod} \to \text{Ab}$ then the assignation $A \mapsto (A \otimes -)$ gives an embedding $\text{Mod}_R \to \mathbf{D}(R)$ which induces an equivalence of categories between Mod_R and the full subcategory of $\mathbf{D}(R)$ consisting of the FP-injective objects (see e.g. [4]). By essentially the same argument of [8, Corollary 3.5(c)], one can show that FP-injective preenvelopes exist in $\mathbf{D}(R)$. Moreover, if $S \in \mathbf{D}(R)$ is a simple functor and $\tau : S \to (A \otimes -)$ is an injective envelope (so that A is pure-injective with local endomorphism ring) then, taking a FP-injective preenvelope of $\text{Coker}\,\tau$, we get a short exact sequence $0 \to S \to (A \otimes -) \xrightarrow{f \otimes -} (B \otimes -)$ in $\mathbf{D}(R)$, and it can be proven that $f : A \to B$ is left almost split, although it is not left minimal unless the FP-injective preenvelope of $\text{Coker}\,\tau$ is indeed an envelope. In fact, by [3, Theorem 2.3], A is the source of a left almost split morphism $A \to B$ in Mod_R if and only if $(A \otimes -)$ is is the injective envelope of a simple functor.

We would like to thank Henning Krause for calling our attention to this remark.

REFERENCES

1. M. Auslander. *Functors and morphisms determined by objects.* In Representation theory of algebras (Proc. Conf. Philadelphia, 1976), 1–244. Dekker, New York (1978).
2. R. Belshoff, E. Enochs and J. Xu. *The existence of flat covers.* Proc. Amer. Math. Soc. **122**, 985-991 (1994).
3. W. Crawley-Boevey. *Modules of finite length over their endomorphism rings.* In Representations of algebras and related topics. London Math. Soc. Lecture Notes **168**, 127-184. Cambridge Univ. Press (1992).
4. W. Crawley-Boevey. *Locally finitely presented additive categories.* Comm. Algebra **22**, 1641-1674 (1994).
5. E. Enochs. *Injective and flat covers, envelopes and resolvents.* Israel J. Math. **39**, 189-209 (1981).
6. H. Herrlich and G. Strecker. *"Category Theory".* Allyn and Bacon, Boston (1973).
7. U. Oberst and H. Rohrl. *Flat and coherent functors.* J. Algebra **14**, 91-105 (1970).
8. J. Rada and M. Saorín. *Rings characterized by (pre)envelopes and (pre)covers of their modules.* Comm. Algebra **26**, 899-912 (1998).
9. J. Rada, M. Saorín and A. del Valle. *Reflective subcategories.* To appear in Glasgow Math. J.
10. B. Stenström. *Purity in functor categories.* J. Algebra **8**, 352-361 (1968).
11. R. Wisbauer. *"Foundations of Module and Ring Theory".* Gordon and Breach, Philadelphia (1991).
12. J. Xu. *"Flat Covers of Modules".* Lecture Notes in Math. **1634**. Springer-Verlag, Berlin (1996).

Faithful Flatness over Hopf Subalgebras: Counterexamples

PETER SCHAUENBURG

MATHEMATISCHES INSTITUT DER UNIVERSITÄT MÜNCHEN, THERESIENSTR. 39, 80333 MÜNCHEN, GERMANY
E-mail address: schauen@rz.mathematik.uni-muenchen.de

1. INTRODUCTION

Throughout the paper we will assume that k is a field, and all tensor products, algebras, coalgebras, etc. are over k.

By the famous Nichols-Zoeller Theorem [10], every finite dimensional Hopf algebra is free over its Hopf subalgebras—a Hopf algebra version of Lagrange's theorem in group theory. In general, the first of Kaplansky's conjectures [3] was that *every* Hopf algebra H is a free module over each Hopf subalgebra $B \subset H$. This is known to be true also when B is semisimple—by another theorem of Nichols and Richmond (Zoeller) [9]—, and when B is finite dimensional and normal by Schneider [13]. However, the conjecture in its general form had already been refuted by Oberst and Schneider [11] when it appeared.

For more general Hopf algebras, a weaker condition, namely, faithful flatness over Hopf subalgebras, seems to be satisfied very often (and is used as a technical tool in many situations [5, 12, 13]). In view of many positive results, the question was asked formally in Montgomery's book [7, Question 3.5.4]: If B is a Hopf subalgebra of a Hopf algebra H, is then H always left and right faithfully flat as a B-module?

For commutative H, this is a classical result on algebraic groups [1, III, 3, no.7]; a purely Hopf algebraic proof including the stronger assertion that H is a projective B-module was given by Takeuchi [16]. Schneider [13] shows that H is B-faithfully flat if B is central in H and H is left or right noetherian.

For the cocommutative case, the assertion is proved in the theory of formal groups [2, Section 2.4.], again, a Hopf algebraic proof is in [16]. Masuoka [4] shows that it is enough to assume that the coradical of H is cocommutative.

Summing up, there are many interesting results showing that a Hopf algebra is faithfully flat over all of its Hopf subalgebras under some additional assumptions; the general question seems to remain open.

In the present paper we discuss a class of pathological counterexamples to faithful flatness over Hopf subalgebras. These are based on the observation that faithful flatness fails when we have certain pathological behavior of the antipodes of the Hopf (sub)algebras in consideration. In fact, we shall show that the Hopf algebra inclusions $B \subset H$ we study do not even have the following property:

(P) If f is an endomorphism of a finitely generated free B-module such that $f \underset{B}{\otimes} H$ is an isomorphism, then f is an isomorphism.

Property (P) is clearly implied by faithful flatness.

Takeuchi [14] constructed the free Hopf algebra $H(C)$ over a coalgebra C and showed that Hopf algebras with non-bijective antipodes exist. Nichols [8] computed a k-basis for $H(M_n(k)^*)$, where $M_n(k)^*$ is the dual coalgebra of the algebra of n-by-n matrices, and showed that this is a Hopf algebra whose antipode is injective but not surjective. We will show that such a Hopf algebra has a Hopf subalgebra over which it is not faithfully flat. This leads us to modify the original question and ask if at least Hopf algebras *with bijective antipode* are always faithfully flat over all their Hopf subalgebras. However, we show that whenever a Hopf algebra with bijective antipode has a Hopf subalgebra whose antipode is not bijective, the inclusion does not have property (P). In view of results of Masuoka and Wigner [6] such an inclusion is also not flat.

This reduces our problem to the first of a pair of dual questions of general Hopf algebra theory: Does every Hopf subalgebra (resp. quotient) of a Hopf algebra with bijective antipode in turn have bijective antipode?

A counterexample to the first question is obtained by a construction, Proposition 2.7, which maps every Hopf algebra H to a universal Hopf algebra $\hat{H}$ with bijective antipode, the map from H to $\hat{H}$ being injective if and only if the antipode of H is.

Obtaining counterexamples to the dual question is slightly more complicated. It seems to have been an open question wether Hopf algebras with surjective, but not injective antipodes exist (but cf. [15]). Note that the results on quotient bialgebras of Hopf algebras obtained in Nichols' paper [8] also imply that under rather mild assumptions on a Hopf algebra H with bijective antipode any of its quotient bialgebras will automatically be a Hopf algebra with bijective antipode. On the other hand, Nichols [8] considers the biideal I in $H(M_2(k)^*)$ generated by the basis element x_{21} of $M_2(k)^*$, uses the diamond lemma to calculate a basis for the quotient and shows that I is not a Hopf ideal, that is, the quotient does not have an antipode. We will modify this example, using the free Hopf algebra with bijective antipode $\hat{H}(C)$ over a coalgebra C, and find a Hopf ideal such that the quotient does not have a skew antipode. Of course, the quotient has surjective, but not bijective antipode.

2. FAITHFUL FLATNESS OVER HOPF SUBALGEBRAS—COUNTEREXAMPLES

Takeuchi [17, Cor.1.5(b)] observed that if H is a Hopf algebra with bijective antipode and $B \subset H$ is a subbialgebra such that H is a faithfully flat left and right B-module, then B is necessarily a Hopf algebra itself. We will be using variants of this observation to provide counterexamples to [7, Question 3.5.4]. In fact, we will use a property of ring inclusions, called property (P) below, which is much weaker than faithful flatness, and give a direct proof that inclusions of bialgebras into Hopf algebras, and of Hopf algebras whose antipode is not bijective into Hopf algebras with bijective antipodes, cannot have property (P).

Definition 2.1. Let $B \subset A$ be a ring inclusion. We say that the inclusion has property (P) if, for each endomorphism f of a finitely generated free right B-module M such that $f \underset{B}{\otimes} A$ is an ismorphism, f is also an isomorphism.

Remark 2.2. Clearly a left faithfully flat ring inclusion has property (P). Moreover, any ring inclusion $B \subset A$ such that B is a left B-direct summand of A has property (P). Also, if A is a generator left B-module, then the inclusion $B \subset A$ has property (P).

In terms of matrices, property (P) can be reformulated as saying that for every square matrix with entries in B which is invertible as a matrix with entries in A, the inverse has entries in B as well. Note that this implies that the condition is left-right symmetric.

Lemma 2.3. *Let $\iota: B \to A$ be an inclusion of algebras having property (P). Let C be a coalgebra and $\phi: C \to B$ a k-linear map. If $\iota\phi: C \to A$ is convolution invertible, then so is ϕ, that is, the convolution inverse of $\iota\phi$ has image in B.*

Proof. Note first that ϕ is convolution invertible if and only if its restriction to every finite dimensional subcoalgebra of C is. Thus we may assume that C is finite dimensional to begin with. By the Lemma on page 91 of [7] we have, for every k-algebra R, an isomorphism of k-algebras

$$T_R^C: \mathrm{Hom}_k(C, R) \cong \mathrm{End}_{-R}^{C-}(C \otimes R),$$

where the left hand side is an algebra by convolution, and the right hand side, which denotes the set of all right R-linear and left C-colinear endomorphisms, is an algebra by composition. The isomorphism is given explicitly by $T_R^C(\phi)(c \otimes r) = c_{(1)} \otimes \phi(c_{(2)})r$. A short computation shows that $T_A^C(\iota\phi) = T_B^C(\phi) \underset{B}{\otimes} A$ as endomorphisms of $C \otimes A = C \otimes B \underset{B}{\otimes} A$. From this, the claim follows immediately. $\qquad\square$

Our first corollary contains a direct proof of [17, Cor.1.5(b)].

Corollary 2.4. *Let H be a Hopf algebra and B a subbialgebra such that $B \subset H$ has property (P). Then B is a Hopf subalgebra. If the antipode of H is bijective, then so is the antipode of B.*

In fact the first claim follows by applying the preceding Lemma with ϕ the inclusion of B into H, the second claim follows by applying the first to $B^{\mathrm{op}} \subset H^{\mathrm{op}}$.

Since we will encounter below examples of Hopf subalgebras with non-bijective antipodes in Hopf algebras with bijective antipodes, the Corollary is sufficient to answer negatively our original question. Still, we list another Corollary which will show that a counterexample is already at hand in the literature:

Corollary 2.5. *Let H be a Hopf algebra and $B \subset H$ a Hopf subalgebra such that the inclusion $S(B) \subset H$ has property (P). Then $S(B) = B$.*

Proof. Consider the map $\phi \colon B \ni b \mapsto S(b) \in S(B)$. Writing $\iota \colon S(B) \to H$ for the inclusion, $\iota\phi$ is convolution invertible, and its inverse is the inclusion $B \to H$, which thus has its image, B, contained in $S(B)$ by Lemma 2.3. $\qquad\square$

Remark 2.6. Nichols [8] showed that there is a Hopf algebra H whose antipode is injective, but not surjective. Applying the preceding Lemma to $B = H$ we see that the inclusion $S(H) \subset H$ does not have property (P); in particular, the answer to [7, Question 3.5.4] is negative.

Next, we want to find a Hopf algebra H *with bijective antipode* and a Hopf subalgebra B in it such that the inclusion $B \subset H$ is not faithfully flat. By Lemma 2.4 it suffices to find an example where the antipode of B is not bijective.

Proposition 2.7. *The underlying functor from the category of Hopf algebras with bijective antipode to the category of Hopf algebras has a left adjoint.*

In other words: Let H be a Hopf algebra. Then there is a Hopf algebra $\hat{H}$ with bijective antipode and a Hopf algebra map $\iota \colon H \to \hat{H}$

such that every Hopf algebra map $H \to F$ to a Hopf algebra F with bijective antipode factors uniquely through ι.

The map $\iota: H \to \hat{H}$ is injective if and only if the antipode of H is injective.

Proof. One can check in general that the colimit of a directed system of bialgebras (Hopf algebras) is again a bialgebra (resp. Hopf algebra) in the obvious way.

Let $\hat{H}$ be the colimit of the directed system of Hopf algebras $(H_i, i \in \mathbb{N}_0)$ and Hopf algebra maps $f_i: H_i \to H_{i+1}$ for $i \in \mathbb{N}_0$, where $H_i = H$ and $f_i = S^2$ for all i. We denote by $[h]_i$ the class in $\hat{H}$ of $h \in H = H_i$. We have a Hopf algebra map $\iota: H \to \hat{H}$ with $\iota(h) = [h]_0$. We define $\overline{S}: \hat{H} \to \hat{H}$ by $\overline{S}([h]_i) = [S(h)]_{i+1}$ (to see that this is well defined, let $t_i: H_i \to \hat{H}^{\mathrm{op}}$ be defined by $t_i(h) = [S(h)]_{i+1}$ and check $t_{i+1}f_i(h) = [S(f_i(h))]_{i+1} = [f_i(S(h))]_{i+1} = [S(h)]_i = t_i(h)$). Then $\overline{S}S([h]_i) = \overline{S}([S(h)]_i) = [S^2(h)]_{i+1} = [h]_i$ and $S\overline{S}([h]_i) = S([S(h)]_{i+1}) = [S^2(h)]_{i+1} = [h]_i$, so that the antipode of $\hat{H}$ is bijective with inverse $\overline{S}$.

To check the universal mapping property of $\hat{H}$, let F be a Hopf algebra with bijective antipode and $\phi: H \to F$. We define Hopf algebra maps $\alpha_i: H_i \to F$ by $\alpha_i(h) = S^{-2i}\phi(h)$ and check $\alpha_{i+1}f_i(h) = S^{-2(i+1)}S^2(h) = S^{-2i}(h) = \alpha_i(h)$, so that there exists a Hopf algebra map $\psi: \hat{H} \to F$ with $\psi([h]_i) = S^{-2i}\phi(h)$ for all $i \in \mathbb{N}_0$ and $h \in H$. In particular, $\psi\iota = \phi$. If $\psi': \hat{H} \to F$ is another Hopf algebra map with $\psi'\iota = \phi$, then $\psi'([h]_i) = \psi'(S^{-2i}S^{2i}([h]_i)) = S^{-2i}\psi'([S^{2i}(h)]_i) = S^{-2i}\psi'([h]_0) = S^{-2i}\phi(h)$ and hence $\psi' = \psi$.

Clearly the antipode of H is injective if ι is injective. On the other hand $\iota(h) = [h]_0 = 0$ holds in $\hat{H}$ if and only if there exists $n \in \mathbb{N}$ with $0 = S^{2n}(h)$ in H. If the antipode of H is injective, this implies $h = 0$. $\qquad\square$

Corollary 2.8. *There exists a Hopf algebra H with bijective antipode and a Hopf subalgebra $B \subset H$ such that the inclusion $B \subset H$ does not satisfy property (P); in particular H is not faithfully flat over B.*

Proof. Let B be a Hopf algebra with injective, but not surjective antipode. Using Proposition 2.7 we view B as a Hopf subalgebra of $H := \hat{B}$. The embedding does not have property (P) by Lemma 2.4. $\qquad\square$

Remark 2.9. Masuoka and Wigner [6, Cor.2.9.] show that when H is a Hopf algebra with bijective antipode and $B \subset H$ is a Hopf subalgebra, then H is left B-flat iff H is left B-faithfully flat iff H is a progenerator left B-module. (Note that while the statement of [6, Cor.2.9] requires

B to also have bijective antipode, the part of the Corollary just cited is proved without that hypothesis.) Hence, we have in fact proved that there exist Hopf algebra inclusions which are not flat.

One should be aware that the counterexample to (faithful) flatness thus obtained still relies entirely upon pathological behaviour of the antipode, in this case of the Hopf subalgebra in question. Thus, an important and certainly very relevant weaker version of the original question still remains open:

Question . Is every Hopf algebra with bijective antipode faithfully flat over all its Hopf subalgebras with bijective antipode?

3. There exists a Hopf algebra whose antipode is surjective, but not bijective

The following simple observation is obtained by combining Takeuchi's construction of the free Hopf algebra generated by a coalgebra [14] and Lemma 2.7:

Lemma 3.1. *The underlying functor from the category of Hopf algebras with bijective antipode to the category of coalgebras has a left adjoint, which we denote by $\hat{H}$. We call $\hat{H}(C)$ the free Hopf algebra with bijective antipode generated by the coalgebra C.*

In fact, we can construct $\hat{H}(C) := \widehat{H(C)}$, where $H(C)$ is the free Hopf algebra generated by C. Alternatively, we can copy the construction of $H(C)$ in [14]: We define coalgebras V_r for $r \in \mathbb{Z}$ by $V_r := C$ if r is even, and $V_r := C^{\mathrm{cop}}$ if r is odd. We let $V := \bigoplus_{r \in \mathbb{Z}} V_r$. We consider the map $S \colon V \to V$ given by $S((v_r)_{r \in \mathbb{Z}}) := (v_{r-1})_{r \in \mathbb{Z}}$. It induces an algebra map $S \colon T(V) \to T(V)^{\mathrm{op}}$. Let N be the ideal of $T(V)$ generated by $\{v_{(1)}S(v_{(2)}) - \varepsilon(v) \cdot 1 | v \in V\} \cup \{S(v_{(1)})v_{(2)} - \varepsilon(v) \cdot 1 | v \in V\}$. One checks that N is a biideal and hence $H(C) := T(V)/N$ is a bialgebra. Moreover, N is stable under both S and its inverse, hence S induces an invertible map $S \colon H(C) \to H(C)^{\mathrm{op}}$. One proves as in [14] that S is an antipode for $H(C)$.

Theorem 3.2. *Let (t_{ij}) be the standard k-basis of $M_4(k)^*$. Then the Hopf ideal I in $H := \hat{H}(M_4(k)^*)$ generated by the images of t_{31}, t_{32}, t_{41} and t_{42} is not stable under the inverse of the antipode. The antipode of H/I is surjective, but not injective.*

Proof. We first align our notations with those of Nichols [8]: We introduce the matrix (t_{ij}) of standard basis elements in $M_n(k)^*$. We denote basis elements in the coalgebras V_r by $x_{ij}^r := t_{ij} \in V_r$ if r is even, and

$x_{ij}^r := t_{ji} \in V_r$ if r is odd. We let $X := \{x_{ij}^r | i, j \in \{1, \dots, n\}, r \in \mathbb{Z}\}$, write $\langle X \rangle$ for the free monoid generated by X and $k\langle X \rangle$ for its monoid algebra (that is, the polynomial ring with noncommuting variables x_{ij}^r). Then we have

$$\hat{H}(M_n(k)^*) \cong k\langle X \rangle / N$$

with comultiplication and antipode given by $\Delta(x_{ij}^r) = \sum_{a=1}^n x_{ia}^r \otimes x_{aj}^r$ and $S(x_{ij}^r) = x_{ji}^{r+1}$, where

$$N = (\sum_{a=1}^n x_{ia}^r x_{ja}^{r+1} - \delta_{ij}, \sum_{a=1}^n x_{ai}^r x_{aj}^{r-1} - \delta_{ij} | i, j \in \{1, \dots, n\}, r \in \mathbb{Z}).$$

Next, we wish to compute a k-basis of $\hat{H}(M_n(k)^*)$ by applying Bergman's diamond lemma. Nichols did the same for $H(M_n(k)^*)$, and our calculations and results are the same as his. Namely, the following four sets of reductions, for all $i, j, k \in \{1, \dots, n\}$ and $r \in \mathbb{Z}$, satisfy the conditions of the diamond lemma:

$$x_{in}^r x_{jn}^{r+1} \to \delta_{ij} - \sum_{a<n} x_{ia}^r x_{ja}^{r+1} \tag{I}$$

$$x_{ni}^r x_{nj}^{r-1} \to \delta_{ij} - \sum_{a<n} x_{ai}^r x_{aj}^{r-1} \tag{II}$$

$$x_{in}^r x_{j,n-1}^{r+1} x_{k,n-1}^{r+2} \to \delta_{jk} x_{in}^r - \delta_{ij} x_{kn}^{r+2} + \sum_{a<n} x_{ia}^r x_{ja}^{r+1} x_{kn}^{r+2} - \sum_{a<n-1} x_{in}^r x_{ja}^{r+1} x_{ka}^{r+2} \tag{III}$$

$$x_{ni}^r x_{n-1,j}^{r-1} x_{n-1,k}^{r-2} \to \delta_{jk} x_{ni}^r - \delta_{ij} x_{nk}^{r-2} + \sum_{a<n} x_{ai}^r x_{aj}^{r-1} x_{nk}^{r-2} - \sum_{a<n-1} x_{ni}^r x_{aj}^{r-1} x_{ak}^{r-2} \tag{IV}$$

We also have to provide a semigroup partial order on $\langle X \rangle$ compatible with these reductions and satisfying the descending chain condition. We define $x_{i_1 j_1}^{r_1} \cdots x_{i_a j_a}^{r_a} \geq x_{k_1 \ell_1}^{s_1} \cdots x_{k_b \ell_b}^{s_b}$ if and only if either $a > b$ or we have $a = b$, $r_\alpha = s_\alpha$ for all α, and $(i_1, j_1, \dots, i_a, j_a) \geq (k_1, \ell_1, \dots, k_a, \ell_a)$ in lexicographical order.

We shall not prove that the conditions of the diamond lemma are really satisfied for (I) through (IV), but rather refer to Nichols' paper [8] where this is asserted for $H(M_n(k)^*)$ instead of $\hat{H}(M_n(k)^*)$ (where we have generators x_{ij}^r and corresponding relations only for $r \geq 0$).

Now to obtain the example in Theorem 3.2, we have to divide out the Hopf ideal I in $H = \hat{H}(M_4(k)^*)$ generated by the elements x_{ij}^0 for $i \geq 3$, $j \leq 2$. This is the ideal generated by all the $S^r(x_{ij}^0) = x_{ij}^r$ for $r \geq 0$ even, $i \geq 3$ and $j \leq 2$, and the $S^r(x_{ji}^0) = x_{ij}^r$ for $r \geq 1$ odd, $i \leq 2$ and $j \geq 3$. The claim of the theorem reduces to $S^{-1}(I) \not\subset I$; we will show, using the diamond lemma, that (for example) $x_{13}^{-1} \neq 0$ in H/I, that is, $x_{13}^{-1} = S^{-1}(x_{31}^0) \notin I$.

The idea for the counterexample comes from Nichols' example that the biideal $(x_{21}^0) \subset H(M_2(k)^*)$ is not a Hopf ideal. Since so far all our computations in $\hat{H}(M_n(k)^*)$ looked exactly the same as the calculations for $H(M_n(k)^*)$ in [8], we would like to point out that the example does not carry over directly:

Remark 3.3. The ideal generated by x_{21}^0 in $\hat{H}(M_2(k)^*)$ is a Hopf ideal stable under the inverse of the antipode.

Proof. To see this, it is sufficient to show that in the quotient $\hat{H}(M_2(k)^*)/(x_{21}^0)$ we have $x_{21}^r = 0$ whenever r is even, and $x_{12}^r = 0$ whenever r is odd. Assume for example that $x_{21}^r = 0$. Then we have $1 = x_{11}^r x_{11}^{r-1} + x_{21}^r x_{21}^{r-1} = x_{11}^r x_{11}^{r-1}$ and $1 = x_{11}^{r+1} x_{11}^r + x_{21}^{r+1} x_{21}^r = x_{11}^{r+1} x_{11}^r$, whence x_{11}^r is a unit. Furthermore $0 = x_{12}^{r+1} x_{11}^r + x_{22}^{r+1} x_{21}^r = x^{r+1} x_{11}^r$ and $0 = x_{11}^r x_{12}^{r-1} + x_{21}^r x_{22}^{r-1} = x_{11}^r x_{12}^{r-1}$, whence $x_{12}^{r+1} = x_{12}^{r-1} = 0$. Similarly, one shows that $x_{12}^r = 0$ implies $x_{21}^{r+1} = x_{21}^{r-1} = 0$. $\qquad \square$

Let us now finish the proof of Theorem 3.2. It will be sufficient that the following system of reductions satisfies the conditions of the diamond lemma (with respect to the order introduced above) and defines $\hat{H}(M_4(k)^*)/I$ as a quotient of $k\langle X \rangle$.

$$x_{i4}^r x_{j4}^{r+1} \to \delta_{ij} - \sum_{a<4} x_{ia}^r x_{ja}^{r+1} \tag{I}$$

$$x_{4i}^r x_{4j}^{r-1} \to \delta_{ij} - \sum_{a<4} x_{ai}^r x_{aj}^{r-1} \tag{II}$$

$$x_{i4}^r x_{j3}^{r+1} x_{k3}^{r+2} \to \delta_{jk} x_{i4}^r - \delta_{ij} x_{k4}^{r+2} + \sum_{a<4} x_{ia}^r x_{ja}^{r+1} x_{k4}^{r+2} - \sum_{a<3} x_{i4}^r x_{ja}^{r+1} x_{ka}^{r+2} \tag{III}$$

$$x_{4i}^r x_{3j}^{r-1} x_{3k}^{r-2} \to \delta_{jk} x_{4i}^r - \delta_{ij} x_{4k}^{r-2} + \sum_{a<4} x_{ai}^r x_{aj}^{r-1} x_{4k}^{r-2} - \sum_{a<3} x_{4i}^r x_{aj}^{r-1} x_{ak}^{r-2} \tag{IV}$$

$$\text{(for } r \geq 0 \text{ even, } i \geq 3, j \leq 2) \quad x_{ij}^r \to 0 \tag{V}$$

$$\text{(for } r \geq 1 \text{ odd, } i \leq 2, j \geq 3) \quad x_{ij}^r \to 0 \tag{VI}$$

$$\text{(for } r \geq 1, i, j \leq 2) \quad x_{2i}^r x_{2j}^{r-1} \to \delta_{ij} - x_{1i}^r x_{1j}^{r-1} \tag{VII}$$

$$\text{(for } i \leq 2) \quad x_{2i}^0 x_{2j}^{-1} \to \delta_{ij} - x_{1i}^0 x_{1j}^{-1} \tag{VII'}$$

$$\text{(for } r \geq 0, i, j \leq 2) \quad x_{i2}^r x_{j2}^{r+1} \to \delta_{ij} - x_{i1}^r x_{j1}^{r+1} \tag{VIII}$$

$$(\text{for } r \geq 2, i, j, k \leq 2) \quad x_{2i}^r x_{1j}^{r-1} x_{1k}^{r-2} \to \delta_{jk} x_{2i}^r - \delta_{ij} x_{2k}^{r-2} + x_{1i}^r x_{1j}^{r-1} x_{2k}^{r-2}$$
$$(\text{IX})$$

$$(\text{for } i, j \leq 2) \quad x_{2i}^1 x_{1j}^0 x_{1k}^{-1} \to \delta_{jk} x_{2i}^1 - \delta_{ij} x_{2k}^{-1} + x_{1i}^1 x_{1j}^0 x_{2k}^{-1}$$
$$(\text{IX'})$$

$$(\text{for } r \geq 0, i, j, k \leq 2) \quad x_{i2}^r x_{j1}^{r+1} x_{k1}^{r+2} \to \delta_{jk} x_{i2}^r - \delta_{ij} x_{k2}^{r+2} + x_{i1}^r x_{j1}^{r+1} x_{k2}^{r+2}$$
$$(\text{X})$$

Note that the first six of these reductions define $\hat{H}(M_4(k)^*)/I$ as a quotient of $k\langle X \rangle$. From our computations, it will be obvious that the rules (VII) to (X) are only introduced to resolve ambiguities, so that the complete set of reductions still defines the same algebra. Since x_{ij}^{-1} are irreducible with respect to (I) through (X), the proof of Theorem 3.2 will be complete when we show that all inclusion and overlap ambiguities between (I) through (X) resolve.

(V) can be contained in (II) as the left factor: For $r \geq 0$ even and $i \leq 2$ the word $x_{4i}^r x_{4j}^{r-1}$ reduces via (V) to zero; on the other hand we have

$$x_{4i}^r x_{4j}^{r-1} \xrightarrow{(\text{II})} \delta_{ij} - \sum_{a<4} x_{ai}^r x_{aj}^{r-1} \xrightarrow{(\text{V})} \delta_{ij} - \sum_{a<3} x_{ai}^r x_{aj}^{r-1}.$$

This reduces to zero by (VI) if $r \geq 2$ and $j \geq 3$. To resolve the ambiguity for $r \geq 2$ and $j \leq 2$ we add one half of (VII), for $r = 0$ we add (VII').

Similarly, if $r \geq 1$ is odd and $j \leq 2$, we have $x_{4i}^r x_{4j}^{r-1} \xrightarrow{(\text{V})} 0$ and

$$x_{4i}^r x_{4j}^{r-1} \xrightarrow{(\text{II})} \delta_{ij} - \sum_{a<4} x_{ai}^r x_{aj}^{r-1} \xrightarrow{(\text{V})} \delta_{ij} - \sum_{a<3} x_{ai}^r x_{aj}^{r-1}$$

which reduces to zero by (VI) if $i \geq 3$ and leads us to add the odd half of (VII) if $i \leq 2$.

(V) can occur as the first factor of (IV): For $r \geq 0$ even and $i \leq 2$ we have

$$x_{4i}^r x_{3j}^{r-1} x_{3k}^{r-2} \xrightarrow{(\text{V})} 0$$

and

$$x_{4i}^r x_{3j}^{r-1} x_{3k}^{r-2} \xrightarrow{(\text{IV})} \delta_{jk} x_{4i}^r - \delta_{ij} x_{4k}^{r-2} + \sum_{a<4} x_{ai}^r x_{aj}^{r-1} x_{4k}^{r-2} - \sum_{a<3} x_{4i}^r x_{aj}^{r-1} x_{ak}^{r-2}$$

$$\xrightarrow{(\text{V})} -\delta_{ij} x_{4k}^{r-2} + \sum_{a<3} x_{ai}^r x_{aj}^{r-1} x_{4k}^{r-2}$$

This reduces to zero by (VII) or (VII') if $j \leq 2$ or $r = 0$. If $r \geq 2$ and $j \geq 3$ it reduces to zero by (VI).

(V) can occur as the second factor of (IV): For $r \geq 1$ odd and $j \leq 2$ we have

$$x_{4i}^r x_{3j}^{r-1} x_{3k}^{r-2} \xrightarrow{\text{(V)}} 0$$

and

$$x_{4i}^r x_{3j}^{r-1} x_{3k}^{r-2} \xrightarrow{\text{(IV)}} \delta_{jk} x_{4i}^r - \delta_{ij} x_{4k}^{r-2} + \sum_{a<4} x_{ai}^r x_{aj}^{r-1} x_{4k}^{r-2} - \sum_{a<3} x_{4i}^r x_{aj}^{r-1} x_{ak}^{r-2}$$

$$\xrightarrow{\text{(V)}} \delta_{jk} x_{4i}^r - \delta_{ij} x_{4k}^{r-2} + \sum_{a<3} x_{ai}^r x_{aj}^{r-1} x_{4k}^{r-2} - \sum_{a<3} x_{4i}^r x_{aj}^{r-1} x_{ak}^{r-2}$$

Using (VI) if $i \geq 3$ or (VII) if $i \leq 2$, this reduces further to

$$\delta_{jk} x_{4i}^r - \sum_{a<3} x_{4i}^r x_{aj}^{r-1} x_{ak}^{r-2}.$$

If $r = 1$ this reduces to zero by (VII'), if $r \geq 3$ and $k \leq 2$ it reduces to zero by (VII), and if $r \geq 3$ and $k \geq 3$, it reduces to zero by (VI).

(V) can occur as the third factor of (IV): For $r \geq 2$ even and $k \leq 2$, we have

$$x_{4i}^r x_{3j}^{r-1} x_{3k}^{r-2} \xrightarrow{\text{(V)}} 0$$

and

$$x_{4i}^r x_{3j}^{r-1} x_{3k}^{r-2} \xrightarrow{\text{(IV)}} \delta_{jk} x_{4i}^r - \delta_{ij} x_{4k}^{r-2} + \sum_{a<4} x_{ai}^r x_{aj}^{r-1} x_{4k}^{r-2} - \sum_{a<3} x_{4i}^r x_{aj}^{r-1} x_{ak}^{r-2}$$

$$\xrightarrow{\text{(V)}} \delta_{jk} x_{4i}^r - \sum_{a<3} x_{4i}^r x_{aj}^{r-1} x_{ak}^{r-2},$$

which reduces to zero by (VII) if $j \leq 2$, and by (VI) if $j \geq 3$.

(VI) can occur as the left factor of (I): If $r \geq 1$ is odd and $i \leq 2$,

$$x_{i4}^r x_{j4}^{r+1} \xrightarrow{\text{(VI)}} 0,$$

while

$$x_{i4}^r x_{j4}^{r+1} \xrightarrow{\text{(I)}} \delta_{ij} - \sum_{a<4} x_{ia}^r x_{ja}^{r+1} \xrightarrow{\text{(VI)}} \delta_{ij} - \sum_{a<3} x_{ia}^r x_{ja}^{r+1}$$

If $j \geq 3$, this reduces to zero by (V), otherwise it forces us to introduce the odd half of (VIII).

(VI) can occur as the right factor of (I), leading in the same way to the even half of (VIII).

(VI) can occur as the leftmost factor of (III): If $r \geq 1$ is odd and $i \leq 2$,

$$x_{i4}^r x_{j3}^{r+1} x_{k3}^{r+2} \xrightarrow{\text{(VI)}} 0,$$

and

$$x_{i4}^r x_{j3}^{r+1} x_{k3}^{r+2} \xrightarrow{\text{(III)}} \delta_{jk} x_{i4}^r - \delta_{ij} x_{k4}^{r+2} + \sum_{a<4} x_{ia}^r x_{ja}^{r+1} x_{k4}^{r+2} - \sum_{a<3} x_{i4}^r x_{ja}^{r+1} x_{ka}^{r+2}$$

$$\xrightarrow{\text{(VI)}} -\delta_{ij} x_{k4}^{r+2} + \sum_{a<3} x_{ia}^r x_{ja}^{r+1} x_{k4}^{r+2}.$$

This reduces to zero by (V) if $j \geq 3$ or by (VIII) if $j \leq 2$.

(VI) can occur as the middle factor of (III): If $r \geq 0$ is even and $j \leq 2$,

$$x_{i4}^r x_{j3}^{r+1} x_{k3}^{r+2} \xrightarrow{\text{(VI)}} 0,$$

and

$$x_{i4}^r x_{j3}^{r+1} x_{k3}^{r+2} \xrightarrow{\text{(III)}} \delta_{jk} x_{i4}^r - \delta_{ij} x_{k4}^{r+2} + \sum_{a<4} x_{ia}^r x_{ja}^{r+1} x_{k4}^{r+2} - \sum_{a<3} x_{i4}^r x_{ja}^{r+1} x_{ka}^{r+2}$$

$$\xrightarrow{\text{(VI)}} \delta_{jk} x_{i4}^r - \delta_{ij} x_{k4}^{r+2} + \sum_{a<3} x_{ia}^r x_{ja}^{r+1} x_{k4}^{r+2} - \sum_{a<3} x_{i4}^r x_{ja}^{r+1} x_{ka}^{r+2}.$$

By applying (VIII) if $i \leq 2$ or (V) if $i \geq 3$ we can reduce this to

$$\delta_{jk} x_{i4}^r - \sum_{a<3} x_{i4}^r x_{ja}^{r+1} x_{ka}^{r+2},$$

which in turn reduces to zero using (VIII) if $k \leq 2$ or (V) if $k \geq 3$.

(VI) can occur as the rightmost factor in (III): If $r \geq -1$ is odd and $k \leq 2$,

$$x_{i4}^r x_{j3}^{r+1} x_{k3}^{r+2} \xrightarrow{\text{(VI)}} 0,$$

and

$$x_{i4}^r x_{j3}^{r+1} x_{k3}^{r+2} \xrightarrow{\text{(III)}} \delta_{jk} x_{i4}^r - \delta_{ij} x_{k4}^{r+2} + \sum_{a<4} x_{ia}^r x_{ja}^{r+1} x_{k4}^{r+2} - \sum_{a<3} x_{i4}^r x_{ja}^{r+1} x_{ka}^{r+2}$$

$$\xrightarrow{\text{(VI)}} \delta_{jk} x_{i4}^r - \sum_{a<3} x_{i4}^r x_{ja}^{r+1} x_{ka}^{r+2}.$$

This reduces to zero by (V) if $j \geq 3$ and by (VIII) if $j \leq 2$.

(VII) cannot overlap any of (I)...(VI).

The right factor of (VII') may overlap (I): For $i \leq 2$

$$x_{2i}^0 x_{24}^{-1} x_{j4}^0 \xrightarrow{\text{(VII')}} -x_{1i}^0 x_{14}^{-1} x_{j4}^0 \xrightarrow{\text{(I)}} -\delta_{1j} x_{1i}^0 + \sum_{a<4} x_{1i}^0 x_{1a}^{-1} x_{ja}^0$$

and

$$x^0_{2\imath} x^{-1}_{24} x^0_{j4} \xrightarrow{(\mathrm{I})} \delta_{2j} x^0_{2\imath} - \sum_{a<4} x^0_{2\imath} x^{-1}_{2a} x^0_{ja}$$

$$\xrightarrow{(\mathrm{VII'})} \delta_{2j} x^0_{2\imath} - \sum_{a<4}(\delta_{\imath a} x^0_{ja} - x^0_{1\imath} x^{-1}_{1a} x^0_{ja})$$

$$= \delta_{2j} x^0_{2\imath} - x^0_{j\imath} + \sum_{a<4} x^0_{1\imath} x^{-1}_{1a} x^0_{ja}$$

The results of the two reductions differ by $x^0_{j\imath} - \delta_{1j} x^0_{1\imath} - \delta_{2j} x^0_{2\imath}$, which vanishes for $j \leq 2$, and equals $x^0_{j\imath}$ for $j \geq 3$. The latter reduces to zero by (V).

The right hand factor of (VII') can overlap (III). For $i \leq 2$

$$x^0_{2\imath} x^{-1}_{24} x^0_{j3} x^1_{k3}$$

$$\xrightarrow{(\mathrm{III})} \delta_{jk} x^0_{2\imath} x^{-1}_{24} - \delta_{2j} x^0_{2\imath} x^1_{k4} + \sum_{a<4} x^0_{2\imath} x^{-1}_{2a} x^0_{ja} x^1_{k4} - \sum_{a<3} x^0_{2\imath} x^{-1}_{24} x^0_{ja} x^1_{ka}$$

$$\xrightarrow{(\mathrm{VII'})} -\delta_{jk} x^0_{1\imath} x^{-1}_{14} - \delta_{2j} x^0_{2\imath} x^1_{k4} + \sum_{a<4}(\delta_{\imath a} - x^0_{1\imath} x^{-1}_{1a}) x^0_{ja} x^1_{k4} + \sum_{a<3} x^0_{1\imath} x^{-1}_{14} x^0_{ja} x^1_{ka}$$

$$= -\delta_{jk} x^0_{1\imath} x^{-1}_{14} - \delta_{2j} x^0_{2\imath} x^1_{k4} + x^0_{j\imath} x^1_{k4} - \sum_{a<4} x^0_{1\imath} x^{-1}_{1a} x^0_{ja} x^1_{k4} + \sum_{a<3} x^0_{1\imath} x^{-1}_{14} x^0_{ja} x^1_{ka}$$

and on the other hand

$$x^0_{2\imath} x^{-1}_{24} x^0_{j3} x^1_{k3} \xrightarrow{(\mathrm{VII'})} -x^0_{1\imath} x^{-1}_{14} x^0_{j3} x^1_{k3}$$

$$\xrightarrow{(\mathrm{III})} -\delta_{jk} x^0_{1\imath} x^{-1}_{14} + \delta_{1j} x^0_{1\imath} x^1_{k4} - \sum_{a<4} x^0_{1\imath} x^{-1}_{1a} x^0_{ja} x^1_{k4} + \sum_{a<3} x^0_{1\imath} x^{-1}_{14} x^0_{ja} x^1_{ka}$$

The results differ by $x^0_{j\imath} x^1_{k4} - \delta_{1j} x^0_{1\imath} x^1_{k4} - \delta_{2j} x^0_{2\imath} x^1_{k4}$, which vanishes if $j \leq 2$, and reduces to zero by (V) if $j \geq 3$.

(VII) can overlap with itself as well as with (VII'): If $r \geq 2$ and $i, j, k \leq 2$ or $r = 1$ and $i, j \leq 2$, we have

$$x^r_{2\imath} x^{r-1}_{2j} x^{r-2}_{2k} \xrightarrow{(\mathrm{VII})} \delta_{\imath j} x^{r-2} - x^r_{1\imath} x^{r-1}_{2k}$$

$$x^r_{2\imath} x^{r-1}_{2j} x^{r-2}_{2k} \xrightarrow{(\mathrm{VII})\ \mathrm{or}\ (\mathrm{VII'})} \delta_{jk} x^r_{2\imath} - x^r_{2\imath} x^{r-1}_{1j} x^{r-2}_{1k}$$

which forces us to introduce rules (IX) and (IX'), respectively.

Similarly, the possible overlaps of (VIII) with itself lead to (X).

Note that (VII), (VIII), (IX) and (X) are precisely the reduction rules that define $H(M_2(k)^*)$. In particular, we need not consider any overlaps involving only those reduction rules.

By a short inspection we see that the only overlaps left to deal with are those involving (IX') with (I), (III) or (VIII). We omit treating the overlap of (the rightmost factor of) (VIII) with (the leftmost factor of) (IX') since it resolves in the same way as that of (VIII) with (IX).

The rightmost factor of (IX') may overlap (I): For $i, j \leq 2$ we have

$$x_{2_i}^1 x_{1_j}^0 x_{14}^{-1} x_{k4}^0 \xrightarrow{(IX')} -\delta_{ij} x_{24}^{-1} x_{k4}^0 + x_{1_i}^1 x_{1_j}^0 x_{24}^{-1} x_{k4}^0$$

$$\xrightarrow{(I)} -\delta_{ij}\delta_{2k} + \delta_{ij}\sum_{a<4} x_{2a}^{-1} x_{ka}^0 + \delta_{2k} x_{1_i}^1 x_{1_j}^0 - \sum_{a<4} x_{1_i}^1 x_{1_j}^0 x_{2a}^{-1} x_{ka}^0,$$

and, on the other hand,

$$x_{2_i}^1 x_{1_j}^0 x_{14}^{-1} x_{k4}^0 \xrightarrow{(I)} \delta_{1k} x_{2_i}^1 x_{1_j}^0 - \sum_{a<4} x_{2_i}^1 x_{1_j}^0 x_{1a}^{-1} x_{ka}^0$$

$$\xrightarrow{(IX')} \delta_{1k} x_{2_i}^1 x_{1_j}^0 - \sum_{a<4}(\delta_{ja}x_{2_i}^1 - \delta_{ij}x_{2a}^{-1} + x_{1_i}^1 x_{1_j}^0 x_{2a}^{-1})x_{ka}^0$$

$$= \delta_{1k} x_{2_i}^1 x_{1_j}^0 - x_{2_i}^1 x_{kj}^0 + \sum_{a<4}\delta_{ij}x_{2a}^{-1}x_{ka}^0 - \sum_{a<4} x_{1_i}^1 x_{1_j}^0 x_{2a}^{-1}x_{ka}^0.$$

The results of the two reductions differ by

$$x_{2_i}^1 x_{kj}^0 - \delta_{1k} x_{2_i}^1 x_{1_j}^0 + \delta_{2k} x_{1_i}^1 x_{1_j}^0 - \delta_{ij}\delta_{2k}.$$

If $k = 1$, this vanishes. If $k = 2$, it equals $x_{2_i}^1 x_{2_j}^0 + x_{1_i}^1 x_{1_j}^0 - \delta_{ij}$ and reduces to zero by (VII). If $k \geq 3$, we have $x_{2_i}^1 x_{kj}^0$ and can reduce to zero by (V).

The last, and most complicated, overlap to deal with is that of the right factor of (XI') with (III). For $i, j \leq 2$ we have the reductions

$$x_{2_i}^1 x_{1_j}^0 x_{14}^{-1} x_{k3}^0 x_{\ell 3}^1$$

$$\xrightarrow{(III)} \delta_{k\ell} x_{2_i}^1 x_{1_j}^0 x_{14}^{-1} - \delta_{1k} x_{2_i}^1 x_{1_j}^0 x_{\ell 4}^1 + \sum_{a<4} x_{2_i}^1 x_{1_j}^0 x_{1a}^{-1} x_{ka}^0 x_{\ell 4}^1 - \sum_{a<3} x_{2_i}^1 x_{1_j}^0 x_{14}^{-1} x_{ka}^0 x_{\ell a}^1$$

$$\xrightarrow{(IX')} -\delta_{k\ell}\delta_{ij}x_{24}^{-1} + \delta_{k\ell} x_{1_i}^1 x_{1_j}^0 x_{24}^{-1} - \delta_{1k} x_{2_i}^1 x_{1_j}^0 x_{\ell 4}^1$$

$$+ \sum_{a<4}(\delta_{ja}x_{2_i}^1 - \delta_{ij}x_{2a}^{-1} + x_{1_i}^1 x_{1_j}^0 x_{2a}^{-1})x_{ka}^0 x_{\ell 4}^1 - \sum_{a<3}(-\delta_{ij}x_{24}^{-1} + x_{1_i}^1 x_{1_j}^0 x_{24}^{-1})x_{ka}^0 x_{\ell a}^1$$

$$= (x_{1_i}^1 x_{1_j}^0 - \delta_{ij})(\delta_{k\ell}x_{24}^{-1} + \sum_{a<4} x_{2a}^{-1}x_{ka}^0 x_{\ell 4}^1 - \sum_{a<3} x_{24}^{-1} x_{ka}^0 x_{\ell a}^1) - \delta_{1k} x_{2_i}^1 x_{1_j}^0 x_{\ell 4}^1 + x_{2_i}^1 x_{kj}^0 x_{\ell 4}^1$$

and

$$x_{2_i}^1 x_{1_j}^0 x_{14}^{-1} x_{k3}^0 x_{\ell 3}^1 \xrightarrow{(IX')} (-\delta_{ij}x_{24}^{-1} + x_{1_i}^1 x_{1_j}^0 x_{24}^{-1})x_{k3}^0 x_{\ell 3}^1$$

$$\xrightarrow{(III)} (x_{1_i}^1 x_{1_j}^0 - \delta_{ij})(\delta_{k\ell}x_{24}^{-1} - \delta_{2k}x_{\ell 4}^1 + \sum_{a<4} x_{2a}^{-1}x_{ka}^0 x_{\ell 4}^1 - \sum_{a<3} x_{24}^{-1} x_{ka}^0 x_{\ell a}^1)$$

The results differ by

$$x^1_{2\imath}x^0_{kj}x^1_{\ell 4} - \delta_{1k}x^1_{2\imath}x^0_{1j}x^1_{\ell 4} - \delta_{2k}\delta_{\imath j}x^1_{\ell 4} + \delta_{2k}x^1_{1\imath}x^0_{1j}x^1_{\ell 4}.$$

This vanishes if $k = 1$. If $k = 2$, it equals

$$(x^1_{2\imath}x^0_{2j} + x^1_{1\imath}x^0_{1j} - \delta_{\imath j})x^1_{\ell 4}$$

and reduces to zero by (VII). For $k \geq 3$ the result is $x^1_{2\imath}x^0_{kj}x^1_{\ell 4}$ and reduces to zero by (V). $\qquad\qquad\qquad\qquad\qquad\qquad\qquad\qquad\qquad\qquad$ $\square$

REFERENCES

[1] DEMAZURE, M., AND GABRIEL, P. *Groupes Algébriques I.* North Holland, Amsterdam, 1970.

[2] GABRIEL, P. *Groupes formels, Exp. VII B, SGA 3, Schémas en Groupes,* vol. 151 of *Lecture Notes in Mathematics.* Springer, 1970.

[3] KAPLANSKY, I. Bialgebras. University of Chicago Lecture Notes. 1975.

[4] MASUOKA, A. On Hopf algebras with cocommutative coradicals. *J. Algebra* *144* (1991), 415–466.

[5] MASUOKA, A. Quotient theory of Hopf algebras. In *Advances in Hopf Algebras,* J. Bergen and S. Montgomery, Eds. Marcel Dekker Inc., 1994, pp. 107–133.

[6] MASUOKA, A., AND WIGNER, D. Faithful flatness of Hopf algebras. *J. Algebra* *170* (1994), 156–164.

[7] MONTGOMERY, S. *Hopf algebras and their actions on rings,* vol. 82 of *CBMS Regional Conference Series in Mathematics.* AMS, Providence, Rhode Island, 1993.

[8] NICHOLS, W. D. Quotients of Hopf algebras. *Comm. Algebra 6* (1978), 1789–1800.

[9] NICHOLS, W. D., AND RICHMOND, M. B. Freeness of infinite dimensional hopf algebras. *Comm. Algebra 20* (1992), 1489–1492.

[10] NICHOLS, W. D., AND ZOELLER, M. B. A Hopf algebra freeness theorem. *Amer. J. Math. 111* (1989), 381–385.

[11] OBERST, U., AND SCHNEIDER, H.-J. Untergruppen formeller Gruppen von endlichem Index. *J. Algebra 31* (1974), 10–44.

[12] SCHNEIDER, H.-J. Normal basis and transitivity of crossed products for Hopf algebras. *J. Algebra 152* (1992), 289–312.

[13] SCHNEIDER, H.-J. Some remarks on exact sequences of quantum groups. *Comm. in Alg. 21* (1993), 3337–3357.

[14] TAKEUCHI, M. Free Hopf algebras generated by coalgebras. *J. Math. Soc. Japan 23* (1971), 561–582.

[15] TAKEUCHI, M. There exists a Hopf algebra whose antipode is not injective. *Sci. Papers Coll. Gen. Ed. Univ. Tokyo 21* (1971), 127–130.

[16] TAKEUCHI, M. A correspondence between Hopf ideals and sub-Hopf algebras. *Manuscripta Math. 7* (1972), 251–270.

[17] TAKEUCHI, M. Quotient spaces for Hopf algebras. *Comm. in Alg. 22* (1994), 2503–2523.

MATHEMATISCHES INSTITUT DER UNIVERSITÄT MÜNCHEN, THERESIENSTR. 39, 80333 MÜNCHEN, GERMANY

E-mail address: schauen@rz.mathematik.uni-muenchen.de

Cohen–Macaulay Modules over Classical Orders

Daniel Simson

Faculty of Mathematics and Informatics
Nicholas Copernicus University, 87-100 Toruń, Poland

This article is an extended and completed version of the second part of the joint survey lecture with Yu. Drozd on Cohen-Macaulay modules over Cohen-Macaulay algebras presented at the Euroconference "Interactions between Ring Theory and Representations of Algebras" in Murcia on 12-17 January 1998. The first part by Yu. Drozd is presented in [13] and is mainly devoted to commutative Cohen-Macaulay algebras.

The present survey article is mainly devoted to non-commutative Cohen-Macaulay algebras of dimension 1, and particularly to representation theory of classical orders.

The paper is divided into eight sections as follows.

1. Classical D-orders, the category of Cohen-Macaulay modules and the Cohen-Macaulay representation types.

2. The Cohen-Macaulay representation type of group orders.

3. A reduction functor.

4. The case $J(\Gamma) \subseteq J(\Lambda) \subseteq \Lambda \subseteq \Gamma$.

5. Socle projective representations of partially ordered sets.

6. Representations of partially ordered sets equipped with an equivalence relation.

7. Tiled D-orders, representations of partially ordered sets and a covering functor.

8. The Cohen-Macaulay type of subamalgams of tiled D-orders and a Tits form.

We shall use the usual rings and modules theory notation introduced in [1]. The reader is referred to [4], [17] and [41] for basic definitions and facts on representation theory of finite dimensional algebras and representation types. For an elementary introduction to Cohen-Macaulay algebras and Cohen-Macaulay modules we refer to [52], and to Cohen-Macaulay modules over classical orders we refer to the survey articles [10] and [36].

1 Classical D-orders, the category of Cohen-Macaulay modules and Cohen-Macaulay representation types

Given an associative ring R with an identity element we denote by $\mathrm{mod}(R)$ the category of finitely generated right R-modules. We denote by $J(R)$ the Jacobson radical of the ring R.

Let D be a complete discrete valuation domain. We denote by $F = D_0$ the field of fractions of D, by $\mathfrak{P} = J(D)$ the unique maximal ideal of D and we set $K = D/\mathfrak{P}$.

We recall that a classical D-**order** Λ in a finite dimensional semisimple F-algebra C is a D-subalgebra Λ of C which is a finitely generated free D-submodule of C and Λ contains an F-basis of C [7].

By a **Cohen-Macaulay algebra** over D we mean a D-algebra Λ which is a finitely generated free D-module.

EXAMPLES 1.1. (a) The D-subalgebra $\Lambda = \begin{pmatrix} D & D \\ \mathfrak{P} & D \end{pmatrix}$ of the simple D_0-algebra $C = \begin{pmatrix} D_0 & D_0 \\ D_0 & D_0 \end{pmatrix}$ is a D-order in C. Note that Λ is a hereditary ring, that is, gl.dim $\Lambda = 1$.

(b) Let $D = \widehat{\mathbb{Z}}_{(p)}$ be the ring of p-adic integers, $D_0 = \widehat{\mathbb{Q}}_{(p)}$ be the fields of p-adic fractions, and let G be a finite group. The group $\widehat{\mathbb{Z}}_{(p)}$-subalgebra $\Lambda = \widehat{\mathbb{Z}}_{(p)}[G]$ is a $\widehat{\mathbb{Z}}_{(p)}$-order in the semisimple $\widehat{\mathbb{Q}}_{(p)}$-algebra $C = \widehat{\mathbb{Q}}_{(p)}[G]$. □

Given a Cohen-Macaulay D-algebra Λ, we denote by $\mathrm{CM}(\Lambda)$ the category of finitely generated right (maximal) **Cohen-Macaulay Λ-modules**, that is, finitely generated right Λ-modules which are free when viewed as D-modules. In case Λ is a classical D-order the category $\mathrm{CM}(\Lambda)$ is also denoted by $\mathrm{latt}(\Lambda)$ and it is called the category of right Λ-lattices (see [7]).

REMARK 1.2. Assume that Λ is a complete noetherian local ring and denote by K.dim Λ the Krull dimension of Λ. Let $T \subseteq \Lambda$ be a noetherian normalization of Λ. Then Λ is a Cohen-Macaulay ring in a classical sense (that is, K.dimΛ equals the depth of Λ) if and only if the T-algebra Λ is a finitely generated free T-module (see [52]). Recall that the maximal Cohen-Macaulay Λ-modules are by definition the finitely generated modules X of depth equals to K.dimΛ.

In case K.dim$\Lambda = 1$, it is easy to show that X is a maximal Cohen-Macaulay Λ-module if and only if X viewed as a T-module is finitely generated free. It is well-known that, in case K.dim$\Lambda = 2$, X is a maximal Cohen-Macaulay Λ-module if and only if X is T-reflexive, that is, the natural T-homomorphism $X \to \operatorname{Hom}_T(\operatorname{Hom}_T(X, T), T)$ is bijective (see [52]). $\qquad\square$

EXAMPLE 1.3. Let Λ be a simple plain curve singularity (see [11])

$$(1.4) \qquad\qquad \Lambda_\Delta = K[[X, Y]]/(f_\Delta)$$

over K, where $K[[X, Y]]$ is the power series K-algebra in two indeterminates X, Y, and f_Δ is the polynomial corresponding to the Dynkin diagram Δ defined as follows

$$f_\Delta = \begin{cases} X^{n+1} + Y^2 & \text{if} \quad \Delta = \mathbb{A}_n, \\ X^{n-1} + XY^2 & \text{if} \quad \Delta = \mathbb{D}_n, \\ X^4 + Y^3 & \text{if} \quad \Delta = \mathbb{E}_6, \\ X^3 Y + Y^3 & \text{if} \quad \Delta = \mathbb{E}_7, \\ X^5 + Y^3 & \text{if} \quad \Delta = \mathbb{E}_8. \end{cases}$$

It is easy to see that Λ_Δ is a Cohen-Macaulay algebra of Krull dimension one and a noetherian normalization of Λ_Δ is isomorphic with the power series algebra $K[[t]]$ in one indeterminate. Since obviously Λ_Δ is a semiprime ring then according to Lemma 1.5 below, Λ_Δ is a $K[[t]]$-order for any Dynkin diagram Δ. It is well-known that Λ_Δ is of finite Cohen-Macaulay type in the sense defined below (see [11]), [36], [41, Sec. 13.4]). $\qquad\square$

LEMMA 1.5. (a) *A D-algebra Λ is a D-order in a finite dimensional semisimple D_0-algebra C if and only if Λ is a semiprime ring which is a Cohen-Macaulay D-algebra.*

(b) *If Λ is a D-order then a right Λ-module X is in $\operatorname{CM}(\Lambda)$ if and only if X is a submodule of a finitely generated free Λ-module.*

Proof. (a) For the proof of (a) we recall the well-known result of A. Goldie that a noetherian ring Λ is semiprime if and only if Λ has a semisimple classical ring of fractions C. Hence (a) easily follows. The proof of (b) is easy and is left as an exercise (see [7] and [31]).   □

We denote by $\Gamma_{CM(\Lambda)} = \Gamma(CM(\Lambda))$ the **Auslander-Reiten quiver** of the category $CM(\Lambda)$, that is, an oriented graph whose vertices are the isomorphism classes $[X]$ of indecomposable modules X in $CM(\Lambda)$. There is an arrow $[X] \to [Y]$ in $\Gamma_{CM(\Lambda)}$ if there is an irreducible morphism $f : X \to Y$ in $\Gamma_{CM(\Lambda)}$, that is, $f : X \to Y$ is a nonisomorphism such that f is not a composition $X \to Z \to Y$ of two nonisomorphisms between indecomposable modules in $\Gamma_{CM(\Lambda)}$, nor a finite sum of such a compositions $X \to Z_j \to Y$ (see [4], [36, Section 2] and [41, Section 11.1]).

We recall from [4] that a morphism $g : Y \to Z$ in $CM(\Lambda)$ is said to be **right almost split**, if g is not a split epimorphism and every morphism $h : U \to Z$ in $CM(\Lambda)$ which is not a split epimorphism factors through g. A morphism $g : Y \to Z$ in $CM(\Lambda)$ is called **right minimal**, if every endomorphism $t : Y \to Y$ such that $gt = g$ is an isomorphism. Dually, a morphism $f : X \to Y$ in $CM(\Lambda)$ is said to be **left almost split** if f is not a split epimorphism and every morphism $h : X \to V$ in $CM(\Lambda)$ which is not a split epimorphism factors through f. A morphism $f : X \to Y$ in $CM(\Lambda)$ is called **left minimal** if every endomorphism $t : Y \to Y$ such that $tg = g$ is an isomorphism. An exact sequence $0 \longrightarrow X \xrightarrow{f} Y \xrightarrow{g} Z \longrightarrow 0$ in $CM(\Lambda)$ is said to be **almost split**, if the morphism f is a minimal left almost split and the morphism g is a minimal right almost split.

The following theorem collects main properties of the category of Cohen-Macaulay modules over classical orders.

THEOREM 1.6. *Let Λ be a D-order in a finite dimensional semisimple D_0-algebra C.*

(a) *Λ is a noetherian non-artinian semiperfect ring and every module in $CM(\Lambda)$ has a projective cover.*

(b) *The category $CM(\Lambda)$ is additive (non-abelian), contains all finitely generated projective Λ-modules and is closed under taking Λ-submodules and extensions in $\text{mod}(\Lambda)$.*

(c) *The endomorphism ring $\text{End}(X)$ of any indecomposable module X in $CM(\Lambda)$ is local.*

(d) *The category $CM(\Lambda)$ has the Krull-Schmidt property in the sense that every object X of $CM(\Lambda)$ has a direct sum decomposition $X \cong X_1 \oplus \ldots \oplus X_m$, where $X_1, \ldots, X_m$ are indecomposable objects of $CM(\Lambda)$, and every such a decomposition is unique up to isomorphism and a permutation of the summands $X_1, \ldots, X_m$.*

(e) *Every module Z in the category* $CM(\Lambda)$ *admits a minimal right almost split morphism* $Y \to Z$ *called a sink morphism* [33, p. 56] *and a minimal left almost split morphism* $Z \to Y'$ *called a source morphism* [33]. *Every indecomposable non-projective module* Z *in the category* $CM(\Lambda)$ *admits an almost split sequence ending at* Z, *and every indecomposable indecomposable non-injective module* Z *in the category* $CM(\Lambda)$ *admits an almost split sequence starting from* Z.

Proof. Assume that D is a complete discrete valuation domain and let p be a generator of the maximal ideal $\mathfrak{P} = J(D)$.

Let A be an arbitrary finitely generated D-algebra. Then $\overline{A} = A/pA$ is a finite dimensional algebra over $K = D/(p)$ and therefore there exists a positive integer m such that $J(A)^m \subseteq pA \subseteq J(A)$. It follows that $A/J(A) \cong \overline{A}/J(\overline{A})$ is a semisimple K-algebra. By applying the completeness of D one shows that idempotents of $A/J(A)$ can be lifted to idempotents of A (see [32]). It follows that any finitely generated D-algebra A is a semiperfect ring.

Applying this fact to a D-order Λ we get (a). If we apply this to the D-algebra $\mathrm{End}(X)$, where X is an indecomposable module in $CM(\Lambda)$, we get (c). Hence the statements (b) and (d) easily follow (see [32]). For the proof of (e) we refer to Roggenkamp [35] and [36]. $\square$

MAIN AIMS 1.7. *In the representation theory of classical D-orders Λ we are mainly interested in the following problems.*

(P1) *Classify the indecomposable modules in* $CM(\Lambda)$, *list them and parameterize them in a suitable way.*

(P2) *Give an explicit description of indecomposable modules X in* $CM(\Lambda)$ *and their endomorphism D-algebras $E_X = \mathrm{End}(X)$ by means of generators and relations.*

(P3) *Determine the E_X-E_Y-bimodule structure of the hom-group* $\mathrm{Hom}_\Lambda(Y, X)$ *and of the extension group* $\mathrm{Ext}^1_\Lambda(Y, X)$ *for any pair of indecomposable modules X and Y in* $CM(\Lambda)$.

(P4) *Given a module in* $CM(\Lambda)$, *find its decomposition into a direct sum of indecomposable modules.*

(P5) *Determine the structure of the category* $CM(\Lambda)$, *the structure of the Auslander-Reiten quiver* $\Gamma_{CM(\Lambda)}$ *of* $CM(\Lambda)$, *and describe the shapes of connected components of the quiver* $\Gamma_{CM(\Lambda)}$.

(P6) *Determine the Cohen-Macaulay representation type of* $CM(\Lambda)$ *(finite, wild, tame, polynomial growth) in the sense defined below.* $\square$

DEFINITION 1.8. A D-order is said to be of **finite Cohen-Macaulay (representation) type** if the category $CM(\Lambda)$ has finitely many isomorphism classes of indecomposable modules [2].

A D-order Λ is said to be of **wild Cohen-Macaulay (representation) type** if there exists a representation embedding functor $\mathrm{mod}\left(\begin{smallmatrix} L & L^3 \\ 0 & L \end{smallmatrix}\right) \to CM(\Lambda)$, where L is a division ring [14], [43], [47].

A D-order Λ is said to be of **tame Cohen-Macaulay (representation) type** if the indecomposable Cohen-Macaulay Λ-modules of any fixed D-rank form a finite set of at most one-parameter families (see [13], [14], [41, Section 14.4], [43, Definition 3.1], [47, Section 7] for details). $\qquad\square$

We recall from [43] (see also [6]) that an additive functor $T : \mathcal{A} \to \mathcal{B}$ between Krull-Schmidt subcategories $\mathcal{A}$ and $\mathcal{B}$ of module categories is said to be a **representation embedding**, if T is exact, carries indecomposable objects to indecomposable ones, and for any pair of objects X and Y in $\mathcal{A}$ the existence of an isomorphism $T(X) \cong T(Y)$ in $\mathcal{B}$ implies the existence of an isomorphism $X \cong Y$ in $\mathcal{A}$.

In the definition of tame Cohen-Macaulay type we usually assume that D is an algebra over an algebraically closed field K such that $D/\mathfrak{p} \cong K$. In this case, according to Drozd and Greuel [14], the tame-wild dichotomy holds, that is, every classical D-order Λ is either of tame Cohen-Macaulay type or of wild Cohen-Macaulay type and these types are mutually exclusive. Furthermore, in this case D-orders of tame Cohen-Macaulay type are divided into two classes: the orders of polynomial growth and the orders of non-polynomial growth (see [43, Definition 3.1] and [47, Section 7]). In case D is an arbitrary discrete complete valuation domain there is no satisfactory definition of tame Cohen-Macaulay type for D-orders.

The following sufficient conditions for the infinite and the wild Cohen-Macaulay type are very useful.

PROPOSITION 1.9. *Let Λ be a D-order in a finite dimensional semisimple D_0-algebra C. Assume that there exists a primitive idempotent e in Λ such that the C-module eC is a direct sum of q simple modules.*

(a) *If $q \geq 4$ then Λ is of infinite Cohen-Macaulay type.*

(b) *If $q \geq 5$ then Λ is of wild Cohen-Macaulay type.*

Proof. See [34, Proposition 2.16]. $\qquad\square$

As a consequence we get the following generalization of an old result of E.C. Dade proved in 1963 (see [34] and [36]).

COROLLARY 1.10. *Let Λ be a local D-order in a finite dimensional semisimple D_0-algebra $C = \mathbb{M}_{m_1}(F_1) \times \cdots \times \mathbb{M}_{m_q}(F_q)$, where $F_1, \ldots, F_q$ are division D_0-algebras.*
   (a) *If Λ is of finite Cohen-Macaulay type then $q \leq 3$.*
   (b) *If Λ is not of wild Cohen-Macaulay type then $q \leq 4$.*   □

The following result is well-known as a test for the finite Cohen-Macaulay type.

PROPOSITION 1.11 (M. AUSLANDER). *If the Auslander-Reiten quiver $\Gamma_{CM(\Lambda)}$ of the category $CM(\Lambda)$ contains a finite connected component $\mathcal{C}$ then Λ is of finite Cohen-Macaulay type and $\mathcal{C} = \Gamma_{CM(\Lambda)}$.*

Proof. Since according to Theorem 1.6 (e) the category $CM(\Lambda)$ has almost split sequences and contains all finitely generated projective Λ-modules then the arguments given for modules over artin algebras (see [4, Section VII.2], [41, Section 11.8]) generalize almost verbatim to the Cohen-Macaulay modules over D-orders.   □

2 The Cohen-Macaulay type of group orders

Let us start with a well-known characterization of integral group orders of finite Cohen-Macaulay type which is due to D. G. Higman, A. Heller, I. Reiner, A. Jones and was proved in the early sixties (see [36, Section 3]).

THEOREM 2.1. *Let G be a finite group. Then Cohen-Macaulay group $\mathbb{Z}$-algebra $\mathbb{Z}[G]$ is of finite Cohen-Macaulay type if and only if for every prime number $p \geq 2$ any Sylow p-subgroup G_p of G is cyclic of order p or p^2.*

The following theorem gives a complete classification of group D-orders $D[G]$ of finite, tame and wild Cohen-Macaulay type, respectively. For a more detailed discussion the reader is referred to Dieterich [8], [9], [10, Section 5] and to Roggenkamp [36, Section 3].

THEOREM 2.2. *Let G be a finite p-group and let D be a complete discrete valuation domain with $J(D) = (p)$, the field of fractions $F = D_0$, and with a valuation $v : D \to \mathbb{N} \cup \{\infty\}$. Let $\Lambda = D[G]$ be the* group D-order. *Assume that* Λ *is not hereditary and the* F-*algebra* $F \otimes_D \Lambda$ *is semisimple.*

(a) *The* D-*order* $\Lambda = D[G]$ *is of finite Cohen-Macaulay type if and only if any of the following conditions is satisfied:*

     (i) $v(p) = 0$.

     (ii) $v(p) = 1$ *and* $G = C_{p^e}$ *is a cyclic group of order* p^e, $e \leq 2$.

     (iii) $v(p) = 2$ *and* $G = C_p$.

     (iv) $p = 3$, $v(3) = 3$ *and* $G = C_3$.

     (v) $p = 2$, $v(2) \geq 4$ *and* $G = C_2$.

(b) *The* D-*order* $\Lambda = D[G]$ *is of tame Cohen-Macaulay type if and only if any of the following conditions is satisfied:*

     (i) $p = 2$, $v(2) = 1$ *and* $G = C_2 \times C_2$.

     (ii) $p = 2$, $v(2) = 1$ *and* $G = C_8$.

     (iii) $p = 2$, $v(2) = 2$ *and* $G = C_4$.

     (iv) $p = 3$, $v(3) = 4$ *and* $G = C_3$.

(c) *The* D-*order* $\Lambda = D[G]$ *is of wild Cohen-Macaulay type in all the remaining cases.*

Proof. See Dieterich [8], [9], [10, Section 5]. $\square$

3   A reduction functor

Let Λ be a D-order in a finite dimensional semisimple D_0-algebra C. It is obvious that the modules in $\mathrm{CM}(\Lambda)$ are noetherian of infinite length. One of the aims of this section is to construct a functor $\mathbb{F}$: $\mathrm{CM}(\Lambda) \to \mathrm{mod}(\overline{\Lambda}_\Gamma)$ (3.11) reducing the study of $\mathrm{CM}(\Lambda)$ to the study of a nice additive subcategory of modules of finite length over an artin algebra $\overline{\Lambda}_\Gamma$ (3.3). For this purpose we fix a D-order Γ such that

$$\Lambda \subseteq \Gamma \subseteq C$$

and Γ is a hereditary D-order, that is, the global dimension of Γ is equal to 1.

Note that such an order Γ do exist, because any maximal over-order Ω of Λ in C is hereditary [31].

Since Γ/Λ is a finitely generated torsion D-module then there exists a positive integer $s \geq 1$ such that

(3.1) $J(\Gamma)^s \subseteq J(\Lambda)$

and consequently there exists a two-sided Γ-ideal I such that the ring Γ/I is artinian,

(3.2) $I \subseteq J(\Lambda)$, $I \subseteq J(\Gamma)$, $\mathrm{length}_D(\Gamma/I) < \infty$ and $I \cap D = (p^c)$, for some $c \geq 1$, where p is a generator of the maximal ideal $\mathfrak{P} = J(D)$ of D. Following Dieterich [10, Section 3] any such a triple (I, Λ, Γ) will be called an **admissible triple**.

Throughout we fix a hereditary D-order Γ and a two-sided ideal I satisfying the conditions in (3.2), and we associate with the admissible triple (I, Λ, Γ) the artinian ring

$$(3.3) \qquad \overline{\Lambda}_\Gamma = \begin{pmatrix} A & {}_A B_B \\ 0 & B \end{pmatrix}$$

where $A = \Lambda/I$, $B = \Gamma/I$ and ${}_A B_B = B$ is viewed as an A-B-bimodule via the ring embedding $A = \Lambda/I \hookrightarrow B = \Gamma/I$ induced by the inclusion $\Lambda \subseteq \Gamma$.

PROPOSITION 3.4. *The ring $\overline{\Lambda}_\Gamma$ (3.3) associated with any admissible triple (I, Λ, Γ) has the following properties.*

(a) *The ring $\overline{\Lambda}_\Gamma$ is an artin algebra. More precisely, $\overline{\Lambda}_\Gamma$ is a finitely generated algebra over the commutative artinian principal ideal ring $D/(p^c)$, where c is as in (3.2).*

(b) *The $D/(p^c)$-algebra $B = \Gamma/I$ is serial in the sense that for every primitive idempotent $e \in B$ the right ideal eB of B has a unique finite composition series of submodules.*

Proof. The statement (a) is an easy observation. For (b) we refer to [36, Section 1B and 1C]. $\qquad\square$

Throughout we shall apply the following convention.

CONVENTION 3.5. Right modules Y over the algebra

$$\overline{\Lambda}_\Gamma = \begin{pmatrix} A & {}_A B_B \\ 0 & B \end{pmatrix}$$

will be identified with triples

$$Y = (Y'_A, Y''_B, \, t : Y' \to Y'')$$

where Y' is the right A-module $Y\left(\begin{smallmatrix} 1 & 0 \\ 0 & 0 \end{smallmatrix}\right)$, Y'' is the right B-module $Y\left(\begin{smallmatrix} 0 & 0 \\ 0 & 1 \end{smallmatrix}\right)$, and t is the A-module homomorphism $y' \mapsto y'\left(\begin{smallmatrix} 0 & 1 \\ 0 & 0 \end{smallmatrix}\right)$. $\qquad\square$

Following [47] we consider two full subcategories

$$(3.6) \qquad \widehat{\mathrm{mod}}^{\mathrm{pr}}_{\mathrm{pr}}(\overline{\Lambda}_\Gamma) \quad \text{and} \quad \widehat{\mathrm{mod}}_{\mathrm{pr}}(\overline{\Lambda}_\Gamma)$$

of $\mathrm{mod}(\overline{\Lambda}_\Gamma)$.

The first one is formed by all $\overline{\Lambda}_\Gamma$-modules $Y = (Y_A', Y_B'', t : Y' \to Y'')$ such that the modules Y_A' and Y_B'' are projective and $(\mathrm{Im}\, t)B = Y_B''$.

The second one is formed by all $\overline{\Lambda}_\Gamma$-modules $Y = (Y_A', Y_B'', t : Y' \to Y'')$ such that the module Y_B'' is B-projective, the A-homomorphism $t : Y' \to Y''$ is injective and $(\mathrm{Im}\, t)B = Y_B''$.

LEMMA 3.7. (a) *Each of the full additive subcategories* $\widehat{\mathrm{mod}}_{\mathrm{pr}}^{\mathrm{pr}}(\overline{\Lambda}_\Gamma)$ *and* $\widehat{\mathrm{mod}}_{\mathrm{pr}}(\overline{\Lambda}_\Gamma)$ *of* $\mathrm{mod}(\overline{\Lambda}_\Gamma)$ *has almost split sequences.*

(b) *The category* $\widehat{\mathrm{mod}}_{\mathrm{pr}}^{\mathrm{pr}}(\overline{\Lambda}_\Gamma)$ *is a hereditary subcategory of* $\mathrm{mod}(\overline{\Lambda}_\Gamma)$, *that is, the group* $\mathrm{Ext}_{\overline{\Lambda}_\Gamma}^2(X,Y)$ *is zero for any pair of modules* X, Y *in* $\widehat{\mathrm{mod}}_{\mathrm{pr}}^{\mathrm{pr}}(\overline{\Lambda}_\Gamma)$. *The projective dimension of every module in* $\widehat{\mathrm{mod}}_{\mathrm{pr}}^{\mathrm{pr}}(\overline{\Lambda}_\Gamma)$ *is at most one.*

(c) *The adjustment functor*

$$(3.8) \qquad \Theta : \widehat{\mathrm{mod}}_{\mathrm{pr}}^{\mathrm{pr}}(\overline{\Lambda}_\Gamma) \longrightarrow \widehat{\mathrm{mod}}_{\mathrm{pr}}(\overline{\Lambda}_\Gamma)$$

$(Y_A', Y_B'', t : Y' \to Y'') \mapsto (\mathrm{Im}\, t, Y_B'', inc : \mathrm{Im}\, t \hookrightarrow Y'')$ *is full, dense and vanishes on modules of the form* $(eA, 0, 0)$, *where* $e \in A$ *is a primitive idempotent.*

(c) Θ *defines a bijection between the set of the isoclasses of indecomposable modules of* $\widehat{\mathrm{mod}}_{\mathrm{pr}}^{\mathrm{pr}}(\overline{\Lambda}_\Gamma)$ *having no direct summands of the form* $(eA, 0, 0)$, *where* $e \in A$ *is a primitive idempotent, and the set of the isoclasses of the indecomposable modules of* $\widehat{\mathrm{mod}}_{\mathrm{pr}}(\overline{\Lambda}_\Gamma)$.

(d) *The category* $\widehat{\mathrm{mod}}_{\mathrm{pr}}^{\mathrm{pr}}(\overline{\Lambda}_\Gamma)$ *is of finite (resp. tame, wild) representation type if and only if the category* $\widehat{\mathrm{mod}}_{\mathrm{pr}}(\overline{\Lambda}_\Gamma)$ *is of finite (resp. tame, wild) representation type.*

Proof. For (a) see [36, p. 469] and [30]. For the proof of (b)–(d) we refer to [47]. □

By [47], the category $\widehat{\mathrm{mod}}_{\mathrm{pr}}^{\mathrm{pr}}(\overline{\Lambda}_\Gamma)$ has nice homological properties and has a useful matrix problem interpretation. In particular $\widehat{\mathrm{mod}}_{\mathrm{pr}}^{\mathrm{pr}}(\overline{\Lambda}_\Gamma)$ is equivalent with an open subcategory [14] of the category $\mathrm{rep}(\mathfrak{B}_\Lambda, K)$ of K-linear representations of a free triangular boks associated with the order Λ (see [14] and [47]).

Let us define the pair of adjoint functors

$$(3.9) \qquad \mathrm{CM}(\Lambda) \underset{\mathrm{res}_\Lambda}{\overset{\mathrm{ind}_\Gamma}{\rightleftarrows}} \mathrm{CM}(\Gamma)$$

where res_Λ is the **restriction functor** associating to any Y_Γ in $\mathrm{CM}(\Gamma)$ the abelian group Y with the natural Λ-module structure. The **induction functor** ind_Γ associates to any module X in $\mathrm{CM}(\Lambda)$ the

Γ-submodule

(3.10) $$\mathrm{ind}_\Gamma(X) = X\Gamma$$

of the C-module $XF = X \otimes_D F$ generated by X. Note that the kernel of the natural epimorphism $\mu : X \otimes_\Lambda \Gamma \to X\Gamma$ is the Γ-submodule of $X \otimes_\Lambda \Gamma$ consisting of all D-torsion elements and

$$X\Gamma \cong (X \otimes_\Lambda \Gamma)/\mathrm{Ker}\,\mu$$

(see [7]). If $f : X \to Y$ is a homomorphism in $\mathrm{CM}(\Lambda)$ we denote by

$$\mathrm{ind}_\Gamma(f) = \hat{f} : \mathrm{ind}(X) \to \mathrm{ind}(Y)$$

the unique Γ-homomorphism making the diagram

$$
\begin{array}{ccc}
X & \hookrightarrow & X\Gamma \\
\downarrow{\scriptstyle f} & & \downarrow{\scriptstyle \hat{f}} \\
Y & \hookrightarrow & Y\Gamma
\end{array}
$$

commutative. It is easy to see that ind_Γ is a covariant additive functor.

Following [15], [18] and [34] we define the reduction functor

(3.11) $$\mathbb{F} : \mathrm{CM}(\Lambda) \longrightarrow \widehat{\mathrm{mod}}_{\mathrm{pr}}(\overline{\Lambda}_\Gamma) \subseteq \mathrm{mod}(\overline{\Lambda}_\Gamma)$$

by the formula

$$\mathbb{F}(X) = (X/XI, X\Gamma/XI,\ u : X/XI \to X\Gamma/XI)$$

where u is the A-module embedding induced by the Λ-monomorphism $X \hookrightarrow X\Gamma$ (see [18], [34]).

THEOREM 3.12. *In the notation and assumption made above the following statements hold.*

(a) *The additive functor $\mathbb{F}$ (3.11) is a representation equivalence, that is, $\mathbb{F}$ is full dense and reflects isomorphisms.*

(b) *If $f : X \to Y$ is a homomorphism in $\mathrm{CM}(\Lambda)$ then $\mathbb{F}(f) = 0$ if and only if $\mathrm{Im}\, f \subseteq YI$.*

(c) *The category $\mathrm{CM}(\Lambda)$ is of finite (resp. tame, wild) representation type if and only if the category $\widehat{\mathrm{mod}}_{\mathrm{pr}}(\overline{\Lambda}_\Gamma)$ is of finite (resp. tame, wild) representation type.*

Proof. See [14], [18], [34], [36]. □

4   The case $J(\Gamma) \subseteq J(\Lambda) \subseteq \Lambda \subseteq \Gamma$

Throughout Λ is a D-order in a semisimple D_0-algebra C. We fix a hereditary D-order $\Gamma \subseteq C$ containing Λ. We recall from (3.1) that there exists a positive integer $s \geq 1$ such that $J(\Gamma)^s \subseteq J(\Lambda)$.

In this section we consider the case when $s = 1$, that is, $J(\Gamma) \subseteq J(\Lambda) \subseteq \Lambda \subseteq \Gamma$. We consider the artin algebra (3.3)

$$\overline{\Lambda}_\Gamma = \begin{pmatrix} A & {}_A B_B \\ 0 & B \end{pmatrix}$$

associated with the admissible triple $(I = J(\Gamma), \Lambda, \Gamma)$, where $A = \Lambda/J(\Gamma) \hookrightarrow B = \Gamma/J(\Gamma)$. In this case the ring B is semisimple and therefore B is Morita equivalent to a product

$$(4.1) \qquad\qquad T = G_1 \times \ldots \times G_t$$

of division rings $G_1, \ldots, G_t$, that is, there exists a Morita equivalence of module categories

$$(4.2) \qquad\qquad \mathrm{mod}(B) \cong \mathrm{mod}(T)$$

The equivalence carries the bimodule ${}_A B_B$ to an A-T-bimodule ${}_A M_T$. We associate with $\overline{\Lambda}_\Gamma$ the new artin algebra

$$(4.3) \qquad\qquad R_\Lambda = \begin{pmatrix} A & {}_A M_T \\ 0 & T \end{pmatrix}$$

where T and M are as above.

We denote by $\mathrm{mod}_{sp}(R_\Lambda)$ the full subcategory of $\mathrm{mod}(R_\Lambda)$ formed by **socle projective modules**, that is, the modules X such that the socle $\mathrm{soc}(X)$ of X is a projective R_Λ-module [39].

We denote by $\mathrm{mod}_{sp}^{\bullet}(R_\Lambda)$ the full subcategory of $\mathrm{mod}_{sp}(R_\Lambda)$ formed by the modules having no simple projective direct summands.

PROPOSITION 4.4.   *Assume that Λ is a D-suborder of a hereditary order Γ such that $J(\Gamma) \subseteq J(\Lambda) \subseteq \Lambda \subseteq \Gamma$. Let $\overline{\Lambda}_\Gamma$ be the artin algebra (3.3) associated with the admissible triple $(I = J(\Gamma), \Lambda, \Gamma)$ and let R_Λ be the artin algebra (4.3).*

(a)   The artin algebra $\overline{\Lambda}_\Gamma$ (4.3) is a right multipeak ring in the sense of [39], that is, the socle $\mathrm{soc}(\overline{\Lambda}_\Gamma)$ of the right $\overline{\Lambda}_\Gamma$-module $\overline{\Lambda}_\Gamma$ is a projective module and the left A-module ${}_A B$ is faithful. In this case

$$\mathrm{soc}(\overline{\Lambda}_\Gamma) = \begin{pmatrix} 0 & {}_A B_B \\ 0 & B \end{pmatrix}$$

(b) *The artin algebra R_Λ is a right multipeak ring Morita equivalent with the algebra $\overline{\Lambda}_\Gamma$.*

(c) *The Morita equivalence* $\mathrm{mod}(\overline{\Lambda}_\Gamma) \cong \mathrm{mod}(R_\Lambda)$ *induces an equivalence of categories*

(4.5) $L : \widehat{\mathrm{mod}}_{\mathrm{pr}}(\overline{\Lambda}_\Gamma) \xrightarrow{\;\approx\;} \mathrm{mod}^\bullet_{\mathrm{sp}}(R_\Lambda) \subseteq \mathrm{mod}_{\mathrm{sp}}(R_\Lambda)$

(d) *The composition*

(4.6) $\widehat{\mathbb{F}} = L \circ \mathbb{F} : \mathrm{CM}(\Lambda) \longrightarrow \mathrm{mod}^\bullet_{\mathrm{sp}}(R_\Lambda) \subseteq \mathrm{mod}_{\mathrm{sp}}(R_\Lambda)$

of the reduction functor $\mathbb{F}$ (3.11) with the Morita equivalence L is a representation equivalence of categories preserving the representation types.

(e) *The category $\mathrm{CM}(\Lambda)$ is of finite (resp. tame, wild) representation type if and only if the category $\mathrm{mod}_{\mathrm{sp}}(R_\Lambda)$ is of finite (resp. tame, wild) representation type.*

Proof. For the proof of (a) and (b) we refer to [39] and to the proof of Lemma 17.39 in [41]. The proof of (c) follows by standard Morita equivalence arguments. The statements (d) and (e) follow from Theorem 3.12.   □

CONCLUSION 4.7. The reduction above and the functor (4.6) reduce the problem of classifying D-suborders Λ of hereditary D-orders Γ such that $J(\Gamma) \subseteq J(\Lambda)$ and the problem of determining the Cohen-Macaulay type of such D-orders Λ to a classification of basic right peak artin algebras R and determining the representation type of their categories $\mathrm{mod}_{\mathrm{sp}}(R)$ consisting of socle projective modules.

Although the problem is not solved for arbitrary right peak artin algebras R a lot is known for large classes of algebras R. There are representation type criteria in the following cases:

(a) R is a hereditary artin algebra [34],

(b) R is an artinian schurian right peak PI-ring (see [41, Section 17.7]),

(c) $R = KI$ is an incidence algebra of a finite poset I over a field K (see Kleiner [23], Nazarova [27], Simson[43], Kasjan-Simson [20] - [22], Kasjan [19] and Dowbor-Kasjan [12]),

(d) $R = K(I, 3)$ is an incidence K-algebra of a finite poset I equipped with a set of zero-relations 3 (see [42], [49], [48]), or

(e) $R = KI_\rho$ is an incidence K-algebra of a finite stratified poset I_ρ having a unique maximal element and equipped with an equivalence relation (see [29], [40], [42], [49], [48]).   □

We discuss the incidence algebras in the following section.

5 Socle projective representations of partially ordered sets

Throughout this section K is a field and $(I, \preceq)$ is a finite poset (i.e. partially ordered set) with respect to a partial order $\preceq$. We shall write $i \prec j$ if $i \preceq j$ and $i \neq j$. For the sake of simplicity we write I instead of $(I, \preceq)$.

We recall that a notion of a matrix K-representation of a partially ordered set S was introduced by Nazarova and Roiter [28] in a connection with the Second Brauer-Thrall Conjecture and with matrix problems arising in the integral representation theory (see [41]). The main aim of their study was to determine indecomposable objects of the additive Krull-Schmidt category Mat_S^{ad} of all matrix K-representations of S (see [41, Chapter 2] and [17]). The objects of Mat_S^{ad} can be viewed as partitioned matrices with coefficients in K and the classification of the indecomposable objects in Mat_S^{ad} is equivalent to the problem of putting any partitioned matrix in Mat_S^{ad} into a normal form by admissible elementary row and column transformations defined by means of the partial order $\preceq$ of S. Indecomposable objects correspond to the indecomposable normal partitioned cells. The finite, tame and wild representation type of Mat_S^{ad} is well defined.

On the other hand, Gabriel [16] has introduced the Krull-Schmidt category S-sp of S-spaces over K. The objects of S-sp are system $\underline{M} = (M_i; M)_{i \in S}$ of finite dimensional K-vector spaces $M_i \subseteq M$ such that $M_i \subseteq M_j$ for all $i \preceq j$ in S. A morphism $f : \underline{M} \to \underline{M}'$ between S-spaces is a K-linear map $f : M \to M'$ such that $f(M_j) \subseteq M'_j$ for all $j \in S$. The direct sum of $\underline{M}$ and $\underline{N}$ in S-sp is defined by formula $\underline{M} \oplus \underline{N} = (M_i \oplus N_i; M \oplus N)_{i \in S}$.

There is a nice functorial connection between Mat_S^{ad} and S-sp given by a full additive dense functor $H : \mathrm{Mat}_S^{ad} \longrightarrow S$-sp (see [41, Section 3.1]) preserving the representation type and vanishing only on a finite set of isomorphism classes of indecomposable objects. Therefore the study of the category Mat_S^{ad} is equivalent to the study of S-sp from the representation theory point of view.

It was proved by Kleiner [23] (see [33], [41, Section 10.1]) that the category S-sp is of finite representation type if and only if the poset S does not contain a subposet isomorphic to one of the critical posets of Kleiner shown in Tables 5.1 below. Moreover, by a well-known theorem of Nazarova [27] (see [33], [41, Section 15.1]) the category S-sp is of tame representation type if and only if the poset S does not contain a subposet isomorphic to one of the hypercritical posets of Nazarova shown in Tables 5.2 below.

TABLES 5.1. Critical Posets of Kleiner

$$\mathcal{K}_1 : (1,1,1,1) = \{\bullet\ \bullet\ \bullet\ \bullet\}, \qquad \mathcal{K}_2 : (2,2,2) = \left\{\begin{array}{ccc} \bullet & \bullet & \bullet \\ \uparrow & \uparrow & \uparrow \\ \bullet & \bullet & \bullet \end{array}\right\},$$

$$\mathcal{K}_3 : (1,3,3) = \left\{\begin{array}{ccc} \bullet & \bullet & \bullet \\ & \uparrow & \uparrow \\ & \bullet & \bullet \\ & \uparrow & \uparrow \\ & \bullet & \bullet \end{array}\right\}, \qquad \mathcal{K}_4 : (N,4) = \left\{\begin{array}{ccc} \bullet & \bullet & \bullet \\ \uparrow\!\!\!\nwarrow & \uparrow & \uparrow \\ & \bullet & \\ & & \uparrow \\ & & \bullet \\ & & \uparrow \\ & & \bullet \end{array}\right\},$$

$$\mathcal{K}_5 : (1,2,5) = \left\{\begin{array}{ccc} \bullet & \bullet & \bullet \\ & \uparrow & \uparrow \\ & \bullet & \bullet \\ & & \uparrow \\ & & \bullet \\ & & \uparrow \\ & & \bullet \\ & & \uparrow \\ & & \bullet \end{array}\right\}.$$

TABLES 5.2. Hypercritical Posets of Nazarova

$$\mathcal{N}_1 : (1,1,1,1,1) = (\bullet\ \bullet\ \bullet\ \bullet\ \bullet), \qquad \mathcal{N}_2 : (1,1,1,2) = \left(\bullet\ \bullet\ \bullet\ \begin{array}{c}\bullet\\\uparrow\\\bullet\end{array}\right)$$

$$\mathcal{N}_3 : (2,2,3) = \left(\begin{array}{ccc} \bullet & \bullet & \bullet \\ \uparrow & \uparrow & \uparrow \\ \bullet & \bullet & \bullet \\ & & \uparrow \\ & & \bullet \end{array}\right), \qquad \mathcal{N}_4 : (1,3,4) = \left(\begin{array}{ccc} & \bullet & \bullet \\ & \uparrow & \uparrow \\ \bullet & \bullet & \bullet \\ & \uparrow & \uparrow \\ & \bullet & \bullet \\ & & \uparrow \\ & & \bullet \end{array}\right)$$

$$\mathcal{N}_5 : (N,5) = \left(\begin{array}{ccc} \bullet & \bullet & \bullet \\ \uparrow\!\!\!\nwarrow & \uparrow & \uparrow \\ \bullet & \bullet & \bullet \\ & & \uparrow \\ & & \bullet \\ & & \uparrow \\ & & \bullet \\ & & \uparrow \\ & & \bullet \end{array}\right), \qquad \mathcal{N}_6 : (1,2,6) = \left(\begin{array}{ccc} \bullet & \bullet \\ \uparrow & \uparrow \\ \bullet & \bullet & \bullet \\ & & \uparrow \\ & & \bullet \\ & & \uparrow \\ & & \bullet \\ & & \uparrow \\ & & \bullet \end{array}\right).$$

A study of a class of matrix problems arising in the theory of Cohen-Macaulay modules over classical orders leads us in [43] to the concept of a peak I-space (explained below), being a natural generalization of the concept of an S-space introduced by Gabriel in [16].

Given a poset I we denote by max I the set of all maximal elements of I (called peaks of I). We suppose that

$$I = \{1, \ldots, n, p_1, \ldots, p_r\}, \quad \max I = \{p_1, \ldots, p_r\}$$

and that the order relation $\prec$ in I is such that $i \prec j$ implies that $i < j$ in the natural order. We can always achieve this by a suitable renumbering of the elements in I.

We recall that a subposet J of I is said to be **a peak subposet** if $J \cap \max I = \max J$.

Usually we view the poset I as a quiver with the commutativity relations induced by the ordering $\prec$ (see [41, Example 10, p. 281]). By the **incidence K-algebra** of I we shall mean the path algebra of I with coefficients in K (see Chapter 14 of [41]). It follows from our assumption on the order $\preceq$ that KI has the following upper triangular $I \times I$-matrix form

$$(5.3) \quad KI = \begin{pmatrix} K & K_{12} & \cdots & K_{1n} & K_{1p_1} & K_{1p_2} & \cdots & K_{1p_r} \\ 0 & K & \cdots & K_{2n} & K_{2p_1} & K_{2p_2} & \cdots & K_{2p_r} \\ \vdots & \vdots & \ddots & \vdots & \vdots & \vdots & \cdots & \vdots \\ 0 & 0 & \cdots & K & K_{np_1} & K_{np_2} & \cdots & K_{np_r} \\ 0 & 0 & \cdots & 0 & K & 0 & \cdots & 0 \\ 0 & 0 & \cdots & 0 & 0 & K & \cdots & 0 \\ \vdots & \vdots & \ddots & \vdots & \vdots & \vdots & \ddots & \vdots \\ 0 & 0 & \cdots & 0 & 0 & 0 & \cdots & K \end{pmatrix}$$

where $K_{ij} = K$ if $i \prec j$ and $K_{ij} = 0$ otherwise. It is easy to see that KI is a right multipeak algebra of finite global dimension.

Given $i \preceq j$ in I we denote by $e_{ij} \in KI$ the matrix having 1 at i-j-th position and zeros elsewhere. Given j in I we denote by $e_j = e_{jj}$ the standard primitive idempotent corresponding to j.

Note that KI can be displayed in the form

$$(5.4) \qquad\qquad KI = \begin{pmatrix} A & {}_A M_B \\ 0 & B \end{pmatrix}$$

where $A = KI^-$, $I^- = I \setminus \max I$, $B = K \times K \times \cdots \times K$ ($|\max I|$-times), the K-vector space $M = \bigoplus\limits_{p \in \max I} \bigoplus\limits_{\substack{i \prec p \\ i \in I^-}} e_{ip}K$ is viewed as a A-B-bimodule in the obvious way and multiplication is given by the matrix multiplication.

Right KI-modules will be identified with the systems

$$X = (X_i;\ _jh_i)_{i,j\in I}$$

where $X_i = Xe_i$ is a finite dimensional vector space over K and $_jh_i :$
$X_i \to X_j$, $i \prec j$, are K-linear maps such that $_ih_i = \mathrm{id}$ and $_th_j\cdot{_j}h_i = {_t}h_i$
for all $i \prec j \prec t$ in I.

The criteria of Kleiner and of Nazarova mentioned above yield the
following result.

THEOREM 5.5. *Let I be a poset with a unique maximal element.*

(a) *The category* $\mathrm{mod}_{sp}(KI)$ *is of finite representation type if and
only if I does not contain any of the five critical posets $\mathcal{K}_1,\dots,\mathcal{K}_5$ of
Kleiner shown in Tables 5.1.*

(b) *The category* $\mathrm{mod}_{sp}(KI)$ *is of tame representation type if and
only if I does not contain any the six hypercritical posets $\mathcal{N}_1,\dots,\mathcal{N}_6$
of Nazarova shown in Tables 5.2.*

Proof. Let p be the unique maximal element of I. Following [43]
we define the category I-spr of filtered socle projective representations
of I (or peak I-spaces) as follows.

The objects of I-spr are systems $\mathbf{M} = (M_i)_{i\in I}$ of finite dimensional
K-vector spaces M_i such that $M_i \subseteq M_j$ if $i \preceq j$ in I. In particular
$M_i \subseteq M_p$ for all $i \in I$.

By a morphism $f : \mathbf{M} \to \mathbf{M}'$ between peak I-spaces $\mathbf{M}$ and $\mathbf{M}'$
we shall mean a K-linear map $f : M \to M'$ such that $f(M_j) \subseteq M'_j$
for all $j \in I$. The direct sum of $\mathbf{M}$ and $\mathbf{N}$ in I-spr is defined by
formula $\mathbf{M} \oplus \mathbf{N} = (M_i \oplus N_i)_{i\in I}$. A peak I-space $\mathbf{M}$ is said to be
indecomposable if $\mathbf{M}$ is non-zero and is not a direct sum of two
non-zero peak I-spaces (see [17], [41, Section 5], [43]). Note that
I-spr $= S$-sp, where $S = I \setminus \{p\}$. It is easy to see that the K-linear
functor

$$(5.6) \qquad \rho : I\text{-spr} \xrightarrow{\ \approx\ } \mathrm{mod}_{sp}(KI)$$

$\mathbf{M} \mapsto \widehat{\mathbf{M}} = (M_j;\ _ju_i)_{i\prec j}$, is an equivalence of categories, where $_ju_i :$
$M_i \to M_j$ is the vector space embedding. The quasi-inverse of ρ is
defined by associating to any $X = (X_i, {_j}h_i)_{i,j\in I,i\prec j}$ in $\mathrm{mod}_{sp}(KI)$ the
peak I-space $\mathbf{M}(X) = (M(X)_j)_{j\in I}$, where

$$M(X)_j = \begin{cases} X_p & \text{for } j = p \in \max I, \\ \mathrm{Im}\,[(_ph_j)_{p\in\max I} : X_j \to M_p = X_p] & \text{for } j \in I \setminus \max I. \end{cases}$$

It follows that our theorem reduces to the context studied by Kleiner
and Nazarova (see [41]) and therefore the theorem follows from the

well-known criteria of Kleiner [23] and of Nazarova [27] mentioned above (see also [17] and [41, Sections 10 and 15]). □

A counterpart of Kleiner's theorem for posets with arbitrary number of maximal elements is the following result proved by the author in [43].

THEOREM 5.7. *Let K be a field, I a finite poset and let KI be the incidence K-algebra (5.3) of I. The following statements are equivalent.*

(a) *The category $\mathrm{mod}_{\mathrm{sp}}(KI)$ is of finite representation type.*

(b) *The Tits quadratic form $q_I : \mathbb{Z}^I \to \mathbb{Z}$ of I defined by the formula*

$$(5.8) \qquad q_I(x) = \sum_{i \in I} x_i^2 + \sum_{\substack{i \prec j \\ j \in I \backslash \max I}} x_i x_j - \sum_{p \in \max I} (\sum_{i \prec p} x_i) x_p.$$

is weakly positive, that is, $q_I(v) > 0$ for any non-zero vector $v \in \mathbb{N}^I$.

(c) *The poset I does not contain as a peak subposet any of the 114 critical posets $\mathcal{P}_1, \ldots, \mathcal{P}_{114}$ presented in [43, Section 5].*

(d) *The set $\mathcal{R}_{q_I}^+ = \{v \in \mathbb{N}^I; q_I(v) = 1\}$ of positive roots of the quadratic form q_I is finite.*

(e) *There is no full faithful exact functor $\mathrm{mod}\Gamma_2(K) \to \mathrm{mod}_{\mathrm{sp}}(KI)$, where $\Gamma_2(K) = \begin{pmatrix} K & K^2 \\ 0 & K \end{pmatrix}$ is the Kronecker K-algebra.*

(f) $\dim_K \mathrm{Ext}_{KI}^1(U_1, U_2) \leq 1$ *for any pair of modules U_1, U_2 in $\mathrm{mod}_{\mathrm{sp}}(KI)$ satisfying the following two conditions:*

(i) $\mathrm{End}(U_1) \cong \mathrm{End}(U_2) \cong K$,

(ii) $\mathrm{Hom}_R(U_1, U_2) = 0$ *and* $\mathrm{Hom}_R(U_2, U_1) = 0$.

Proof. The equivalence of the statements (a)-(c) is proved in [43, Section 3]. For the proof of the equivalences of (a), (d), (e) and (f) we refer to [24], [25], [48] and [49]. □

REMARK. Criteria for tame representation type of the category $\mathrm{mod}_{\mathrm{sp}}(KI)$ analogous to that one in Theorem 5.7 are given for a relatively large class of posets by S. Kasjan and the author in [21], [22], [19] (see also [12]). □

EXAMPLE 5.9. Let I be the following poset

$$
\begin{array}{ccccccccc}
5 & \to & p_1 & & & & & & \\
\uparrow & & & & & & & & \\
I: & & 4 & & \longrightarrow & & p_2 & & \\
\uparrow & & & & & & \uparrow & & \\
1 & \to & 2 & \to & 3 & \to & p_3 & &
\end{array}
$$

having exactly three maximal elements p_1, p_2, p_3. The incidence K-algebra KI of I has the form

$$
KI = \begin{pmatrix}
K & K & K & 0 & 0 & K & K & K \\
  & K & K & 0 & 0 & 0 & K & K \\
  &   & K & 0 & 0 & 0 & K & K \\
  &   &   & K & 0 & K & K & 0 \\
  &   &   &   & K & K & 0 & 0 \\
  &   & 0 &   &   & K & 0 & 0 \\
  &   &   &   &   &   & K & 0 \\
  &   &   &   &   &   &   & K
\end{pmatrix} \subseteq \mathrm{M}_8(K)
$$

The Tits quadratic form $q_I : \mathbb{Z}^8 \to \mathbb{Z}$ is defined by the formula

$$
\begin{aligned}
q_I(x) &= x_1^2 + x_2^2 + x_3^2 + x_5^2 + x_5^2 + x_{p_1}^2 + x_{p_2}^2 + x_{p_3}^2 \\
&\quad + x_1(x_2 + x_3 + x_4) + x_2 x_3 \\
&\quad - (x_1 + x_4 + x_5) x_{p_1} - x_4 x_{p_2} - (x_1 + x_2 + x_3)(x_{p_2} + x_{p_3}) \\
&= (x_5 - \tfrac{1}{2} x_{p_1})^2 + \tfrac{3}{4}(x_{p_1} - \tfrac{2}{3} x_1 - \tfrac{2}{3} x_4)^2 + \tfrac{2}{3}(\tfrac{1}{4} x_1 + x_4 - \tfrac{3}{4} x_{p_2})^2 \\
&\quad + \tfrac{5}{8}(-\tfrac{3}{5} x_1 - \tfrac{4}{5} x_2 - \tfrac{4}{5} x_3 + x_{p_2})^2 \\
&\quad + (-\tfrac{1}{2} x_1 - \tfrac{1}{2} x_2 - \tfrac{1}{2} x_3 + x_{p_3})^2 \\
&\quad + \tfrac{3}{20}(x_1 - \tfrac{1}{3} x_2 - \tfrac{1}{3} x_3)^2 + \tfrac{1}{3}(x_2 - x_3)^2 + \tfrac{1}{3} x_2 x_3
\end{aligned}
$$

It is easy to see that q_I is weakly positive, and according to Theorem 5.7 the category $\mathrm{mod}_{sp}(KI)$ is of finite representation type.

Consider the following three D-suborders

$$
\Lambda_1 = \begin{pmatrix} D & D & \mathfrak{P} \\ \mathfrak{P} & D & \mathfrak{P} \\ \mathfrak{P} & \mathfrak{P} & D \end{pmatrix}, \quad
\Lambda_2 = \begin{pmatrix} D & D & D & D \\ \mathfrak{P} & D & \mathfrak{P} & \mathfrak{P} \\ \mathfrak{P} & \mathfrak{P} & D & D \\ \mathfrak{P} & \mathfrak{P} & \mathfrak{P} & D \end{pmatrix}, \quad
\Lambda_3 = \begin{pmatrix} D & D & D \\ \mathfrak{P} & D & D \\ \mathfrak{P} & \mathfrak{P} & D \end{pmatrix}
$$

of $\Gamma_1 = \mathrm{M}_3(D)$, $\Gamma_2 = \mathrm{M}_4(D)$ and $\Gamma_3 = \mathrm{M}_3(D)$, respectively. Denote by Λ the suborder of $\Lambda_1 \times \Lambda_2 \times \Lambda_3$ consisting of all triples $(a, b, c) \in \Lambda_1 \times \Lambda_2 \times \Lambda_3$ of matrices a, b, c satisfying the following conditions:

(i) $a_{ij} \equiv b_{ij}$ for all $1 \le i \le j \le 2$,

(ii) $b_{11} \equiv c_{11}$, $b_{13} \equiv c_{12}$, $b_{14} \equiv c_{13}$, $b_{33} \equiv c_{22}$, $b_{44} \equiv c_{33}$, $b_{34} \equiv c_{23}$,

where $x \equiv y$ means $x - y \in \mathfrak{P}$.

We note that Λ is contained in the hereditary D-order $\Gamma = \Gamma_1 \times \Gamma_2 \times \Gamma_3$ and the Jacobson radical $J(\Gamma) = \mathrm{M}_3(\mathfrak{P}) \times \mathrm{M}_4(\mathfrak{P}) \times \mathrm{M}_3(\mathfrak{P})$ of Γ is contained in the Jacobson radical of Λ. Then $(J(\Gamma), \Lambda, \Gamma)$ is an admissible triple and Proposition 4.4 applies to $\Lambda \subseteq \Gamma$.

We claim that the algebra R_Λ (4.3) is isomorphic with the incidence K-algebra KI and therefore $\overline{\Lambda}_\Gamma = \left(\begin{smallmatrix} A & B \\ 0 & B \end{smallmatrix} \right)$ (3.3) is Morita equivalent with

KI, where $A = \Lambda/J(\Gamma)$ and $B = \Gamma/J(\Gamma)$. For this we note that A is isomorphic with KJ, where J is the subposet of I consisting of elements $\{1,2,3,4,5\}$. The algebra B is Morita equivalent wit the product $K \times K \times K$ of three copies of K. Hence our claim easily follows (see [46] for other examples of a similar type).

Consequently there is a K-algebra isomorphism $R_\Lambda \cong KI$ and according to Proposition 4.4 there exists a representation equivalence

$$\widehat{\mathbb{F}}' : \mathrm{CM}(\Lambda) \longrightarrow \mathrm{mod}^\bullet_{\mathrm{sp}}(KI) \subseteq \mathrm{mod}_{\mathrm{sp}}(KI)$$

induced by the functor (4.6). The functor $\widehat{\mathbb{F}}'$ establishes a bijection between the set of isoclasses of indecomposable modules in $\mathrm{CM}(\Lambda)$ and the set of the isoclasses of indecomposable objects in the category $\mathrm{mod}_{\mathrm{sp}}(KI)$ being not isomorphic to the simple projective modules $\mathbf{S}(p_1), \mathbf{S}(p_2), \mathbf{S}(p_3)$ corresponding to the maximal elements p_1, p_2, p_3 of I.

Since $R_\Lambda \cong KI$ and the category $\mathrm{mod}_{\mathrm{sp}}(KI)$ is of finite representation type then according to Proposition 4.4 the order Λ is of finite Cohen-Macaulay type.

By constructing the preprojective component of $\mathrm{mod}_{\mathrm{sp}}(KI)$ like in [41, Section 11.10] or in [49, Example 6.6], or by applying the Tits quadratic form argument given in [43] (and in Theorem 5.7) and by determining the cardinality of the set $\mathcal{R}^+_{q_I}$ of positive roots of q_I, one can show that the category $\mathrm{mod}_{\mathrm{sp}}(KI)$ has 100 isoclasses of indecomposable objects and therefore each of the categories $\mathrm{mod}^\bullet_{\mathrm{sp}}(KI)$ and $\mathrm{CM}(\Lambda)$ has 97 isoclasses of indecomposable modules. □

6 Representations of posets equipped with an equivalence relation

Let J be a poset. Consider the set of pairs

$$\blacktriangle J^- := \{(i,j);\ i \preceq j \text{ in } J^-\} \subseteq J^- \times J^-$$

where $J^- = J \backslash \max J$. Following [29] and [40], by a **stratified poset** (or a **poset with an equivalence relation**) we shall mean a pair

(6.1) $$J_\rho = (J, \rho)$$

where J is a poset and $\rho \subseteq \blacktriangle J^- \times \blacktriangle J^-$ is a binary equivalence relation such that the subset

(6.2)
$$KJ_\rho = \{\lambda = [\lambda_{pq}]_{p,q \in J} \in KJ;\ \lambda_{ij} = \lambda_{st} \text{ if } (i,j)\rho(s,t) \text{ holds in } \blacktriangle J^-\}$$

is a K-subalgebra $KJ\rho$ of the incidence K-algebra KJ of J. In this case we call $KJ\rho$ the **incidence K-algebra** of the stratified poset $KJ\rho$.

The reader is referred to [40] and [42] for examples and an elementary introduction to stratified posets. Note that the definition of a stratified poset given here differs from the original one given in [40], and extends it.

The following lemma is an easy observation (see [40]).

LEMMA 6.3. (a) *If $J\rho = (J, \rho)$ is a stratified poset then the incidence algebra $KJ\rho$ is a right multipeak algebra.*

(b) *Assume that J is a poset and ρ is a binary equivalence relation on $\blacktriangle J^-$ such that if $(i,j)\rho(p,q)$ then there is a poset isomorphism $\sigma :$ $[i,j] \to [p,q]$ with the property that $(i,t)\rho(p,\sigma(t))$ and $(t,j)\rho(\sigma(t),q)$ for all $t \in [i,j]$, where $[s,p] = \{j \in J; s \preceq j \preceq p\}$ (see [29], [40]). Then $J\rho = (J, \rho)$ is a stratified poset.* □

The following two important problems are not solved yet.

PROBLEM 6.4. *Does the converse to the implication in Lemma 6.3 (b) hold?* □

PROBLEM 6.5. *Given a stratified poset $J\rho$, determine the representation type of the category $\mathrm{mod}_{\mathrm{sp}}(KJ\rho)$ in terms of $J\rho$.* □

REMARK 6.6. A motivation for determining the representation type of the category $\mathrm{mod}_{\mathrm{sp}}(KJ\rho)$, where $J\rho$ is a stratified posets, follows from the fact that for any stratified poset $J\rho$ there exists a classical D-order Λ such that the right multipeak algebra R_Λ (4.3) associated with Λ is isomorphic with $KJ\rho$, and according to Proposition 4.4 the functor (4.6) induces a representation equivalence of categories

$$\widehat{\mathbb{F}}' : \mathrm{CM}(\Lambda) \longrightarrow \mathrm{mod}^{\bullet}_{\mathrm{sp}}(KJ\rho) \subseteq \mathrm{mod}_{\mathrm{sp}}KJ\rho)$$

preserving the representation types. In particular, the category $\mathrm{CM}(\Lambda)$ is of finite (resp. tame, wild) representation type if and only if the category $\mathrm{mod}_{\mathrm{sp}}(KJ\rho)$ is of finite (resp. tame, wild) representation type. □

A combinatorial characterization of stratified posets $J\rho = (J, \rho)$ for which the category $\mathrm{mod}_{\mathrm{sp}}(KJ\rho)$ is of finite or tame representation type remains an open problem and seems to be rather difficult.

A criterion for $J\rho$ to have the category $\mathrm{mod}_{\mathrm{sp}}(KJ\rho)$ of finite representation type is given in [29] in case $J\rho$ is a bipartite completed poset, and in [42] in case $J\rho$ is a bipartite stratified poset. In [42] for bipartite stratified posets a Galois covering technique is developed. This allows an efficient study of $\text{mod}_{\text{sp}}(KJ\rho)$ in the tame representation type case. By applying these results we get in [45], [49] and [48] a characterization of three-partite subamalgams of tiled D-orders of finite and tame Cohen-Macaulay type (see Section 8).

In [5] and [53] a characterization of weakly completed posets with an equivalence relation for which the category $\text{mod}_{\text{sp}}(KJ\rho)$ is of tame representation type is given. A polynomial growth criterion is also presented there.

7   Tiled D-orders, representations of partially ordered sets and a covering functor

Let D be a complete discrete valuation domain with the field of fractions $F = D_0$. Let $\mathfrak{P}$ be the unique maximal ideal of D and $K = D/\mathfrak{P}$.

7.1   0-1-orders

Let us start by recalling a useful observation of Drozd, Zavadski and Kirichenko [15], [54] (see also [41, Section 13.]).

Following [54] a D-order Λ in the simple D_0-algebra $C = \mathbb{M}_n(D_0)$ is called a 0-1-order if Λ is a D-suborder of the hereditary $n \times n$ matrix order $\Gamma = \mathbb{M}_n(D)$ of the following form

$$(7.1) \qquad \Lambda = \begin{pmatrix} D & D_{12} & \cdots & D_{1n} \\ \mathfrak{P} & D & \cdots & D_{2n} \\ \vdots & \vdots & \ddots & \vdots \\ \mathfrak{P} & \mathfrak{P} & \cdots & D \end{pmatrix}$$

where D_{ij} is either $\mathfrak{P}$ or D. We associate with Λ the set

$$I_\Lambda = \{1, \ldots, n\}$$

and the relation $\preceq$ defined by the formula $i \preceq j \quad \Leftrightarrow \quad D_{ij} = D$. It is easy to see that $(I_\Lambda, \preceq)$ is a poset.

The following criterion follows from [54].

THEOREM 7.2. *Let Λ be a 0-1-order of the form* (7.1) *and let I_Λ be the poset associated with Λ.*

(a) *The D-order Λ is of finite Cohen-Macaulay type if and only if the poset I_Λ does not contain any of the critical posets $\mathcal{K}_1, \ldots, \mathcal{K}_5$ of Kleiner shown in Tables 5.1.*

(b) *The D-order Λ is of tame Cohen-Macaulay type if and only if the poset I_Λ does not contain any of the critical posets $\mathcal{N}_1, \ldots, \mathcal{N}_6$ of Nazarova shown in Tables 5.2.*

Proof. It is clear that $\Gamma = \mathbb{M}_n(D)$ is a hereditary order containing Λ, $J(\Gamma) = \mathbb{M}_n(\mathfrak{P})$ is contained in Λ. Therefore $(J(\Gamma), \Lambda, \Gamma)$ is an admissible triple and Proposition 4.4 applies. A straightforward analysis shows that the algebra R_Λ (4.3) associated to the triple $(J(\Gamma), \Lambda, \Gamma)$ is isomorphic with the incidence algebra KI_Λ^* of the poset

$$I_\Lambda^* = I_\Lambda \cup \{p\}$$

obtained from I_Λ by adding a unique maximal element p. By Proposition 4.4 there exist a representation equivalence functor

$$\widehat{\mathbb{F}}' : \mathrm{CM}(\Lambda) \longrightarrow \mathrm{mod}_{\mathrm{sp}}^\bullet(KI_\Lambda^*) \subseteq \mathrm{mod}_{\mathrm{sp}}(KI_\Lambda^*) \cong I_\Lambda^*\text{-spr}$$

induced by the functor (4.6). The functor $\widehat{\mathbb{F}}'$ establishes a bijection between the set of the isoclasses of indecomposable modules in $\mathrm{CM}(\Lambda)$ and the set of the isoclasses of indecomposable objects in the category $\mathrm{mod}_{\mathrm{sp}}(KI_\Lambda^*)$ being not isomorphic to the simple projective module $\mathbf{S}(p)$ corresponding to the unique maximal element p of I_Λ^*. Then the theorem is a consequence of the criteria of Kleiner [23] and Nazarova [27] rephrased in Theorem 5.5. $\square$

7.2   Tiled orders and associated covering posets

By a **tiled D-order** we mean the order of the form

$$(7.3) \qquad \Lambda = \begin{pmatrix} D & \mathfrak{P}^{m_{12}} & \cdots & \mathfrak{P}^{m_{1n}} \\ \mathfrak{P}^{m_{21}} & D & \cdots & \mathfrak{P}^{m_{2n}} \\ \vdots & \vdots & \ddots & \vdots \\ \mathfrak{P}^{m_{n1}} & \mathfrak{P}^{m_{n2}} & \cdots & D \end{pmatrix} \subseteq \mathbb{M}_n(D_0)$$

in the simple matrix algebra $A = \mathbb{M}_n(D_0)$, where $m_{ij} \in \mathbb{Z}$ are integers such that $m_{ij} + m_{jt} \geq m_{it}$ for all $i, j, t = 1, \ldots, n$. We set $\mathfrak{P}^0 = D$.

Throughout we suppose that Λ is basic, or equivalently, the inequality $m_{ij} + m_{ji} > 0$ holds for all $i, j = 1, \ldots, m, i \neq j$.

Following Zavadski and Kirichenko [55] we associate to any basic tiled order Λ of the form (7.3) the infinite poset $\mathcal{I}(\Lambda)$ consisting of all indecomposable projective right ideals of Λ contained in a fixed simple right ideal **S** of the simple algebra $C = \mathbb{M}_n(D_0)$. We take for the partial order in $\mathcal{I}(\Lambda)$ the inclusion.

It is shown in [55] that the poset $\mathcal{I}(\Lambda)$ is isomorphic to the infinite poset

$$(7.4) \qquad I(\Lambda) = \{1,\ldots,n\} \times \mathbb{Z} = \{1\} \times \mathbb{Z} \cup \cdots \cup \{n\} \times \mathbb{Z}$$

which is a disjoint union of n (= the number of rows in Λ) infinite chains

$$
\begin{array}{ccccccccc}
\cdots & \longrightarrow & (1,-1) & \longrightarrow & (1,0) & \longrightarrow & (1,1) & \longrightarrow & (1,2) & \longrightarrow & \cdots \\
\cdots & \longrightarrow & (2,-1) & \longrightarrow & (2,0) & \longrightarrow & (2,1) & \longrightarrow & (2,2) & \longrightarrow & \cdots \\
& & \vdots & & \vdots & & \vdots & & \vdots & & \\
\cdots & \longrightarrow & (n,-1) & \longrightarrow & (n,0) & \longrightarrow & (n,1) & \longrightarrow & (n,2) & \longrightarrow & \cdots
\end{array}
$$

connected by the following additional relations

$$(i,t) \prec (j,s) \quad \Leftrightarrow \quad m_{ij} \leq s - t.$$

The additive group $\mathbb{Z}$ of integers acts on the poset $I(\Lambda)$ by the shift $\sigma(i,t) = (i,t+1)$ as an automorphism group (see also [41, Section 13.2]).

The following surprising result was proved by Zavadski and Kirichenko in [55].

THEOREM 7.5. *Let D be a complete discrete valuation domain with the unique maximal ideal $\mathfrak{P}$ and $K = D/\mathfrak{P}$. A **tiled D-order** Λ of the form (7.3) is of finite Cohen-Macaulay type if and only if the infinite poset $I(\Lambda)$ does not contain a poset isomorphic to any of the critical posets of Kleiner $\mathcal{K}_1,\ldots,\mathcal{K}_5$ of Kleiner shown in Tables 5.1.* $\square$

EXAMPLE 7.6. The D-order $\Lambda = \begin{pmatrix} D & \mathfrak{P} & \mathfrak{P}^2 \\ \mathfrak{P}^3 & D & \mathfrak{P} \\ \mathfrak{P}^2 & \mathfrak{P}^3 & D \end{pmatrix}$ is of finite Cohen-Macaulay type, because the infinite poset $I(\Lambda)$ has the form

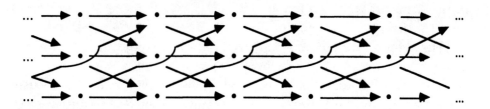

and obviously it does not contain any of the critical posets $\mathcal{K}_1, \ldots, \mathcal{K}_5$ of Kleiner. □

In case D is arbitrary, a connection between peak $I(\Lambda)$-spaces and Cohen-Macaulay modules over Λ is studied in the proof of the main theorem of [55] and by Rump [38, Theorems 4.2, 5.1 and 6.1]. In case $D = K[[t]]$ is a power series K-algebra there exists a covering type functor

$$(7.7) \qquad \qquad \mathbb{G} : \widetilde{I_\Lambda}\text{-}\widetilde{\mathrm{spr}} \longrightarrow \mathrm{CM}(\Lambda)$$

defined by Roggenkamp and Wiedemann in [37] (see also [41, Section 13.3]), where

$$(7.8) \qquad \qquad \widetilde{I_\Lambda} = I(\Lambda) \cup \{p\}$$

is the infinite one-peak poset obtained from $I(\Lambda)$ (4.2) (with the natural $\mathbb{Z}$-action) by attaching a unique maximal element p.

Note that $\mathbb{G}$ is a covering type functor providing us with a simple tool for the study of $\mathrm{CM}(\Lambda)$ by means of peak $\widetilde{I_\Lambda}$-spaces, because of the following result proved in [37] (see also [41, Chapter 13]).

THEOREM 7.9. *Let* Λ *be a tiled* $K[[t]]$*-order* (7.3) *and let* $\widetilde{I_\Lambda} = I(\Lambda) \cup \{p\}$ *be the infinite poset defined above. Denote by* $\widetilde{I_\Lambda}\text{-}\widetilde{\mathrm{spr}} \cong \mathrm{mod}^\bullet_{\mathrm{sp}}(K\widetilde{I_\Lambda})$ *the full subcategory of* $\widetilde{I_\Lambda}\text{-}\mathrm{spr} \cong \mathrm{mod}_{\mathrm{sp}}(K\widetilde{I_\Lambda})$ *formed by the objects* $\mathbf{M} = (M_\beta; M_p)$ *such that* $M_\beta = M_p$ *for* β *sufficiently large. Then the following hold.*

(a) *If* $\mathbf{M}$ *is an indecomposable object in* $\widetilde{I_\Lambda}\text{-}\widetilde{\mathrm{spr}}$*, then the Cohen-Macaulay* Λ*-module* $\mathbb{G}(\mathbf{M})$ *is indecomposable and* $\mathbb{G}(\mathbf{M}) \cong \mathbb{G}(\sigma\mathbf{M})$*, where* $\sigma\mathbf{M}$ *is the shift of* $\mathbf{M}$*.*

(b) *If* $\mathbf{f} : \mathbf{M} \to \mathbf{N}$ *is an irreducible morphism, then* $\mathbb{G}(\mathbf{f})$ *is irreducible. If* $\mathbb{X}$ *is an Auslander-Reiten sequence in* $\widetilde{I_\Lambda}\text{-}\widetilde{\mathrm{spr}}$*, then* $\mathbb{G}(\mathbb{X})$ *is an Auslander-Reiten sequence in* $\mathrm{CM}(\Lambda)$*.*

(c) *Suppose that* $\widetilde{I_\Lambda}$ *does not contain the critical posets of Kleiner shown in Tables 5.1. Then the functor* $\mathbb{G}$ *is dense. Moreover, if* $\mathbf{M}$ *and* $\mathbf{N}$ *are indecomposable peak* $\widetilde{I_\Lambda}$*-spaces and* $\mathbb{G}(\mathbf{M}) \cong \mathbb{G}(\mathbf{N})$*, then* $\mathbf{M} \cong \sigma^t\mathbf{N}$ *for some* $t \in \mathbb{Z}$*.*

(d) *The functor* $\mathbb{G}$ *induces an isomorphism of Auslander-Reiten quivers* $\Gamma(\widetilde{I_\Lambda}\text{-}\widetilde{\mathrm{spr}})/\sigma^{\mathbb{Z}} \cong \Gamma_{CM(\Lambda)}$*.*

The following result was proved in [41, Section 15.12].

THEOREM 7.10. *Let* K *be an algebraically closed field,* Λ *a tiled* $K[[t]]$*-order* (7.3) *and let* $I(\Lambda)$ *be the infinite poset* (7.4) *associated* *with Λ above. If the category* $\mathrm{CM}(\Lambda)$ *is of tame representation type then the poset* $I(\Lambda)$ *does not contain any of the hypercritical posets* $\mathcal{N}_1,\ldots,\mathcal{N}_6$ *of Nazarova.*   □

PROBLEM 7.11 [41, Section 15.12]. *Does the converse of Theorem 7.10 hold?*   □

The answer is positive for Cohen-Macaulay tame tiled $K[t]]$-orders of polynomial growth.

THEOREM 7.12 (Simson 1997). *If K is an algebraically closed field then the tiled $K[[t]]$-order Λ (7.3) is of tame Cohen-Macaulay type and of polynomial growth if and only if the infinite poset $I(\Lambda)$ (7.4) associated with Λ does not contain any of the hypercritical posets $\mathcal{N}_1,\ldots,\mathcal{N}_6$ of Nazarova and does not contain the following poset of Nazarova and Zavadski*

$$\mathcal{NZ}: \quad \overset{\circ\ \ \ \circ}{\underset{\circ\ \ \ \circ\ \ \ \circ\ \ \ \circ}{\downarrow\!\times\!\downarrow}}$$

In this case the covering functor $\mathbb{G}$ (7.7) is dense.

The proof will be published in a subsequent paper.   □

Now we give an example of a tiled D-order Λ for which we are not able to decide if Λ is of tame or wild Cohen-Macaulay type.

EXAMPLE 7.13. Let $D = K[[t]]$ and let Λ be the tiled D-order of the form

$$\Lambda = \begin{pmatrix} D & \mathfrak{P} & \mathfrak{P}^2 & \mathfrak{P}^3 \\ \mathfrak{P} & D & \mathfrak{P} & \mathfrak{P}^2 \\ \mathfrak{P}^3 & \mathfrak{P}^2 & D & \mathfrak{P} \\ \mathfrak{P}^4 & \mathfrak{P}^3 & \mathfrak{P} & D \end{pmatrix}$$

The infinite poset $I(\Lambda)$ (7.4) of Λ has the form

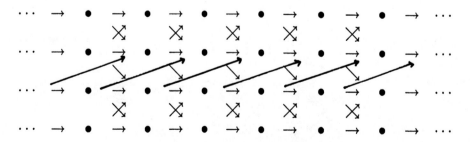

Note that $I(\Lambda)$ contains the poset $\mathcal{NZ}$, but it does not contains the hypercritical posets of Nazarova (Tables 5.2). It follows from [41,

Chapter 15] that every finite subposet of $I(\Lambda)$ is representation-tame, and $\widetilde{I_\Lambda}\text{-}\widetilde{\mathrm{spr}}$ is tame of non-polynomial growth. One can show that the covering functor (7.7) for our order Λ is not dense. Unfortunately we are not able to prove that the order Λ is of tame Cohen-Macaulay type. $\qquad\qquad\qquad\qquad\qquad\qquad\qquad\qquad\qquad\qquad\qquad\qquad\qquad\Box$

Let us finish this section by recalling from [26] a useful connection between the covering functor (7.7) and the completion functor (see also [41, Section 13.4]).

For this purpose we suppose that $D = K[[t]]$ and Λ is the tiled D-order (7.3). We associate with Λ the $\mathbb{Z}$-graded $K[t]$-subalgebra

$$(7.14) \qquad T_\Lambda = \begin{pmatrix} K[t] & t^{m_{12}}K[t] & \cdots & t^{m_{1n}}K[t] \\ t^{m_{21}}K[t] & K[t] & \cdots & t^{m_{2n}}K[t] \\ \vdots & \vdots & \ddots & \vdots \\ t^{m_{n1}}K[t] & t^{m_{n2}}K[t] & \cdots & K[t] \end{pmatrix}$$

of $\mathrm{M}_n(L)$, where $L = K[t, t^{-1}]$ is the Laurent polynomial algebra. This means that $T_\Lambda \subseteq \Lambda$ has the form (7.3) with $D = K[[t]]$, $\mathfrak{P}^{m_{ij}}$ and $K[t]$, $t^{m_{ij}}K[t]$ interchanged. The $\mathbb{Z}$-grading of T_Λ is defined as follows. We consider the Laurent polynomial algebra $L = K[t, t^{-1}] = \bigoplus_{r\in\mathbb{Z}} Kt^r$ as a $\mathbb{Z}$-graded algebra by viewing t as a homogeneous element of degree one. Then we view the matrix algebra $\mathrm{M}_n(L) = \bigoplus_{r\in\mathbb{Z}} \mathrm{M}_n(K)t^r$ as a $\mathbb{Z}$-graded algebra with the grading induced by the grading of L. The $\mathbb{Z}$-grading of T_Λ is induced by the $\mathbb{Z}$-grading of $\mathrm{M}_n(L)$. The graded $K[t]$-algebra

$$T_\Lambda = \bigoplus_{m\in\mathbb{Z}} (T_\Lambda)_m$$

is called the **$\mathbb{Z}$-graded tiled $K[t]$-algebra associated with** Λ.

If we denote by e_{ij}, $i, j = 1, \ldots, n$, the elementary standard unit matrices of $\mathrm{M}_n(L)$ and we set $e_j = e_{jj}$, then $\varepsilon = \{e_1, \ldots, e_n\} \subseteq (T_\Lambda)_0$ is a complete set of primitive orthogonal idempotents of the algebra $(T_\Lambda)_0$. If we view $K[t]$ as a subalgebra of T_Λ embedded diagonally then the homogeneous elements $e_{ij}t^{r+m_{ij}}$ of degree $r+m_{ij}$ form a $K[t]$-basis of T_Λ. The discussion above together with the arguments given in [26] yields.

THEOREM 7.15. *Let Λ be a tiled $K[[t]]$-order (7.3) and $T_\Lambda \subseteq \Lambda$ be the associated $\mathbb{Z}$-graded tiled $K[t]$-algebra (7.14).*

(a) T_Λ is a $\mathbb{Z}$-graded Cohen-Macaulay algebra of graded Krull dimension one and there is an isomorphism $\widehat{T}_\Lambda \cong \Lambda$ of $K[[t]]$-algebras, where $\widehat{T}_\Lambda$ is the completion of T_Λ with respect to the (t)-topology.

(b) *If* $\mathrm{CM}^{\mathbb{Z}}(T_\Lambda)$ *is the category of* $\mathbb{Z}$-*graded Cohen-Macaulay* G_Λ-*modules and* $\mathbb{G}$ *is the covering functor* (7.7) *then there are equivalences of categories* $\mathrm{CM}^{\mathbb{Z}}(T_\Lambda) \cong \widetilde{I_\Lambda}\text{-}\overline{\mathrm{spr}}$ *and* $\mathrm{CM}(\widehat{T}_\Lambda) \cong \mathrm{CM}(\Lambda)$ *such that the diagram*

$$
\begin{array}{ccc}
\mathrm{CM}^{\mathbb{Z}}(T_\Lambda) & \xrightarrow{\;\widehat{(\cdot)}\;} & \mathrm{CM}(\widehat{T}_\Lambda) \\
\Big\downarrow{\scriptstyle\cong} & & \Big\downarrow{\scriptstyle\cong} \\
\widetilde{I_\Lambda}\text{-}\widetilde{\mathrm{spr}} & \xrightarrow{\;\mathbb{G}\;} & \mathrm{CM}(\Lambda)
\end{array}
$$

is commutative, where $\widehat{(\cdot)}$ *is the completion functor with respect to the* (t)-*topology* [3]. ∎

The idea presented in Theorem 7.15 applies also to a class of pull-back or subamalgams of tiled $K[[t]]$-orders. We outline it in the following example.

EXAMPLE 7.16. Let $D = K[[t]]$ and let Λ be the tiled $K[[t]]$-order

$$
\Lambda = \begin{pmatrix} D & \mathfrak{P} \\ \mathfrak{P}^2 & D \end{pmatrix}
$$

The infinite one-peak poset $\widetilde{I_\Lambda}$ (7.8) of Λ has the form

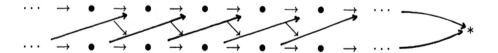

It is clear that $\widetilde{I_\Lambda}$ does not contain the critical posets of Kleiner shown in Tables 5.1, and according to Theorem 7.5 the order Λ is of finite Cohen-Macaulay type.

Consider now the push-out $K[[t]]$-suborder Ω of $\Lambda \times \Lambda$ consisting of all pairs (λ, λ') of two by two matrices $\lambda, \lambda' \in \Lambda$ such that $\lambda_{22} - \lambda'_{11} \in (t)$. Let T_Λ be the $\mathbb{Z}$-graded Cohen-Macaulay $K[t]$-algebra (7.14) associated with Λ. Denote by $T_\Omega \subseteq \Omega$ the $\mathbb{Z}$-graded $K[t]$-subalgebra of $T_\Lambda \times T_\Lambda \subseteq \Lambda \times \Lambda$ consisting of all pairs (λ, λ') of two by two matrices $\lambda, \lambda' \in \Lambda$ such that $\lambda_{22} - \lambda'_{11} \in (t)$.

One can show that T_Ω is a $\mathbb{Z}$-graded Cohen-Macaulay algebra of graded Krull dimension one and there is an isomorphism $\widehat{T}_\Omega \cong \Omega$ of $K[[t]]$-algebras, where $\widehat{T}_\Omega$ is the completion of T_Ω with respect to the

(t)-topology. Moreover there is a commutative diagram

$$
\begin{array}{ccc}
\mathrm{CM}^{\mathbb{Z}}(T_{\Omega}) & \xrightarrow{\widehat{(\cdot)}} & \mathrm{CM}(\widehat{T_{\Omega}}) \\
\Big\downarrow{\cong} & & \Big\downarrow{\cong} \\
\mathrm{mod}^{\bullet}_{sp}(K\widetilde{I_{\Omega}}) & \longrightarrow & \mathrm{CM}(\Omega)
\end{array}
$$

where $\widehat{(\cdot)}$ is the completion functor with respect to the (t)-topology [3], $\mathrm{mod}^{\bullet}_{sp}(K\widetilde{I_{\Omega}})$ is the category of socle projective modules defined in Section 4, the vertical functors are equivalences of categories and $K\widetilde{I_{\Omega}}$ is the path K-algebra of the two-peak infinite quiver $\widetilde{I_{\Omega}}$ obtained from the infinite two-peak poset

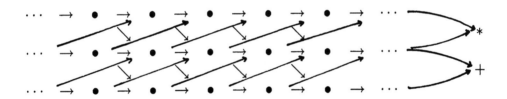

by attaching all cross zero-relations. Note that the quiver $\widetilde{I_{\Omega}}$ is obtained from the disjoint union $\widetilde{I_{\Lambda}} \uplus \widetilde{I_{\Lambda}}$ of two copies of the infinite one-peak poset $\widetilde{I_{\Lambda}}$ by making the identification of the bottom infinite chain of one copy of $\widetilde{I_{\Lambda}}$ with the top infinite chain of the second copy of $\widetilde{I_{\Lambda}}$. It follows from [42, Proposition 2.8] that $K\widetilde{I_{\Omega}}$ is the incidence algebra $K(\widetilde{I_{\Lambda}} \uplus \widetilde{I_{\Lambda}})_{\rho}$ (in the sense of Section 5) of the poset $\widetilde{I_{\Lambda}} \uplus \widetilde{I_{\Lambda}}$ equipped with the equivalence relation ρ defined by the identification described above.

It is easy to see that the two-peak quiver $\widetilde{I_{\Omega}}$ with zero-relations does not contain any of the two-peak critical forms presented in [42, pp.3570–3571]. Then by applying to $K\widetilde{I_{\Omega}}$ the arguments used in [42, Theorem 4.30] one can show that the category $\mathrm{mod}^{\bullet}_{sp}(K\widetilde{I_{\Omega}})$ has finitely many indecomposable modules up to $\mathbb{Z}$-shift, and consequently, the category $\mathrm{CM}^{\mathbb{Z}}(T_{\Omega})$ has finitely many indecomposable $\mathbb{Z}$-graded modules up to $\mathbb{Z}$-shift. From this, by applying a covering type properties of the completion functor [3], we can conclude that the order $\Omega \cong \widehat{T_{\Omega}}$ is of finite Cohen-Macaulay type. $\qquad\square$

8   Cohen-Macaulay type of subamalgams of tiled D-orders and a Tits quadratic form

Throughout K is an algebraically closed field and D is a complete discrete valuation domain which is a K-algebra such that $D/\mathfrak{P} \cong K$.

Our aim of this section is to define for a class of D-orders $\Lambda^\bullet$ a positive integer $m_{\Lambda^\bullet}$, a Tits reduced quadratic form $q_{\Lambda^\bullet} : \mathbb{Z}^{m_{\Lambda^\bullet}} \longrightarrow \mathbb{Z}$ and a system

$$* : G_{\Lambda^\bullet}(v) \times \mathcal{M}_{\Lambda^\bullet}(v) \longrightarrow \mathcal{M}_{\Lambda^\bullet}(v), \quad v \in \mathbb{Z}^{m_{\Lambda^\bullet}},$$

of algebraic groups $G_{\Lambda^\bullet}(v)$ acting on irreducible algebraic K-varieties $\mathcal{M}_{\Lambda^\bullet}(v)$ in such a way that $\Lambda^\bullet$ is of tame Cohen-Macaulay type if and only if $\dim G_{\Lambda^\bullet}(v) \geq \dim \mathcal{M}_{\Lambda^\bullet}(v)$, or equivalently, if and only if the quadratic form $q_{\Lambda^\bullet}$ is weakly non-negative, that is, $q_{\Lambda^\bullet}(v) \geq 0$ for all vectors $v \in \mathbb{N}^{m_{\Lambda^\bullet}}$.

In particular we do it for three-partite subamalgams of tiled D-orders $\Lambda^\bullet$ defined as follows. Denote by $\mathbb{M}_n(D)$ the full $n \times n$-matrix ring with coefficients in D. Suppose that $n, n_1, n_2 > 0$ and $n_3 \geq 0$ are natural numbers, and Λ is a tiled D-suborder of $\mathbb{M}_n(D)$ of the form

$$(8.1) \qquad \Lambda = \left. \begin{pmatrix} D & {}_1D_2 & \cdots & {}_1D_{n-1} & {}_1D_n \\ \mathfrak{P} & D & \cdots & {}_2D_{n-1} & {}_2D_n \\ \vdots & \vdots & \ddots & \vdots & \vdots \\ \mathfrak{P} & \mathfrak{P} & \cdots & D & {}_{n-1}D_n \\ \mathfrak{P} & \mathfrak{P} & \cdots & \mathfrak{P} & D \end{pmatrix} \right\} n$$

such that
   (a) $_iD_j$ is either D or $\mathfrak{P}$,
   (b) Λ admits a three-partition

$$(8.2) \qquad \Lambda = \left(\begin{array}{c|c|c} \Lambda_1 & \mathcal{X} & \mathbb{M}_{n_1}(D) \\ \hline \mathbb{M}_{n_3 \times n_1}(\mathfrak{P}) & \Lambda_3 & \mathcal{Y} \\ \hline \mathbb{M}_{n_1}(\mathfrak{P}) & \mathbb{M}_{n_1 \times n_3}(\mathfrak{P}) & \Lambda_2 \end{array} \right) \begin{array}{l} {\scriptstyle \}n_1} \\ {\scriptstyle \}n_3} \\ {\scriptstyle \}n_2} \end{array}$$

where $\Lambda_2 = \Lambda_1$, $n_1 = n_2$, $n_1 + n_2 + n_3 = n$ and Λ_3 is a hereditary

$n_3 \times n_3$-matrix D-order

$$(8.3) \qquad \Lambda_3 = \left.\begin{pmatrix} D & D & \dots & D & D \\ \mathfrak{P} & D & \dots & D & D \\ \vdots & \ddots & \ddots & \vdots & \vdots \\ \mathfrak{P} & \mathfrak{P} & \dots & D & D \\ \mathfrak{P} & \mathfrak{P} & \dots & \mathfrak{P} & D \end{pmatrix}\right\} n_3$$

In particular $_iD_j = D$ holds in Λ for $1 \le i \le n_1$ and $n_1 + n_3 < j \le n$.

Note that $1 = \varepsilon_1 + \varepsilon_3 + \varepsilon_2$, where ε_1, ε_3 and ε_2 are the matrix idempotents of Λ corresponding to the identity elements of Λ_1, Λ_3 and Λ_2, respectively. By a **three-partite subamalgam** of Λ we shall mean the D-suborder

$$\Lambda^\bullet = \{\lambda = [\lambda_{ij}] \in \Lambda; \quad \varepsilon_1 \lambda \varepsilon_1 - \varepsilon_2 \lambda \varepsilon_2 \in \mathbb{M}_{n_1}(\mathfrak{P})\}$$

of Λ consisting of all matrices $\lambda = [\lambda_{ij}]$ of Λ such that the left upper corner $n_1 \times n_1$ submatrix $\varepsilon_1 \lambda \varepsilon_1$ of λ is congruent modulo $\mathbb{M}_{n_1}(\mathfrak{P})$ to the right lower corner $n_1 \times n_1$ submatrix $\varepsilon_2 \lambda \varepsilon_2$ of λ.

To any such a D-order $\Lambda^\bullet$ we associate $m_{\Lambda^\bullet} = n_1 + 2n_3 + 2$ and the reduced Tits quadratic form

$$(8.4) \qquad q_{\Lambda^\bullet} : \mathbb{Z}^{n_1 + 2n_3 + 2} \longrightarrow \mathbb{Z}$$

in the indeterminates $x_*, x_+, x_1, \dots, x_{n_1+n_3}, \overline{x}_{n_1+1}, \dots, \overline{x}_{n_1+n_3}$ defined by the formula

$$q_{\Lambda^\bullet}(x_1, \dots, x_{n_1+n_3}, \overline{x}_{n_1+1}, \dots, \overline{x}_{n_1+n_3}, x_*, x_+) =$$

$$= x_*^2 + x_+^2 + \sum_{j=1}^{n_1+n_3} x_j^2 + \sum_{j=n_1+1}^{n_1+n_3} \overline{x}_j^2$$

$$+ \sum_{\substack{iD_j=D \\ 1 \le i < j \le n_1+n_3}} x_i x_j + \sum_{s<t} \overline{x}_s \overline{x}_t + \sum_{\substack{tD_s=D \\ n_1 < t \le n_1+n_3 < s}} x_{s-n_1-n_3} \overline{x}_t$$

$$- x_+\Big(\sum_{j=1}^{n_1+n_3} x_j\Big) - x_*\Big(\sum_{j=1}^{n_1} x_j + \sum_{j=n_1+1}^{n_1+n_3} \overline{x}_j\Big).$$

In [49], [48] we also associate with $\Lambda^\bullet$ a poset (i.e. a partially ordered set) $I_{\Lambda^\bullet}$ consisting of $m_{\Lambda^\bullet} = n_1 + 2n_3 + 2$ elements and having precisely two maximal elements p_1, p_2, a set $3_{\Lambda^\bullet} \subseteq \{(i,j); i \prec j \text{ in } I_{\Lambda^\bullet}\}$ of zero-relations in $I_{\Lambda^\bullet}$ and an algebraic group action

$$(8.5) \qquad * : G_{\Lambda^\bullet}(v) \times \mathcal{M}_{\Lambda^\bullet}(v) \longrightarrow \mathcal{M}_{\Lambda^\bullet}(v)$$

$v \in \mathbb{N}^{m_{\Lambda^\bullet}} = \mathbb{N}^{n_1+2n_3+2}$, defined as follows. The set $\mathcal{M}_{\Lambda^\bullet}(v)$ is the irreducible K-variety of all block matrices of the form (compare with [43])

$$
A = \begin{array}{|c|c|c|c|}
\hline
A_{1p_1} & A_{2p_1} & \cdots & A_{mp_1} \\
\hline
A_{1p_2} & A_{2p_2} & \cdots & A_{mp_2} \\
\hline
\end{array}
\begin{array}{l} {\scriptstyle\}}v(p_1) \\ {\scriptstyle\}}v(p_2) \end{array}
$$
$$
\underbrace{\phantom{A_{1p_1}}}_{v(1)} \underbrace{\phantom{A_{2p_1}}}_{v(2)} \quad \cdots \quad \underbrace{\phantom{A_{mp_1}}}_{v(m)}
$$

with coefficients in K, where $m = n_1 + 2n_3$, $A_{ip} = 0$ if either $i \not\prec p$, or $(i,p) \in 3_{\Lambda^\bullet}$, $p \in \{p_1, p_2\}$. We define $G_{\Lambda^\bullet}(v)$ to be the algebraic group

$$
G_{\Lambda^\bullet}(v) = H_{\Lambda^\bullet}(v) \times \mathrm{Gl}(v(p_1), K) \times \mathrm{Gl}(v(p_2), K)
$$

where $H_{\Lambda^\bullet}(v)$ is a group consisting of all matrices of the form

$$
h = \begin{pmatrix}
h_{11} & h_{12} & \cdots & h_{1m} \\
0 & h_{22} & \cdots & h_{2m} \\
\vdots & \vdots & \ddots & \vdots \\
0 & 0 & \cdots & h_{mm}
\end{pmatrix} \in \mathrm{Gl}(v(1) + \cdots + v(m), K)
$$

where $h_{ii} \in \mathrm{Gl}(v(i), K)$ and $h_{ij} = 0$ if either i and j are not comparable in $I_{\Lambda^\bullet}$, or $(i,j) \in 3_{\Lambda^\bullet}$. We suppose that the partial order in $I_{\Lambda^\bullet}$ is such that $i \prec j$ implies that $i < j$ in the natural order.

The multiplication of two matrices $h = [h_{ij}]$ and $h' = [h'_{ij}]$ in $H_{\Lambda^\bullet}(v)$ is the matrix $h'' = [h''_{ij}]$, where

$$
(h \cdot h')_{ij} = h''_{ij} = \begin{cases} \sum_{i \preceq s \preceq j} h_{is} h'_{sj} & \text{if } i \preceq j \text{ and } (i,j) \notin 3_{\Lambda^\bullet} \\ 0 & \text{if } i \not\preceq j \text{ or } (i,j) \in 3_{\Lambda^\bullet}. \end{cases}
$$

It is clear that the group $H_{\Lambda^\bullet}(v)$ is parabolic. Note that in case the set $3_{\Lambda^\bullet}$ is not empty the group $H_{\Lambda^\bullet}(v)$ is not a subgroup of the group $\mathrm{Gl}(v(1) + \cdots + v(m), K)$. The action (8.5) is defined by the formula $(h, (g_1, g_2)) * A = \mathrm{diag}(g_1, g_2) \cdot A \circ h^{-1}$, where $h \in H_{\Lambda^\bullet}(v)$, $g_j \in \mathrm{Gl}(v(p_j), K)$, $A \circ h^{-1} = A'$ is a partitioned matrix with $A'_{sp} = (Ah^{-1})_{sp}$, if $s \preceq p$ and $(s,p) \notin 3$, and $A'_{sp} = 0$, if $s \not\preceq p$ or $(s,p) \in 3$.

Our main results of this section are the following two theorems.

THEOREM 8.6. *Let $K, D, \mathfrak{P}, \Lambda, \Lambda^\bullet, q_{\Lambda^\bullet}$ and the group action $*:$ $G_{\Lambda^\bullet}(v) \times \mathcal{M}_{\Lambda^\bullet}(v) \to \mathcal{M}_{\Lambda^\bullet}(v)$, $v \in \mathbb{N}^{m_{\Lambda^\bullet}} = \mathbb{N}^{n_1+2n_3+2}$, be as above. If the part $\mathcal{X}$ or the part $\mathcal{Y}$ of the D-order Λ in (8.2) consists of matrices with coefficients in $\mathfrak{P}$ then the following conditions are equivalent.*

(a) The D-order $\Lambda^\bullet$ is of tame Cohen-Macaulay type.

(b) The Tits reduced quadratic form $q_{\Lambda^\bullet} : \mathbb{Z}^{n_1+2n_3+2} \to \mathbb{Z}$ is weakly non-negative, that is, $q_{\Lambda^\bullet}(v) \geq 0$ for all vectors $v \in \mathbb{N}^{m_{\Lambda^\bullet}}$.

(c) $\dim G_{\Lambda^\bullet}(v) \geq \dim \mathcal{M}_{\Lambda^\bullet}(v)$ *for every* $v \in \mathbb{N}^{n_1+2n_3+2}$, *where* $\dim$ *is the variety dimension.*

(d) *For every vector* $v \in \mathbb{N}^{n_1+2n_3+2}$ *the constructible subset* $\mathrm{ind}\mathcal{M}_{\Lambda^\bullet}(v)$ *of* $\mathcal{M}_{\Lambda^\bullet}(v)$ *consisting of indecomposable* $G_{\Lambda^\bullet}(v)$-*orbits contains a constructible subset* $C(v)$ *such that* $\mathcal{M}_{\Lambda^\bullet}(v) = G_{\Lambda^\bullet}(v) * C(v)$ *and* $\dim C(v) \leq 1$.

(e) *If* $R = KI_{\Lambda^\bullet}/\mathfrak{Z}_{\Lambda^\bullet}$ *is the factor of the incidence* K-*algebra* $KI_{\Lambda^\bullet}$ *of the poset* $I_{\Lambda^\bullet}$ *modulo the ideal* $\mathfrak{Z}_{\Lambda^\bullet}$ *of zero-relations then*

$$\dim_K \mathrm{Ext}^1_R(U_1, U_2) \leq 2$$

for any pair of modules U_1, U_2 *in* $\mathrm{mod}(R)$ *satisfying the following conditions:*

(i) $\mathrm{End}(U_1) \cong \mathrm{End}(U_2) \cong K$,

(ii) $\mathrm{Hom}_R(U_1, U_2) = 0$ *and* $\mathrm{Hom}_R(U_2, U_1) = 0$,

(iii) *the kernel of the projective cover* $P(U_j) \to U_j$ *of* U_j *is a semisimple projective* R-*module for* $j = 1, 2$.

(f) *Either* $n_3 = 0$ *and the* D-*order* Λ_1 *in* (8.2) *does not contain minor* D-*suborders of one of the forms*

$$\Delta_0 = \begin{pmatrix} D & \mathfrak{P} & \mathfrak{P} \\ \mathfrak{P} & D & \mathfrak{P} \\ \mathfrak{P} & \mathfrak{P} & D \end{pmatrix}, \ \Delta_1 = \begin{pmatrix} D & \mathfrak{P} & D \\ \mathfrak{P} & D & \mathfrak{P} \\ \mathfrak{P} & \mathfrak{p} & D \end{pmatrix}$$

$$\Delta_2 = \begin{pmatrix} D & D & \mathfrak{P} \\ \mathfrak{P} & D & \mathfrak{P} \\ \mathfrak{P} & \mathfrak{p} & D \end{pmatrix}, \ \Delta_3 = \begin{pmatrix} D & \mathfrak{P} & \mathfrak{P} \\ \mathfrak{P} & D & D \\ \mathfrak{P} & \mathfrak{p} & D \end{pmatrix}$$

or else $n_3 \geq 1$, Λ_1 *is hereditary of the form* (8.3) *and the* D-*order* $\Lambda^\bullet$ *does not contain (up to domination in the sense of* [49], [50], [51]*) as a three-partite minor* D-*suborders any of* 17 *three-partite defined forms* $\Omega_1, \ldots, \Omega_{17}$ *of size at most* 15×15 *presented in* [50, Section 1] *and* [51, Section 7]*).*

In this case the structure of the Auslander-Reiten quiver $\Gamma_{CM(\Lambda)}$ of the category $\mathrm{CM}(\Lambda^\bullet)$ is described (see [49]).

THEOREM 8.7. *Under the notation and assumption of Theorem 8.6 the following conditions are equivalent.*

(a) *The* D-*order* $\Lambda^\bullet$ *is of tame Cohen-Macaulay type and of polynomial growth.*

(b) *Either* $n_3 \geq 1$, *the* D-*order* Λ_1 *in* (8.2) *is hereditary of the form* (8.3) *and the* D-*order* $\Lambda^\bullet$ *does not contain as a three-partite minor* D-*suborder any of the* 17 *three-partite orders* $\Omega_1, \ldots, \Omega_{17}$ *(see* [50, Section 1] *and* [51, Section 7]*) up to domination, or else* $n_3 = 0$

and there exists at most one pair (i,j) *such that* $1 \leq i < j \leq n_1$ *and* $_iD_j = \mathfrak{P}$ *in* Λ (8.1).

A structure of three-partite subamalgam D-orders $\Lambda^\bullet$ which are of tame Cohen-Macaulay type and of non-polynomial growth is completely described in [51, Section 6].

Our main tools in the proof of Theorems 8.6 and 8.7 are the reduction functors defined in [18], [34], [42], [49], the Galois covering technique developed in [40] and [42], elementary algebraic geometry arguments (see [48, Section 3]) and the tameness criterion for posets given in [21]. For details we refer to [48], [50] and [51].

References

[1] F.W. Anderson and K.R. Fuller, *"Rings and Categories of Modules"*, Graduate Texts in Mathematics 13, Springer, Berlin, 1974.

[2] M. Auslander and I. Reiten, The Cohen-Macaulay type of Cohen-Macaulay rings, *Adv. in Math.* **73**(1989), 1-23.

[3] M. Auslander and I. Reiten, Graded modules and their completions, *in "Topics in Algebra, Part I: Rings and Representations of Algebras"*, Banach Center Publications, Vol. 26, PWN Warszawa 1990, pp. 181-192.

[4] M. Auslander, I. Reiten and S. Smalø, *"Representation Theory of Artin Algebras"*, Cambridge Studies in Advanced Mathematics 36, Cambridge University Press, 1995.

[5] V. M. Bondarenko and A. G. Zavadski, Posets with an equivalence relation of tame type and of finite growth, *in* Proceedings of the Tsukuba International Conference on Representations of Finite Dimensional Algebras, *Canadian Mathematical Society Conference Proceedings,* **11**(1991), pp. 67-88.

[6] W. Crawley-Boevey, Modules of finite length over their endomorphism rings, in *"Representations of Algebras and Related Topics"*, London Math. Soc. Lecture Notes **168** (1992), 127-184.

[7] C.W. Curtis and I. Reiner, *"Methods of Representation Theory"*, Vol. I, Wiley Classics Library Edition, New York, 1990.

[8] E. Dieterich, Group rings of wild representation type, *Math. Ann.*, **266**(1983), 1-22.

[9] E. Dieterich, Representation types of group rings over complete discrete valuation rings, *in* Lecture Notes in Math., 1142(1985), pp. 112-125.

[10] E. Dieterich, Tame orders, *in "Topics in Algebra, Part I: Rings and Representations of Algebras"*, Banach Center Publications, Vol. 26, PWN Warszawa, 1990, pp. 233-261.

[11] E. Dieterich and A. Wiedemann, The Auslander-Reiten quiver of a simple curve singularity, *Trans. Amer. Math. Soc.*, **294**(1986), 455-475.

[12] P. Dowbor and S. Kasjan, Galois covering technique and tame non-simply connected posets of polynomial growth, *J. Pure Appl. Algebra* 1999, in press.

[13] Y. A. Drozd, Cohen-Macaulay modules and vector bundles, Proc. Euroconference "Interactions between Ring Theory and Representations of Algebras", Murcia, 12-17 January 1998, this volume.

[14] Y. A. Drozd and M. G. Greuel, Tame-wild dichotomy for Cohen-Macaulay modules, Math. Ann. **294**(1992), 387-394.

[15] J. Drozd, A. G. Zavadski and V. V. Kirichenko, Matrix problems and integral representations, *Izv. Akad. Nauk SSSR* **38**(1974) 291-293 (in Russian).

[16] P. Gabriel, Indecomposable representations I, *Manuscripta Math.* **6**(1972), 71-103.

[17] P. Gabriel and A. V. Roiter, *"Representations of Finite Dimensional Algebras"*, Algebra VIII, Encyclopedia of Math. Sc., Vol 73, Springer-Verlag, 1992.

[18] E. L. Green and I. Reiner, Integral representations and diagrams, *Michigan Math. J.* **25**(1978), 53-84. (1988), 312-336.

[19] S. Kasjan, A criterion for polynomial growth of $\tilde{A}_n$-free two-peak posets of tame prinjective type, *preprint,* Toruń, 1997.

[20] S. Kasjan and D. Simson, Fully wild prinjective type of posets and their quadratic forms, *J. Algebra* **172**(1995), 506-529.

[21] S. Kasjan and D. Simson, Tame prinjective type and Tits form of two-peak posets I, *J. Pure Appl. Algebra* **106**(1996), 307-330.

[22] S. Kasjan and D. Simson, Tame prinjective type and Tits form of two-peak posets II, J. Algebra **187**(1997), 71-96.

[23] M. Kleiner, Partially ordered sets of finite type, *in* Zap. Nauchn. Sem. Leningrad. Otdel. Mat. Inst. Steklov. (LOMI), **28**(1972), 32-41 (in Russian).

[24] J. Kosakowska and D. Simson, On Tits form and prinjective representations of posets of finite prinjective type, *Comm. Algebra*, **26**(1998), 1613-1623.

[25] J. Kosakowska and D. Simson, Posets of infinite prinjective type and embeddings of Kronecker modules into the category of prinjective peak I-spaces, Preprint, Toruń, 1998.

[26] H. Lenzing and D. Simson, Cohen-Macaulay modules, the completion functor, representations of posets and rational tiled orders, preprint.

[27] L. A. Nazarova, Partially ordered sets of infinite type, *Izv. Akad. Nauk SSSR* **39**(1975), 963-991 (in Russian).

[28] L. A. Nazarova and A.V. Roiter, Representations of partially ordered sets, *in* Zap. Nauchn. Sem. Leningrad. Otdel. Mat. Inst. Steklov. (LOMI), **28**(1972), 5-31 (in Russian).

[29] L. A. Nazarova and V. A. Roiter, Representations of completed posets, *Comment. Math. Helv.* **63**(1988), 498-526.

[30] J. A. de la Peña and D. Simson, Prinjective modules, reflection functors, quadratic forms and Auslander-Reiten sequences, *Trans. Amer. Math. Soc.* **329**(1992), 733-753.

[31] I. Reiner, *"Maximal Orders"*, Academic Press, London - New York - San Francisco, 1975.

[32] I. Reiner, Topics in integral representation theory, *in* Lecture Notes in Math., Springer-Verlag, Vol. 744, 1979.

[33] C. M. Ringel, *"Tame Algebras and Integral Quadratic Forms"*, Lecture Notes in Math. Vol. 1099, Springer-Verlag, Berlin-Heidelberg-New York-Tokyo, 1984.

[34] C. M. Ringel and K. W. Roggenkamp, Diagrammatic methods in the representation theory of orders, *J. Algebra* **60**(1979), 11-42.

[35] K. W. Roggenkamp, The construction of almost split sequences for integral group rings and orders, *Commun. Algebra* **5**(1977), 1363-1373.

[36] K. W. Roggenkamp, Indecomposable representations of orders, in *"Topics in Algebra, Part I: Rings and Representations of Algebras"*, Banach Center Publications, Vol. 26, PWN Warszawa, 1990, pp. 449-491.

[37] K. W. Roggenkamp and A. Wiedemann, Auslander-Reiten quivers of Schurian orders, *Comm. Algebra* **12**(1984), 2525-2578.

[38] W. Rump, Enlacements and representation theory of completely reducible orders, Lecture Notes in Math., 1178 (1986), pp. 272-308

[39] D. Simson, Socle reductions and socle projective modules, *J. Algebra*, **103**(1986), 18-68.

[40] D. Simson, Representations of bounded stratified posets, coverings and socle projective modules, in *"Topics in Algebra, Part I: Rings and Representations of Algebras"*, Banach Center Publications, Vol. 26, PWN Warszawa, 1990, pp. 499-533.

[41] D. Simson, *"Linear Representations of Partially Ordered Sets and Vector Space Categories"*, Algebra, Logic and Applications, Vol. 4, Gordon & Breach Science Publishers, New York, 1992.

[42] D. Simson, Right peak algebras of two-separate stratified posets, their Galois coverings and socle projective modules, *Comm. Algebra* **20**(1992), 3541-3591.

[43] D. Simson, On representation types of module subcategories and orders, *Bull. Pol. Acad. Sci., Math.*, **41**(1993), 77-93.

[44] D. Simson, Posets of finite prinjective type and a class of orders, *J. Pure Appl. Algebra* **90**(1993), 77-103.

[45] D. Simson, A reduction functor, tameness and Tits form for a class of orders, *J. Algebra* **174**(1995), 430-452.

[46] D. Simson, Socle projective representations of partially ordered sets and Tits quadratic forms with application to lattices over orders, in Proceedings of the Conference on Abelian Groups and Modules, Colorado Springs, August 1995, *Lecture Notes in Pure and Appl. Math.*, Vol. 182, 1996, pp. 73-111.

[47] D. Simson, Prinjective modules, propartite modules, representations of bocses and lattices over orders, J. Math. Soc. Japan, **49**(1997), 31-68.

[48] D. Simson, Representation types, Tits reduced quadratic forms and orbit problems for lattices over orders, *Proc. AMS-IMS-SIAM Joint Summer Conference "Trends in the Representation Theory of Finite Dimensional Algebras", University of Washington, Seattle, July 1997*, Contemporary Math. 229(1998), 307-342.

[49] D. Simson, Three-partite subamalgams of tiled orders of finite lattice type, *J. Pure Appl. Algebra* **138**(1999), 151-184.

[50] D. Simson, A reduced Tits quadratic form and tameness of three-partite subamalgams of tiled orders, Trans. Amer. Math. Soc., in press.

[51] D. Simson, Tame three-partite subamalgams of tiled orders of polynomial growth, Colloq. Math. **82**(1999), in press.

[52] Y. Yoshino, *"Cohen-Macaulay Modules over Cohen-Macaulay Rings"*, London Math. Soc. Lecture Notes Series, Vol 146, Cambridge University Press, 1990.

[53] A. G. Zavadski, An algorithm for posets with an equivalence relation, *in* Proceedings of the Tsukuba International Conference on Representations of Finite Dimensional Algebras, *Canadian Mathematical Society Conference Proceedings,* **11**(1991), pp. 299-322.

[54] A. G. Zavadski and V. V. Kirichenko, Torsion-free modules over prime rings, *in* Zap. Naychn. Sem. Leningrad. Otdel. Mat. Inst. Steklov. (LOMI), **57**(1976), 100-116 (in Russian).

[55] A. G. Zavadski and V. V. Kirichenko, Semimaximal rings of finite type, *Mat. Sbornik* **103**(1977), 323-345 (in Russian).

On Associated Primes and Weakly Associated Primes

BORIS ŠIROLA

DEPARTMENT OF MATHEMATICS, UNIVERSITY OF ZAGREB, BIJENIČKA 30, 10000 ZAGREB, CROATIA
E-mail address: sirola@math.hr

INTRODUCTION

Let $\mathcal{R}$ be a ring and M an $\mathcal{R}$-module. If $\mathcal{R}$ is commutative, the importance of the associated prime ideals of M for studying of the relationship between the prime ideal structure of $\mathcal{R}$ and the structure of M (subquotients, composition series, characteristic cycles, etc.) is quite clear. As usual, for $\mathcal{R}$ noncommutative the situation becomes more complicated. One possibility is to consider the well known associated primes Ass M of M, which generalizes the above "commutative" notion in case when the ring is (left) Noetherian and the module is finitely generated. But then the above mentioned relationship cannot be investigated in a satisfactory way via these prime ideals. In order to obtain a better treatment of this problem some new prime ideals were introduced, such as e.g. the affiliated primes Aff M of M; see J. T. Stafford [St1, St2]. Here we propose a new definition for associated primes as a natural generalization of the classical definition from commutative theory. We call these primes w-associated (weakly assoc.) to M and denote them by w-Ass M. Let us emphasize that in the most interesting situation when $\mathcal{R}$ is (left) Noetherian and M finitely generated it turns out that the w-associated primes are nothing else but the annihilator primes Annspec M of M. Also, in [Š1] we introduce a set of primes which is larger and more accessible (easier for computing) than w-Ass M. These primes are called the foundation primes of M and denoted by Fnd M. Although larger, this set is not too big and so we can successfully study w-Ass M via Fnd M; for some interesting rings we can completely determine the set w-Ass M within Fnd M (see [Š1, Thm.B]). Besides, the set Fnd M is interesting in its own right (cf. [Š3, Sect. 5]). It is also worthy to note the following very pleasant fact which also concerns the uniqueness question for foundation primes. Namely, we introduce the foundation filtration of a module which is unique, if it exists, and so produces the unique set of foundation primes. This uniqueness turns out to be a far reaching technical convenience in our approach.

This note is closely related to [Š1]. The main purpose is to give some remarks and examples which will better explain the mutual relationships among the sets Ass M, w-Ass M, Annspec M, Aff M, and Fnd M. In short, we show that all these sets are in general mutually nonequal and that Ass $M \subseteq$ w-Ass $M \subseteq$ Annspec $M \cap$ Fnd M. Example 2.6 provides a module, from the very interesting category of highest weight modules for a semisimple Lie algebra, having a w-associated prime (equiv. annihilator prime) which is not associated; for another example see [GW,

---

Research supported in part by the Ministry of Science and Technology, Republic of Croatia.

Ex. 2ZE]. Example 2.7 shows that Aff M and Fnd M are generally nonequal; let it be said that in all known examples it holds Aff $M \subseteq$ Fnd M, but we do not know whether this is always true. Finally, Example 2.8 shows that the sets Aff M and w-Ass M are generally nonequal, too (see also [GW, Ex. 2O]). Section 1 of the note serves for introducing the main definitions and explaining some basic facts and results, while Section 2 contains the examples.

Further results concerning all the mentioned sets of prime ideals and in particular the relationships between these for (some) left and for (some) right modules can be found in [Š2].

The author thanks Dragan Miličić for many stimulating discussions about the subject. Actually, Example 2.6 arose from those discussions. He also thanks Ken Goodearl for turning our interest to study the relationships of the w-associated and foundation primes with the affiliated primes. The simple fact that a w-associated prime equals an annihilator prime when the basic ring is left Noetherian and a module is finitely generated was brought to our attention by the referee of the previous version of the paper [Š1]. He also gave some other valuable remarks about the subject. The author express his sincere gratefulness to him.

Some of the main results of our work were presented at the Euroconference "Interactions between Ring Theory and Representations of Algebras", Murcia, España, January 1998.

1. PRELIMINARIES

Throughout this note every ring will have an identity and every ideal will be two-sided. Also, the notion "module" always means a unital left module. For a ring $\mathcal{R}$ and an $\mathcal{R}$-module M, with $M^\times$ denote the set of all nonzero elements of M. For $x \in M^\times$ let $\mathfrak{A}_x$ denote the greatest ideal of $\mathcal{R}$ which annihilates x.

Recall the following classical definition from commutative theory.

Definition 1.1. Let $\mathcal{R}$ be a commutative ring and M an $\mathcal{R}$-module. A prime ideal $\mathcal{P}$ of $\mathcal{R}$ is *associated* to M if $\mathcal{P} = \text{ann}(x)$ for some nonzero $x \in M$.

The next definition introduces the well known "noncommutative analogue" of this notion.

Definition 1.2. Let $\mathcal{R}$ be a ring and M an $\mathcal{R}$-module. A prime ideal of $\mathcal{R}$ which equals the annihilator of some nonzero submodule of M is called an *annihilator prime* for M. By Annspec M denote the set of annihilator primes for M. Given an ideal I of $\mathcal{R}$, a nonzero submodule $N \leq M$ is called an *I-prime* module if $I = \text{ann}(N')$ for any nonzero submodule $N' \leq N$. (It is immediate that then I must be a prime ideal.) A prime ideal $\mathcal{P}$ of $\mathcal{R}$ is *associated* to M if there exists a $\mathcal{P}$-prime submodule of M. By Ass M denote the set of primes associated to M.

Although the above definitions of an associated prime are not quite similar it turns out that they are equivalent when $\mathcal{R}$ is a commutative Noetherian ring and M is a finitely generated $\mathcal{R}$-module [GW, Ex. 4ZB]. Therefore, in that case one can consider the latter definition as a generalization of Definition 1.1. But we can also introduce the following definition of an associated prime in the noncommutative case, which is actually a direct generalization of Definition 1.1 (see Lemma 1.4(i) below for justification of the terminology).

Definition 1.3. Let $\mathcal{R}$ be a ring and M an $\mathcal{R}$-module. A prime ideal $\mathcal{P}$ of $\mathcal{R}$ is *w-associated* (*weakly* associated) to M if $\mathcal{P}$ is the annihilator of some nonzero cyclic submodule of M; i.e. $\mathcal{P} = \mathrm{ann}(\mathcal{R}x)$, or equivalently $\mathcal{P} = \mathfrak{A}_x$, for some $x \in M^\times$. Denote

$$\text{w-Ass } M := \text{the set of primes w-associated to } M.$$

Note that a w-associated prime is a special kind of an annihilator prime. While these two notions are generally nonequal in case of infinitely generated modules (e.g., for $\mathcal{R} = \mathbb{Z}$ and $M = \bigoplus_{n=1}^{\infty} \mathbb{Z}/n\mathbb{Z}$, the ideal (0) is an annihilator prime for M but not a w-associated prime), the part (ii) of the next lemma says that they are equivalent in the most interesting case.

Lemma 1.4. *Let $\mathcal{R}$ be a ring and M an $\mathcal{R}$-module.*

  (i) *If a prime ideal $\mathcal{P}$ of $\mathcal{R}$ is associated to M then it is w-associated to M.*

  (ii) *Suppose that moreover $\mathcal{R}$ is left Noetherian and M is finitely generated. Then the notions of an annihilator prime and a w-associated prime are equivalent.*

Proof. (i) Let N be a $\mathcal{P}$-prime submodule of M and take an arbitrary nonzero element $x \in N$. Then clearly $\mathcal{P} \subseteq \mathfrak{A}_x$. Consider now the cyclic module $N' := \mathcal{R}x$. Since $\mathfrak{A}_x N' = 0$ we have $\mathfrak{A}_x \subseteq \mathcal{P}$, thereby proving the claim.

(ii) Let a prime $\mathcal{P}$ be the annihilator of some nonzero submodule $N \leq M$. Write $N = \mathcal{R}x_1 + \cdots + \mathcal{R}x_n$, $x_i \in N$. Then $\mathcal{P} = \mathrm{ann}(\mathcal{R}x_1) \cap \cdots \cap \mathrm{ann}(\mathcal{R}x_n)$. Since $\mathcal{P}$ is a prime, it follows $\mathcal{P} = \mathrm{ann}(\mathcal{R}x_i)$ for some i. $\square$

The main purpose of the paper [Š1] is to study w-Ass M as an abstract set, and as a subset of the prime spectrum $\mathrm{Spec}\,\mathcal{R}$ equipped with Jacobson topology. For this we introduce a larger and more accessible set of prime ideals which we call the foundation prime ideals, and then study the w-associated primes via these new primes. For convenience of the reader and further needs we will now introduce the notation and terminology from [Š1], which will be freely used in the sequel, and also present some results.

Let $\mathcal{R}$ be a ring and M an $\mathcal{R}$-module; we fix $\mathcal{R}$ and M throughout. For any ideal $\mathfrak{A}$ of $\mathcal{R}$ and a submodule $N \leq M$ the set

$$M(\mathfrak{A}; N) := \{x \in M \mid \mathfrak{A}x \subseteq N\}$$

is obviously a submodule of M, which we call the *submodule of $\mathfrak{A}$-invariants modulo N*. In particular, the module $M(\mathfrak{A}; 0) = \mathrm{ann}_M \mathfrak{A}$ we will write as $M(\mathfrak{A})$. Now denote

$$M_x := M(\mathfrak{A}_x),$$

and then consider the lattice of submodules

$$\Lambda(M) := \{M_x \mid x \in M^\times\},$$

and the set

$$\mathcal{E}(M) := \text{the set of minimal modules in } \Lambda(M).$$

The following proposition collects some simple facts on the modules M_x and the ideals $\mathfrak{A}_x$.

Proposition 1.5. *Let M be an $\mathcal{R}$-module.*

  (i) *If x, y are elements of $M^\times$ then $M_x \subseteq M_y$ if and only if $\mathfrak{A}_x \supseteq \mathfrak{A}_y$. In particular, $\mathcal{E}(M)$ corresponds to the maximal elements of the set of ideals $\{\mathfrak{A}_x \mid x \in M^\times\}$. Further, if αx is an element of $M^\times$ for some $\alpha \in \mathcal{R}$ then $\mathfrak{A}_x \subseteq \mathfrak{A}_{\alpha x}$.*

Now suppose that the set $\mathcal{E}(M)$ is nonempty.

(ii) *For an arbitrary module $N \in \mathcal{E}(M)$ and every element $x \in N^\times$ we have*

$$(1) \qquad\qquad M_x = N \quad \text{and} \quad \mathfrak{A}_x = \mathrm{ann}(N).$$

Further, the ideal $\mathrm{ann}(N)$ *is prime.*

(iii) *If $N \in \mathcal{E}(M)$ possesses a simple submodule then the ideal* $\mathrm{ann}(N)$ *is primitive.*

Proof. (i) Clear.

(ii) Let $m \in M^\times$ such that $M_m = N$. For an arbitrary $x \in N^\times$ we have $\mathfrak{A}_m x = 0$, and then $M_x \subseteq N$. Hence, by the minimality of N in $\Lambda(M)$ it follows $M_x = N$. Also, since $\mathfrak{A}_m N = 0$ we have $\mathfrak{A}_m = \mathrm{ann}(N)$. Let us show that the ideal $\mathrm{ann}(N)$ is prime. Suppose $\mathcal{P}$ and $\mathcal{Q}$ are ideals of $\mathcal{R}$ such that $\mathcal{P}\mathcal{Q} \subseteq \mathrm{ann}(N)$. If $\mathcal{Q} \not\subseteq \mathrm{ann}(N)$, then there exist elements $x \in N$ and $q \in \mathcal{Q}$ such that $qx \neq 0$. The fact $\mathcal{P}qx = 0$, by (1), implies $\mathcal{P} \subseteq \mathrm{ann}(N)$, which completes the proof of (ii).

(iii) Suppose that E is a simple submodule of N and take some element $e \in E^\times$. By the inclusion $\mathrm{ann}(E) \subseteq \mathfrak{A}_e$ and (1) we have $\mathrm{ann}(E) = \mathrm{ann}(N)$. Hence, by the definition of a primitive ideal, (iii) follows . $\qquad\square$

The following lemma enables to define important notions of the foundation and the foundation filtration for a given module.

Lemma 1.6. *The sum $\sum_{N \in \mathcal{E}(M)} N$ is actually a direct sum.*

Proof. See [Š1, Lemma 2.2]. $\qquad\square$

Definition 1.7. If $\mathcal{E}(M)$ is a nonempty set, the submodule $\bigoplus_{N \in \mathcal{E}(M)} N$ is called the *foundation* of M and denoted by $\mathrm{fnd}(M)$. Further, if $(\mathcal{F}) = (\mathcal{F}_i)_{i \geq 0}$ is a (finite or infinite) exhaustive filtration of M such that $\mathcal{F}_i$ are submodules of M satisfying the equalities $\mathrm{fnd}(M/\mathcal{F}_i) = \mathcal{F}_{i+1}/\mathcal{F}_i$ for every $i \geq 0$, we call $(\mathcal{F})$ the *foundation filtration* of M and then say that M has the foundation filtration. In this case we define the sets of prime ideals (see Prop. 1.5(ii))

$$\mathrm{Fnd}_i\, M := \{\mathrm{ann}(N) \mid N \in \mathcal{E}(M/\mathcal{F}_{i-1})\} \qquad \text{for } i > 0,$$
$$\mathrm{Fnd}\, M := \bigcup_{i>0} \mathrm{Fnd}_i\, M.$$

The primes of $\mathrm{Fnd}\, M$ are called the *foundation primes* of M. We say that the primes from $\mathrm{Fnd}_i\, M$ belong to the i-th *foundation level*. If i is the minimal foundation level to which a prime $\mathcal{P} \in \mathrm{Fnd}\, M$ belongs we write $\mathrm{mfl}(\mathcal{P}) = i$.

Now we have the following theorem (see [Š1, Thm. 2.5).

Theorem 1.8. *Let $\mathcal{R}$ be a left Noetherian ring and M a finitely generated $\mathcal{R}$-module.*

(a) *Then the module M has the finite foundation filtration and every $\mathrm{Fnd}_i\, M$ is a finite set of primes satisfying the incomparability condition. Moreover, the (finite) set $\mathrm{Fnd}\, M$ contains w-Ass M (cf. Lemma 1.4).*

(b) *Let $\mathcal{P} \in$ w-Ass M, $\mathcal{Q} \in \mathrm{Fnd}\, M$, $p = \mathrm{mfl}(\mathcal{P})$, and $q = \mathrm{mfl}(\mathcal{Q})$. If $q < p$ then $\mathcal{Q} \not\subseteq \mathcal{P}$.*

Note that the set of foundation primes is nicely organized by foundation levels. This fact together with the preceding theorem enable us the mentioned studying of w-Ass M via $\mathrm{Fnd}\, M$. Using the fact that $\mathrm{Fnd}_1\, M \subseteq$ w-Ass M and the technique of "climbing over foundation levels", the set w-Ass M can be recognized within $\mathrm{Fnd}\, M$

for, e.g., $\mathcal{R}$ commutative or equal to a quotient ring of the enveloping algebra of a finite dimensional nilpotent Lie algebra [Š1, Thm. B].

We also note the following (cf. [Š1, Cor. and Def. 2.6]). Having the foundation filtration of a module M such that every $\text{Fnd}_i\, M$ is a finite set we can construct in a natural way (simply, "climb" over all foundation levels begining with the first one) an exhaustive filtration $(M_i)_{i\geq 0}$ of M which refines the foundation filtration of M. Related with this filtration we can order the set $\text{Fnd}\, M$ as $(\mathcal{P}_1, \mathcal{P}_2, \dots)$. Now we have the inclusions

$$(2) \qquad\qquad M_i \subseteq M(\mathcal{P}_i; M_{i-1}) \qquad \text{for } i \geq 1.$$

A filtration $(M_i)_{i\geq 0}$ which is a refinement of the foundation filtration constructed as above is called a *full filtration* of M.

Finally, for further needs recall the definition of affiliated series and affiliated primes. (The affiliated primes were first introduced by Stafford [St1, St2], partly in order to find a set of primes which would contain more information than the set of associated primes does.) We give a definition which will be convenient for our purposes. It is immediate that this definition is equivalent to the usual one (see, e.g., [GW, p. 33] and [LW, p. 166]).

Definition 1.9. Let $0 = M_0 \subseteq M_1 \subseteq \cdots \subseteq M_n = M$ be a sequence of submodules of an $\mathcal{R}$-module M, and $(\mathcal{P}_1, \dots, \mathcal{P}_n)$ an ordered set of prime ideals of $\mathcal{R}$. Then this sequence of submodules and this set of primes are called an *affiliated series* for M and *affiliated primes* of M corresponding to the given affiliated series, respectively, if the following two conditions hold:

(A1) M_i/M_{i-1} is a $\mathcal{P}_i$-prime module; and

(A2) $M_i = M(\mathcal{P}_i; M_{i-1})$, for every i (cf. (2)).

In general, an affiliated prime of M is a prime ideal which is an affiliated prime corresponding to some affiliated series for M. The set of all affiliated primes of M denote by $\text{Aff}\, M$.

Remark 1.10. An unsatisfactory fact is that a module may have different affiliated series which produce different sets of corresponding affiliated primes. Assuming that the basic ring is moreover an algebra of finite GK-dimension, Lenagan and Warfield [LW] defined a "standard" affiliated series which again are not necessarily unique but always give the same affiliated primes. Note that while dealing with full filtrations of modules no such problems arise (without any restrictions on the basic ring); the reason is that our definitions of the foundation filtration and a full filtration are canonical, i.e. no particular choice of affiliated series is involved.

2. Examples

Keep the notation and terminology from Section 1 in force.

As we mentioned in the introduction one of our goals is to show that the converse of the claim (i) in Lemma 1.4 does not always hold. An example which we offer to see this is a certain highest weight module for a semisimple Lie algebra; more precisely, the dual of a Verma module of length two. First we give one more, general and simple, fact.

Lemma 2.1. *Let M be an $\mathcal{R}$-module. Then $\text{ann}(M) \subseteq \mathfrak{A}_x$ for any $x \in M^\times$. Moreover, the latter inclusion is an equality if M is a cyclic module and x is a cyclic generator of M.*

Proof. If M is a cyclic module with a cyclic generator x then $\mathfrak{A}_x M = (\mathfrak{A}_x \mathcal{R})x = 0$. Hence, the lemma follows.   □

Now we need some preparations. If $\mathfrak{a}$ is a Lie algebra then its enveloping algebra will be denoted as usual by $\mathcal{U}(\mathfrak{a})$. Let now $\mathfrak{g}$ be a semisimple Lie algebra over an algebraically closed field of characteristic zero. Fix a Borel subalgebra $\mathfrak{b}$ in $\mathfrak{g}$ and let $\mathfrak{h}$ be a Cartan subalgebra of $\mathfrak{g}$ contained in $\mathfrak{b}$. Consider the *category of highest weight modules*, which will be denoted by $\mathcal{HW}$-mod. Recall that a $\mathfrak{g}$-module M is a highest weight module, with respect to $\mathfrak{b}$, if M is finitely generated and $\dim \mathcal{U}(\mathfrak{b})x < \infty$, for every $x \in M$. Let $M(\lambda)$ denote the Verma module for $\lambda \in \mathfrak{h}^*$, and $L(\lambda)$ its unique irreducible quotient (see, e.g., [D]). It is well known that the modules $L(\lambda)$, for λ running over $\mathfrak{h}^*$, determine all the irreducible objects in $\mathcal{HW}$-mod.

Henceforth we use without special references the results from [D, Chap. 7].

The following result must be well known but we do not know any appropriate reference.

Lemma 2.2. *Keeping the above notation let V be a Verma module. Then* $\operatorname{ann}(V) = \mathfrak{A}_v$ *for any $v \in V^\times$. (As an immediate consequence we get* $\operatorname{ann}(V) = \operatorname{ann}(W)$ *for any nonzero submodule $W \leq V$.)*

Proof. By denoting $\mathcal{P} := \operatorname{ann}(V)$ and using Lemma 2.1 we have $\mathcal{P} \subseteq \mathfrak{A}_v$ for any $v \in V^\times$. Now we consider the lattice $\Lambda(V)$ and let $w \in V$ be such that $V_w \in \mathcal{E}(V)$; or equivalently, $\mathfrak{A}_w$ is a maximal ideal in the set of ideals $\{\mathfrak{A}_v \mid v \in V^\times\}$. (Since the enveloping algebra is Noetherian, the set $\Lambda(V)$ is nonempty.) For a highest weight vector v_0 of V it holds $\mathcal{P} = \mathfrak{A}_{v_0}$, again by Lemma 2.1. We would like to show that $\mathcal{P} = \mathfrak{A}_w$. To see this suppose the contrary, which is by Proposition 1.5 further equivalent to $V_w \neq V$. Let E be the least nonzero submodule of V; it is well known that such E exists and moreover that it is isomorphic to a Verma module $M(\mu)$, for some $\mu \in \mathfrak{h}^*$. Then clearly $E \leq V_w$, and hence $\mathfrak{A}_w = \operatorname{ann}(E)$; the latter equality by using Proposition 1.5 again. Thus, we obtain $\mathcal{P} \subset \operatorname{ann}(E)$, which is impossible since both the ideals $\mathcal{P}$ and $\operatorname{ann}(E)$ are minimal primitives by a famous result of Duflo [D, Thm. 8.4.4]. This finishes the proof.   □

Corollary 2.3. *Let V be a Verma module. Then*

$$\operatorname{Ass} V = \text{w-Ass}\, V = \operatorname{Annspec} V = \operatorname{Fnd} V = \operatorname{Fnd}_1 V = \operatorname{Aff} V = \{\operatorname{ann}(V)\}.$$

In particular, V has a unique affiliated series, i.e. $0 = M_0 < M_1 = V$.

Proof. Clear.   □

Let us proceed toward the announced example. For that purpose recall some basic facts concerning the dual modules of highest weight modules [BGG] (see also [M]). Let p be the principal antiautomorphism of $\mathcal{U}(\mathfrak{g})$ and ι the automorphism of $\mathcal{U}(\mathfrak{g})$ which acts as $\iota_{|\mathfrak{h}} = -\operatorname{id}_\mathfrak{h}$ and by $\iota(\mathfrak{g}_\alpha) = \mathfrak{g}_{-\alpha}$ on the root subspaces. Define the Chevalley antiautomorphism $\tau := \iota \circ p$. Let now M be a highest weight module and M^* its linear dual. Define the $\mathfrak{g}$-module structure on M^* by

$$(uf)(x) := f(\tau(u)x) \qquad \text{for } u \in \mathcal{U}(\mathfrak{g}), \ f \in M^*, \ x \in M,$$

and set

$$M^\sim := \{f \in M^* \mid \dim \mathcal{U}(\mathfrak{h})f < \infty\}.$$

It is immediate that $M^\sim$ is also a $\mathfrak{g}$-module, and we call it the *dual* of M. Denote $I(\lambda) = M(\lambda)^\sim$ for $\lambda \in \mathfrak{h}^*$. The following basic facts about dual modules will be freely used below; for a proof see, e.g., [BGG, M].

Lemma 2.4. (i) *If M is a highest weight module then $M^\sim$ is such a module, too, and $(M^\sim)^\sim = M$. Hence, $M \mapsto M^\sim$ defines an exact contravariant functor from the category $\mathcal{HW}$-mod into itself.*

(ii) *For any $\lambda \in \mathfrak{h}^*$ it holds $L(\lambda)^\sim = L(\lambda)$; therefore, $I(\lambda)$ has a unique irreducible $\mathfrak{g}$-submodule, which is moreover isomorphic to $L(\lambda)$.*

The following lemma was brought to our attention by D. Miličić. Although this result must be also well known, for convenience of the reader we include his argument.

Lemma 2.5. *Let V be a Verma module. Then $\mathrm{ann}(V) = \mathrm{ann}(V^\sim)$.*

Proof. The ideal $p(\mathrm{ann}(V))$ annihilates V^* considered as the contragredient module of V, and then clearly $\tau(\mathrm{ann}(V)) \subseteq \mathrm{ann}(V^\sim)$. By replacing $V \leftrightarrow V^\sim$, the same reasoning combined with the latter inclusion yields $\tau(\mathrm{ann}(V)) = \mathrm{ann}(V^\sim)$. (Note that the last equality is also valid under the weaker assumption that V is a highest weight module.) Now, the already noted fact that V has a (unique) irreducible submodule, which is itself isomorphic to some Verma module, together with Lemma 2.2 complete the proof. □

Example 2.6. Let $M(\lambda)$, $\lambda \in \mathfrak{h}^*$, be a Verma module of length two having a composition series $0 < M(\mu) < M(\lambda)$, for a certain $\mu \in \mathfrak{h}^*$ (e.g., let $\mathfrak{g} = \mathfrak{sl}(2)$ and $\lambda - \rho$ a dominant weight; 2ρ is the corresponding positive root). Denote the annihilators $\mathcal{P} := \mathrm{ann}(M(\lambda))$ and $\mathcal{Q} := \mathrm{ann}(L(\lambda))$. By Lemma 2.2 we have $\mathcal{P} \subseteq \mathcal{Q}$ but we can moreover take $\mathcal{P} \subset \mathcal{Q}$.

Claim 1. Annspec $I(\lambda) = \{\mathcal{P}, \mathcal{Q}\}$.

[Take an arbitrary $f \in I(\lambda) \setminus L(\lambda)$ and put $\overline{f} := f + L(\lambda)$. Then for f and $\overline{f}$ we have the inclusions $\mathcal{P} \subseteq \mathfrak{A}_f \subseteq \mathfrak{A}_{\overline{f}} = \mathcal{P}$ (use Lemmas 2.5 and 2.2), and so $\mathcal{P} = \mathfrak{A}_f$. Since $\mathcal{Q}$ is obviously an annihilator prime the claim is proved.]

Claim 2. Ass $I(\lambda) = \{\mathcal{Q}\}$.

[We only have to observe that the existence of a nonzero submodule N of $I(\lambda)$ satisfying $\mathrm{ann}(N) = \mathcal{P}$ would imply $L(\lambda) \leq N$. Therefore, $\mathcal{P}$ is not a prime associated to $I(\lambda)$.]

It is also instructive to formulate the following claim; an easy proof is left to the reader.

Claim 3. $I(\lambda)$ has a unique affiliated series, i.e. the series $0 < L(\lambda) < I(\lambda)$, with the corresponding affiliated primes $(\mathcal{Q}, \mathcal{P})$.

The following example shows that the sets of affiliated primes and foundation primes, of a given module, are generally nonequal.

Example 2.7. This example was used in [Š1, Example 4.5] for computing the sets Fnd M and w-Ass M for the module M defined below. For convenience of the reader we recall the main features and give some more details.

Let $\mathbb{F}$ be an algebraically closed field and $\mathcal{A} := \mathbb{F}[X_1, \ldots, X_n]$ the polynomial ring in $n > 1$ variables. Consider the ideals (notation: the ideal of any ring $\mathcal{R}$ generated with some subset $X \subseteq \mathcal{R}$ is denoted by $\langle X \rangle$) $\mathcal{I} := \langle X_i X_j \mid i \neq j \rangle$, and $\mathcal{P}_i := \langle X_1, \ldots, \hat{X}_i, \ldots, X_n \rangle$, $\mathcal{B}_i := \langle X_i, X_j X_k \mid j \neq i, k \neq i, \text{ and } j \neq k \rangle$ for $i = 1, \ldots, n$. Define the torsion $\mathcal{A}$-module $M := \mathcal{A}/\mathcal{I}$ and its submodules $N_i := \mathcal{B}_i/\mathcal{I}$.

Then $\mathcal{E}(M) = \{N_1, \ldots, N_n\}$.Denote $\mathcal{F}_1 := \mathrm{fnd}(M)$, and observe that $M/\mathcal{F}_1 \simeq \mathbb{F}$ (an $\mathcal{A}$-module isomorphism) so that $\mathcal{E}(M/\mathcal{F}_1) = \{M/\mathcal{F}_1\}$. Put $\mathcal{Q} := \mathrm{ann}(M/\mathcal{F}_1)$. Clearly, we have $\mathcal{Q} = \langle X_1, \ldots, X_n \rangle$. Now, $(\mathcal{F}_0 := 0, \mathcal{F}_1, \mathcal{F}_2 := M)$ is the foundation filtration of M, and the refinement $(M_0 := 0, M_i := \bigoplus_{j \leq i} N_j$ for $i \leq n, M_{n+1} := M)$ is a full filtration of M.

Claim. It holds the following:

   (a) $M(\mathcal{P}_i; M_{i-1}) \subseteq \mathcal{F}_1$ for every $i < n$; and
   (b) $M(\mathcal{P}_n; M_{n-1}) = M$.

   [(a) Note that $M_i = (\sum_{j \leq i} \mathcal{B}_j)/\mathcal{I}$ for $i \leq n$, and in particular $M_n = \mathcal{F}_1 = \mathcal{Q}/\mathcal{I}$. Suppose that there exists $x \in M \setminus \mathcal{F}_1$ such that $\mathcal{P}_i x \subseteq M_{i-1}$. With no loss of generality we can take $x = 1 + \mathcal{I}$. Now the latter inclusion implies that the ideal $\sum_{j \leq i-1} \mathcal{B}_j$ contains $\mathcal{P}_i$, and in particular the element X_n, which is obviously not true.

   (b) To prove the inclusion ($\supseteq$) take $x \in M \setminus \mathcal{F}_1$; the case $x \in \mathcal{F}_1$ being clear. Since $\mathfrak{A}_{x+\mathcal{F}_1} = \mathcal{Q}$ we have $\mathcal{Q}x \subseteq \mathcal{F}_1$, and so $\mathcal{P}_n x \subseteq \mathcal{F}_1$. We want to see that $\mathcal{P}_n x \subseteq M_{n-1}$. Supposing the contrary, there exists $p \in \mathcal{P}_n$ such that $px = \sum_{j<n} \gamma_j + \gamma$ where $\gamma_j \in N_j$ and $\gamma \in (N_n)^\times$. Hence, by putting $\mathcal{S} := \prod_{i=1}^{n-1} \mathcal{P}_i$ and using the fact that $\mathcal{S}$ kills every γ_j, we have

$$\mathcal{P}_1 \mathcal{S} \gamma = \mathcal{P}_1 \mathcal{S} px \subseteq \mathcal{Q} \mathcal{S} \mathcal{P}_n x = 0$$

(the latter equality since $\mathcal{P}_1 \cdots \mathcal{P}_n \mathcal{Q}$ kills every element of M). Thus $\mathcal{P}_1 \mathcal{S} \subseteq \mathfrak{A}_\gamma = \mathcal{P}_n$, and then $\mathcal{P}_j \subseteq \mathcal{P}_n$ for some $j < n$, which is impossible by Theorem 1.8(i). This finishes the proof.]

   Now, using (a) of the above claim it follows $M(\mathcal{P}_i; M_{i-1}) = M_i$ for $i < n$ (see [Š1]). Having all this it is easy to conclude that M has $n!$ mutually different affiliated series, but each of these gives the same set of affiliated primes $\mathrm{Aff}\, M = \{\mathcal{P}_1, \ldots, \mathcal{P}_n\}$. At the same time we have $\mathrm{Fnd}_1 M = \text{w-Ass}\, M = \{\mathcal{P}_1, \ldots, \mathcal{P}_n\}$ and $\mathrm{Fnd}_2 M = \{\mathcal{Q}\}$.

   In general, the sets of w-associated primes and affiliated primes of a finitely generated module over a commutative Noetherian ring are equal [GW, Ex. 4ZB]. We will finish this note with an easy example which shows that the latter does not always hold (see [Š1, Example 4.4]).

Example 2.8. Let $\mathbb{F}$ be a field and $\mathcal{R} := \left(\begin{smallmatrix} \mathbb{F} & \mathbb{F} \\ 0 & \mathbb{F} \end{smallmatrix} \right)$ a matrix ring. Then $\mathcal{R}$ has only two prime ideals, namely $\mathcal{P} := \left(\begin{smallmatrix} 0 & \mathbb{F} \\ 0 & \mathbb{F} \end{smallmatrix} \right)$ and $\mathcal{P}' := \left(\begin{smallmatrix} \mathbb{F} & \mathbb{F} \\ 0 & 0 \end{smallmatrix} \right)$. It is an easy exercise to see that for the $\mathcal{R}$-module $M = {}_\mathcal{R}\mathcal{R}$ we have w-Ass $M = \mathrm{Ass}\, M = \{\mathcal{P}\}$, and also that M has the unique affiliated series, namely $0 < \mathcal{P}' < M$, with the corresponding set of affiliated primes $(\mathcal{P}, \mathcal{P}')$. Note that the latter series equals the foundation filtration of M, and so $\mathrm{Aff}\, M = \mathrm{Fnd}\, M$.

REFERENCES

[BGG] I. N. Bernstein, I. M. Gelfand and S. I. Gelfand, *Ob ednoj kategorii g-modulej*, Funkcional. Anal. i Priložen. **10** (1976), 1–8.

[D]   J. Dixmier, *Enveloping Algebras*, North-Holland Mathematical Library, vol. 14, North-Holland, Amsterdam, 1977.

[GW] K. R. Goodearl and R. B. Warfield, Jr., *An Introduction to Noncommutative Noetherian Rings*, "London Math. Soc. Stud. Texts", vol. 16, Cambridge Univ. Press, Cambridge, 1989.

[LW]   T. H. Lenagan and R. B. Warfield, Jr., *Affiliated series and extensions of modules*, J. Algebra **142** (1991), 164–187.

[M]   D. Miličić, *Localization and representation theory of reductive Lie groups*, (mimeographed notes), to appear (available electronically from http://www.math.utah.edu/~milicic).

[St1]   J. T. Stafford, *On the regular elements of Noetherian rings*, in Ring Theory, Proceedings of the 1978 Antwerp Conference (F. Van Oystaeyen, Ed.), New York (1979), Dekker, pp. 257–277.

[St2]   ———, *Noetherian full quotient rings*, Proc. London Math. Soc. **44** (1982), 385–404.

[Š1]   B. Širola, *On annihilator primes and foundation primes*, preprint.

[Š2]   ———, *The ϑ-transfer technique*, preprint.

[Š3]   ———, *On module supports and Casselman's subrepresentation theorem*, preprint.

On Kaplansky's Conjectures

Yorck Sommerhäuser

MATHEMATISCHES INSTITUT DER UNIVERSITÄT MÜNCHEN, THERESIENSTR. 39, 80333 MÜNCHEN, GERMANY

1   Introduction

In the autumn of 1973, I. Kaplansky gave a course on bialgebras in Chicago. For this course, he prepared some lecture notes that he originally intended to turn into a comprehensive account on the subject. Later, he changed his mind, and therefore in 1975 these lecture notes were published without larger additions at the University of Chicago Press. These lecture notes contain, besides a fairly comprehensive bibliography of the literature available at that time, two appendices. The first of these appendices is concerned with bialgebras of low dimension, whereas the second one contains a list of ten conjectures on Hopf algebras which are known today as Kaplansky's conjectures.

Kaplansky's conjectures did not arise as the product of a long investigation in the field of Hopf algebras; also, Kaplansky did not make many contributions to the solution of his conjectures. He only intended to list a number of interesting problems at the end his lecture notes - lecture notes that he himself called informal. Because of this, it happened that one conjecture in the list was already solved at the time of publication, another one is very simple. That the conjectures nevertheless gained considerable importance for the field is due to the fact that Kaplansky achieved to touch upon a number of questions of fundamental character.

This article tries to summarize the present knowledge about Kaplansky's conjectures. Brief surveys can be found in [52] and [63], some conjectures are also discussed in [37] and [42]. Here, the exposition shall be more detailed, but nevertheless not comprehensive. Since Kaplansky's lecture notes are not always easily accessible, we have reproduced the conjectures in their original formulation in an appendix. The reader should note that, except for the appendix, the formulation of results does not follow literally the quoted sources. In addition, usually not all important results of a quoted article are mentioned. We also note that Kaplansky posed various other conjectures concerning different fields of mathematics; these are not discussed here.

2   The first conjecture

2.1   Kaplansky's first conjecture states that a Hopf algebra is a free module over any Hopf subalgebra. This assertion should be compared with the group ring situation, where a system of representatives for the cosets forms a basis of the group ring over the group ring of a subgroup. At the time of its publication, it was already known that the conjecture is false, because U. Oberst and H. J. Schneider had, in an article quoted by Kaplansky, constructed a counterexample (cf. [50], Prop. 10, p. 31, see also [37], Example 3.5.2, p. 38). In this counterexample, Oberst and Schneider construct an extension of real Hopf algebras via Galois descent from the extension

$$\mathbb{C}[2\mathbb{Z}] \subset \mathbb{C}[\mathbb{Z}]$$

of the complex group ring of the integers over the group ring of the even numbers. In fact, instead of the extension $\mathbb{R} \subset \mathbb{C}$, they consider more generally arbitrary quadratic Galois extensions of fields. Later, Schneider gave a counterexample over arbitrary fields (cf. [64]): Suppose that H is the Hopf algebra representing the affine group scheme of the special linear group of degree n and $A = K[Z_n]$ the group ring of the cyclic group of order n, representing the affine group scheme of n-th roots of unity. The roots of unity are realized via diagonal matrices as a normal subgroup of the special linear group. In this situation, we can look at the quotient scheme, which is represented by a Hopf algebra B. Now Schneider proves that H is not free over B if n is even.

2.2   However, several positive results were proved in the following years, showing that the conjecture holds under additional assumptions. Kaplansky already mentioned that W. D. Nichols proved that the conjecture holds if the coradical of the large Hopf algebra is contained in the Hopf subalgebra, this is also a direct consequence of a result by D. E. Radford (cf. [57], Cor. 2.3, p. 146). In the same year, Radford proved that the conjecture holds for a large class of Hopf algebras, namely the pointed ones (cf. [56], p. 271):

Theorem   Suppose that A is a Hopf subalgebra of the Hopf algebra H. Suppose that H is pointed, i. e., every simple subcoalgebra is one-dimensional. Then H is a free left and right A-module.

Furthermore, Radford proved in [58] that commutative Hopf algebras are free over finite-dimensional Hopf subalgebras. He also proved that here the assumption that the Hopf subalgebra be finite-dimensional could be weakend to the assumption that only the coradical of the Hopf subalgebra be finite-dimensional by imposing the additional assumption that the coradical of the large Hopf algebra is a Hopf subalgebra, which is the case over fields of characteristic zero by a result of M. Takeuchi (cf. [78]). He also proved there that Hopf algebras are free over finite-dimensional Hopf subalgebras if the coradical of the large Hopf algebra is cocommutative.

2.3 W. D. Nichols had been concerned with Kaplansky's conjectures from the very beginning. In the eighties, he and his former Ph. D. student M. B. Zoeller worked in particular on Kaplansky's first conjecture. They summarized their efforts in a series of articles (cf. [86], [43], [45]) of subsequent generalizations, culminating in what is now known as the Nichols-Zoeller theorem:

Theorem A finite-dimensional Hopf algebra is free over every Hopf subalgebra.

In fact, Nichols and Zoeller prove the more general theorem that relative Hopf modules are free. Their theorem implies that the dimension of a Hopf subalgebra divides the dimension of the large Hopf algebra, which is the Hopf algebra version of Lagrange's theorem on the order of subgroups. This theorem has many consequences and is commonly considered as one of the most important theorems of the theory of Hopf algebras. Nichols and Zoeller also exhibited an example that infinite dimensional Hopf algebras need not be free over finite dimensional grouplike Hopf subalgebras (cf. [44]). These authors have also shown in a more recent article that a Hopf algebra is free over a semisimple Hopf subalgebra (cf. [46]). (Due to marriage, the name 'Zoeller' has changed to 'Richmond'.)

The work of Nichols and Zoeller has been amplified by H. J. Schneider. Given a Hopf subalgebra of a finite-dimensional Hopf algebra, it is possible to form a quotient which is not always a Hopf algebra, but always is a coalgebra. The original Hopf algebra then becomes a comodule over this quotient. Now Schneider shows (cf. [65]) that the original Hopf algebra can be decomposed, as a module and a comodule, into a tensor product of the Hopf subalgebra and the quotient. In a different article, Schneider proves that Hopf algebras are free over finite dimensional normal Hopf subalgebras (cf. [66]).

The Nichols-Zoeller theorem has been extended to coideal subalgebras that are quasi-Frobenius algebras by A. Masuoka (cf. [28]). There, he also determines sufficient conditions for a coideal subalgebra to be a quasi-Frobenius algebra (cf. also [29]).

2.4 Although nowadays quite a lot is known about Kaplansky's first conjecture, the fact that it does not hold in the infinite dimensional case has given rise to the question whether weaker properties than freeness hold in all cases. One possible variant of this question has been stated by S. Montgomery in [37], Question 3.5.4, p. 39:

Is H always left and right faithfully flat over any subHopfalgebra K?

This question might be considered as a perpetuation of Kaplansky's first conjecture. It plays an important role for the quotient theory of Hopf algebras.

Several things are known about this question today. It has been known for quite a time from the theory of algebraic groups that this is the case if the Hopf algebra is commutative (cf. [7], Chap. III, § 3, no. 7, Theorem 7.2, [82], sec. 14.1). This result can also be found in an article by M. Takeuchi (cf. [77], Theorem 3.1), where it is also shown that a Hopf algebra with cocommutative coradical is faithfully flat over every Hopf subalgebra. Later, Takeuchi proved that commutative Hopf algebras are projective over their Hopf subalgebras (cf. [80], Cor. 1, p. 460). However, as was shown later by A. Masuoka and D. Wigner (cf. [30]), faithful flatness and projectivity are essentially equivalent conditions in this situation:

Theorem Suppose that H is a Hopf algebra with bijective antipode, and that A is a right coideal subalgebra of H. Then the following assertions are equivalent:

1. H is faithfully flat as a left A-module.

2. H is a projective generator as a left A-module.

This theorem applies in particular if A is a Hopf subalgebra of H. Of course, the first statement can easily be deduced from the second. In addition, H. J. Schneider has proved that a left noetherian Hopf algebra is faithfully flat over central Hopf subalgebras (cf. [66]). S. H. Ng has shown that Hopf algebras are faithfully flat over grouplike Hopf subalgebras if the characteristic of the base field is zero (cf. [40]); he also considers the case of positive characteristic.

These results have been partially generalized to coideal subalgebras: A. Masuoka and D. Wigner prove that commutative Hopf algebras are flat over right coideal subalgebras (cf. [30]), they are not faithfully flat in general. Moreover, Masuoka has shown that Hopf algebras with cocommutative coradical are faithfully flat over certain coideal subalgebras (cf [27]).

However, the answer to the above question is negative in general: P. Schauenburg has constructed a counterexample (cf. [61]) by modifying constructions of W. D. Nichols and M. Takeuchi (cf. [41], [79]). His example relies on the fact that the Hopf subalgebra can be chosen to have a non-bijective antipode, although the antipode of the large Hopf algebra may be bijective. Therefore, as Schauenburg points out, the remaining question is whether Hopf algebras with bijective antipodes are faithfully flat over Hopf subalgebras with bijective antipodes. This question may be considered as the current form of Kaplansky's first conjecture.

3   The third and the ninth conjecture

3.1   Kaplansky's third conjecture states that a Hopf algebra over a field of characteristic zero does not contain nonzero central nilpotent elements. Although is seems that no literature on this conjecture exists, it has been obvious to most researchers for some time that this is false.

Consider a left integral Λ_H in a finite-dimensional non-semisimple Hopf algebra. By the generalized Maschke theorem due to R. G. Larson and M. E. Sweedler (cf. [25]), we then have $\epsilon_H(\Lambda_H) = 0$, and therefore also $\Lambda_H^2 = 0$. Λ_H is therefore a nilpotent element. Now, if Λ_H is also a right integral, i. e., if H is unimodular, then Λ_H is obviously a central element. Kaplansky's third conjecture would therefore imply that, over fields of characteristic zero, all finite-dimensional unimodular Hopf algebras are semisimple.

This is, however, not the case. There exist finite-dimensional Hopf algebras that are not semisimple, for example the Taft algebra (cf. [75]), which is, however, not unimodular. But every finite-dimensional Hopf algebra may be embedded in the Drinfel'd double $D(H)$, and this is a unimodular Hopf algebra by a result of D. E. Radford (cf. [60], Theorem 4, [69], Theorem 5.4). By another result of Radford (cf. [60], Prop. 7), $D(H)$ is semisimple if and only if H is semisimple and cosemisimple. Therefore, the Drinfel'd double of the Taft algebra is a finite-dimensional unimodular Hopf algebra which is not semisimple.

3.2 We now consider another counterexample to Kaplansky's third conjecture, namely the Frobenius-Lusztig kernel of $U_q(sl(2))$, which is slightly more explicit than the example considered above and, as we shall see below, also provides a counterexample for Kaplanksky's ninth conjecture. In fact, it can be shown that it is a quotient of the Drinfel'd double of the Taft algebra.

We work over an algebraically closed field of characteristic zero. Suppose that q is a primitive k-th root of unity, where $k > 2$ is odd - the even case is similar. The Frobenius-Lusztig kernel of $U_q(sl(2))$ is defined as the algebra U with generators K, K^{-1}, E and F and relations

$$KK^{-1} = 1 = K^{-1}K$$
$$KE = q^2 EK \qquad KF = q^{-2}FK$$
$$EF - FE = \frac{K - K^{-1}}{q - q^{-1}}$$
$$K^k = 1 \qquad E^k = 0 \qquad F^k = 0$$

The coproduct is determined on the generators by the formulas:

$$\Delta(K) = K \otimes K \quad \Delta(K^{-1}) = K^{-1} \otimes K^{-1}$$
$$\Delta(E) = E \otimes 1 + K \otimes E \quad \Delta(F) = F \otimes K^{-1} + 1 \otimes F$$

This Hopf algebra has a basis consisting of the elements $E^i K^j F^m$ for $i, j, m = 0, \ldots, k-1$. If $T := \sum_{i=1}^{k} q^{2i(k-1)}K^i$, it is easy to see that $E^{k-1}TF^{k-1}$ is a left and right integral, and therefore a central element (cf. [72], p. 368, [69],

Prop. 6.3, note a slight difference in the definitions). Since the counit vanishes on the integral, its square is zero.

The integral is not the only nilpotent central element of U. Its center can be described completely (cf. [62]).

3.3 As noted by H. J. Schneider (cf. [63]), the Frobenius-Lusztig kernel considered above also provides a counterexample for Kaplansky's ninth conjecture. This conjecture states that for a finite dimensional Hopf algebra over an algebraically closed field, the dimension of the Jacobson radical of the Hopf algebra equals the dimension of the Jacobson radical of the dual if the characteristic of the base field does not divide the dimension. In particular, semisimple Hopf algebras should be cosemisimple, a problem that will be considered below in conjunction with the fifth conjecture.

We now explain briefly why Kaplansky's ninth conjecture is refuted by the Frobenius-Lusztig kernel U. The simple modules of U are known (cf. [17], Theorem VI.5.7, p. 137, [14], sec. 2.13, p. 24); in the odd case considered above, there is one isomorphism class of simple modules for each of the dimensions $1, \ldots, k$. Since the Jacobson radical J of U consists, by definition, precisely of the elements that annihilate every simple modules, U/J has the same simple modules as U. Since U/J is a semisimple algebra, its dimension is the sum of the squares of the dimensions of the simple modules, i. e., we have:

$$dim\, U/J = \sum_{m=1}^{k} m^2 = \frac{1}{6}k(k+1)(2k+1)$$

On the other hand, the Jacobson radical of U^* is precisely orthogonal to the coradical C of U (cf. [37], Remark 5.1.7, p. 58). Kaplansky's conjecture therefore would imply that the coradical has the same dimension as U/J. But, as can be shown as in [37], Lemma 5.5.5, p. 76, the coradical of U is spanned by the powers of K, and therefore we have:

$$dim\, C = k$$

Note that the proof in [37] needs a slight modification. Further details on the Jacobson radical of U can be found in [72], Cor. 3.8, p. 367; note that the setup there is slightly different.

4   The fifth and the seventh conjecture

4.1 It is easy to see that the antipode of a Hopf algebra is an involution if the Hopf algebra is commutative or cocommutative. Since the notion of an antipode in a Hopf algebra generalizes the notion of an inverse in a group, it is reasonable to expect that this is a more general feature. In fact, it was not known for some time whether there exist Hopf algebras whose antipodes were not involutions - at least in the finite-dimensional case. Such a Hopf algebra was then constructed by E. J. Taft (cf. [75], [76]) and also by D. E. Radford (cf. [53]). Before that, R. G. Larson as well as M. E. Sweedler already had constructed examples of infinite dimensional Hopf algebras with antipodes of infinite order (cf. [18], [74], Chap. IV, Exercises), whereas Radford proved later that antipodes of finite-dimensional Hopf algebras have finite order (cf. [54], [55]).

When Kaplansky wrote down his conjectures, it was therefore known that the antipode was not always an involution. Kaplansky's fifth conjecture now states that the antipode of a finite-dimensional Hopf algebra over an algebraically closed field is an involution if the Hopf algebra is semisimple or cosemisimple. Here, a Hopf algebra is called cosemisimple if the dual Hopf algebra is semisimple.

It should first be observed that the question of algebraic closure is not important here, because the order of the antipode remains unchanged under base field extension. As will be explained below, the characteristic of the base field is of greater importance.

One of the first results on this conjecture was obtained by R. G. Larson, who proved it for Hopf algebras over fields of characteristic zero or sufficiently large characteristic, provided that the Hopf algebra is semisimple and the irreducible modules have dimension one or two (cf. [21]). His results were generalized by D. E. Radford, who proved that it suffices to assume that the two-sided ideals corresponding to the simple modules of dimension one or two generate the Hopf algebra H as a coalgebra (cf. [59]).

Kaplansky's fifth conjecture is better understood if it is split into two partial problems:

1. Is every semisimple Hopf algebra cosemisimple?

2. Is the antipode of a semisimple, cosemisimple Hopf algebra an involution?

In what follows, we shall treat these problems separately.

4.2   The most substantial progress on Kaplansky's fifth conjectures was made by R. G. Larson and D. E. Radford in two closely related articles published in 1988. In the first one of these they answered affirmatively the first question above under the assumption that the base field has characteristic zero (cf. [23]). Their proof relies on a trace formula that expresses the trace of the squared antipode in terms of the integrals of the Hopf algebra:

$$Tr(S_H^2) = \epsilon_H(\Gamma_H)\rho_H(1)$$

Here, ϵ_H and S_H denote the counit and the antipode of the Hopf algebra H, whereas $\Gamma_H \in H$ and $\rho_H \in H^*$ denote right integrals satisfying $\rho_H(\Gamma_H) = 1$. As H. J. Schneider points out, the immediate consequence $dim\, H = \epsilon_H(\Gamma_H)\rho_H(1)$ if $S_H^2 = id_H$ was already observed in [49]. He has later used the methods of this article to give simplified proofs of the first trace formula stated above as well as the second trace formula discussed below, and also for Radford's formula for the fourth power of the antipode (cf. [63]).

Recently, the first problem mentioned above has also been solved in large positive characteristic (cf. [68]). The techniques there rely on the one hand on the adjunction of a grouplike element making the square of the antipode an inner automorphism, on the other hand on the Perron-Frobenius theorem to prove the invertibility of the character of the adjoint representation.

4.3 We now discuss the second problem mentioned above. It should be observed that for this problem, we already know by Radford's formula that the fourth power of the antipode is the identity (cf. [55]). The most important progress on the second problem was also made by Larson and Radford in their second article, which was also published in 1988. There, they proved that the antipode of a semisimple, cosemisimple Hopf algebra is an involution if the characteristic of the base field is zero or larger than the square of the dimension of the Hopf algebra. Their technique relies on a second trace formula:

$$Tr(S_H^2) = (dim\, H)Tr(S_H^2 \mid_{x_R H})$$

Here, $x_R \in H$ denotes the character of the regular representation of H^*, i. e., the element satisfying $\phi(x_R) = Tr(L_\phi)$, where $L_\phi : H^* \to H^*$ denotes left multiplication by ϕ. It should be noted that it is reasonable to expect that this character is an integral of H. If it were possible to prove this in general, the above trace formula would read $Tr(S_H^2) = dim\, H$, and it would be easy to derive strong conclusions on Kaplansky's fifth conjecture.

4.4 The methods of Larson and Radford described so far also enabled them to solve Kaplansky's seventh conjecture. This conjecture states that for a semisimple, cosemisimple Hopf algebra over an algebraically closed field, the characteristic of the field does not divide the dimension of the Hopf algebra. Once again, the assumption that the base field be algebraically closed is not necessary. Using the above trace formulas, it is possible to give their complete proof of this conjecture in a few lines:

By Larson and Sweedler's generalization of Maschke's theorem, a Hopf algebra is semisimple if and only if the counit does not vanish on a nonzero integral. Therefore the right hand side of the first trace formula is nonzero if the Hopf algebra is semisimple and cosemisimple. Therefore, also the left hand side is nonzero, and by the second trace formula this implies that the dimension of the Hopf algebra is nonzero as an element of the base field, i. e., the characteristic of the base field does not divide the dimension.

4.5 The results on Kaplansky's fifth conjecture mentioned so far can be summarized in the following theorem:

Theorem Suppose that H is a semisimple Hopf algebra. Suppose that the characteristic p of the base field is zero or satisfies $p > m^{m-4}$ where $m = 2\,(dim\,H)^2$. Then H is cosemisimple and the antipode of H is an involution.

There are, however, additional results on this conjecture. According to W. D. Nichols, M. Eberwein has proved that the antipode of a semisimple Hopf algebra is an involution, provided that the characteristic of the base field is larger than the squared dimension of the Hopf algebra and the irreducible modules are of dimension 1, 2 or 3 (cf. [42], [8]). This generalizes the above mentioned result of Larson (cf. [21]). Larson and Radford have also proved a number of additional results; they have studied the question whether the property that the antipode is an involution can be extended from a Hopf subalgebra to a larger Hopf algebra, and they have carried out computer-based analysis of the algebra structures for a potential counterexample to Kaplansky's fifth conjecture (cf. [24]).

5 The sixth conjecture

5.1 Kaplansky's sixth conjecture may be interpreted as follows: For a Hopf algebra over an algebraically closed field, the dimension of every simple module divides the dimension of the Hopf algebra. In this generality, it was known at that time that this is false even for group rings. For example, the special linear groups $SL(2,p)$ of 2×2-matrices over a field with p elements, where p is an odd prime, admit simple modules over algebraically closed fields of characteristic p with a dimension that does not divide the order of the group (cf. [6], Example (17.17), p. 426). However, it was an open question at that time whether, for an absolutely irreducible representation in characteristic $p > 0$, the p-component of the dimension of the irreducible module divides the order of the group; this was Problem 17 in a list of open problems compiled by R. Brauer in 1963 (cf. [3], [11], p. 166). Even this conjecture was refuted when J. G. Thackray showed in 1981 that McLaughlin's simple group of order $2^7 \cdot 3^6 \cdot 5^3 \cdot 7 \cdot 11$ has a simple module of dimension $2^9 \cdot 7$ in characteristic 2 (cf. [81]). Of course, the result holds for group rings over algebraically closed fields of characteristic zero, or if the characteristic of the field does not divide the order of the group. More generally, it holds for p-solvable groups (cf. [6], Cor. (22.5), p. 518).

The first result on the conjecture for arbitrary Hopf algebras is due to R. G. Larson. This result also shows why the problem is more difficult for Hopf algebras than for groups. As noted above, the conjecture is true for group rings of finite groups over algebraically closed fields of characteristic zero (cf. [6], Prop. (9.32), p. 216). The proof of this result exploits properties of algebraic integers, and, in doing so, exploits the fact that the group ring $K[G]$ has an obvious integral form, namely $\mathbb{Z}[G]$. Larson's result says that the same holds if the Hopf algebra under consideration is defined over $\mathbb{Z}$. More precisely, Larson proves (cf. [20], Prop. 4.2, p. 208):

Theorem Suppose that R is a Dedekind domain with fraction field K and that H is a Hopf algebra over R that is finitely generated projective over R. Assume that $H \otimes_R K$ is split semisimple and involutory. Denote the set of left ideals in H by L. Then, for every simple $H \otimes_R K$-module V, the principal ideal $(dim\, V)$ divides the ideal $\epsilon_H(L)$, i. e., we have $(dim\, V) \supseteq \epsilon_H(L)$.

5.2 One of the most important results on Kaplansky's sixth conjecture is a theorem of W. D. Nichols and M. B. Richmond that anwers the question affirmatively if the simple module is two-dimensional (cf. [47], Cor. 12):

Theorem The dimension of a semisimple Hopf algebra over an algebraically closed field is even if the Hopf algebra has a simple module of dimension 2.

The technique of the authors is to analyse the decomposition of the tensor product of the two-dimensional simple module and its dual. More precisely, the argument proceeds in the way that either this decomposition behaves similar to the decomposition of the regular module of the Lie algebra $sl(2)$, in which case the Hopf algebra considered is infinite-dimensional, or the corresponding Hopf algebra admits certain Hopf algebra quotients of dimension 2, 12, 24 or 60 whose representations are similar to those of the dihedral groups D_n, the tetrahedral group A_4, the octahedral (resp. hexahedral) group S_4, or the icosahedral (resp. dodecahedral) group A_5. The authors have continued their investigation in [48].

5.3 Kaplansky's sixth conjecture has also been considered by S. Montgomery and S. J. Witherspoon (cf. [39]). They prove that, in certain situations, Kaplansky's conjecture holds for a Hopf algebra if it holds for a subalgebra. More precisely, they prove:

Theorem Suppose that A is a finite-dimensional algebra over an algebraically closed field such that the dimension of the simple modules of A divide the dimension of A. Suppose that G is a finite group such that the characteristic of the base field does not divide the order of G. If B is a crossed product of A and the group ring $K[G]$, then the dimensions of the simple modules of B divide the dimension of B. This also holds if B is a crossed product of A and the dual group ring $K[G]^*$.

Using results of G. I. Kac and A. Masuoka (cf. [15], [33]) Montgomery and Witherspoon then show that Hopf algebras of prime power dimension over algebraically closed fields of characteristic zero can be constructed by iterating the crossed product constructions described in the theorem. Therefore, Kaplansky's sixth conjecture holds for these Hopf algebras.

5.4 Kaplansky's sixth conjecture has also been verified for other classes of Hopf algebras. Y. Zhu has proved that, for a semisimple Hopf algebra H over an algebraically closed field of characteristic zero, the dimension of the simple $D(H)$-submodules of H divides the dimension of H (cf. [85]). Here, $D(H)$ denotes the Drinfel'd double of H. His argument relies, besides the so-called class equation that will be discussed below, on the fact that the action of the Drinfel'd double on H precisely centralizes the action of the character algebra on H. Similar results and additional consequences were obtained independently, but later in the author's dissertation, among them the semisimplicity of the character ring in positive characteristic. Zhu's result has been generalized by P. Etingof and S. Gelaki to arbitrary simple modules of the Drinfel'd double $D(H)$. As a consequence, they were able to obtain the following theorem (cf. [9], Theorem 1.5):

Theorem Suppose that H is a quasitriangular semisimple Hopf algebra over an algebraically closed field of characteristic zero. Then the dimension of every simple H-module divides the dimension of H.

6 The eighth conjecture

6.1 Kaplansky's eighth conjecture states that a Hopf algebra of prime dimension over an algebraically closed field is commutative and cocommutative. Since commutativity properties remain unchanged under extensions of the base field, the assumption of algebraic closure is again not relevant here. However, in the algebraically closed case, something more can be said. As observed by P. Cartier and D. Harrison (cf. [37], Theorem 2.3.1) finite-dimensional cocommutative cosemisimple Hopf algebras over algebraically closed fields are group rings. As a consequence of the Nichols-Zoeller theorem, it was shown by R. G. Larson and D. E. Radford that a Hopf algebra of prime dimension is semisimple and cosemisimple if the characteristic of the base field is zero or greater than the dimension of the Hopf algebra (cf. [24], Theorem 2.3). Therefore, the conjecture would imply in these cases that the only example of a Hopf algebra of prime dimension is the group ring of the cyclic group of prime order. Of course, this does not hold if the characteristic of the base field coincides with the dimension of the Hopf algebra, because then the restricted enveloping algebra of the base field, considered an abelian restricted Lie algebra via the Frobenius homomorphism, is another example of a commutative and cocommutative Hopf algebra of dimension p.

The history of Kaplansky's eighth conjecture is slightly strange. Kaplansky himself had considered the dimensions 2, 3 and 5 in the first appendix of his notes. He added as a remark to the conjecture that G. I. Kac had partial results on this conjecture, and his bibliography also contains the corresponding article [15]. The conjecture was then considered as open for about twenty years; the only contribution was given by C. R. Cai and H. X. Chen, who proved the conjecture for dimension 7 and 13 over fields of characteristic zero (cf. [4]). A year later, Y. Zhu was able to prove the conjecture completely over fields of characteristic zero, using a rather different argument based on the Drinfel'd double construction (cf. [83]). According to Zhu, E. Effros then pointed out Kac's work to him, and upon reading it he discovered that Kac' argument could be extended comparatively easily to the general case. Zhu then decided to follow Kac' argument in the published version of his manuscript (cf. [84]).

The argument of Kac and Zhu is based on a theorem which is now called the class equation, and may be considered as even more important than the solution of the conjecture itself:

Theorem Suppose that H is a semisimple Hopf algebra over an algebraically closed field of characteristic zero. Suppose that $e_1, \ldots, e_m$ is a complete set of primitive orthogonal idempotents of the character ring $Ch(H)$, where e_1 is an integral of H^*. Then we have

$$dim\, H = \sum_{i=1}^{m} dim\, e_i H^*$$

$dim\, e_1 H^* = 1$ and $dim\, e_i H^*$ divides $dim\, H$.

Note that the nontrivial part is the divisibility statement. The theorem reduces to the ordinary class equation if applied to the group ring of a finite group, because the idempotents above then become the characteristic functions of the conjugacy classes. If applied to the dual group ring, the theorem implies that the dimensions of the simple modules divide the order of the group. A simplified proof of the class equation has been given by M. Lorenz (cf. [26]).

6.2   Nowadays, Kaplansky's eighth conjecture may be understood as the first step in the program to understand the structure of semisimple Hopf algebras in terms of the prime factors of their dimension, similar to the situation in finite group theory. This program, also called the classification program for semisimple Hopf algebras, has recently experienced considerable progress, the class equation being an important instrument in this investigation. One of the first consequences of the class equation is the observation that a Hopf algebra of prime power order contains a nontrivial central grouplike element (cf. [15], [33]). Here, the proof is analogous to the proof that a p-group has nontrivial center. As a consequence, A. Masuoka has proved that over a field of characteristic zero,

Hopf algebras of dimension p^2 are commutative and cocommutative. A. Masuoka
has also contributed several other results (cf. [31], [32], [34], [35], [36]). Further
results can be found in [13], [10], [70]. Since this development is reviewed in [38],
we shall not give further details here.

7   The tenth conjecture

7.1   Kaplansky's tenth conjecture states that, over an algebraically closed field,
the number of isomorphism classes of Hopf algebras of a given dimension is
finite, provided that the characteristic of the base field does not divide the dimension. This was proved by D. Ştefan for isomorphism classes of semisimple
and cosemisimple Hopf algebras (cf. [71]). Note that, due to the positive solution of the seventh conjecture, for a semisimple and cosemisimple Hopf algebra
the characteristic does not divide the dimension. H. J. Schneider has given a
simplified proof of this theorem (cf. [67]).

7.2   However, the conjecture is false in general. This was realized more or
less simultaneously by three groups of researchers in 1997, namely N. Andruskiewitsch and H. J. Schneider on the one hand, M. Beattie, S. Dăscălescu
and L. Grünenfelder on the other hand, and also by S. Gelaki (cf. [1], [2], [12]).
In the approach of the first two groups, the conjecture is refuted by constructing pointed Hopf algebras of dimension p^4 for a prime number p, whereas in
Gelaki's approach infinite families of Hopf algebras of dimension mn^2 are constructed, where $m > 2$ and $n > 1$ are natural numbers such that n divides m.
The approach of Andruskiewitsch and Schneider also leads to the classification
of pointed Hopf algebras of dimension p^3 over an algebraically closed field of
characteristic zero, where p is a prime. This has also be obtained by S. Caenepeel
and S. Dăscălescu (cf. [5]).

8   The second and the fourth conjecture

8.1   Kaplansky's second conjecture states that a coalgebra is admissible if and
only if every finite subset is contained in a finite-dimensional admissible subcoalgebra. Here, a coalgebra is called admissible if it admits an algebra structure
making it a Hopf algebra. This question should be viewed as a variant of the theorem that every finite subset of a coalgebra is contained in a finite-dimensional
subcoalgebra (cf. [16], Theorem 2, p. 7, [37], Theorem 5.1.1, p. 56). To the best
of the author's knowledge, this conjecture of Kaplansky has not been considered
in the literature.

8.2   Kaplansky's fourth conjecture states that if an element x in a Hopf algebra H satisfies

$$a_{(1)}xS(a_{(2)}) = \epsilon(a)x$$

for every $a \in H$, then x is contained in the center of H. Here, we have used the following variant of the Heyneman-Sweedler sigma notation for the coproduct:

$$\Delta(a) = a_{(1)} \otimes a_{(2)}$$

This is obvious, because then we have:

$$ax = a_{(1)}xS(a_{(2)})a_{(3)} = \epsilon(a_{(1)})xa_{(2)} = xa$$

This calculation can certainly be found in many places, among them [37], Lemma 5.7.2, p. 83.

9   Appendix

In this appendix, we reproduce literally appendix 2 from [16], which contains Kaplansky's conjectures. It may be helpful for the understanding of the conjectures to know that Kaplansky's lecture notes contain a bibliography which, among many other references, lists the items labeled [15], [19] and [50] of the present article. Kaplansky's conjectures are:

1. If C is a Hopf subalgebra of the Hopf algebra B then B is a free left C-module. (Remark. Nichols has proved this if B contains the coradical of C.)

2. Call a coalgebra C <u>admissible</u> if it admits an algebra structure making it a Hopf algebra. The conjecture states that C is admissible if and only if every finite subset of C lies in a finite-dimensional admissible subcoalgebra. (Remarks. 1. Both implications seem hard. 2. There is a corresponding conjecture where "Hopf algebra" is replaced by "bialgebra". 3. There is a dual conjecture for locally finite algebras.)

3. A Hopf algebra of characteristic 0 has no non-zero central nilpotent elements.

4. (Nichols). Let x be an element in a Hopf algebra H with antipode S. Assume that for any a in H we have

$$\sum b_i x S(c_i) = \epsilon(a)x$$

where $\Delta a = \sum b_i \otimes c_i$. Conjecture: x is in the center of H.

In the remaining six conjectures H is a finite-dimensional Hopf algebra over an algebraically closed field.

5. If H is semisimple on either side (i.e. either H or the dual H^* is semisimple as an algebra) the square of the antipode is the identity.

6. The size of the matrices occuring in any full matrix constituent of H divides the dimension of H.

7. If H is semisimple on both sides the characteristic does not divide the dimension.

8. If the dimension of H is prime then H is commutative and cocommutative.

Remark. Kac, Larson, and Sweedler have partial results on 5 - 8.

In the two final conjectures assume that the characteristic does not divide the dimension of H.

9. The dimension of the radical is the same on both sides.

10. There are only a finite number (up to isomorphism) of Hopf algebras of a given dimension.

References

[1] N. Andruskiewitsch/H. J. Schneider: Lifting of quantum linear spaces and pointed Hopf algebras of order p^3, Preprint, 1997

[2] M. Beattie/S. Dăscălescu/L. Grünenfelder: On the number of types of finite dimensional Hopf algebras, Preprint, 1997

[3] R. Brauer: Representations of finite groups. In: Lectures on modern mathematics, Vol. I, 133-175, Wiley, New York, 1963

[4] C. R. Cai/H. X. Chen: Prime dimensional Hopf algebras. In: Proceedings of the first China-Japan international symposium on ring theory, Okayama Univ., Okayama, 1992

[5] S. Caenepeel/S. Dăscălescu: Pointed Hopf algebras of dimension p^3, Preprint, 1997

[6] C. W. Curtis/I. Reiner: Methods of representation theory with applications to finite groups and orders, Vol. I, Wiley Interscience, New York, 1981

[7] M. Demazure/P. Gabriel: Groupes algébriques I, North Holland, Amsterdam 1970

[8] M. Eberwein: Cosemisimple Hopf algebras, Ph. D. Dissertation, Florida State University, Tallahassee, 1992

[9] P. Etingof/S. Gelaki: Some properties of finite-dimensional semisimple Hopf algebras, Preprint, 1997

[10] P. Etingof/S. Gelaki: Semisimple Hopf algebras of dimension pq are trivial, Preprint, 1998

[11] W. Feit: The representation theory of finite groups, North-Holland, Amsterdam, 1982

[12] S. Gelaki: On pointed Hopf algebras and Kaplansky's 10th conjecture, Preprint, 1998

[13] S. Gelaki/S. Westreich: On semisimple Hopf algebras of dimension pq, Preprint, 1998

[14] J. C. Jantzen: Lectures of quantum groups, Grad. Stud. in Math. 6, Amer. Math. Soc., Providence, 1996

[15] G. I. Kac: Certain arithmetic properties of ring groups, Funktsional. Anal. i Prilozhen. 6 (1972), 88-90. English translation: Funct. Anal. Appl. 6 (1972), 158-160

[16] I. Kaplansky: Bialgebras, Lecture Notes in Mathematics, University of Chicago, Chicago, 1975

[17] C. Kassel: Quantum groups, Grad. Texts in Math. 155, Springer Verlag, Berlin, 1995

[18] R. G. Larson: The order of the antipode of a Hopf algebra, Proc. Amer. Math. Soc. 21 (1969), 167-170

[19] R. G. Larson: Characters of Hopf algebras, J. Algebra 17 (1971), 352-368

[20] R. G. Larson: Orders in Hopf algebras, J. Algebra 22 (1972), 201-210

[21] R. G. Larson: Cosemisimple Hopf algebras with small simple subcoalgebras are involutory, Comm. Algebra 11 (1983), 1175-1186

[22] R. G. Larson/D. E. Radford: Semisimple cosemisimple Hopf algebras, Amer. J. Math. 110 (1988), 187-195

[23] R. G. Larson/D. E. Radford: Finite dimensional cosemisimple Hopf algebras in characteristic 0 are semisimple, J. Algebra 117 (1988), 267-289

[24] R. G. Larson/D. E. Radford: Semisimple Hopf algebras, J. Algebra 171 (1995), 5-35

[25] R. G. Larson/M. E. Sweedler: An associative orthogonal bilinear form for Hopf algebras, Amer. J. Math. 91 (1969), 75-93

[26] M. Lorenz: On the class equation for Hopf algebras, Preprint, 1996, to appear in Proc. Amer. Math. Soc.

[27] A. Masuoka: On Hopf algebras with commutative coradicals, J. Algebra 144 (1991), 415-466

[28] A. Masuoka: Freeness of Hopf algebras over coideal subalgebras, Comm. Algebra 20 (1992), 1353-1373

[29] A. Masuoka: Coideal subalgebras in finite Hopf algebras, J. Algebra 163 (1994), 819-831

[30] A. Masuoka/D. Wigner: Faithful flatness of Hopf algebras, J. Algebra 170 (1994) 156-164

[31] A. Masuoka: Semisimple Hopf algebras of dimension 6, 8, Israel J. Math. 92 (1995), 361-373

[32] A. Masuoka: Semisimple Hopf algebras of dimension $2p$, Comm. Algebra 23 (1995), 1931-1940

[33] A. Masuoka: The p^n-theorem for semisimple Hopf algebras, Proc. Amer. Math. Soc. 124 (1996), 735-737

[34] A. Masuoka: Semisimple Hopf algebras of dimension p^3 obtained by an extension, J. Algebra 178 (1995), 791-806

[35] A. Masuoka: Some further classification results on semisimple Hopf algebras, Comm. Algebra 24 (1996), 307-329

[36] A. Masuoka: Calculation of some groups of Hopf algebra extensions, J. Algebra 191 (1997), 568-588

[37] S. Montgomery: Hopf Algebras and their Actions on Rings, CBMS Regional Conf. Ser. in Math., Vol. 82, Amer. Math. Soc., Providence, R. I., 1993

[38] S. Montgomery: Classifying finite-dimensional semisimple Hopf algebras, Preprint, 1997

[39] S. Montgomery/S. J. Witherspoon: Irreducible representations of crossed products, J. Pure Appl. Algebra, in press

[40] S. H. Ng: On the projectivity of module coalgebras, to appear in: Proc. Amer. Math. Soc.

[41] W. D. Nichols: Quotients of Hopf algebras, Comm. Algebra 6 (1978), 1789-1800

[42] W. D. Nichols: Cosemisimple Hopf algebras. In: Advances in Hopf algebras, Marcel Dekker, New York, 1994

[43] W. D. Nichols/M. B. Zoeller: Finite dimensional Hopf algebras are free over grouplike subalgebras, J. Pure Appl. Algebra 56 (1989), 51-57

[44] W. D. Nichols/M. B. Zoeller: Freeness of infinite dimensional Hopf algebras over grouplike subalgebras, Comm. Algebra 17 (1989), 413-424

[45] W. D. Nichols/M. B. Zoeller: A Hopf algebra freeness theorem, Amer. J. Math. 111 (1989), 381-385

[46] W. D. Nichols/M. B. Richmond: Freeness of infinite dimensional Hopf algebras, Comm. Algebra 20 (1992), 1489-1492

[47] W. D. Nichols/M. B. Richmond: The Grothendieck group of a Hopf algebra, J. Pure Appl. Algebra 106 (1996), 297-306

[48] W. D. Nichols/M. B. Richmond: The Grothendieck algebra of a Hopf algebra I, Comm. Algebra 26 (1998), 1081-1095

[49] U. Oberst/H.-J. Schneider: Über Untergruppen endlicher algebraischer Gruppen, Manuscripta Math. 8 (1973), 217-241

[50] U. Oberst/H.-J. Schneider: Untergruppen formeller Gruppen von endlichem Index, J. Algebra 31 (1974), 10-44

[51] B. Pareigis: Endliche Hopf-Algebren, Algebra-Ber., Uni-Druck, Munich, 1973

[52] B. Pareigis: Lectures on quantum groups and non-commutative geometry, Lecture notes, Munich, 1998

[53] D. E. Radford: A free rank 4 Hopf algebra with antipode of order 4, Proc. Nat. Acad. Sci. USA 30 (1971), 55-58

[54] D. E. Radford: The antipode of a finite-dimensional Hopf algebra over a field has finite order, Bull. Amer. Math Soc. 81 (1975), 1103-1105

[55] D. E. Radford: The order of the antipode of a finite-dimensional Hopf algebra is finite, Amer. J. Math. 98 (1976), 333-355

[56] D. E. Radford: Pointed Hopf algebras are free over Hopf subalgebras, J. Algebra 45 (1977), 266-273

[57] D. E. Radford: Operators on Hopf algebras, Amer. J. Math. 99 (1977), 139-158

[58] D. E. Radford: Freeness (Projectivity) criteria for Hopf algebras over Hopf subalgebras, J. Pure Appl. Algebra 11 (1977), 15-28

[59] D. E. Radford: On the antipode of a cosemisimple Hopf algebra, J. Algebra 88 (1984), 68-88

[60] D. E. Radford: Minimal quasitriangular Hopf algebras, J. Algebra 157 (1993), 285-315

[61] P. Schauenburg: Faithful flatness over Hopf subalgebras - Counterexamples. In: Proceedings of the conference 'Interactions between Ring theory and Representations of algebras', Murcia, Spain, 1998

[62] S. Schmidt-Samoa: Ein Quotient der Quantengruppe $U_q(sl_2)$ im Einheitswurzelfall, Diplomarbeit, Univ. Göttingen, Göttingen, 1995

[63] H.-J. Schneider: Lectures on Hopf algebras, Universidad de Cordoba Trabajos de Matematica, Serie "B", No. 31/95, Cordoba, Argentina, 1995

[64] H. J. Schneider: Zerlegbare Untergruppen affiner Gruppen, Math. Ann. 255 (1981), 139-158

[65] H.-J. Schneider: Normal basis and transitivity of crossed products for Hopf algebras, J. Algebra 152 (1992), 196-231

[66] H.-J. Schneider: Some remarks on exact sequences of quantum groups, Comm. Algebra 21 (1993), 3337-3357

[67] H.-J. Schneider: In preparation

[68] Y. Sommerhäuser: On Kaplansky's fifth conjecture, Preprint gk-mp-9702/50, to appear in: J. Algebra

[69] Y. Sommerhäuser: Ribbon transformations, Integrals, and Triangular Decompositions, Preprint gk-mp-9707/52, to appear in: J. Algebra

[70] Y. Sommerhäuser: Yetter-Drinfel'd Hopf algebras over groups of prime order, in preparation

[71] D. Ştefan: The set of types of n-dimensional semisimple and cosemisimple Hopf algebras is finite, J. Algebra 193 (1997), 571-580

[72] R. Suter: Modules over $U_q(sl_2)$ Comm. Math. Phys. 163 (1994) 359-393

[73] M. E. Sweedler: Integrals for Hopf algebras, Ann. of Math. (2) 89 (1969), 323-335

[74] M. E. Sweedler: Hopf algebras, W. A. Benjamin, New York, 1969

[75] E. J. Taft: The order of the antipode of finite-dimensional Hopf algebra, Proc. Nat. Acad. Sci. USA 68 (1971), 2631-2633

[76] E. J. Taft/R. L. Wilson: There exist finite-dimensional Hopf algebras with antipodes of arbitrary even order, J. Algbra 62 (1980), 283-291

[77] M. Takeuchi: A correspondence between Hopf ideals and sub-Hopf algebras, Manuscripta Math. 7 (1972), 251-270

[78] M. Takeuchi: On a semidirect product decomposition of affine groups over a field of characteristic 0, Tôhoku Math. J. 24 (1972), 453-456

[79] M. Takeuchi: Free Hopf algebras generated by coalgebras, J. Math. Soc. Japan 23 (1971), 561-582

[80] M. Takeuchi: Relative Hopf modules - equivalences and freeness criteria, J. Algebra 60 (1979), 452-471

[81] J. G. Thackray: Modular representations of some finite groups, Dissertation, Univ. Cambridge, Cambridge, 1981

[82] W. Waterhouse: Introduction to affine group schemes, Grad. Texts in Math. 66, Springer Verlag, Berlin, 1979

[83] Y. Zhu: Quantum double construction of quasitriangular Hopf algebras and Kaplansky's conjecture, Preprint, 1993

[84] Y. Zhu: Hopf algebras of prime dimension, Internat. Math. Res. Notices 1 (1994), 53-59

[85] Y. Zhu: A commuting pair in Hopf algebras, Proc. Amer. Math. Soc. 125 (1997), 2847-2851

[86] M. B. Zoeller: Freeness of Hopf algebras over semisimple grouplike subalgebras, J. Algebra 118 (1988), 102-108

The Duality Theorem for Coactions of Multiplier Hopf Algebras

Alfons Van Daele and Yinhuo Zhang[*]
Dept. of Mathematics
University of Leuven
Celestijnenlaan 200B, B-3001 Heverlee, Belgium

Abstract

In this note we give a proof of the duality theorem for a coaction of a multiplier Hopf algebra pair claimed in [5]. Let $\langle A, B \rangle$ be a multiplier Hopf algebra pair in the sense of [4], R a non-degenerate algebra on which B coacts. Then we can form the double smash product algebra $(R \# A) \# B$. If the pair $\langle A, B \rangle$ satisfies a sort of RL-condition, then we have an algebra isomorphism:

$$(R \# A) \# B \cong R \otimes (A \# B).$$

As a special case, if A is an algebraic quantum group and $B = \widehat{A}$, we obtain

$$(R \# A) \# \widehat{A} \cong R \otimes (A \diamond \widehat{A}),$$

where $A \diamond \widehat{A}$ is isomorphic with the finite rank operator algebra on the vector space A. In case $\langle A, B \rangle$ is a dual pair of Hopf algebras, we get the Blattner-Montgomery duality theorem [3, 2.1].

1   Preliminaries

Throughout we fix a ground field k. Algebras, tensor products and vector spaces without indications are over k. First let us recall the definition of a multiplier Hopf algebra (see [6] for details). Let A be an algebra with or without unit, but with a *non-degenerate* product (i.e., $aA = 0$ or $Aa = 0 \Rightarrow a = 0$). The multiplier algebra $M(A)$ may be viewed as the largest algebra with unit in which A is a dense ideal. Consider the tensor product $A \otimes A$ of A which is again an algebra with a non-degenerate product. The embedding $A \otimes A \hookrightarrow M(A \otimes A)$ factors through $M(A) \otimes M(A)$, i.e. $A \otimes A \hookrightarrow M(A) \otimes M(A) \hookrightarrow M(A \otimes A)$.

A comultiplication (or a coproduct) on A is a homomorphism $\Delta : A \to M(A \otimes A)$ such that $\Delta(a)(1 \otimes b)$ and $(a \otimes 1)\Delta(b)$ are elements of $A \otimes A$ for all $a, b \in A$. We must have Δ *coassociative* in the sense that

$$(a \otimes 1 \otimes 1)(\Delta \otimes \iota)(\Delta(b)(1 \otimes c)) = (\iota \otimes \Delta)((a \otimes 1)\Delta(b))(1 \otimes 1 \otimes c)$$

for all $a, b, c \in A$ (where ι denotes the identity map).

---

[*]Supported by Research Council of K U Leuven

1.1 Definition [6] A pair (A, Δ) of an algebra A with a non-degenerate product and a comultiplication Δ on A is called a *multiplier Hopf algebra* if the linear maps from $A \otimes A$ to itself given by

$$a \otimes b \to \Delta(a)(1 \otimes b)$$
$$a \otimes b \to (a \otimes 1)\Delta(b)$$

are bijective.

The bijectivities of the above two maps are equivalent with the existence of a *counit* ϵ and an *antipode* S satisfying (and defined by)

$$(\epsilon \otimes \iota)(\Delta(a)(1 \otimes b)) = ab$$
$$(\iota \otimes \epsilon)((a \otimes 1)\Delta b)) = ab$$
$$m(S \otimes \iota)(\Delta(a)(1 \otimes b)) = \epsilon(a)b$$
$$m(\iota \otimes S)((a \otimes 1)\Delta(b)) = \epsilon(b)a,$$

where $\epsilon : A \to k$ is a homomorphism, $S : A \to M(A)$ is an anti-homomorphism and m is the multiplication map which can be extended to $A \otimes M(A)$ and $M(A) \otimes A$.

A multiplier Hopf algebra is said to be *regular* if (A, Δ^{op}) is a multiplier Hopf algebra (or equivalently if the antipode S is bijective from A to A). Any Hopf algebra is a multiplier Hopf algebra. Conversely a multiplier Hopf algebra with unit is a Hopf algebra. So the notion of multiplier Hopf algebra is a natural generalization of the notion of a Hopf algebra for algebras without unit.

Throughout this paper we only consider regular multiplier Hopf algebra, and all the algebras considered have non-degenerate products. Let (A, Δ) be a multiplier Hopf algebra.

A non-zero linear functional φ on A is called a *left integral* if $(\iota \otimes \varphi)\Delta(a) = \varphi(a)1 \in M(A)$ for all $a \in A$. Similarly a right integral can be defined. Such an integral does not always exists. But is easy to see that the antipode will convert a left integral to a right integral. In general, left and right integrals are not necessarily the same.

In [9] a regular multiplier Hopf algebra with a non-trivial integral is called an *algebraic quantum group* (see also [7]).

1.2 Theorem [9] Let (A, Δ) be an algebraic quantum group with a left integral φ. Set $\hat{A} = \{\varphi(\cdot A) \mid a \in A\}$. Considering $\hat{A}$ as a subspace of A^*, the duals of the coproduct and the product of A give a product and a coproduct $\hat{\Delta}$ of $\hat{A}$ and $(\hat{A}, \hat{\Delta})$ is again algebraic quantum group such that $(\hat{\hat{A}}, \hat{\hat{\Delta}}) \cong (A, \Delta)$.

Let (A, Δ) be an algebraic quantum group. Given elements $(a_1, \cdots, a_n)$ in A. There exists an element $e \in A$ such that $a_i e = ea_i = a_i$. If A has a cointegral (i.e. an element $t \in A$ such that $ah = \epsilon(a)h$ for all $a \in A$), then the element e can be chosen to be an idempotent, cf. [10]. In this case A is said to have local units.

Recall from [4] that the comultiplication Δ of A allows us to write $\Delta(a) = \sum a_{(1)} \otimes a_{(2)}$ in the sense that

$$\Delta(a)(1 \otimes b) = \sum a_{(1)} \otimes a_{(2)}b$$
$$\text{and} \quad (b \otimes 1)\Delta(a) = \sum ba_{(1)} \otimes a_{(2)}$$

in $A \otimes A$ for all $a, b \in A$. In this case, we say that $a_{(2)}$ (resp. $a_{(1)}$) is *covered* by the element b.

Now let us recall from [5] some notions about actions of multiplier Hopf algebras.

1.3 Definition Let (A, Δ) be a multiplier Hopf algebra. An algebra R is called an *A-module algebra* if R is a left unital A-module (i.e. $AR = R$) and satisfies the compatibility conditions :

$$a \cdot (xy) = \sum (a_{(1)} \cdot x)(a_{(2)} \cdot y)$$

for all $a \in A$ and $x, y \in R$.

The above identity makes sense because the element y may be replaced by cy and

$$a \cdot (x(c \cdot y)) = m(\Delta(a)(1 \otimes c) \cdot (x \otimes y)).$$

In this case, we say that $a_{(2)}$ is *covered* by y. Similarly we could also have replaced x by bx because A is assumed to be regular.

1.4 Definition Let R be an A-module algebra. The *smash product* algebra $R\#A$ is defined as follows. We let $R\#A = R \otimes A$ as a vector space while the multiplication is given by

$$(x\#a)(y\#b) = \sum x(a_{(1)} \cdot y)\#a_{(2)}b$$

for all $x, y \in R$ and $a, b \in A$. The product makes sense because $\sum a_{(1)} \otimes a_{(2)}b \in A \otimes A$. Remark that $R\#A$ is a non-degenerate associative algebra. There is a natural action of $R\#A$ on R given by

$$(x\#a) \cdot y = x(a \cdot y)$$

where $x, y \in R$ and $a \in A$.

1.5 Definition [11] Let (A, Δ) be a multiplier Hopf algebra, and R an algebra. By a coaction of A on R, we mean a monomorphism $\Gamma : R \to M(R \otimes A)$ satisfying :

i). $\Gamma(R)(1 \otimes A) \subseteq R \otimes A$ and $(1 \otimes A)\Gamma(R) \subseteq R \otimes A$,
ii). $(\Gamma \otimes \iota)\Gamma = (\iota \otimes \Delta)\Gamma$.

In this case R is called an *A-comodule algebra*. If, in addition, $(R \otimes 1)\Gamma(R) \subseteq R \otimes A$, then Γ is called a *reduced* coaction.

1.6 Examples i) Let (A, Δ) be a multiplier Hopf algebra. Then $\Gamma = \Delta$ is a reduced coaction of A on itself.

ii) Let R be an algebra, G the group of automorphisms of R. Let $K(G)$ be the dual multiplier Hopf algebra of the group Hopf algebra kG. Then R is a $K(G)$-comodule algebra with coaction Γ given by $\Gamma(x) = \sum_{g \in G} g(x) \otimes p_g \in M(R \otimes K(G))$, where $p_g(\sigma) = \delta_{g\sigma}$. It is easy to see that Γ is reduced if and only if for any two elements $x, y \in R$, $x\sigma(y) = 0$ except for finitely many element $\sigma \in G$.

For a coaction $\Gamma : R \longrightarrow M(R \otimes A)$, the injectivity of Γ is equivalent to the counitary property:

$$(\iota \otimes \epsilon)\Gamma(x) = x$$

for all $x \in R$.

We are allowed to use the sigma notations to express a coaction Γ. Write $\Gamma(x) = \sum x_{(0)} \otimes x_{(1)}$ and $(\Gamma \otimes \iota)\Gamma(x) = \sum x_{(0)} \otimes x_{(1)} \otimes x_{(2)}$. Then we may rewrite the conditions i) and ii) of (1.5) in sigma notations :

For all $x \in R$ and $a, b \in A$, we have

i). $\sum x_{(0)} \otimes x_{(1)}a = \Gamma(x)(1 \otimes a) \in R \otimes A$ and $\sum x_{(0)} \otimes ax_{(1)} = (1 \otimes a)\Gamma(x) \in R \otimes A$,

ii). $\sum x_{(0)} \otimes x_{(1)}a \otimes x_{(2)}b = \sum_i x_{(0)} \otimes (x_{(1)}a_i)_{(1)} \otimes (x_{(1)}a_i)_{(2)}b_i$,

iii). $\sum x_{(0)}\epsilon(x_{(1)}) = 1 \in M(R)$,

where $a \otimes b = \sum_i \Delta a_i(1 \otimes b_i)$.

Note that whenever one uses the Sweedler notation one has to make sure that $x_{(1)}$ is covered on one side by an element of A.

1.7 Theorem. [11, 2.2]. Let (A, Δ) be an algebraic quantum group, and R an algebra. Then R is an A-module algebra if and only if R is $\hat{A}$-comodule algebra.

In particular, if A is an algebraic quantum group of discrete type, then $\hat{A}$ is a Hopf algebra with integral. In this case, the coaction of A (or $\hat{A}$) arising from the action of $\hat{A}$ (or A) can be easily expressed as follows (see [14] for the details).

Let (A, Δ) be an algebraic quantum group of discrete type, $\hat{A}$ the dual Hopf algebra. Let t be a left cointegral of A, φ a left integral of A such that $\varphi(t) = 1$. Then

i). In case R is an A-module algebra, the coaction of $\hat{A}$ on R is

$$\Gamma(x) = \sum S_A^{-1}(t_{(1)}) \cdot x \otimes \varphi(\cdot t_{(2)}),$$

ii). In case R is an $\hat{A}$-module algebra, the coaction of A on R is

$$\Gamma(x) = \sum \varphi(\cdot t_{(2)}) \cdot x \otimes S_A^{-1}(t_{(1)}).$$

For further details about corepresentations of multiplier Hopf algebras, one may refer to [12].

2   The duality theorem

Let us first recall the notion of a pairing between two regular multiplier Hopf algebras (as introduced in [4]). Start with two regular multiplier Hopf algebras A and B and a non-degenerate bilinear from $\langle\,,\,\rangle$ from $A \times B$ to k. For all $a \in A$ and $b \in B$ we can define an element $a \rightharpoonup b$ in $M(B)$ by

$$
\begin{aligned}
(a \rightharpoonup b)b' &= \sum \langle a, b_{(2)}\rangle b_{(1)}b' \\
b'(a \rightharpoonup b) &= \sum \langle a, b_{(2)}\rangle b'b_{(1)}
\end{aligned}
$$

whenever $b' \in B$. Let us now assume that the pairing is such that $a \rightharpoonup b \in B$ for all $a \in A$ and $b \in B$. Then, it makes sense to require that

$$\langle a', a \rightharpoonup b\rangle = \langle a'a, b\rangle$$

for all $a, a' \in A$ and $b \in B$. This essentially means that the product in A is dual to the coproduct in B. Then, for all $a, a', a'' \in A$ and $b \in B$ we will have

$$
\begin{aligned}
\langle a, (a'a'') \rightharpoonup b\rangle &= \langle a(a'a''), b\rangle \\
&= \langle aa', a'' \rightharpoonup b\rangle \\
&= \langle a, a' \rightharpoonup (a'' \rightharpoonup b)\rangle
\end{aligned}
$$

and because we have taken our pairing to be non-degenerate, we get

$$(a'a'') \rightharpoonup b = a' \rightharpoonup (a'' \rightharpoonup b).$$

Therefore, B is a left A-module.

This observation led to the notion of a *multiplier Hopf algebra pairing* in [4] which consists of two regular multiplier Hopf algebras A and B together with a non-degenerate bilinear form $\langle \, , \, \rangle : A \times B \longrightarrow k$ satisfying the following conditions :

$$\begin{array}{ll} \sum \langle a_{(1)}, b \rangle a_{(2)} \in A & \sum \langle a, b_{(1)} \rangle b_{(2)} \in B \\ \sum \langle a_{(2)}, b \rangle a_{(1)} \in A & \sum \langle a, b_{(2)} \rangle b_{(1)} \in B \end{array} \qquad (1)$$

for all $a \in A$ and $b \in B$, and

$$\begin{array}{ll} \langle \sum \langle a_{(1)}, b \rangle a_{(2)}, b' \rangle = \langle a, bb' \rangle & \langle a', \sum \langle a, b_{(1)} \rangle b_{(2)} \rangle = \langle aa', b \rangle \\ \langle \sum \langle a_{(2)}, b \rangle a_{(1)}, b' \rangle = \langle a, b'b \rangle & \langle a', \sum \langle a, b_{(2)} \rangle b_{(1)} \rangle = \langle a'a, b \rangle, \end{array} \qquad (2)$$

for all $a, a' \in A$ and $b, b' \in B$.

It follows that four modules are involved. We have that B is a left and a right A-module (in fact, an A-bimodule) and that A is a left and a right B-module (also a B-bimodule). We will denote the left actions by $\rightharpoonup$ and the right actions by $\leftharpoonup$. All of these modules are unital for a multiplier Hopf algebra pairing (see [5, 2.8]).

Remark that the fact that these actions are unital also implies that e.g. in the expression

$$\sum \langle a_{(1)}, b \rangle a_{(2)},$$

the element $a_{(1)}$ is also *covered* by b in the following sense. Take an element $e \in A$ such that $e \rightharpoonup b = b$ (this is possible because B is a unital left A-module and A has one-sided local units). Then we will have

$$\sum \langle a_{(1)}, b \rangle a_{(2)} = \sum \langle a_{(1)} e, b \rangle a_{(2)}$$

so that $a_{(1)}$ is covered in the usual sense. This covering of $a_{(1)}$ by b through the pairing in turn implies that

$$\sum \langle a_{(1)}, b \rangle a_{(2)} \in A$$

for all $a \in A$ and $b \in B$. This shows that the axioms for a pairing are not independent from each other. These observations are important for the formulas and calculations further in this paper.

It is shown in [5, 6.1] that the two actions:

$$\begin{array}{ll} A \otimes B \longrightarrow B, & a \rightharpoonup b = \sum \langle a, b_{(2)} \rangle b_{(1)}, \\ B \otimes A \longrightarrow A, & b \rightharpoonup a = \sum \langle a_{(2)}, b \rangle a_{(1)} \end{array}$$

make A (or B) a left B- (or A-) module algebra in the sense of definition 1.4. Because we have an action of A on B, we can define the smash product $B \# A$. Similarly, the action of B on A yields another smash product $A \# B$. These two smash products are anti-isomorphic as algebras, cf.[5, 6.4].

Remark that in case $\langle A, B \rangle$ is a Hopf algebra pairing, we do not require the bilinear form $\langle \, , \, \rangle$ to be non-degenerate. This is because we do not have any problem with the unitary property.

Now let R be a left A-module algebra. Consider the smash product $R \# A$. The canonical action of B on A induces a natural action of B on $R \# A$ given by

$$b(x \# a) = x \#(b \rightharpoonup a) = \sum \langle a_{(2)}, b \rangle \, x \# a_{(1)}$$

whenever $a \in A$, $b \in B$ and $x \in R$.

It easily follows that this map makes $R \# A$ into a left B-module. It is a unital module because A is a unital B-module. But just as A is a B-module algebra, the same is true for $R \# A$. Namely, the action defined above makes $R \# A$ into a B-module algebra.

Now, we consider the *bismash product* $(R \# A) \# B$. The product of two elements $(x \# a) \# b$ and $(x' \# a') \# b'$ in $(R \# A) \# B$ is given by

$$
\begin{aligned}
((x \# a) \# b)((x' \# a') \# b') &= \sum ((x \# a)(x' \#(b_{(1)} \rightharpoonup a'))) \# b_{(2)} b' \\
&= \sum \langle a'_{(2)}, b_{(1)} \rangle ((x \# a)(x' \# a'_{(1)})) \# b_{(2)} b' \\
&= \sum \langle a'_{(2)}, b_{(1)} \rangle ((x(a_{(1)} \cdot x')) \# a_{(2)} a'_{(1)}) \# b_{(2)} b'.
\end{aligned}
$$

Remark that in this last expression, b' covers $b_{(2)}$ and x' covers $a_{(1)}$. This implies that indirectly, also both $a'_{(1)}$ and $a'_{(2)}$ are covered.

In general, R can be made into a left $(R \# A)$-module (see (1.4)). Here, we can make $R \# A$ into a left $((R \# A) \# B)$-module. We have the following formula. Whenever $(x \# a) \# b$ is an element in $(R \# A) \# B$ and $x' \# a'$ is in $R \# A$, we get

$$
\begin{aligned}
((x \# a) \# b)(x' \# a') &= (x \# a)(x' \#(b \rightharpoonup a')) \\
&= \sum \langle a'_{(2)}, b \rangle (x \# a)(x' \# a'_{(1)}) \\
&= \sum \langle a'_{(2)}, b \rangle x(a_{(1)} \cdot x') \# a_{(2)} a'_{(1)}.
\end{aligned}
$$

Here again $a_{(1)}$ is covered by x' and $a'_{(2)}$ is covered by b. Remark that $R \# A$ is a faithful $((R \# A) \# B)$-module for the action defined above (see [5, 7.4].

Now, let us assume that the action of A on R comes from a coaction Γ of B on R. Let us use the Sweedler notation also for these coactions. So, write $\Gamma(x) = \sum x_{(0)} \otimes x_{(1)}$ and make sure that $x_{(1)}$ is always covered by an element of B. If $a \in A$ and $x \in R$, then

$$a \cdot x = \sum x_{(0)} \langle a, x_{(1)} \rangle.$$

In what follows, we will use the following notations:

$$
\begin{aligned}
\pi(a)a' &= aa' \\
\pi(b)a' &= b \rightharpoonup a' = \sum \langle a'_{(2)}, b \rangle a'_{(1)} \\
\pi'(a)a' &= a'a \\
\pi'(b)a' &= a' \leftharpoonup b = \sum \langle a'_{(1)}, b \rangle a'_{(2)},
\end{aligned}
$$

whenever $a, a' \in A$ and $b \in B$. Now we are able to prove our main result:

2.1 Theorem. Let $\langle A, B \rangle$ be a multiplier Hopf algebra pairing, and R a left A-module algebra such that the A-action comes from a B-coaction on R. If $\pi(A)\pi(B)\pi'(B) = \pi(A)\pi(B)$, then we have an isomorphism of algebras:

$$(R\#A)\#B \longrightarrow R \otimes (A\#B).$$

Proof. We will show this isomorphism in two steps. First we can view the algebra $(R\#A)\#B$ (or $R \otimes (A\#B)$) as a subalgebra of the operator algebra on vector space $R\#A$ (or $R \otimes A$). This is assured by the fact that the canonical action of $(R\#A)\#B$) (or $R\otimes(A\#B)$) on $R\#A$ (or $R\otimes A$) is faithful. Then we construct an algebra monomorphism from $(R\#A)\#B$ into the operator algebra on $R\otimes A$, and prove that the image of $(R\#A)\#B$ coincides with $R \otimes (A\#B)$. This method is essentially the same as the one given in [3].

Define a linear map $W : R \otimes A \longrightarrow R\#A$ by

$$W(x \otimes a) = \sum a_{(1)} \cdot x\#a_{(2)}.$$

This is well-defined because $a_{(1)}$ is covered by x. Remark that W is surjective because $\Delta(A)(A \otimes 1) = A \otimes A$ and R is unital. In fact, W is a bijection and the inverse is given by

$$W^{-1}(x\#a) = \sum S^{-1}(a_{(1)}) \cdot x \otimes a_{(2)}$$

whenever $a \in A$ and $x \in R$.

Now, we can use this vector space isomorphism W to construct a new action of $(R\#A)\#B$ on $R \otimes A$. It will still be faithful and we will use it to examine the bismash product $(R\#A)\#B$. For $a, a' \in A$, $b \in B$ and $x, x' \in R$, we get

$$W^{-1}(x\#a)W(x' \otimes a') = \sum((S^{-1}a'_{(1)})(S^{-1}a_{(1)}) \cdot x)x' \otimes a_{(2)}a'_{(2)}$$
$$W^{-1}bW(x' \otimes a') = x' \otimes (b \rightharpoonup a').$$

If we combine these two formulas, we get

$$W^{-1}((x\#a)\#b)W(x' \otimes a') = \sum \langle a'_{(3)}, b \rangle((S^{-1}a'_{(1)})(S^{-1}a_{(1)}) \cdot x)x' \otimes a_{(2)}a'_{(2)} \qquad (3)$$

for all $a, a' \in A$, $b \in B$ and $x, x' \in R$.

Now the linear isomorphism W induces an monomorphism γ from the algebra $(R\#A)\#B$ into the operator algebra on the vector space $R \otimes A$.

Let P be the algebra of linear maps on $R \otimes A$ spanned by the operators of form (3), i.e., the image of γ. We have seen that the map

$$x \otimes a \to \sum S^{-1}(a_{(1)}) \cdot x \otimes a_{(2)}$$

is a bijection of $R \otimes A$ (in fact, it is the map W^{-1}). It follows that the algebra P is the span of all operators of the form

$$x' \otimes a' \to \sum \langle a'_{(3)}, b \rangle((S^{-1}a'_{(1)}) \cdot x)x' \otimes aa'_{(2)}$$

where $a \in A$, $b \in B$ and $x \in R$. We can recognize the action of $A\#B$ on A in the last factor. Indeed, the above operator can be rewritten as

$$x' \otimes a' \to \sum((S^{-1}a'_{(1)}) \cdot x)x' \otimes (a\#b)a'_{(2)},$$

where in the last factor, we use the natural left action of $B\#A$ on A.

If we examine the above formulas, we see that we have some kind of a *twisting* of the algebras R and $A\#B$. The action of R on $R \otimes A$ here is given by

$$x' \otimes a' \to \sum ((S^{-1}a'_{(1)}) \cdot x) x' \otimes a'_{(2)}.$$

Since the action of A on R comes from the coaction, say Γ, of B on R, we may rewrite the forgoing formula by using the Sweedler notation. Write $\Gamma(x) = \sum x_{(0)} \otimes x_{(1)}$. Then the algebra P of operators on $R \otimes A$ considered before, is spanned by linear operators of the form

$$x' \otimes a' \to \sum x_{(0)} x' \otimes \langle S^{-1}a'_{(1)}, x_{(1)} \rangle (a\#b) a'_{(2)}. \tag{4}$$

In this formula, the element $a'_{(2)}$ is covered by $a\#b$ and so, also $x_{(1)}$ will be covered through the pairing with $S^{-1}a'_{(1)}$ and this expression has a meaning.

Now we can rewrite the formula (4) in terms of representations π and π'. That is

$$x' \otimes a' \to \sum x_{(0)} x' \otimes \pi(a)\pi(b)\pi'(S^{-1}(x_{(1)})) a'. \tag{5}$$

This makes sense because a' is in $\pi'(B)A$ and it then covers $x_{(1)}$. It follows that the operator (5) is of form

$$\sum x_{(0)} \otimes \pi(a)\pi(b)\pi'(S^{-1}(x_{(1)}))).$$

Now applying the condition that $\pi(A)\pi(B) = \pi(A)\pi(B)\pi'(B)$, we get

$$\sum x_{(0)} \otimes \pi(a)\pi(b)\pi'(S^{-1}(x_{(1)}))) \in R \otimes \pi(A)\pi(B).$$

Thus we have shown that the algebra P is a subalgebra of $R \otimes \pi(A)\pi(B)$ which is identified with the algebra $R \otimes (A\#B)$ by means of operators on the vector space $R \otimes A$. In fact, P is equal to $R \otimes \pi(A)\pi(B)$ since

$$R \otimes B = (1 \otimes B)\Gamma(R)$$

and

$$\begin{aligned} R \otimes \pi(A)\pi(B) &= R \otimes \pi(A)\pi(B)\pi'(B) \\ &= (1 \otimes \pi(A)\pi(B))[(\iota \otimes \pi')((1 \otimes B)\Gamma(R))] \\ &= P. \end{aligned}$$

Therefore, the proof is completed.   □

Remark that the condition $\pi(A)\pi(B)\pi'(B) = \pi(A)\pi(B)$ is fulfilled in the case of $\langle A, \hat{A} \rangle$ because in this case $\pi(A)\pi'(\hat{A}) = \pi(A)\pi(\hat{A})$ are finite rank operators. Indeed, let φ and ψ be a left and a right integral of A. We have $\hat{A} = \varphi(A \cdot) = \psi(A \cdot)$ [9]. Given $a \in A$ and $b \in \hat{A}$, there are unique elements c and d of A such that $b = \varphi(c \cdot) = \psi(d \cdot)$. It is easy to compute that, for $x \in A$,

$$\begin{aligned} \pi(a)\pi(b)(x) &= a\varphi(cx) \\ &= \sum aS(c_{(1)})\varphi(c_{(2)}x). \end{aligned}$$

This implies that $\pi(a)\pi(b)$ is a finite rank operator

$$\sum aS(c_{(1)}) \otimes \varphi(c_{(2)} \cdot).$$

Similarly, $\pi(a)\pi'(b)$ is the finite rank operator

$$\sum aS^{-1}(d_{(2)}) \otimes \psi(d_{(1)} \cdot).$$

Since $\Delta(A)(1 \otimes A) = \Delta(A)(A \otimes 1) = A \otimes A$ and S is bijective, we get

$$\pi(A)\pi(\widehat{A}) = \pi(A)\pi'(\widehat{A}) = A \otimes \widehat{A}$$

where we identify $A \otimes \widehat{A}$ with the finite rank operators on vector space A. As a consequence, we obtain

2.2 Corollary [5, Thm.7.6] Let A be an algebraic quantum group, R an A-module algebra. Then

$$(R\#A)\#\widehat{A} \cong R \otimes (A \diamond \widehat{A})$$

where $A \diamond \widehat{A}$ is the finite rank operator algebra on A.

An interesting application of corollary 2.3 is the duality theorem for the coaction of the quantum group $SL_q(2)$ on the algebra $R = \mathbb{C}\langle x, y, x^{-1} \rangle / \langle yx - qxy \rangle$, where q is a non-zero number in $\mathbb{C}$.

If we require that q is not a root of unit, then $SL_q(2)$ is a cosemisimple Hopf algebra, and hence an algebraic quantum group with a left integral φ. The dual algebraic quantum group $\widehat{A}$ is discrete in the sense of [8]. The reader may refer to [13] for the complete description of $\widehat{A}$, which is denoted by $D_q(sl_2)$. Thus we have an isomorphism of algebras:

$$(R\#D_q(sl_2))\#SL_q(2) \cong R \otimes (D_q(sl_2)\#SL_q(2)),$$

where $R = \mathbb{C}\langle x, x^{-1}, y \rangle / \langle yx - qxy \rangle$ and $D_q(sl_2)\#SL_q(2)$ is the finite rank operator algebra on the vector space $D_q(sl_2)$, or the Heisenberg double of $SL_q(2)$.

As a consequence of the theorem 2.1, we also obtain the Blattner-Montgomery duality theorem.

2.3 Corollary [3, Thm. 2.1] Let $\langle A, B \rangle$ be a Hopf algebra pairing, and R a left A-module algebra such that the action of A comes from a coaction of B on R. If $\pi'(B)$ is contained in $\pi(A)\pi(B)$, then we have an isomorphism of algebras:

$$(R\#A)\#B \cong R \otimes (A\#B).$$

Proof. By assumption $\pi'(B) \subseteq \pi(A)\pi(B)$, we get

$$\pi(A)\pi(B)\pi'(B) = \pi(A)\pi'(B)\pi(B) \subseteq \pi(A)\pi(B)\pi(B) = \pi(A)\pi(B).$$

Since $\pi'(B)$ has the identity, we get $\pi(A)\pi(B)\pi'(B) = \pi(A)\pi(B)$. So theorem 2.1 can be applied to this case. $\qquad\square$

References

[1] E. Abe, Hopf algebras, Cambridge University Press (1977).

[2] S. Baaj and E. Skandalis, Unitaires Multiplicatifs et Dualité pour les Produits Croisés de C*-Algèbres, Ann. Scient. Éc. Norm. Sup. 4e Série, t.26 (1993), 425-488.

[3] R.J. Blattner and S. Montgomery, A Duality Theorem for Hopf Module Algebras, J.Alg. 95(1985),153-172.

[4] B. Drabant and A. Van Daele, Pairing and Quantum Double of Multiplier Hopf Algebras, preprint K.U. Leuven (1996).

[5] B. Drabant, A. Van Daele and Y.H. Zhang, Actions of Multiplier Hopf Algebras. preprint K.U. Leuven (1997).

[6] A. Van Dacle, Multiplier Hopf Algebras, Trans. AMS 342 (1994), 917-932.

[7] A. Van Daele, Multiplier Hopf Algebras and Duality, Quantum Groups and Quantum Spaces, Banach Center Publications, 40, Warsaw (1997), 51-58.

[8] A. Van Daele, Discrete Quantum Groups, J.Alg. 180(1996), 431-444.

[9] A. Van Daele, An algebraic framework for group duality, preprint University of Oslo, to appear in Adv. in Math.

[10] A. Van Daele and Y.H. Zhang, Multiplier Hopf Algebras of Discrete Type, preprint K.U. Lcuven (1996).

[11] A. Van Daele and Y.H. Zhang, Galois Theory for Multiplier Hopf Algebra with Intcgrals, preprint K.U Leuven (1997)

[12] A.Van Dacle and Y.H.Zhang, Corepresentation Theory of Multiplier Hopf Algebra I, prepreint, K.U.Leuven (1997).

[13] A.Van Dacle and Y.H. Zhang, Discrete Quantum Group $D_q(sl_2)$, preprint, K.U.Leuven, 1998.

[14] Y.H. Zhang, The Quantum Double of a CoFrobenius Hopf Algebra, preprint K.U. Leuven (1997).

Static Modules and Equivalences

Robert Wisbauer

Mathematical Institute of the University

40225 Düsseldorf, Germany

Abstract

By a well known theorem of K. Morita, any equivalence between full module categories over rings R and S, are given by a bimodule ${}_RP_S$, such that ${}_RP$ is a finitely generated projective generator in R-*Mod* and $S = \mathrm{End}_R(P)$. There are various papers which describe equivalences between certain subcategories of R-*Mod* and S-*Mod* in a similar way with suitable properties of ${}_RP_S$. Here we start from the other side: Given any bimodule ${}_RP_S$ we ask for the subcategories which are equivalent to each other by the functor $\mathrm{Hom}_R(P, -)$. In R-*Mod* these are the *P-static (= P-solvable) modules*. In this context properties of s-Σ-quasi-projective, w-Σ-quasi-projective and (self-) tilting modules ${}_RP$ are reconsidered as well as Mittag-Leffler properties of P_S. Moreover for any ring extension $R \to A$ related properties of the A-module $A \otimes_R P$ are investigated.

1 Introduction

It was noticed by K. Morita that an R-Module P is a finitely generated, projective generator P in R-*Mod*, if and only if the functor

$$\mathrm{Hom}_R(P, -) : R\text{-}Mod \to S\text{-}Mod,$$

defines a category equivalence, where $S = \mathrm{End}_R(P)$.

Many authors have worked on generalizations of this setting by looking at representable equivalences between proper subcategories. Imposing various conditions on these subcategories, such as closure under submodules, factor modules or extensions, the problem was to find a module P with suitable properties to characterize the equivalence under consideration.

For this purpose notions like quasi-projective, s-Σ-quasi-projective, w-Σ-quasi-projective modules were introduced, and the generator property was replaced by weaker conditions. We refer to the papers of U. Albrecht, R. Colpi, T.G. Faticoni,

K.R. Fuller, A.I. Kashu, T. Kato, C. Menini, T. Onodera, A. Orsatti, M. Sato and others for this approach. It should be mentioned that the importance of *tilting modules* in representation theory gave a new impact to this kind of investigation.

Here we suggest to put the question the other way round. We do not assume the subcategories to be given but we start with any R-module P and $S = \mathrm{End}_R(P)$. Then we ask if there are any non-trivial subcategories $\mathcal{C} \subset R\text{-}Mod$ and $\mathcal{D} \subset S\text{-}Mod$ for which $\mathrm{Hom}_R(P, -)$ provides an equivalence. Since the functor $P \otimes_S -$ is left adjoint to $\mathrm{Hom}_R(P, -)$ we know that the modules in $\mathcal{C}$ must be "invariant" under $P \otimes_S \mathrm{Hom}_R(P, -)$. Following Nauman [20] and Alperin [3] we call these modules *P-static*. Other names in the literature are *reflexive* (see [6]) or *P-solvable* or *P-coreflexive modules* (see [12, p. 75]), and in Ulmer [26] the class of static modules is called the *fixpoint category*.

Of course P itself and every finite direct sum of copies of P are P-static. We will say P is $\sum$-*self-static* if any direct sum of copies of P is P-static. Under this condition we obtain a straightforward characterization of P-static modules (in 3.7) which shows the importance of the projectivity notions mentioned above.

Section 4 will mainly be concerned with the interplay of conditions imposed on the category of static modules (or on the image of $\mathrm{Hom}_R(P, -)$) and properties of the module P. This includes characterizations of self-tilting modules and related equivalences.

It is well known that generators in a full module category can be characterized by properties over their endomorphism and biendomorphism rings. In Section 5 we provide similar characterizations for w-$\sum$-quasi-projective and self-tilting modules. In particular it turns out that, for a faithful $\sum$-self-static self-tilting R-module P, the ring R is dense in the biendomorphism ring of P, and P_S has P-dcc in the sense of Zimmermann [31], a property which makes P_S a certain Mittag-Leffler module. Similar Mittag-Leffler properties for P_S are observed for the case that the category of P-static modules is closed under products in the category of P-generated modules.

There is another topic considered in the paper of Nauman. Ring extensions $R \to A$ are studied and the transfer of properties from an R-module P to the A-module $A \otimes_R P$. Exploiting his basic ideas we give an account of this relationship in Section 6 thus extending results of Fuller [13] in this direction.

The main concern of this note is to generalize known results and provide simple proofs by relating papers which were written independently. Various results scattered around in the literature are gathered under a common point of view. In particular it should be mentioned that our techniques also apply to modules P which are not self-small thus including interesting examples from abelian group theory (e.g., [30, 4.12, 5.8]).

2 Preliminaries

Let R be an associative ring with unit and $R\text{-}Mod$ the category of unital left R-modules. Homomorphisms of modules will be written on the opposite side of the scalars. For unexplained notation we refer to [27].

Throughout the paper P will be a left R-module and $S := \mathrm{End}_R(P)$.

An R-module N is P-generated if there exists an exact sequence

$$0 \to K \to P^{(\Lambda)} \to N \to 0, \quad \Lambda \text{ some set,}$$

and N is P-presented if there exists such a sequence where K is P-generated.

Gen(P), Pres(P) and $\sigma[P]$ will denote the full subcategories of $R\text{-}Mod$ whose objects are P-generated, P-presented or submodules of P-generated modules, respectively. $\sigma[P]$ is closed under direct sums, factor modules and submodules in $R\text{-}Mod$ and hence is a Grothendieck category. Recall that for $Q \in \sigma[P]$, $Q|_P^\Lambda$ denotes the product of Λ copies of Q in $\sigma[P]$ (e.g., [30]). If Q is P-injective then $Q|_P^\Lambda = \mathrm{Tr}(P, Q^\Lambda)$ (the trace of P in Q^Λ).

An R-module N is P-cogenerated if there exists an exact sequence

$$0 \to N \to P^\Lambda \to L \to 0, \quad \Lambda \text{ some set,}$$

and N is P-copresented if there exists such a sequence where L is P-cogenerated.

By Cog(P) and Cop(P) we denote the full subcategories of $R\text{-}Mod$ consisting of P-cogenerated, resp., P-copresented modules.

Add(P) (resp. add(P)) stands for the class of modules which are direct summands of (finite) direct sums of copies of P.

2.1 Canonical functors. Related to $_R P_S$ we have the adjoint pair of functors

$$\mathrm{Hom}_R(P, -) : R\text{-}Mod \to S\text{-}Mod, \quad P \otimes_S - : S\text{-}Mod \to R\text{-}Mod,$$

and for any $N \in R\text{-}Mod$ and $X \in S\text{-}Mod$, the canonical morphisms

$$\mu_N : P \otimes_S \mathrm{Hom}_R(P, N) \to N, \quad p \otimes f \mapsto (p)f,$$
$$\nu_X : X \to \mathrm{Hom}_R(P, P \otimes_S X), \quad x \mapsto [p \mapsto p \otimes x].$$

We recall the following useful properties (e.g., [27, 45.8]).

2.2 Proposition. *Consider any* $N \in R\text{-}Mod$ *and* $X \in S\text{-}Mod$.

(1) *Each of the following compositions of maps yield the identity:*

$$\mathrm{Hom}_R(P, N) \xrightarrow{\nu_{\mathrm{Hom}(P,N)}} \mathrm{Hom}_R(P, P \otimes_S \mathrm{Hom}_R(P, N)) \xrightarrow{\mathrm{Hom}(P, \mu_N)} \mathrm{Hom}_R(P, N),$$
$$P \otimes_S X \xrightarrow{id \otimes \nu_X} P \otimes_S \mathrm{Hom}_R(P, P \otimes_S X) \xrightarrow{\mu_{P \otimes X}} P \otimes_S X.$$

(2) $\operatorname{Coke}(\nu_{\operatorname{Hom}(P,N)}) \simeq \operatorname{Hom}_R(P, \operatorname{Ke}(\mu_N))$ *and* $\operatorname{Ke}(\mu_{P\otimes X}) \simeq P \otimes_S \operatorname{Coke}(\nu_X)$.

2.3 Static and adstatic modules. An R-module N is called P-*static* if μ_N is an isomorphism and the class of all P-static R-modules is denoted by $\operatorname{Stat}(P)$.

An S-module X is called P-*adstatic* if ν_X is an isomorphism and we denote the class of all P-adstatic S-modules by $\operatorname{Adst}(P)$.

The name P-*static* was used in Alperin [3] and Nauman [20] and the name P-*adstatic* should remind that we have an adjoint situation. It is easy to see that, for every P-static module N, $\operatorname{Hom}_R(P, N)$ is P-adstatic, and for any P-adstatic module X, $P \otimes_R X$ is P-static. In fact we have the following (e.g., Onodera [21, Theorem 1], Alperin [3, Lemma], Nauman [20, Theorem 2.5], Faticoni [12, Proposition 6.3.2]):

2.4 Basic equivalence. *For any R-module P, the functor*

$$\operatorname{Hom}_R(P, -) : \operatorname{Stat}(P) \to \operatorname{Adst}(P)$$

defines an equivalence with inverse $P \otimes_S -$.

3 Σ-self-static and pseudo-finite modules

Clearly the module P and finite direct sums P^k are P-static. Moreover, for P finitely generated any direct sum $P^{(\Lambda)}$ is P-static. This also holds more generally when P is *self-small*, i.e., if for any set Λ, the canonical map

$$\operatorname{Hom}_R(P, P)^{(\Lambda)} \to \operatorname{Hom}_R(P, P^{(\Lambda)})$$

is an isomorphism. However this condition is not necessary for $P^{(\Lambda)}$ to be P-static. Because of the importance of this property we give it its own name.

3.1 Definition. We say that P is Σ-*self-static* if, for any set Λ, $P^{(\Lambda)}$ is P-static, i.e., we have an isomorphism

$$\mu_{P^{(\Lambda)}} : P \otimes_S \operatorname{Hom}_R(P, P^{(\Lambda)}) \to P^{(\Lambda)}.$$

As we will see soon Σ-self-static modules can be far from being finitely generated. Nevertheless many examples of Σ-self-static modules have a property which is familiar from finitely generated modules. Again we suggest a name for this.

3.2 Definition. We call a module P *pseudo-finite* if, for any set Λ, and any morphisms

$$P \xrightarrow{g} P^{(\Lambda)} \xrightarrow{h} N,$$

where $gh \neq 0$, there exists a morphism $\bar{g} : P \to \text{Im}(g) \cap P^{\Lambda_o}$, for some finite subset $\Lambda_o \subset \Lambda$, such that $\bar{g}h \neq 0$. These maps are displayed in the diagram

$$P \xrightarrow{\bar{g}} P^{\Lambda_o}$$
$$\downarrow \varepsilon$$
$$P \xrightarrow{g} P^{(\Lambda)} \xrightarrow{h} N,$$

where ε denotes the canonical inclusion.

Clearly every self-small module is pseudo-finite, and it is easy to see that any direct summand of a direct sum of finitely generated modules is pseudo-finite. In particular P is pseudo-finite provided it is projective in $\sigma[P]$. Moreover, if P is a generator in $\sigma[P]$ it is also pseudo-finite.

We do not expect that every pseudo-finite module is $\sum$-self-static. However the last two examples mentioned share a generalized projectivity condition which makes them $\sum$-self-static as we will prove in our next propositon.

Recall that P is *self-pseudo-projective* in $\sigma[P]$ if any diagram with exact sequence

$$P \cdot\overset{\alpha}{\cdots}\cdot P$$
$$\beta \vdots \qquad \downarrow$$
$$0 \to K \to L \to N \to 0,$$

where $K \in \text{Gen}(P)$ and $L \in \sigma[P]$, can be non-trivially commutatively extended by some $\alpha : P \to P$, $\beta : P \to L$. This condition is equivalent to $\text{Gen}(P)$ being closed under extensions in $\sigma[P]$, and also to the fact that $\text{Hom}_R(P, -)$ respects exactness of sequences of the form (see [17, Proposition 2.2])

$$0 \to \text{Tr}(P, L) \to L \to L/\text{Tr}(P, L) \to 0, \text{ for any } L \in \sigma[P].$$

3.3 Pseudo-finite self-pseudo-projective modules. *Let P be pseudo-finite and self-pseudo-projective. Then:*

(1) *For any $N \in R\text{-Mod}$,* $\quad \text{Hom}_R(P, \text{Ke}(\mu_N)) = 0.$

(2) *P is $\sum$-self-static.*

Proof. (1) Let $\{f_\lambda\}_\Lambda$ be a generating set of the S-module $\text{Hom}_R(P, N)$ and consider the canonical map

$$S^{(\Lambda)} \to \text{Hom}_R(P, N), \quad s_\lambda \mapsto s_\lambda f_\lambda.$$

Tensoring with P_S we obtain the morphism

$$h : P^{(\Lambda)} \simeq P \otimes S^{(\Lambda)} \to P \otimes_S \text{Hom}_R(P, N), \quad p_\lambda \mapsto p_\lambda \otimes f_\lambda,$$

where the kernel of h is P-generated. By our projectivity condition, for every non-zero map $g : P \to P \otimes_S \operatorname{Hom}_R(P, N)$, we may construct a commutative diagram

$$
\begin{array}{ccc}
P & \xrightarrow{\alpha} & P \\
\beta \downarrow & & \downarrow g \\
P^{(\Lambda)} & \xrightarrow{h} & P \otimes_S \operatorname{Hom}_R(P, N) \to 0 \\
& \bar{h} \searrow & \downarrow \mu_N \\
& & N,
\end{array}
$$

where $\alpha g = \beta h \neq 0$. By our finiteness condition we may assume that $\operatorname{Im} \beta$ is contained in a finite partial sum of $P^{\Lambda_o} \subset P^{(\Lambda)}$ and $\beta h \neq 0$. With the canonical projections π_λ related to $P^{(\Lambda)}$, and $\Lambda_o = \{\lambda_1, \ldots, \lambda_k\}$, we have for any $p \in P$,

$$
(p)\beta h = \sum_{\imath=1}^{k} (p)\beta\pi_{\lambda_\imath} \otimes f_{\lambda_\imath} = p \otimes \sum_{\imath=1}^{k} \beta\pi_{\lambda_\imath} f_{\lambda_\imath} = p \otimes \beta\bar{h}.
$$

Now assume $0 \neq \operatorname{Im} g \subset \operatorname{Ke} \mu_N$. Then $\beta\bar{h} = 0$ and hence $\beta h = 0$, contradicting our assumption. So we have $\operatorname{Hom}_R(P, \operatorname{Ke}(\mu_N)) = 0$.

The map $\mu_{P^{(\Lambda)}} : P \otimes_S \operatorname{Hom}_R(P, P^{(\Lambda)}) \to P^{(\Lambda)}$ is surjective and is split by the map

$$
P^{(\Lambda)} \to P \otimes_S \operatorname{Hom}_R(P, P^{(\Lambda)}), \quad p_\lambda \mapsto p_\lambda \otimes \varepsilon_\lambda.
$$

Hence $\operatorname{Ke}(\mu_{P^{(\Lambda)}})$ is a direct summand and so it is P-generated. Now (1) implies that $\mu_{P^{(\Lambda)}}$ is injective. $\qquad\square$

The above proposition subsumes several well known results:

3.4 Corollary. *Let P be an R-module and $T = \operatorname{Tr}(P, R)$.*

(1) *Assume P is projective in $\sigma[P]$ or $P = TP$. Then $\operatorname{Hom}_R(P, \operatorname{Ke}(\mu_N)) = 0$ and P is $\sum$-self-static.*

(2) *If P is a generator in $\sigma[P]$ then $\sigma[P] = \operatorname{Stat}(P)$ (and P is $\sum$-self-static).*

Proof. (1) Assume P is projective in $\sigma[P]$. Then obviously P is self-pseudo-projective and is a direct summand of a direct sum of finitely generated modules, hence pseudo-finite.

If $P = TP$ then every P-generated module is T-generated and vice versa. From this it is easy to see that $\operatorname{Gen}(P)$ is closed under extensions in $\sigma[P]$ (even in R-Mod) and so P is self-pseudo-projective. Consider any morphisms

$$
P \xrightarrow{g} P^{(\Lambda)} \xrightarrow{h} N,
$$

where $gh \neq 0$. Choose some $t \in T$, $a \in P$ with $(ta)gh \neq 0$. Then by restriction we have a non-zero map

$$Ta \to Ra \xrightarrow{g} R(a)g \subset (P)g \cap P^{\Lambda_o} \xrightarrow{h} N,$$

for some finite $\Lambda_o \subset \Lambda$. Since Ta is P-generated, there exists $\bar{g} : P \to (P)g \cap P^{\Lambda_o}$ with $\bar{g}h \neq 0$, showing that P is pseudo-finite.

Now the assertion follows from 3.3.

(2) Let P be a generator in $\sigma[P]$. Then trivially P is self-pseudo-projective. Arguments similar to those used in the proof of (1) show that P is pseudo-finite. Hence by 3.3, for any R-module N, $\mathrm{Hom}_R(P, \mathrm{Ke}(\mu_N)) = 0$ and hence $\mathrm{Ke}(\mu_N) = 0$ $\square$

Remarks. The second case in 3.4(1) was shown in Onodera [21, Theorem 2]. Notice that for P projective in R-*Mod*, $TP = P$. Assertion (2) was proved in Zimmermann-Huisgen [32, Lemma 1.3].

More examples of $\sum$-self-static and pseudo-finite modules are provided by our next result.

3.5 Non-singular noetherian rings. *Let R be a left noetherian ring with injective hull $E(R)$.*

(1) *If R is left non-singular, then $E(R)$ is $\sum$-self-static.*

(2) *If R is left hereditary, then $E(R)$ is pseudo-finite and self-pseudo-projective.*

Proof. (1) By our assumptions we have the S-module isomorphisms

$$E(R) \simeq \mathrm{Hom}_R(R, E(R)) \simeq \mathrm{End}_R(E(R)) =: S,$$

and for every non-singular injective R-module V,

$$E(R) \otimes_S \mathrm{Hom}_R(E(R), V) \simeq \mathrm{Hom}_R(E(R), V) \simeq V,$$

showing that V is $E(R)$-static. In particular $E(R)$ is $\sum$-self-static.

(2) Let $E(R) \xrightarrow{g} E(R)^{(\Lambda)} \xrightarrow{h} N$ be any morphisms with $gh \neq 0$. Choose any $a \in E(R)$ with $(a)gh \neq 0$. Then $\mathrm{Im}\, g$ contains an injective hull L of Ra. By the uniqueness of maximal essential extensions in $E(R)^{(\Lambda)}$, L is contained in a finite partial sum of $E(R)^{(\Lambda)}$. Since L is generated by $E(R)$ there exists some $\bar{g} : E(R) \to L$ with $\bar{g}h \neq 0$. $\square$

As a special case we conclude from (1) that the rationals Q are $\sum$-self-static. This is not a surprise since it is shown in Arnold-Murley [4, Corollary 1.4] that any module with countable endomorphism ring is in fact self-small.

The following observation is due to D.K. Harrison (see [14, Proposition 2.1]):

3.6 Properties of $\mathbb{Q}/\mathbb{Z}$. *Put $S = \text{End}_{\mathbb{Z}}(\mathbb{Q}/\mathbb{Z})$ and let V be any divisible torsion $\mathbb{Z}$-module. Then we have an isomorphism*

$$\mu_V : \mathbb{Q}/\mathbb{Z} \otimes_S \text{Hom}_{\mathbb{Z}}(\mathbb{Q}/\mathbb{Z}, V) \to V.$$

So in particular $\mathbb{Q}/\mathbb{Z}$ is $\sum$-self-static. However $\mathbb{Q}/\mathbb{Z}$ is not a self-small $\mathbb{Z}$-module since self-small torsion $\mathbb{Z}$-modules are finite (by [4, Proposition 3.1]).

For $\sum$-self-static R-modules we have the following.

3.7 Characterization of static modules. *For R-modules N, P consider the following statements:*

(i) *N is P-static;*

(ii) *there exists an exact sequence $P^{(\Lambda')} \to P^{(\Lambda)} \to N \to 0$ in R-Mod, which stays exact under $\text{Hom}_R(P, -)$;*

(iii) *there exists an exact sequence $0 \to K \to P^{(\Lambda)} \to N \to 0$ in R-Mod with $K \in \text{Gen}(P)$, which stays exact under $\text{Hom}_R(P, -)$.*

For any P, $(i) \Rightarrow (ii) \Leftrightarrow (iii)$.

If P is a $\sum$-self-static R-module, then $(i) \Leftrightarrow (ii) \Leftrightarrow (iii)$.

Proof. Recall that for any P-generated N and $\Lambda = \text{Hom}_R(P, N)$, the canonical exact sequence $P^{(\Lambda)} \to N \to 0$ remains exact under $\text{Hom}_R(P, -)$.

$(ii) \Rightarrow (iii)$ This is obvious since $\text{Hom}_R(P, -)$ is left exact.

$(iii) \Rightarrow (ii)$ Given the sequence in (iii), put $\Lambda' = \text{Hom}_R(P, K)$. Now the assertion follows by the preceding remark.

$(i) \Rightarrow (iii)$ From any exact sequence $0 \to K \to P^{(\Lambda)} \to N \to 0$ we construct the commutative diagram

$$
\begin{array}{ccccccc}
P \otimes_S \text{Hom}_R(P, K) & \to & P \otimes_S \text{Hom}_R(P, P^{(\Lambda)}) & \to & P \otimes_S \text{Hom}_R(P, N) & \to & 0 \\
\downarrow \mu_K & & \downarrow \mu_{P^{(\Lambda)}} & & \downarrow \mu_N & & \\
0 \to \qquad K & \to & P^{(\Lambda)} & \to & N & & \to 0.
\end{array}
$$

Now assume (i), put $\Lambda = \text{Hom}_R(P, N)$ and consider the canonical epimorphism. Then the upper sequence in the diagram is exact and $\mu_{P^{(\Lambda)}}$ is an epimorphism. Hence $\text{Coke}(\mu_K) \simeq \text{Ke}(\mu_N) = 0$ and so K is P-generated.

$(iii) \Rightarrow (i)$ Let P be $\sum$-self-static. For the diagram above, assume that K is P-generated and $\text{Hom}_R(P, -)$ is exact on the given sequence. Then again the upper sequence in the diagram is exact and since $\mu_{P^{(\Lambda)}}$ is an isomorphism, $0 = \text{Coke}(\mu_K) \simeq \text{Ke}(\mu_N)$. Hence μ_N is an isomorphism. $\qquad \square$

Remarks. In Alperin [3], modules with property 3.7(ii) are called *Auslander with respect to P* and the equivalence $(i) \Leftrightarrow (ii)$ is asserted in Lemma 2 without any further condition. However the proof given there only holds for self-small P.

The implication $(i) \Rightarrow (iii)$ was also observed in Faticoni [12, Corollary 6.1.9].

4 Static modules and equivalences

4.1 Classes of modules related to P. From the preceding definitions we have the following chain of subclasses

$$\text{add}\,(P) \subset \text{Stat}(P) \subset \text{Pres}(P) \subset \text{Gen}(P) \subset \sigma[P] \subset R\text{-}Mod.$$

Since any direct summand of a P-static module is again P-static we have that

$$\text{Add}\,(P) \subset \text{Stat}(P) \text{ if and only if } P \text{ is } \Sigma\text{-self-static.}$$

Our investigations will be concerned with the problem when some of these classes coincide. For example, P is a generator in $\sigma[P]$ if and only if $\text{Stat}(P) = \sigma[P]$. If P is a semisimple module, then clearly $\text{Add}\,(P) = \sigma[P]$, and if P is locally noetherian and cohereditary in $\sigma[P]$, then $\text{Add}\,(P) = \text{Pres}(P)$ (see [30]).

For classes $\mathcal{C} \subset R\text{-}Mod$ and $\mathcal{D} \subset S\text{-}Mod$ we use the notation

$$H_P(\mathcal{C}) = \{X \in S\text{-}Mod \mid X \simeq \text{Hom}_R(P, C) \text{ for some } C \in \mathcal{C}\};$$
$$P_S(\mathcal{D}) = \{N \in R\text{-}Mod \mid N \simeq P \otimes_S D \text{ for some } D \in \mathcal{D}\}.$$

Throughout this section let Q be any injective cogenerator in $\sigma[P]$ and $P^* = \text{Hom}_R(P, Q)$. Then we have the chain of classes of S-modules

$$\text{add}(_S S) \subset \text{Adst}(P) \subset H_P(\text{Gen}(P)) \subset \text{Cop}(P^*) \subset \text{Cog}(P^*) \subset S\text{-}Mod.$$

With this notation we collect some elementary properties.

4.2 Fundamental relationships.

(1) $H_P(R\text{-}Mod) = H_P(\text{Gen}(P)) \subset \text{Cop}(P^*)$.

(2) *If $(P^*)^\Lambda \in \text{Adst}(P)$, for any set Λ, then $H_P(\text{Gen}(P)) = \text{Cop}(P^*)$.*

(3) *If P is Σ-self-static, then $P_S(S\text{-}Mod) = \text{Pres}(P)$.*

(4) *P is self-small if and only if $\text{Add}(_S S) \subset \text{Adst}(P)$.*
    Then $\text{Hom}_R(P, -) : \text{Add}(P) \to \text{Add}(_S S)$ is an equivalence.

(5) *If P is Σ-self-static, then P is self-small provided $\text{Adst}(P)$ is closed under submodules.*

Proof. (1) The first equality follows from $\text{Hom}_R(P, L) = \text{Hom}_R(P, \text{Tr}(P, L))$. Since Q is a cogenerator in $\sigma[M]$, for any $N \in \sigma[P]$ we have an exact sequence

$$0 \to N \to Q|_P^\Lambda \to Q|_P^{\Lambda'}.$$

Applying $\text{Hom}_R(P, -)$ we obtain $\text{Hom}_R(P, N) \in \text{Cop}(P^*)$.

(2) For $X \in \text{Cop}(P^*)$ we have an exact sequence in $S\text{-}Mod$,

$$0 \to X \to (P^*)^\Lambda \xrightarrow{g} (P^*)^{\Lambda'}.$$

Applying $P \otimes_S -$ and $\text{Hom}_R(P, -)$ we obtain the commutative exact diagram

$$
\begin{array}{ccccccc}
0 \to & X & \to & (P^*)^\Lambda & \xrightarrow{g} & (P^*)^{\Lambda'} \\
& \downarrow \alpha & & \downarrow \simeq & & \downarrow \simeq \\
0 \to & \text{Hom}_R(P, P \otimes_S K) & \to & \text{Hom}_R(P, P \otimes_S (P^*)^\Lambda) & \to & \text{Hom}_R(P, P \otimes_S (P^*)^{\Lambda'}),
\end{array}
$$

where $K = \text{Ke}\,(id_P \otimes g)$ and the isomorphisms are given by our assumption. Hence α is an isomorphisms showing $X \in H_P(\text{Gen}(P))$.

(3) $P_S(S\text{-}Mod) \subset \text{Pres}(P)$ always holds.

Let $P^{(\Lambda')} \xrightarrow{f} P^{(\Lambda)} \to N \to 0$ be a P-presentation for $N \in \text{Pres}(P)$. Applying $\text{Hom}_R(P, -)$ we obtain an exact sequence

$$\text{Hom}_R(P, P^{(\Lambda')}) \to \text{Hom}_R(P, P^{(\Lambda)}) \to X \to 0,$$

where $X = \text{Coke}(\text{Hom}_R(P, f))$. Tensoring with P_S we obtain the commutative exact diagram

$$
\begin{array}{ccccccc}
P \otimes_S \text{Hom}_R(P, P^{(\Lambda')}) & \to & P \otimes_S \text{Hom}_R(P, P^{(\Lambda)}) & \to & P \otimes_S X & \to 0 \\
\downarrow \simeq & & \downarrow \simeq & & \downarrow \\
P^{(\Lambda')} & \to & P^{(\Lambda)} & \to & N & \to 0,
\end{array}
$$

from which we see that $N \simeq P \otimes_S X$ and hence $\text{Pres}(P) \subset P_S(S\text{-}Mod)$.

(4) One implication follows directly from the definition. If $S^{(\Lambda)} \in \text{Adst}(P)$ then

$$\text{Hom}_R(P, P)^{(\Lambda)} \simeq S^{(\Lambda)} \simeq Hom_R(P, P \otimes_S S^{(\Lambda)}) \simeq \text{Hom}_R(P, P^{(\Lambda)}),$$

showing that P is self-small.

(5) Clearly $S^{(\Lambda)} \subset \text{Hom}_R(P, P^{(\Lambda)})$ and $\text{Hom}_R(P, P^{(\Lambda)}) \in \text{Adst}(P)$. If $\text{Adst}(P)$ is closed under submodules then $S^{(\Lambda)} \in \text{Adst}(P)$ and hence $\text{Add}(_S S) \subset \text{Adst}(P)$. $\quad \square$

The next result considers the case when the image of $\text{Hom}_R(P, -)$ is contained in $\text{Adst}(P)$ and was proved in Sato [25, Theorem] and Kashu [15, Proposition 9.5].

4.3 Conditions on the image of $\text{Hom}_R(P, -)$. *The following are equivalent:*

(a) $H_P(\text{Gen}(P)) = \text{Adst}(P)$;

(b) $\text{Cop}(P^*) = \text{Adst}(P)$;

(c) $P_S(S\text{-}Mod) = \text{Stat}(P)$;

(d) $\operatorname{Pres}(P) = \operatorname{Stat}(P)$;

(e) $\operatorname{Hom}_R(P, \operatorname{Ke} \mu_N) = 0$, *for every* $N \in \operatorname{Gen}(P)$;

(f) $P \otimes_S \operatorname{Coke} \nu_X = 0$, *for every* $X \in S\text{-Mod}$;

(g) $\operatorname{Hom}_R(P, -) : \operatorname{Pres}(P) \to \operatorname{Cop}(P^*)$ *is an equivalence (with inverse* $P \otimes_S -$).

Proof. $(a) \Leftrightarrow (b)$ Under the given conditions we have for any set Λ,

$$(P^*)^\Lambda \simeq \operatorname{Hom}_R(P, Q^\Lambda) \simeq \operatorname{Hom}_R(P, \operatorname{Tr}(P, Q^\Lambda)) \in \operatorname{Adst}(P).$$

Hence by 4.2, $H_P(\operatorname{Gen}(P)) = \operatorname{Cop}(P^*)$.

$(c) \Leftrightarrow (d)$ Under the given conditions, P is $\sum$-self-static and hence $P_S(S\text{-Mod}) = \operatorname{Pres}(P)$ (by 4.2).

$(a) \Leftrightarrow (e)$ Assume for any $N \in \operatorname{Gen}(P)$ that $\operatorname{Hom}_R(P, N)$ is P-adstatic. Then by 2.2, $0 = \operatorname{Coke}(\nu_{Hom(P,N)}) \simeq \operatorname{Hom}_R(P, \operatorname{Ke}(\mu_N))$.

The same formula yields the converse conclusion.

$(c) \Leftrightarrow (f)$ Assume (c). Then for an $X \in S\text{-Mod}$, $\mu_{P\otimes X}$ is an isomorphism, and by 2.2, $0 = \operatorname{Ke}(\mu_{P\otimes X}) \simeq P \otimes_S \operatorname{Coke}(\nu_X)$.

Again the same formula yields the converse conclusion.

$(e) \Rightarrow (f)$ We have $0 = \operatorname{Hom}_R(P, \operatorname{Ke}(\mu_{P\otimes X}) \simeq \operatorname{Hom}_R(P, P \otimes_S \operatorname{Coke}(\nu_X))$, and hence $P \otimes_S \operatorname{Coke}(\nu_X) = 0$.

$(f) \Rightarrow (e)$ By assumption, $0 = P \otimes_S \operatorname{Coke}(\nu_{Hom(P,N)}) \simeq P \otimes_S \operatorname{Hom}_R(P, \operatorname{Ke}(\mu_N))$, and so $\operatorname{Hom}_R(P, \operatorname{Ke}(\mu_N)) = 0$.

$(a) \Leftrightarrow (g)$ This follows from the basic equivalence 2.4. $\qquad \square$

It was shown in 3.3 that pseudo-finite self-pseudo-projective modules satisfy the conditions in 4.3.

Interesting cases arise imposing conditions on the categories $\operatorname{Stat}(P)$ and $\operatorname{Adst}(P)$.

4.4 Conditions on $\operatorname{Stat}(P)$.

(1) *The following are equivalent for the R-module P:*

    (a) *P is $\sum$-self-static and $\operatorname{Stat}(P)$ is closed under factor modules;*

    (b) *$\operatorname{Stat}(P) = \operatorname{Gen}(P)$;*

    (c) *$\operatorname{Hom}_R(P, -) : \operatorname{Gen}(P) \to \operatorname{Cop}(P^*)$ is an equivalence.*

(2) *The following are equivalent for the R-module P:*

    (a) *$\operatorname{Stat}(P)$ is closed under submodules;*

    (b) *$\operatorname{Gen}(P) = \sigma[P]$;*

    (c) *$\operatorname{Stat}(P) = \sigma[P]$;*

(d) $\mathrm{Hom}_R(P,-) : \sigma[P] \to \mathrm{Cop}(P^*)$ *is an equivalence.*

Proof. (1) $(a) \Leftrightarrow (b)$ is clear by the fact that the $P^{(\Lambda)}$'s belong to $\mathrm{Stat}(P)$.

$(b) \Leftrightarrow (c)$ From 4.3 we know that $H_P(\mathrm{Gen}(P)) = \mathrm{Cop}(P^*)$. Now the assertion follows from the basic equivalence 2.4.

(2) $(a) \Rightarrow (b) \Rightarrow (c)$ Since any $P^k \in \mathrm{Stat}(P)$, for any $k \in \mathbb{N}$, (a) implies that all submodules of P^k are P-generated and hence $\sigma[P] = \mathrm{Gen}(P)$. By 3.4(2) this implies $\mathrm{Stat}(P) = \sigma[P]$.

The other implications are obvious (by (1)).   □

Remarks. The modules described in 4.4(1) were named W_o-*modules* and those in 4.4(2) are called W-*modules* in Orsatti [22]. In fact 4.4 is a refinement of Teorema 3.2 and Proposizione 5.1 given there. Moreover it is pointed out in [22, 3.5] that $P := \mathbb{Z}_{p^\infty}$ (Prüfer p-group) is a W_o-module over $\mathbb{Z}$ with $\mathrm{Cop}(P^*) \neq \mathrm{Cog}(P^*)$.

Recall that a module P is *s-Σ-quasi-projective* if $\mathrm{Hom}_R(P,-)$ respects exactness of sequences

$$P^{(\Lambda')} \to P^{(\Lambda)} \to N \to 0, \text{ where } \Lambda', \Lambda \text{ are any sets.}$$

P is *w-Σ-quasi-projective* if $\mathrm{Hom}_R(P,-)$ respects exactness of sequences

$$0 \to K \to P^{(\Lambda)} \to N \to 0, \text{ where } K \in \mathrm{Gen}(P) \text{ and } \Lambda \text{ is any set.}$$

P is called *self-tilting* (in [30]) if P is w-Σ-quasi-projective and $\mathrm{Gen}(P) = \mathrm{Pres}(P)$.

4.5 Implications from projectivity. *Let P be Σ-self-static.*

(1) *If P is w-Σ-quasi-projective or s-Σ-quasi-projective, then $\mathrm{Pres}(P) = \mathrm{Stat}(P)$.*

(2) *If P is self-tilting, then $\mathrm{Gen}(P) = \mathrm{Stat}(P)$ and $\mathrm{Adst}(P) = \mathrm{Cop}(P^*)$.*

Proof. (1) This is obvious from 3.7.

(2) By [30, 3.2 and 3.3], self-tilting modules are self-pseudo-projective and hence $\mathrm{Adst}(P) = H_P(\mathrm{Gen}(P)) = \mathrm{Cop}(P^*)$ follows from 4.3.   □

The following results are shown in Colpi [6], Sato [24] and Faticoni [12].

4.6 Conditions on $\mathrm{Adst}(P)$.

(1) *The following are equivalent for an R-module P:*

(a) $\mathrm{Adst}(P) = \mathrm{Cog}(P^*)$;

(b) *P is self-small and w-Σ-quasi-projective;*

(c) $\mathrm{Hom}_R(P,-) : \mathrm{Pres}(P) \to \mathrm{Cog}(P^*)$ *is an equivalence.*

(2) *The following are equivalent for P:*

   (a) $\mathrm{Adst}(P) = S\text{-}Mod;$

   (b) *P is self-small and s-Σ-quasi-projective;*

   (c) $\mathrm{Hom}_R(P, -) : \mathrm{Pres}(P) \to S\text{-}Mod$ *is an equivalence.*

Proof. (1) See Colpi [6, Proposition 3.7] and Faticoni [12, Theorem 6.1.9],

   (2) See Sato [24, Theorem 2.1] and Faticoni [12, Theorem 6.1.14]. $\square$

Self-small self-tilting modules are also known as $*$-modules (see [30]). Combining the preceding propositions we obtain a characterization of $*$-modules given in Colpi [6, Theorem 4.1]:

4.7 Corollary. *The following are equivalent for an R-module P:*

   (a) $\mathrm{Gen}(P) = \mathrm{Stat}(P)$ *and* $\mathrm{Adst}(P) = \mathrm{Cog}(P^*);$

   (b) *P is self-small and self-tilting;*

   (c) $\mathrm{Hom}_R(P, -) : \mathrm{Gen}(P) \to \mathrm{Cog}(P^*)$ *is an equivalence.*

Remarks. s(emi)-Σ-quasi-projective modules were defined in Sato [24] and the notion of w(eakly)-Σ-quasi-projective modules was introduced in the study of $*$-modules (see Colpi [6]). Notice that the condition $\mathrm{Gen}(P) = \mathrm{Pres}(P)$ in the definition of self-tilting modules was already considered by Onodera in [21]. However, he combined it with projectivity of P such yielding a projective self-generator (see [21, Theorem 5]).

Obviously if P is a generator in $\sigma[P]$ then P is self-pseudo-projective in $\sigma[P]$ but need neither be s-Σ-quasi-projective nor w-Σ-quasi-projective.

5   Properties over the (bi)endomorphism ring

Let P be an R-module, $S = \mathrm{End}_R(P)$ and $B = \mathrm{End}_S(P) = \mathrm{Biend}_R(P)$, the biendomorphism ring. There is a remarkable interplay between the properties of P as a module over R, B and S, and we begin with considering properties of P as an S-module.

Recall that P is said to be *direct projective* if for every direct summand $X \subset P$, any epimorphism $P \to X$ splits (see [27, 41.18]), and P is Σ-*direct projective* if any direct sum of copies of P is direct projective. The latter means that P is a projective object in $\mathrm{Add}(P)$ and can be characterized by the fact that for $L \in \mathrm{Gen}(P)$ and $N \in \mathrm{Add}(P)$, any epimorphism $L \to N$ splits (see [11, Lemma 11.2]). For abelian groups this is known as the *Baer splitting property* (see [12, 7.1]).

The following is a variation and extension of a result in Ulmer [26] and the Theorems 2.2 and 2.5 in Albrecht [1].

5.1 P_S (faithfully) flat. *We keep the notation above.*

(1) *The following are equivalent:*

    (a) *P_S is a flat module;*

    (b) *for any $k, l \in \mathbb{N}$, the kernel of any $f : P^k \to P^l$ is P-generated;*

    (c) *$\mathrm{Stat}(P)$ is closed under kernels.*

In this case $\mathrm{Stat}(P)$ is also closed under P-generated submodules.

(2) *If P is self-small the following are equivalent:*

    (a) *P_S is a faithfully flat module;*

    (b) *$\mathrm{Stat}(P)$ is closed under kernels, and $\mathrm{Hom}_R(P, -)$ is exact on short exact sequences in $\mathrm{Stat}(P)$;*

    (c) *$\mathrm{Stat}(P)$ is closed under kernels, and P is $\sum$-direct projective;*

    (d) *$\mathrm{Stat}(P)$ is closed under kernels, and for any left ideal $I \subset S$, $I = \mathrm{Hom}_R(P, PI)$;*

    (e) *$\mathrm{Stat}(P)$ is closed under kernels and for any left ideal $I \subset S$, $PI \neq P$.*

Proof. (1) $(a) \Leftrightarrow (b)$ is well known (e.g., [27, 15.9]).

$(a) \Rightarrow (c)$ Consider an exact sequence $0 \to K \to L \to N$, where $L, N \in \mathrm{Stat}(P)$. By our assumptions we may construct an exact commutative diagram

$$
\begin{array}{ccccc}
0 \to & P \otimes_S \mathrm{Hom}_R(P, K) & \to & P \otimes_S \mathrm{Hom}_R(P, L) & \to & P \otimes_S \mathrm{Hom}_R(P, N) \\
& \downarrow \mu_K & & \downarrow \simeq & & \downarrow \simeq \\
0 \to & K & \to & L & \to & N
\end{array}
$$

showing that μ_K is an isomorphism.

$(c) \Rightarrow (b)$ is obvious.

Assume (a) holds. Let K be a P-generated submodule of some $L \in \mathrm{Stat}(P)$. An argument similar to that in $(a) \Rightarrow (c)$ shows $K \in \mathrm{Stat}(P)$.

(2) $(a) \Rightarrow (b)$ Consider an exact sequence $L \xrightarrow{f} N \to 0$, where $L, N \in \mathrm{Stat}(P)$. We construct an exact commutative diagram

$$
\begin{array}{ccccc}
P \otimes \mathrm{Hom}_R(P, L) & \to & P \otimes \mathrm{Hom}_R(P, N) & \to & P \otimes_S X & \to 0 \\
\downarrow \simeq & & \downarrow \simeq & & & \\
L & \to & N & & \to & 0,
\end{array}
$$

where $X = \mathrm{Coke}\,(\mathrm{Hom}_R(P, f))$. This implies $P \otimes_S X = 0$ and hence $X = 0$.

$(b) \Rightarrow (c)$ For every $N \in \mathrm{Add}(P)$, $\mathrm{Hom}_R(N, -)$ is exact on short exact sequences in $\mathrm{Stat}(P)$ and so every epimorphism $P^{(\Lambda)} \to N$ splits, i.e., P is $\sum$-direct projective.

$(c) \Rightarrow (b)$ Let $L \to N \to 0$ be an exact sequence, where $L, N \in \mathrm{Stat}(P)$. For any $P \to N$ we obtain, by a pullback construction, a commutative diagram

$$
\begin{array}{ccc}
U & \to & P & \to & 0 \\
\downarrow & & \downarrow & & \\
L & \to & N & \to & 0.
\end{array}
$$

As a kernel of a morphism $L \oplus P \to N$, U is P-generated and hence $U \to P$ splits (since P is $\sum$-direct projective).

$(b) \Rightarrow (d)$ Let $I \subset S$ be a left ideal. By (1), $PI \subset P$ is P-static and for a generating set $\{\gamma_\lambda\}_\Lambda$ of I, we have an exact sequence of modules in $\mathrm{Stat}(P)$,

$$
0 \to K \to P^{(\Lambda)} \xrightarrow{\sum \gamma_\lambda} PI \to 0,
$$

and so $\mathrm{Hom}_R(P, -)$ is exact on this sequence. By standard arguments this implies $I = \mathrm{Hom}_R(P, PI)$.

$(d) \Rightarrow (e) \Rightarrow (a)$ is clear. $\qquad\square$

Notice that for P finitely generated, $I = \mathrm{Hom}_R(P, PI)$ for every left ideal $I \subset S$, if and only if P is *intrinsically projective* (see [28, 5.7]).

Next we recall some well known cases of special interest.

5.2 Proposition. *We keep the notation above.*

(1) *P is a generator in R-Mod if and only if $R \simeq B$ and P_S is finitely generated and projective.*

(2) *P is a progenerator in R-Mod if and only if $R \simeq B$ and P_S is a progenerator in Mod-S.*

(3) *P is self-small and tilting in R-Mod if and only if $R \simeq B$ and P_S is self-small and tilting in Mod-S.*

(4) *If P is faithful and a generator in $\sigma[P]$, then R is dense in B and P_S is flat.*

(5) *If P is self-tilting and self-small, then $P \otimes_S -$ is exact on short exact sequences with modules from $H_P(\mathrm{Gen}(P))$.*

Proof. For (1),(2) and (4) we refer to [27, 18.8 and 15.7]. (3) is proved in Colby-Fuller [5, Proposition 1.1].

(5) By 4.7 we have that $\mathrm{Hom}_R(P, -) : \mathrm{Gen}(P) \to \mathrm{Cog}(P^*) = H_P(\mathrm{Gen}(P))$ is an equivalence. Now the assertion follows from Colpi-Menini [9, Proposition 1.1]. $\qquad\square$

We have a striking left right symmetry in (2) and (3) for the properties in R-*Mod* but there are only implications in one direction for the properties in $\sigma[P]$ considered in (4) and (5). The question arises which property of P_S would guarantee the converse implication in (4). To answer the corresponding question for (5) also some form of density property is needed. Our next result relates to the latter problem.

5.3 Proposition. *Let P be an R-module and $B = \mathrm{Biend}_R(P)$.*

(1) *Every P-presented R-module is a (P-presented) B-module.*

(2) *If P is w-Σ-quasi-projective then $\mathrm{Hom}_R(P, N) = \mathrm{Hom}_B(P, N)$, for every $N \in \mathrm{Pres}(P)$.*

(3) *If P is Σ-self-static then $\mathrm{Hom}_R(P, N) = \mathrm{Hom}_B(P, N)$, for every $N \in \mathrm{Stat}(P)$.*

(4) *If P is Σ-self-static the following are equivalent:*

   (a) *$_RP$ is w-Σ-quasi-projective;*

   (b) *$_BP$ is w-Σ-quasi-projective and $\mathrm{Hom}_R(P, N) = \mathrm{Hom}_B(P, N)$, for every $N \in \mathrm{Pres}(P)$.*

(5) *If P is faithful and Σ-self-static the following are equivalent:*

   (a) *$_RP$ is self-tilting;*

   (b) *$_BP$ is self-tilting and R is dense in B.*

Proof. (1) For $N \in \mathrm{Pres}(P)$ we have a short exact sequence

$$P^{(\Lambda')} \xrightarrow{f} P^{(\Lambda)} \xrightarrow{g} N \to 0, \text{ where } \Lambda', \Lambda \text{ are any sets.}$$

Since every P-generated submodule of $P^{(\Lambda)}$ is a B-submodule (see [27, 15.6]), $\mathrm{Ke}\, g = \mathrm{Im}\, f$ is a B-module. Hence N is a B-module and g is a B-morphism.

(2) We have an exact sequence

$$\mathrm{Hom}_R(P, P^{(\Lambda')}) \to \mathrm{Hom}_R(P, P^{(\Lambda)}) \to \mathrm{Hom}_R(P, N) \to 0.$$

It is easy to see that $\mathrm{Hom}_R(P, P^{(\Lambda)}) = \mathrm{Hom}_B(P, P^{(\Lambda)})$ and this implies $\mathrm{Hom}_R(P, N) = \mathrm{Hom}_B(P, N)$.

(3) In view of 3.7 the same proof as in (2) applies.

(4) This follows easily by (1) and (2).

(5) $(a) \Rightarrow (b)$ Let P be self-tilting. Then for any submodule $K \subset P^n$, $n \in \mathbb{N}$, the factor module $N = P^n/K \in \mathrm{Gen}(P) = \mathrm{Pres}(P)$ is a B-module and by (4), $\mathrm{Hom}_R(P, N) = \mathrm{Hom}_B(P, N)$. In particular the canonical projection $P^n \to N$ is a B-morphism and hence its kernel K is a B-submodule of P^n. This implies that R is dense in B (e.g., [27, 15.7]).

$(b) \Rightarrow (a)$ is obvious since R dense in B implies $\sigma[_RP] = \sigma[_BP]$ (see [27, 15.8]). $\square$

Let $\mathrm{Inj}\,(P)$ denote the class of all injectives in $\sigma[P]$. Since $\mathrm{Inj}\,(P) \subset \mathrm{Gen}(P)$, for any self-small self-tilting module P, $\mathrm{Inj}\,(P) \subset \mathrm{Stat}(P)$. More generally we may ask for which P the latter inclusion holds. Before answering this let us recall the canonical map

$$\alpha_{L,P} : L \otimes_S Hom_R(P, V) \to \mathrm{Hom}_R(\mathrm{Hom}_S(L, P), V), \quad l \otimes f \mapsto [g \mapsto (g(l))f],$$

which is an isomorphism provided L_S is finitely presented and V is P-injective (e.g., [27, 25.5]).

Following Zimmermann [31, 3.2], we say that P_S has L-dcc if $\alpha_{L,P}$ is a monomorphism for all $V \in \text{Inj}(P)$. The notation was chosen to indicate that the condition is related to descending chain conditions on certain matrix subgroups of P.

Now let Q be an injective cogenerator in $\sigma[P]$. Then the above condition is equivalent to

$$\alpha_{L,P} : L \otimes_S Hom_R(P, Q^\Lambda) \to Hom_R(Hom_S(L, P), Q^\Lambda)$$

being a monomorphism. This indicates the relationship to certain Mittag-Leffler modules (e.g., Albrecht [2], Rothmaler [23]).

Let $\mathcal{X}$ be a class of left S-modules. P_S is called an $\mathcal{X}$-*Mittag-Leffler* or $\mathcal{X}$-*ML module* if, for any family $\{X_\lambda\}_\Lambda$ of modules in $\mathcal{X}$, the canonical map

$$P \otimes_S \prod_\Lambda X_\lambda \to \prod_\Lambda (P \otimes_S X_\lambda),$$

is injective. In particular, for $\mathcal{X} = \{X\}$, P_S is X-*Mittag-Leffler* or X-*ML*, if the canonical map

$$P \otimes_S X^\Lambda \to (P \otimes_S X)^\Lambda,$$

is injective for any index set Λ.

The connection of these notions with static modules becomes obvious if we put $L = P$ and assume that P is balanced (i.e. $R \simeq B$). Then P_S has P-dcc implies an isomorphism

$$P \otimes_S Hom_R(P, V) \to Hom_R(Hom_S(P, P), V) \simeq V,$$

for all injective $V \in \sigma[P]$, since they are P-generated. For the injective cogenerator $Q \in \sigma[P]$ and $P^* := Hom_R(P, Q)$ this corresponds to the condition that

$$P \otimes_S (P^*)^\Lambda \simeq (P \otimes_S P^*)^\Lambda,$$

is a monomorphism for any set Λ (i.e., P_S is P^*-ML), and this is equivalent to

$$P \otimes_S (P^*)^\Lambda \simeq P \otimes_S Hom_R(P, Q|_P^\Lambda) \to Q|_P^\Lambda,$$

being an isomorphism for any set Λ.

Summarizing these remarks and referring to the basic equivalence 2.4 we have (see [31, Corollary 3.10]):

5.4 P balanced with P_S-dcc. *For a balanced bimodule $_RP_S$, the following are equivalent:*

(a) $\mathrm{Inj}\,(P) \subset \mathrm{Stat}(P)$;

(b) P_S *has P_S-dcc;*

(c) $Q \in \mathrm{Stat}(P)$ *and P_S is a P^*-ML module;*

(d) *for every set Λ, $Q|_P^\Lambda \in \mathrm{Stat}(P)$;*

(e) $\mathrm{Hom}_R(P, -) : \mathrm{Inj}\,(P) \to H_P(\mathrm{Inj}\,(P))$ *is an equivalence, (and*

$$H_P(\mathrm{Inj}\,(P)) = \{X \in S\text{-}Mod\,|\, X \text{ is a direct summand of } (P^*)^\Lambda,\ \Lambda \text{ some set}\}).$$

Combining this with our previous observations on density properties (in 5.3) we are now able to describe when P-injectives are P-static.

5.5 Injective and static modules. *For a $\sum$-self-static R-module P, the following are equivalent:*

(a) $\mathrm{Inj}\,(P) \subset \mathrm{Stat}(P)$;

(b) *for every set Λ, $Q|_P^\Lambda \in \mathrm{Stat}(P)$;*

(c) $\mathrm{Hom}_R(P, -) : \mathrm{Inj}\,(P) \to H_P(\mathrm{Inj}\,(P))$ *is an equivalence (with inverse $P \otimes_S -$);*

(d) P_S *has P_S-dcc, $\mathrm{Inj}\,(P) \subset \mathrm{Pres}(P)$ and $\mathrm{Hom}_R(P,V) = \mathrm{Hom}_B(P,V)$, for all $V \in \mathrm{Inj}\,(P)$.*

Proof. $(a) \Leftrightarrow (b) \Leftrightarrow (c)$ is clear by the observations preceding 5.4.

$(a) \Rightarrow (d)$ Clearly $\mathrm{Inj}\,(P) \subset \mathrm{Pres}(P)$, and by 5.3(3), for every injective $V \in \sigma[P]$, $\mathrm{Hom}_R(P,V) = \mathrm{Hom}_B(P,V)$. The last equality implies

$$\mathrm{Hom}_R(L,V) = \mathrm{Hom}_B(L,V), \text{ for any } P\text{-generated } B\text{-module } L.$$

Indeed, for $f \in \mathrm{Hom}_R(L,V)$ let $l \in L$, $b \in B$. There exists $g \in \mathrm{Hom}_B(P,L)$ and $p \in P$ with $(p)g = l$, implying $(bl)f = (bp)gf = b((p)gf) = b(l)f$. This shows $f \in \mathrm{Hom}_B(L,V)$.

Now let W be an injective module in $\sigma[_BP] \subset \sigma[P]$ and $\alpha : W \to V$ an R-monomorphism for some injective $V \in \sigma[P]$. Then $\alpha \in \mathrm{Hom}_B(W,V)$ and hence it splits proving that W is injective in $\sigma[P]$.

This implies $\mathrm{Inj}\,(_BP) \subset \mathrm{Stat}(_BP)$ and P_S has P_S-dcc by 5.4.

$(d) \Rightarrow (a)$ Any injective $V \in \sigma[P]$ is a B-module and hence there exists a B-monomorphism $\beta : V \to V'$, for some injective B-module V' in $\sigma[_BP]$. This is R-split by some $\beta' : \mathrm{Hom}_R(V',V) = \mathrm{Hom}_B(V',V)$ (see proof above) and hence V is injective in $\sigma[_BP]$. Now 5.4 implies $\mathrm{Inj}\,(P) = \mathrm{Inj}\,(_BP) \subset \mathrm{Stat}(_BP) = \mathrm{Stat}(P)$. $\quad\square$

If P is a cogenerator in $\sigma[P]$ then it satisfies the density property (see [27, 15.7]) and so we have:

5.6 Corollary. *If P is an injective cogenerator in $\sigma[P]$ the following are equivalent:*

(a) $\mathrm{Inj}\,(P) \subset \mathrm{Stat}(P)$;

(b) P_S *has* P_S-*dcc;*

(c) P_S *is an S-ML module;*

(d) *for every set* Λ, $\mathrm{Tr}(P, P^\Lambda) \in \mathrm{Stat}(P)$.

If P satisfies (a) and is Σ-pure-injective in $\sigma[P]$, then P is Σ-self-static.

Notice that the case $\mathrm{Inj}\,(P) = \mathrm{Stat}(P)$ is described in [30, 6.5]. This condition characterizes locally noetherian cohereditary modules P for which every (injective) module in $\sigma[P]$ is embedded in some $P^{(\Lambda)}$.

Since the inclusion functor $\mathrm{Gen}(P) \to R\text{-}Mod$ is left adjoint to the trace functor $\mathrm{Tr}(P, -) : R\text{-}Mod \to \mathrm{Gen}(P)$, the product of any family $\{N_\lambda\}_\Lambda$ in $\mathrm{Gen}(P)$ is just $\mathrm{Tr}(P, \prod_\Lambda N_\lambda)$ (see [27, 45.11]). By this we may describe when $\mathrm{Stat}(P)$ is closed under certain products. A special case of the next proposition is considered in Albrecht [2, Theorem 3.2].

5.7 $\mathrm{Stat}(P)$ closed under products in $\mathrm{Gen}(P)$.

(1) *The following are equivalent:*

(a) $\mathrm{Stat}(P)$ *is closed under products in* $\mathrm{Gen}(P)$;

(b) *for any family* $\{N_\lambda\}_\Lambda$ *in* $\mathrm{Stat}(P)$, $P \otimes_S \mathrm{Hom}_R(P, \prod_\Lambda N_\lambda) \simeq \mathrm{Tr}(P, \prod_\Lambda N_\lambda)$;

(c) P_S *is an* $\mathrm{Adst}(P)$-*ML module.*

(2) *Assume that P_S is flat and an $\mathrm{Adst}(P)$-ML module. Then $\mathrm{Stat}(P)$ has inverse limits.*

Proof. (1) $(a) \Leftrightarrow (b)$ is clear by the definitions.

$(b) \Rightarrow (c)$ Consider a family $\{X_\lambda\}_\Lambda$ of modules in $\mathrm{Adst}(P)$. Then $P \otimes_S X_\lambda \in \mathrm{Stat}(P)$ and

$$
\begin{aligned}
P \otimes_S \prod_\Lambda X_\lambda &\simeq P \otimes_S \prod_\Lambda \mathrm{Hom}_R(P, P \otimes_S X_\lambda) \\
&\simeq P \otimes_S \mathrm{Hom}_R(P, \prod_\Lambda(P \otimes_S X_\lambda)) \simeq \mathrm{Tr}(P, \prod_\Lambda(P \otimes_S X_\lambda)),
\end{aligned}
$$

showing that P_S is an $\mathrm{Adst}(P)$-ML module.

$(c) \Rightarrow (b)$ Take any family $\{N_\lambda\}_\Lambda$ of modules in $\mathrm{Stat}(P)$. Then $\mathrm{Hom}_R(P, N_\lambda) \in \mathrm{Adst}(P)$ and therefore we have a monomorphism

$$P \otimes_S \mathrm{Hom}_R(P, \prod_\Lambda N_\lambda) \simeq P \otimes_S \prod_\Lambda \mathrm{Hom}_R(P, N_\lambda) \to \prod_\Lambda N_\lambda,$$

and this implies $P \otimes_S \mathrm{Hom}_R(P, \prod_\Lambda N_\lambda) \simeq \mathrm{Tr}(P, \prod_\Lambda N_\lambda)$.

(2) Any category has inverse limits provided it has products and kernels. By (1), the ML-property implies that $\mathrm{Stat}(P)$ has products. As shown in 5.1, P_S flat implies that $\mathrm{Stat}(P)$ has kernels. □

Notice that in particular $\mathrm{Stat}(P) = \mathrm{Gen}(P)$ implies that P_S is an $\mathrm{Adst}(P)$-ML module (by 5.7(1)).

We are now in a position to get new characterizations for generators and tilting modules in $\sigma[M]$.

5.8 Corollary. *Let P be a faithful R-module.*

(1) *The following are equivalent:*

   (a) *P is a generator in $\sigma[P]$;*

   (b) *R is dense in B, P_S is flat and has P_S-dcc.*

(2) *If P is finitely generated the following are equivalent:*

   (a) *P is a projective generator in $\sigma[P]$;*

   (b) *R is dense in B, P_S is faithfully flat and has P_S-dcc.*

(3) *If P is self-small then the following are equivalent:*

   (a) *P is self-tilting;*

   (b) *R is dense in B, $_BP$ is w-Σ-quasi-projective, P_S has P_S-dcc, and $P \otimes_S -$ is exact on short exact sequences with modules from $H_P(\mathrm{Gen}(P))$.*

(4) *For P self-small the following are equivalent:*

   (a) *P is self-tilting and P_S is flat;*

   (b) *P is a projective generator in $\sigma[P]$.*

Proof. (1) $(a) \Rightarrow (b)$ follows from 5.2(4) and 5.5.

$(b) \Rightarrow (a)$ By 5.4, $\mathrm{Inj}(P) \subset \mathrm{Stat}(P)$. For any $K \in \sigma[P]$, there exists an exact sequence $0 \to K \to Q_1 \to Q_2$, where Q_1, Q_2 are injectives in $\sigma[P]$. By 5.1 this implies that K is P-generated.

(2) By [27, 18.5], any finitely generated generator in $\sigma[P]$ is projective in $\sigma[P]$ if and only if it is faithfully flat over its endomorphism ring.

(3) $(a) \Rightarrow (b)$ follows from 5.2(5), 5.3(5) and 5.5.

$(b) \Rightarrow (a)$ By 5.5, Inj $(P) \subset$ Stat(P). Let K be P-generated and consider an exact sequence $0 \to K \to \widehat{K} \to N \to 0$, where $\widehat{K}$ is the P-injective hull of K. $_B P$ being w-Σ-quasi-projective the functor Hom$_B(P, -)$ is exact on this sequence and we obtain an exact commutative diagram

$$
\begin{array}{ccccccccc}
0 \to & P \otimes_S \text{Hom}_B(P, K) & \to & P \otimes_S \text{Hom}_B(P, \widehat{K}) & \to & P \otimes_S \text{Hom}_B(P, N) & \to 0 \\
& \downarrow \mu_K & & \downarrow \simeq & & \downarrow \mu_N & \\
0 \to & K & \to & \widehat{K} & \to & N & \to 0,
\end{array}
$$

showing that μ_K is an isomorphism and hence $K \in$ Stat$(_B P)$. So $_B P$ is self-tilting and by density $_R P$ is self-tilting.

(4) $(a) \Rightarrow (b)$ Let P be self-tilting with P_S flat. Then by (1) and (3), P is a generator in $\sigma[P]$. By [30, Proposition 4.1], any self-tilting module which is a generator in $\sigma[P]$ is projective in $\sigma[P]$.

$(b) \Rightarrow (a)$ is trivial. $\qquad\qquad\qquad\qquad\qquad\qquad\qquad\qquad\qquad\qquad\qquad\qquad$ $\square$

Remark. Abelian group theorists have been mainly interested in modules which are (faithfully) flat over their endomorphism rings while in representation theory (self-) tilting modules have received much attention. From 5.8(4) we see that these notions generalize projective generators in different directions.

6 Ring extensions and equivalences

In this section we investigate the behaviour of equivalences as considered in the previous sections under ring extensions.

6.1 Ring extensions. Let $\alpha : R \to A$ be a morphism of associative rings with units. Related to it we have the *induction functor*

$$R\text{-Mod} \to A\text{-Mod}, \quad M \mapsto A \otimes_R M,$$

and the *restriction functor*

$$A\text{-Mod} \to R\text{-Mod}, \quad {}_A N \mapsto {}_R N.$$

For a given R-module P, putting $S = \text{End}_R(P)$ and $T = \text{End}_A(A \otimes_R P)$ we have the ring morphism

$$\beta : S \to T, \quad f \mapsto id \otimes f.$$

We will be interested in A-modules (resp., T-modules) which have certain properties as R-modules (resp., S-modules). Refining the notations introduced before we set

$$
\begin{aligned}
\mathrm{add}_R^A(P) &= \{V \in A\text{-}Mod \mid {}_RV \in \mathrm{add}(P)\}, \\
\mathrm{Add}_R^A(P) &= \{V \in A\text{-}Mod \mid {}_RV \in \mathrm{Add}(P)\}, \\
\mathrm{Gen}_R^A(P) &= \{V \in A\text{-}Mod \mid {}_RV \in \mathrm{Gen}(P)\}, \\
\mathrm{Pres}_R^A(P) &= \{V \in A\text{-}Mod \mid {}_RV \in \mathrm{Pres}(P)\}, \\
\mathrm{Stat}_R^A(P) &= \{V \in A\text{-}Mod \mid {}_RV \in \mathrm{Stat}_R(P)\}, \\
\mathrm{Adst}_S^T(P) &= \{X \in T\text{-}Mod \mid {}_SX \in \mathrm{Adst}(P)\} .
\end{aligned}
$$

It is easy to see (e.g., [13, Lemma 1.2]) that

$$A \otimes_R P \in \mathrm{Gen}(P) \text{ if and only if } \mathrm{Gen}_A(A \otimes_R P) = \mathrm{Gen}_R^A(P).$$

From Hom-tensor relations (e.g., [28, 15.6]) we have the

6.2 Basic isomorphisms.

(1) *For any $V \in A\text{-}Mod$ there is a functorial S-module isomorphism*

$$\varphi : \mathrm{Hom}_R(P, V) \to \mathrm{Hom}_A(A \otimes_R P, V), \quad f \mapsto [a \otimes p \mapsto a \cdot (p)f].$$

(2) *For $V = A \otimes_R P$ we get $\mathrm{Hom}_R(P, A \otimes_R P) \simeq T$ and the R-module isomorphism*

$$id \otimes \varphi : P \otimes_S \mathrm{Hom}_R(P, A \otimes_R P) \to P \otimes_S T.$$

Our main interest is to transfer properties of the R-module P to the A-module $A \otimes_R P$. It turns out that the conditions we are looking at can be transferred provided $A \otimes_R P$ is P-static as an R-module. With the above preparations we can prove the following crucial result (see [20, Theorem 4.9, 5.5]):

6.3 Related equivalences. *Assume $A \otimes_R P \in \mathrm{Stat}_R^A(P)$. Then:*

(1) $\mathrm{Stat}_R^A(P) = \mathrm{Stat}_A(A \otimes_R P)$.

(2) $\mathrm{Adst}_S^T(P) = \mathrm{Adst}(A \otimes_R P)$ *and there is an equivalence*

$$\mathrm{Hom}_A(A \otimes_R P, -) : \mathrm{Stat}_R^A(P) \to \mathrm{Adst}_S^T(P).$$

(3) *If P is self-small then we have an equivalence*

$$\mathrm{Hom}_A(A \otimes_R P, -) : \mathrm{Add}_R^A(P) \to \mathrm{Add}_S^T(S).$$

In each case the inverse functor is $(A \otimes_R P) \otimes_T -$.

Proof. $A \otimes_R P \in \text{Stat}_R^A(P)$ means $P \otimes_S \text{Hom}_R(P, A \otimes_R P) \simeq A \otimes_R P$.

(1) Combined with an isomorphism from 6.2 we get the R-module isomorphism

$$P \otimes_S T \simeq A \otimes_R P.$$

This implies the isomorphisms for any $V \in A\text{-}Mod$,

$$\begin{aligned}(A \otimes_R P) \otimes_T \text{Hom}_A(A \otimes_R P, V) &\simeq (P \otimes_S T) \otimes_T \text{Hom}_A(A \otimes_R P, V) \\ &\simeq P \otimes_S \text{Hom}_A(A \otimes_R P, V) \\ &\simeq P \otimes_S \text{Hom}_R(P, V).\end{aligned}$$

Now $V \in \text{Stat}_R^A(P)$ means by definition that the last expression in this chain is isomorphic to V, whereas $V \in \text{Stat}_A(A \otimes_R P)$ means that the first expression is isomorphic to V.

(2) First notice the R-module isomorphisms for any T-module V,

$$P \otimes_S V \simeq (P \otimes_S T) \otimes_T V \simeq (A \otimes_R P) \otimes_T V.$$

Assume $V \in \text{Adst}_S^T(P)$. Then we have the isomorphisms

$$\begin{aligned}V \simeq \text{Hom}_R(P, P \otimes_S V) &\simeq \text{Hom}_R(P, (A \otimes_R P) \otimes_T V) \\ &\simeq \text{Hom}_A(A \otimes_R P, (A \otimes_R P) \otimes_T V),\end{aligned}$$

proving $V \in \text{Adst}_T(A \otimes_R P)$.

The same chain of isomorphisms shows $\text{Adst}_T(A \otimes_R P) \subset \text{Adst}_S^T(P)$.

In view of (1) and the first part of (2) the final assertion about the equivalence follows from the basic equivalence 2.4 applied to $A \otimes_R P$.

(3) We know that $\text{Add}_R^A(P) \subset \text{Stat}_R^A(P) = \text{Stat}_A(A \otimes_R P)$, and it is easy to verify that $H_P(\text{Add}_R^A(P)) \subset \text{Add}_S^T(S) \subset \text{Adst}_S^T(P)$. $\square$

Combining the preceding observations we obtain:

6.4 w-Σ-quasi-projective modules. *Let P be a w-Σ-quasi-projective R-module and assume $A \otimes_R P \in \text{Pres}(P)$. Then:*

(1) *$A \otimes_R P$ is a w-Σ-quasi-projective A-module.*

(2) *If P is Σ-self-static we have an equivalence*

$$\text{Hom}_A(A \otimes_R P, -) : \text{Pres}_R^A(P) \to \text{Adst}_S^T(P),$$

with inverse functor $(A \otimes_R P) \otimes_T -$.

(3) *If P is a self-small R-module then $A \otimes_R P$ is a self-small A-module.*

Proof. (1) By [30, 3.2], factor modules of P-presented modules by P-generated modules are P-presented. This implies

$$\operatorname{Pres}_A(A \otimes_R P) \subset \operatorname{Pres}_R^A(P),$$

and by the functorial isomorphism in 6.2 we conclude that $A \otimes_R P$ is w-Σ-quasi-projective as an A-module.

(2) By 4.5, P-presented modules are P-static and so $A \otimes_R P \in \operatorname{Pres}(P)$ implies $A \otimes_R P \in \operatorname{Stat}_R^A(P)$. Hence by 6.3, we have the inclusions

$$\operatorname{Pres}_A(A \otimes_R P) \subset \operatorname{Pres}_R^A(P) = \operatorname{Stat}_R^A(P) = \operatorname{Stat}_A(A \otimes_R P) \subset \operatorname{Pres}_A(A \otimes_R P).$$

From this and 2.4 we obtain the equivalence as given.

(3) By [30, 5.1], for P self-small and w-Σ-quasi-projective, $\operatorname{Hom}_R(P, -)$ commutes with direct limits of P-presented modules. Since $A \otimes_R P \in \operatorname{Pres}(P)$, any infinite direct sum $(A \otimes_R P)^{(\Lambda)}$ is the direct limit of its finite partial sums and hence

$$\begin{aligned}
\operatorname{Hom}_A(A \otimes_R P, (A \otimes_R P)^{(\Lambda)}) &\simeq \operatorname{Hom}_R(P, (A \otimes_R P)^{(\Lambda)}) \\
&\simeq \operatorname{Hom}_R(P, A \otimes_R P)^{(\Lambda)} \\
&\simeq \operatorname{Hom}_A(A \otimes_R P, A \otimes_R P)^{(\Lambda)}.
\end{aligned}$$

$\square$

6.5 Self-tilting modules. *Let P be a self-tilting and $\sum$-self-static R-module, and assume $A \otimes_R P \in \operatorname{Gen}(P)$. Then $A \otimes_R P$ is a self-tilting A-module and we have an equivalence*

$$\operatorname{Hom}_A(A \otimes_R P, -) : \operatorname{Gen}_R^A(P) \to \operatorname{Adst}_S^T(P),$$

with inverse functor $(A \otimes_R P) \otimes_T -$.

Proof. By 6.4, we have the equalities

$$\operatorname{Gen}_A(A \otimes_R P) = \operatorname{Gen}_R^A(P) = \operatorname{Pres}_R^A(P) = \operatorname{Pres}_A(A \otimes_R P) = \operatorname{Stat}_A(A \otimes_R P).$$

$\square$

Recall that P is a tilting module in *R-Mod* if it is self-tilting and a subgenerator in *R-Mod*. We consider some special cases of this.

6.6 Corollary. *Assume $A \otimes_R P \in \operatorname{Gen}(P)$. Then:*

(1) *If P is a projective generator in $\sigma[P]$, then $A \otimes_R P$ is a projective generator in $\sigma_A[A \otimes_R P]$.*

(2) *If P is self-tilting and $\operatorname{Gen}(P)$ is closed under extensions in R-Mod, then $A \otimes_R P$ is self-tilting and $\operatorname{Gen}_A(A \otimes_R P)$ is closed under extensions in A-Mod.*

(3) *Assume that P is $\sum$-self-static and (i) A_R is flat or (ii) $_A A$ is finitely cogenerated and $A \otimes_R P$ is a faithful A-module.*

If P is tilting in R-Mod, then $A \otimes_R P$ is tilting in A-Mod.

Proof. (1) Recall that P is a projective generator in $\sigma[P]$ if and only if it is self-tilting and a generator in $\sigma[P]$.

Let P be a projective generator in $\sigma[P]$. Then P is $\sum$-self-static by 3.4, and by 6.5, $A \otimes_R P$ is self-tilting. Since $\sigma_A[A \otimes_R P] \subset \mathrm{Gen}_R^A(P) = \mathrm{Gen}_A(A \otimes_R P)$ we see that $A \otimes_R P$ is a generator in $\sigma_A[A \otimes_R P]$.

(2) Suppose $\mathrm{Gen}(P)$ is closed under extensions in R-Mod. Since $\mathrm{Gen}_A(A \otimes_R P) = \mathrm{Gen}_R^A(P)$ this implies that $\mathrm{Gen}_A(A \otimes_R P)$ is closed under extensions in A-Mod.

(3) Suppose $R \in \sigma[P]$, i.e., there is a monomorphism $R \to P^k$, for some $k \in \mathbb{N}$. If A_R is flat, then $A \simeq A \otimes_R R \to A \otimes_R P^k$ is a monomorphism and hence $A \otimes_R P$ is a subgenerator in A-Mod. If $_A A$ is finitely cogenerated and $A \otimes_R P$ is faithful, then $A \subset (A \otimes_R P)^k$, for some $k \in \mathbb{N}$. Now the assertions follow from 6.5. $\qquad \square$

Remarks. Let P be finitely generated. If P is self-tilting then it is a $*$-module and 6.5 implies Fuller [13, Theorem 2.2]. Condition (1) in 6.6 corresponds to [13, Corollary 2.4]. If P satisfies the condition in 6.6(2) then P is called *quasi-tilting in R-Mod* (see [8]) and we obtain [13, Corollary 2.5].

Acknowledgements. The author is very indebted to John Clark for helpful discussions on the subject.

References

[1] U. ALBRECHT, Endomorphism rings of faithfully flat abelian groups, *Results in Math.* 17, 179-201 (1990).

[2] U. ALBRECHT, The construction of A-solvable abelian groups, *Czech. Math. J.* 44(119), 413-430 (1994).

[3] J.L. ALPERIN, Static modules and non-normal subgroups, *J. Austral. Math. Soc. (Series A)* 49, 347-353 (1990).

[4] D.M. ARNOLD, C.E. MURLEY, Abelian groups, A, such that $\mathrm{Hom}(A,\text{-})$ preserves direct sums of copies of A, *Pacific J. Math.* 56, 7-20 (1975).

[5] R.R. COLBY, K.R. FULLER, Tilting, cotilting, and serially tilted rings, *Comm. Algebra* 18 (5), 1585-1615 (1990).

[6] R. COLPI, Some remarks on equivalences between categories of modules, *Comm. Algebra* 18 (6), 1935-1951 (1990).

[7] R. COLPI, Tiltings in Grothendieck categories, *Preprint* (1997).

[8] R. COLPI, G. D'ESTE, A. TONOLO, Quasi-tilting modules and counter equivalences, *J. Algebra* 191, 461-494 (1997).

[9] R. COLPI, C. MENINI, On the structure of *-modules,
J. Algebra 158, 400-419 (1993).

[10] R. COLPI AND J. TRLIFAJ, Tilting modules and tilting torsion theories,
J. Algebra 178 (2), 614-634 (1995).

[11] N.V. DUNG, D.V. HUYNH, P. SMITH, R. WISBAUER, Extending modules,
Pitman Research Notes Math. 313 (1994).

[12] T.G. FATICONI, Categories of modules over endomorphism rings,
Memoirs AMS 492 (1993).

[13] K.R. FULLER, *-Modules over Ring Extensions,
Comm. Algebra 25(9), 2839-2860 (1997).

[14] D.K. HARRISON, Infinite abelian groups and homological methods,
Ann. Math 69, 366-391 (1959).

[15] A.I. KASHU, Radicals and torsions in modules (Russian),
Shtiinza, Kishinev (1983).

[16] A.I. KASHU, Duality between localization and colocalization in adjoint situations (Russian), *Mat. Issled.* 65, 71-87 (1982).

[17] A.I. KASHU, Module classes and torsion theories in Morita contexts (Russian),
Mat. Issled.: Strongly regular algebras and PI-algebras, 3-14 (1987).

[18] T. KATO, Duality between colocalization and localization,
J. Algebra 55, 351-374 (1978).

[19] C. MENINI AND A. ORSATTI, Representable equivalences between categories of modules and applications, *Rend. Sem. Mat. Univ. Padova* 82, 203-231 (1989).

[20] S.K. NAUMAN, Static modules and stable Clifford theory,
J. Algebra 128, 497-509 (1990).

[21] T. ONODERA, Codominant dimensions and Morita equivalences,
Hokkaido Math. J. 6, 169-182 (1977).

[22] A. ORSATTI, Equivalenze rappresentabili tra categorie di moduli,
Rend. Sem. Mat. Fiz. 60, 243-260 (1990).

[23] P. ROTHMALER, Mittag-Leffler modules and positive atomicity, manuscript (1994).

[24] M. SATO, Fuller's theorem on equivalences, *J. Algebra* 52, 274-284 (1978).

[25] M. SATO, On equivalences between module categories, *J. Algebra* 59, 412-420 (1979).

[26] F. ULMER, Localizations of endomorphism rings and fixpoints, *J. Algebra* 43, 529-551 (1976).

[27] R. WISBAUER, Foundations of Module and Ring Theory, *Gordon and Breach, Reading* (1991).

[28] R. WISBAUER, Modules and algebras: Bimodule structure and group actions on algebras, *Longman, Pitman Monographs* 81 (1996).

[29] R. WISBAUER, On module classes closed under extensions, *Rings and radicals*, B. Gardner, Liu Shaoxue, R. Wiegandt (ed.), Pitman RN 346, 73-97 (1996).

[30] R. WISBAUER, Tilting in module categories, *Abelian groups, module theory, and topology*, D. Dikranjan, L. Salce (ed.), Marcel Dekker LNPAM 201, 421-444 (1998).

[31] W. ZIMMERMANN, Modules with chain conditions for finite matrix subgroups, *J. Algebra* 190, 68-87 (1997).

[32] B. ZIMMERMANN-HUISGEN, Endomorphism rings of self-generators, *Pacific J. Math.* 61, 587-602 (1975).